**Computational Methods in Electromagnetic Compatibility**

# Computational Methods in Electromagnetic Compatibility: Antenna Theory Approach versus Transmission line Models

*Dragan Poljak and Khalil El Khamlichi Drissi*

WILEY

This edition first published 2018


*Registered Offices*
John Wiley & Sons, Inc., 111 River Street, Hoboken, NJ 07030, USA

*Editorial Office*
111 River Street, Hoboken, NJ 07030, USA

For details of our global editorial offices, customer services, and more information about Wiley products visit us at www.wiley.com.

Wiley also publishes its books in a variety of electronic formats and by print-on-demand. Some content that appears in standard print versions of this book may not be available in other formats.



*Library of Congress Cataloging-in-Publication Data*

Names: Poljak, D. (Dragan), author. | Drissi, Khalil E. (Khalil El Khamlichi), author.
Title: Computational methods in electromagnetic compatibility / by Dragan Poljak, Khalil El Khamlichi Drissi.
Description: Hoboken, NJ : John Wiley & Sons, 2018. | Includes bibliographical references and index. |
Identifiers: LCCN 2017054464 (print) | LCCN 2017059021 (ebook) | ISBN 9781119337195 (pdf) | ISBN 9781119337072 (epub) | ISBN 9781119337171 (cloth)
Subjects: LCSH: Electromagnetic compatibility–Data processing. | Electromagnetism–Mathematics.
Classification: LCC TK7867.2 (ebook) | LCC TK7867.2 .P653 2018 (print) | DDC 621.38–dc23
LC record available at https://lccn.loc.gov/2017054464

Cover Design: Wiley
Cover Image: © shuoshu/Gettyimages

Set in 10/12pt WarnockPro by SPi Global, Chennai, India

Printed in the United States of America

10 9 8 7 6 5 4 3 2 1

*To my lifetime inspiration, to my beloved daughters, my wife, my sister, my mother, and to the everlasting memory of my father who recently passed away and who will never be forgotten...*

Dragan Poljak

*To my dear parents, for all their sacrifices, their love, their tenderness, their support and their prayers. A special thought to my mother whom I miss terribly, I think of you everyday and I will probably never come to terms with the way your life ended on this earth.*

Khalil El Khamlichi Drissi

# Contents

# Preface

Electromagnetic Compatibility (EMC) as a topic has become very important in the last few decades. The vitality of EMC nowadays can be seen in many academic activities, as there are many universities worldwide offering undergraduate or graduate EMC courses, either obligatory or optional. Moreover, today, in the world of wireless communication and Internet of Things (IoT), many electronic products, devices, or systems are required to pass immunity and emission testing regarding EMC standards. Accordingly, there are dozens of books related to various EMC aspects currently available from major scientific publishers. Nevertheless, books rarely deal with EMC computational models and related numerical methods.

The previous book by Dragan Poljak, *Advanced Modeling in Computational Electromagnetic Compatibility*, was published by Wiley in February 2007. The present book authored by Dragan Poljak and Khalil El Khamlichi Drissi provides an overview of the further advances in the area of computational electromagnetics arising from a decade of very close and highly intensive collaboration between the Dragan research group from the University of Split, Croatia, and the Khalil group from Universitė Clermont Auvergne, France.

This rather fruitful collaboration resulted in successful joint projects and numerous journal and conference papers. The beauty of this collaboration reflects in merging two research teams tackling similar problems with different approaches related to antenna theory models (Dragan group) and transmission line methods (Khalil group). Furthermore, there is the benefit of discussing different solution methods related to boundary integral equation techniques and finite difference techniques. Moreover, throughout the book a trade-off between the different formulations and numerical solution methods is provided.

While the previous Wiley book by Dragan was primarily focused on academic examples, the present book by Dragan and Khalil deals with many practical engineering problems. The most significant topics covered in the book are related to realistic antenna systems, such as antennas for air traffic control or ground penetrating radar (GPR) antennas, grounding systems, such as grounding systems for wind turbines and biomedical applications of electromagnetic fields, such as transcranial magnetic stimulation. The book includes a large number of illustrative computational examples and reference list at the end of each chapter. Rigorous theoretical background and mathematical details of various formulations and solution methods being used throughout the book are presented in detail.

The authors hope that the present book gives not only a useful description of their expertise related to computational EMC but also updated information on the latest advances in this area.

The book is divided in two parts. The first part deals with electromagnetic field coupling to thin wire configurations of an arbitrary shape covering the following topics: introductory aspects of computational electromagnetics, antenna theory versus transmission line approximation, electromagnetic field coupling to overhead and buried wires, transient analysis of grounding systems and lightning channel modeling. An important goal of this part of the book is to provide a trade-off between a highly efficient transmission line approach, rather widely used by EMC community researchers and engineers, and antenna theory models providing the most rigorous analysis of high frequency (HF) and transient phenomena.

The second part of the book deals with advanced modeling of bioelectromagnetics phenomena featuring the method of moments (MoM), boundary element method (BEM) and hybrid finite element method (FEM)/ BEM, respectively. Of particular interest is not only human exposure to low frequency (LF) and HF electromagnetic fields but also some biomedical applications of electromagnetic fields.

We hope that this book will be useful material for undergraduate, graduate and postdoc students to learn about advanced EMC computational models and that it will also enable engineers in industry to solve some demanding practical problems. We also think that the book could be used for various university courses involving not only computational EMC models but also computational electromagnetics in general or numerical modeling in engineering itself.

The book requires a general background in electrical engineering, involving mainly basic electromagnetics. Fundamental EMC concepts such as numerical modeling principles are given in this book. Thus, the book is convenient for students, specialists, researchers and engineers.

To sum up, we are glad we have managed to compose this material stemming from more than a decade of very intensive collaboration in the areas of EMC and bioelectromagnetics. Of course, there are many rather challenging problems we plan to deal with together in days to come.

Split, Croatia–Clermont-Ferrand
France, June 2017

*Dragan Poljak*
*Khalil El Khamlichi Drissi*

# Part I

# Electromagnetic Field Coupling to Thin Wire Configurations of Arbitrary Shape

# 1

# Computational Electromagnetics – Introductory Aspects

This introductory section deals with the character of a physical model and the corresponding mathematical method to solve the problem of interest. The models are characterized to be simplified imaginary simulations of the real-world systems one attempts to understand. However, models include only those properties and relationships required to understand aspects of real systems that are of interest at the given moment, i.e. those aspects of real systems one knows, or those one is aware of after all. The rest of the information about a real system is simply neglected. Furthermore, this introductory section discusses the fundamental framework to describe electromagnetic phenomena – Maxwell's equations, wave equations, and conservation laws.

## 1.1 The Character of Physical Models Representing Natural Phenomena

### 1.1.1 Scientific Method, a Definition, History, Development ... ?

*Scientists create tools, that's what they do...*
*C.P. Snow*

Science could be considered as an entire set of facts, definitions, theorems, techniques, and relationships, and is tested on phenomena in the *real*, *objective*, and *external* world and, itself, has many elements of imagination, logic, creativity, judgment, metaphor, and instrumentations.

The essence of science is definitely more in *research methods* and specific *way of reasoning*, and less in particular facts and results.

Scientific insight starts with observing a certain phenomenon, and then organizing the collected observations in a sort of hypothesis that is tested on additional observations, and if necessary, modified. Then, predictions based on these modified hypotheses are carried out, and some experiments are

*Computational Methods in Electromagnetic Compatibility: Antenna Theory Approach versus Transmission line Models*, First Edition. Dragan Poljak and Khalil El Khamlichi Drissi.

performed to test the predictions. When the range of predictions provided by the hypothesis is considered to be satisfactory for the scientific community, the hypothesis is referred to as a *scientific theory* or *natural law*.

This rather successful methodology, more than four centuries old, is called *science* or *scientific method*.

Scientific method was born in the beginning of the seventeenth century with Galileo having abandoned Aristotle's theory of motion. It was Galileo who came up with the principle of the relativity of motion and with the statement that only change in motion required force.

At the same time, a separation of science from philosophy began in the form of shift from consideration of the *nature of phenomenon* (*essence*) to explanation of the *behavior of a phenomenon*. Namely, the *Aristotelian essentialistic approach* to the explanation of natural phenomena was replaced by the *mathematical predictive approach*. Instead of asking the question *why* scientists started to ask *how* [1]. As once Kelvin pointed out – *to know something about phenomena means to measure them and express them in terms of numbers*.

What is considered to be one of the crucial issues in the analysis of a natural phenomenon is related to the development and application of a physical model enabling one to predict the behavior of a system with a certain level of accuracy.

One of the crucial aspects of the scientific method and related technological progress is definitely *the physical model* of a natural phenomenon of interest.

### 1.1.2 Physical Model and the Mathematical Method to Solve the Problem – The Essence of Scientific Theories

> *Therefore, the goal of the scientific method is to establish the model of a physical phenomenon and to develop related mathematical methods for the analysis of the given problem.*

Various theoretical and experimental procedures are used while developing a model. *Models* are simplified imaginary simulations of the natural systems one attempts to understand and include only those properties and relationships required to understand aspects of real systems that are currently of interest, i.e. those aspects one knows, or, those one is aware of after all. The rest of the details about a real system are simply neglected from a model.

The concept of physical model represents the essence of reductionistic approach within the scientific method. How much the model of a given physical phenomenon is satisfactory depends on what is required from the model. In the language of mathematics, almost all problems arising in electromagnetics can be formulated in terms of differential, integral, or variational equations.

Generally, there are two basic approaches to solving problems in electromagnetics – the *differential (field) approach* and the *integral (source) approach*.

The field approach deals with a solution of a corresponding differential equation with associated initial and boundary conditions, specified at a boundary of a computational domain. Solving some differential equation type one obtains the spatial and temporal distribution of the corresponding field or potential.

Historically, this approach has been developed by Boskovic, Faraday, Maxwell, and others, and is generally very useful for handling the problems with closed domains and clearly specified boundary conditions, the so-called *interior field problems.*

The source or integral approach is based on the solution of a corresponding integral equation that yields the distribution of electromagnetic filed sources in terms of charge or current distribution, respectively.

In the past, this approach was promoted by Franklin, Cavendish, and Ampere, among others, and is convenient for the treatment of the *exterior* (unbounded) *field problems.*

Thus, a classical boundary-value problem can be formulated in terms of the operator equation:

$$L(u) = p \tag{1.1}$$

on the domain $\Omega$ with conditions

$$F(u) = q|_{\Gamma} \tag{1.2}$$

prescribed on the boundary $\Gamma$.

$L$ is the linear differential operator, $u$ is the solution of the problem, and $p$ is the excitation function representing the known sources inside the domain. Note that $u$ usually represents potentials (such as scalar potential $\varphi$) or fields (such as electric field $E$).

The character of the differential approach is depicted in Figure 1.1 [2].

Methods for the solution of the interior field problem are generally referred to as *differential methods* or *field methods.*

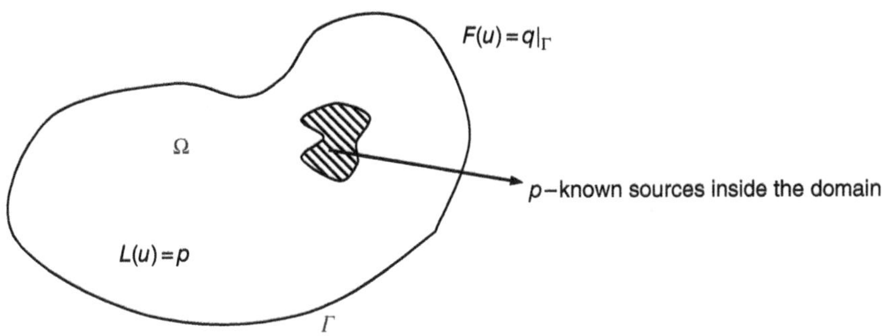


**Figure 1.1** Differential approach concept.

**Figure 1.2** Integral approach concept.

Essentially, a differential approach isolates the calculation domain from the rest of the world. The interaction of the domain of interest with the rest is expressed (*de facto* replaced) by a set of prescribed boundary conditions.

If instead of differential operator $L$ one considers an integral operator $g$, then unknowns are related to field sources (charge or current densities, respectively), distributed along the boundary $\Gamma'$. Namely, it can be written as

$$g(u) = h, \tag{1.3}$$

where $h$ denotes the excitation function.

Figure 1.2 illustrates the character of the integral approach [2].

Solution methods for the exterior field problem are generally referred to as integral methods or source methods. In this case, the domain of interest is unbounded (infinite). However, the source distribution represents all that exists, i.e. all interactions coming from the outer world are neglected. For example, when a basic electromagnetic model is developed for a dipole antenna in engineering electromagnetics the antenna is assumed to be insulated in free space [2].

Finally, for dynamic phenomena the initial condition of a physical system has to be considered. Basically, any law of nature represents physical states in a mathematical form (written in terms of differential, integral, or variational equations). Thus, a prescribed initial condition (behavior of the considered physical quantity at $t = 0$) by definition implies that nothing exists earlier than $t = 0$. This is also considered to be the origin of time asymmetry in physical laws.

Generally, techniques for the solution of operator equations can be referred to as analytical, numerical, or hybrid methods. Analytical solution methods provide *exact solutions* but are, on the other hand, limited to a narrow range of applications, mostly related to canonical problems.

Unfortunately, there are not many realistic scenarios in physics or practical engineering problems that can be worked out using these techniques.

Numerical techniques are applicable to almost all scientific engineering problems, but the main drawbacks are related to the limits governed by the approximation contained in the model itself.

Moreover, the criteria for accuracy, stability, and convergence are not always straightforward and clear to the researcher in a particular area [2].

### 1.1.3 Philosophical Aspects Behind Scientific Theories

One of the crucial questions in the philosophy of science is how physical models really work, or, more generally, how scientific theories are developed or "upgraded."

Looking back into the history of physics, the development of Maxwell's kinetic theory of gases and electromagnetic field theory was not motivated by experimental findings that were not compatible with the existing paradigm (in the sense of Kuhn [3]), as was the case with relativity and quantum mechanics.

In the case of electromagnetism almost all facts, known in Maxwell's time, were interpreted satisfactorily within the Newton paradigm and incorporated into a powerful theoretical frame.

This theory was intensively in use till Hertz experimentally verified one of the main goals of the Maxwell theory – the existence of electromagnetic waves.

However, the principal motivation in the background of Maxwell's work was essentially philosophical, or even metaphysical in nature, i.e. the consequence of his own point of view. The origin of Maxwell's ideas came from Michael Faraday and his study of electromagnetic induction. Faraday, together with Boskovic, was one of the first scientists who came up with the idea of *field* versus action at a distance concept.

Some rather old, but still important questions from philosophy of science are as follows:

- Is scientific insight into the absolute truth possible, taking into account limitations of our conceptual, language, and mathematical tools?
- Is the rise of knowledge cumulative in nature and is it clearly directed to the objective truth?

According to *logical positivism*, *Popper's falsificationism*, and *Kuhn's social relativism* the objective truth is out of reach for the human mind. Furthermore, for Ernst Mach and Vienna circle followers, theories are systems of quantitative relationships between measurable phenomena, and are not directed toward the absolute and objective truth. Moreover, Mach and other empiricists claim that only theories directly testable with experiments should be accepted [4, 5]. For Niels Bohr, theory is a tool to explain various experimental data.

No universal theory exists for Popper that would be conclusively proved in an inductive sense. Theory is alive while its disadvantages are not found. For Kuhn, the *natural selection* of scientific theories is driven by the request for problem solving. For Wittgenstein, the origin of scientific triumph is aspiration for generality.

*Gödel incompleteness theorem* has destroyed the basis of *the axiomatic method*. Stephen Hawking lost faith in the existence of *Theory of Everything* as *Gödel theorem* had convinced him that any system could not be complete if it was consistent. There can always be a proposition that cannot be proved or refuted.

Einstein's reasoning was also affected by the strong philosophical background, classical education, and culture of dialogue. For Einstein, many professional scientists of his time have seen thousands of trees, but have never seen a wood [4, 5]. Einstein esteems that the knowledge of historical and philosophical background of science could set one free of prejudices of which most of the generation suffers. Thus, after initial respect for the Vienna circle, Einstein's attitudes began to differ from the circle ideas. The circle refused any element of a theory, as metaphysical, if there was no clear connection with an experience.

Einstein claims that veracity of a theory can never be proved, as it is never known if future experience will contradict its conclusions. Einstein moves aside Schlick and Reichenbach as new empirical philosophy, according to Einstein, turns science into something like engineering. Einstein's own experience leads him to a strong attitude that creative theoretical thinking cannot be replaced with algorithm for building and testing theories. Passion for knowledge, according to Einstein, creates the illusion that the objective world can be comprehended rationally, without any empirical foundation – in short, by means of metaphysics.

Therefore, the old question still of interest in both philosophy and science is this: Does scientific knowledge come from out-of-mind reality, or it is necessarily just a reflection of the mind and is it limited by its own insight abilities?

### 1.1.4 On the Character of Physical Models

Physical model represents the fundamental concept within the framework of the scientific method for the representation and understanding of natural phenomena. Physical models are simplified imaginary simulations of the real-world systems one attempts to understand, including only those properties and relationships required to understand the aspects of real systems one considers, i.e. those aspects of real systems one knows, or is generally aware of. The rest of the facts about a real system are simply neglected from the model. As a matter of fact, how much the model of a given physical phenomenon is satisfactory then strongly depends on what is required from the particular model. Thus, one draws conclusions from an incomplete information set.

Therefore, models are tools for capturing particular insights of the phenomena and they do not represent a full proof for a system behavior under all circumstances. Moreover, mathematically described physical models are abstractions of the natural world, while the related computational models, convenient for implementation on a digital computer, are eventually abstraction of the physical world.

Therefore, physical models and related solution methods are *problem dependent*.

## 1.2 Maxwell's Equations

*In his hands electricity first became a mathematically exact science and the same might be said of other larger parts of physics.*
*Sir James H. Jeans*

It was not possible to incorporate an increasing knowledge on electricity and magnetism through the nineteenth century into Newton's physics framework. Thus, contributions of Faraday, Maxwell, Heaviside, Hertz, and others led to the revolutionary concept of field in classical physics. The field was shown not to be just mathematical abstract entity, but pure physical reality. Consequently, by adopting the field notion the *action at a distance* concept was abandoned [5].

With James C. Maxwell, not only a rigorous electromagnetic field theory came along but also a grand unification of electricity, magnetism, and light. Namely, almost a quarter century before the Hertz experimental verification Maxwell theoretically anticipated the existence of electromagnetic waves. Light is just an electromagnetic wave, visible to the human eye propagating through ether. After Maxwell, H. A. Lorentz extended Maxwell's theory with electrodynamics of charged particle.

### 1.2.1 Original Form of Maxwell's Equations

Maxwell's equations were modified a few times [6] in the last 150 years since they were originally formulated by Maxwell and published for the first time [7]. The changes were regarding the physical interpretation, mathematical expression, and general approach to the solution methods for different problems. In the mid-1860s Maxwell originally derived 20 scalar equations. What is today considered to be modern Maxwell's equations are a set of vector equations independently derived by Heaviside and Hertz by the end of the nineteenth century.

Historically speaking, with Maxwell's equations, a rigorous mathematical basis has been established for a proper description of electromagnetic phenomena. Moreover, the appearance of Maxwell's equations has provided the paradigm shift from the old *action at a distance* concept to the *field approach.* Maxwell's equations have undergone significant changes twice [8]. First, it was when Heaviside reduced the scalar form into the vector notation. He also made important modifications, having abandoned the potentials in favor of fields. Next time, a significant improvement was initiated by Larmor due to the discovery of the electron.

Important advancements of Maxwell's theory in the mid-1880s were carried out by Poynting, FitzGerald, and Heaviside. Lorentz's contribution is related to the development of microscopic theory by means of Maxwell's equations and inclusion of the force acting on a charged particle arising from the existence of fields.

### 1.2.2 Modern Form of Maxwell's Equations

The laws of classical electromagnetism can be expressed very concisely by a set of four differential equations. There is also an equivalent integral form of these equations. The differential form of Maxwell's equations is commonly used in solving engineering problems while their integral forms are convenient in providing a deeper insight into the underlying physical laws.

The *first Maxwell equation* is the differential form of Faraday law (the time-varying magnetic flux density $\vec{B}$causes the curl of electric field $\vec{E}$) given by

$$\nabla \times \vec{E} = -\frac{\partial \vec{B}}{\partial t}. \tag{1.4}$$

Hence, the time-varying magnetic fields are vortex sources of electric fields.

The *second Maxwell equation* is the differential form of generalized Ampere's law stating that either a current density $\vec{J}$ or a time-varying electric flux density $\vec{D}$ gives rise to a magnetic field $\vec{H}$. This can be expressed as

$$\nabla \times \vec{H} = \vec{J} + \frac{\partial \vec{D}}{\partial t}. \tag{1.5}$$

It is worth noting that the term $\partial\vec{D}/\partial t$ was originally added by Maxwell to the original expression for Ampere's law, thus making the law consistent with the electric charge conservation. This term is usually referred to as a displacement current density.

The *third Maxwell equation* states that electric monopoles exist, so that

$$\nabla \cdot \vec{D} = \rho, \tag{1.6}$$

i.e. charge densities $\rho$ are the monopole sources of the electric field.

Finally, the *fourth Maxwell equation* states that magnetic poles always occur in pairs and are due to electric currents; no free poles can exist. This is expressed by the divergence Maxwell equation:

$$\nabla \cdot \vec{B} = 0, \tag{1.7}$$

which implies that the magnetic field is always solenoidal.

The integral form of the Faraday law states that any change of magnetic flux density $B$ through any closed loop induces an electromotive force around the loop. Taking the surface integration over (1.4) and applying the Stokes theorem yields

$$\oint_c \vec{E}\,\mathrm{d}\vec{s} = -\int_S \frac{\partial \vec{B}}{\partial t} \cdot \mathrm{d}\vec{S}, \tag{1.8}$$

where the line integral is taken around the loop and with $\mathrm{d}\vec{S} = \vec{n}\mathrm{d}S$.

The voltage induced by a varying flux has a polarity such that the induced current in a closed path gives rise to a secondary magnetic flux, which opposes the change in time-varying source magnetic flux.

The integral form of the Ampere law is derived by integrating (1.5) and applying the Stokes theorem:

$$\oint_c \vec{H}\,d\vec{s} = \int_S \vec{J}\,d\vec{S} + \int_S \frac{\partial \vec{D}}{\partial t} \cdot d\vec{S}. \tag{1.9}$$

Equation (1.6) is the Ampere circular rule with Maxwell addition of the second term on the right-hand side (the displacement current). The generalized Ampere law states that either an electric current or a time-varying electric flux gives rise to magnetic field.

Taking the volume integral over (1.6) and applying the Gauss divergence theorem results in

$$\oint_S \vec{D}\,d\vec{S} = \int_V \rho\,dV, \tag{1.10}$$

where the right-hand side represents the total charge within the volume $V$.

Equation (1.6) is the Gauss flux law for the electric field stating that the flux of $D$ vector corresponds to the total electric charge within the domain.

The Gauss flux law for the magnetic field can be derived by taking the volume integral of (1.7) and applying the Gauss divergence theorem, i.e.

$$\oint_S \vec{B}\,d\vec{S} = 0, \tag{1.11}$$

stating that the flux of $B$ vector over any closed surface $S$ is identically zero.

What is also necessary for a description of electromagnetic phenomena in a linear medium are the constitutive equations:

$$\vec{D} = \varepsilon\vec{E}, \tag{1.12}$$

$$\vec{J} = \sigma\vec{E}, \tag{1.13}$$

$$\vec{B} = \mu\vec{H}, \tag{1.14}$$

and the Lorentz force equation

$$\vec{F} = q(\vec{v} \times \vec{B}), \tag{1.15}$$

where $q$ denotes the charged particle, $v$ is the particle velocity, $\varepsilon$ is permittivity, $\sigma$ is conductivity, and $\mu$ is permeability of a medium, respectively.

To solve Maxwell's equations for a given problem the continuity conditions at the interface of two media with different electrical properties must be specified [2]:

$$\vec{n} \times (\vec{E}_1 - \vec{E}_2) = 0, \tag{1.16}$$

$$\vec{n} \times (\vec{H}_1 - \vec{H}_2) = \vec{J}_s, \tag{1.17}$$

$$\vec{n}(\vec{D}_1 - \vec{D}_2) = \rho_s, \tag{1.18}$$

$$\vec{n}(\vec{B}_1 - \vec{B}_2) = 0, \tag{1.19}$$

where $\vec{n}$ is a unit normal vector directed from medium 1 to medium 2, and subscripts 1 and 2 denote fields in regions 1 and 2. Equations (1.16) and (1.19) state that the tangential components of $E$ and the normal components of $B$ are continuous across the boundary. Equation (1.17) represents that the tangential component of $H$ is discontinuous due to the surface current density $J_s$ induced on the boundary. Equation (1.18) means that the discontinuity in the normal component of $D$ is the same as the surface charge density $\rho_s$ on the boundary.

In the case of a perfect conductor, the electric field $E$ and magnetic field $H$ vanish within the perfectly conducting medium. These fields are replaced by the surface charge density $\rho_s$ and surface current density $J_s$. At higher frequencies, there is a well-known effect that confines current largely to surface regions. The so-called skin depth in common situations is often sufficiently small for the surface phenomenon to be an accurate representation. Therefore, the familiar rules for the behavior of time-varying fields at a boundary defined by good conductors follow directly from consideration of the limit condition, i.e. when the conductor is perfect.

As no time-varying field exists in a perfect conductor, the electric flux density is entirely normal to the conductor and supported by a surface charge density at the interface.

$$D_n = \rho_{s.} \tag{1.20}$$

The magnetic field is entirely tangential to the perfect conductor and is equilibrated by a surface current density:

$$H_s = J_s. \tag{1.21}$$

Conditions at the extremes of the boundary value problem are obtained by extending the interface conditions.

### 1.2.3 From the Corner of Philosophy of Science

Essentially, development of Maxwell kinetic theory of gases and electromagnetic field theory was not motivated by experimental findings which were not compatible with existing paradigm (in a sense of T.S. Kuhn [3]), as was the case with relativity and quantum mechanics.

As already explained in *1.1.3*, almost all facts from electrodynamics, known in Maxwell time, were interpreted relatively satisfactory within the Newton paradigm. That approach was standardly used till Hertz experimental verification of the very existence of electromagnetic waves.

The origin of Maxwell's ideas to replace the *action at a distance* concept with the concept of *physical field*, was essentially philosophical, or a pure abstract, mathematical thought, and came from previous works of Faraday and Boskovic.

Maxwell introduced a revolution not only in electromagnetics but also in thermodynamics [9]. His approach to represent a physical phenomenon in terms of statistical function was a remarkable improvement in science generally. Such an approach led not only to the statistical nature of the second law of thermodynamics, but also provided the development of mathematical description of quantum mechanics.

Furthermore, his introduction of today's famous Maxwell's demon having questioned the second law of thermodynamics contributed to the development of information theory in the twentieth century.

No doubt, Maxwell, himself, was and is one of the greatest scientists of all times.

### 1.2.4 FDTD Solution of Maxwell's Equations

One of the widely used approaches for a direct solution of Maxwell's equations in the last few decades is the use of the finite difference time domain (FDTD) method. FDTD solution of Maxwell's equations is based on discretizing the differential equations by means of pulse basis approximation and converting them into a finite difference equation. FDTD is a highly versatile method enabling the analysis of objects with a wide range of size and complexity, from a microstrip circuit to a helicopter or the human body. The method discretizes a domain of interest into unit cells, leading to the use of the so-called staircase approximation for smoothly curved surfaces or volumes. The unit cell, usually called "the Yee cell" after Yee who introduced the first FDTD algorithm [10], is shown in Figure 1.3.

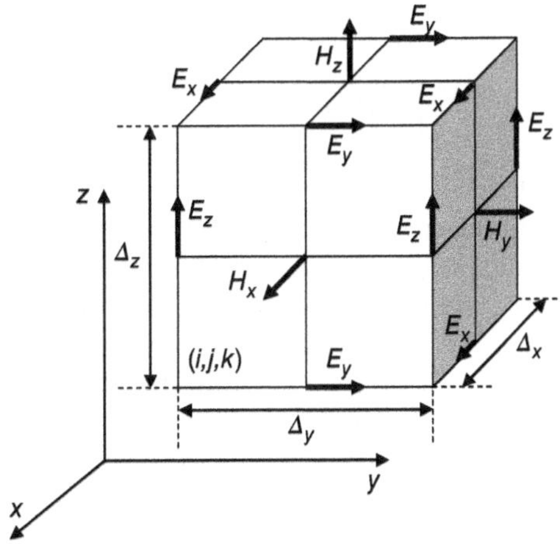


**Figure 1.3** Configuration of electric and magnetic fields in a cell.

Assuming isotropic and nondispersive media, Maxwell's equations (1.4) and (1.5) can be written as

$$\nabla \times \vec{E}(r,t) = -\mu \frac{\partial \vec{H}(r,t)}{\partial t}, \tag{1.22}$$

$$\nabla \times \vec{H}(r,t) = \sigma \vec{E}(r,t) + \varepsilon \frac{\partial \vec{E}(t,r)}{\partial t}. \tag{1.23}$$

Using the Yee cell [10] and performing the space–time finite difference discretization one obtains [11]

$$E_x^{n+1}\left(i+\frac{1}{2},j,k\right) = A \cdot E_x^n\left(i+\frac{1}{2},j,k\right) + B \cdot \left[\frac{H_z^{n+\frac{1}{2}}\left(i+\frac{1}{2},j+\frac{1}{2},k\right) - H_z^{n+\frac{1}{2}}\left(i+\frac{1}{2},j-\frac{1}{2},k\right)}{\Delta y} - \frac{H_y^{n+\frac{1}{2}}\left(i+\frac{1}{2},j,k+\frac{1}{2}\right) - H_y^{n+\frac{1}{2}}\left(i+\frac{1}{2},j,k-\frac{1}{2}\right)}{\Delta z}\right], \tag{1.24}$$

$$H_x^{n+\frac{1}{2}}\left(i,j+\frac{1}{2},k+\frac{1}{2}\right) = H_x^{n-\frac{1}{2}}\left(i,j+\frac{1}{2},k+\frac{1}{2}\right) - \frac{\Delta t}{\mu} \cdot \left[\frac{E_z^n\left(i,j+1,k+\frac{1}{2}\right) - E_z^n\left(i,j,k+\frac{1}{2}\right)}{\Delta y} - \frac{E_y^n\left(i,j+\frac{1}{2},k+1\right) - E_y^n\left(i,j+\frac{1}{2},k\right)}{\Delta z}\right], \tag{1.25}$$

where $A$ and $B$ are given by

$$A = \frac{1 - \dfrac{\sigma \cdot \Delta t}{2 \cdot \varepsilon}}{1 + \dfrac{\sigma \cdot \Delta t}{2 \cdot \varepsilon}}, \quad B = \frac{\dfrac{\Delta t}{\varepsilon}}{1 + \dfrac{\sigma \cdot \Delta t}{2 \cdot \varepsilon}}. \tag{1.26}$$

Other electromagnetic field components are obtained by a simple circular permutation of the space variables. It is possible to improve the original FDTD formulation to provide more accurate modeling of smooth curves, but with the price of increasing the complexity of the algorithms. Also, the FDTD method encounters some difficulties in modeling thin wires and the related feed-gap concept [9]. Moreover, within the FDTD method $E$ and $H$ fields are not computed exactly at a distance of half a cell from each other.

The thin wire itself is defined as a conductive object with radius smaller than the size of an FDTD cell [9, 11].

In many engineering applications, such as antenna analysis and design and grounding systems studies, it is necessary to model electrically thin conducting cylinders requiring the radii $r_0$ of such conducting structures to be smaller than the smallest Yee cell dimensions. Thus, a special formulation must be implemented to accurately represent these radii. The details can be found elsewhere, e.g. in [11].

Contour integral approach [9], derived from Maxwell's equations in integral form, is used to account for the air–ground interface. It is convenient to treat the problems including inhomogeneous media and a nonuniform spatial discretization by using Maxwell's equations in their integral form [11].

Figure 1.4 shows a grid of FDTD discretization of the air–ground interface. Thus, from the generalized Ampere's law (1.9) one obtains

$$\oint_c \vec{H}\cdot d\vec{s} = \left[H_z^{n+\frac{1}{2}}(i+1/2,j+1/2,k) - H_z^{n+\frac{1}{2}}(i+1/2,j-1/2,k)\right]\cdot\Delta z + \left[H_y^{n+\frac{1}{2}}(i+1/2,j,k-1/2) - H_y^{n+\frac{1}{2}}(i+1/2,j,k+1/2)\right]\cdot\Delta y, \quad (1.27)$$

$$\iint_s \left(\sigma\vec{E}_x + \varepsilon\frac{\partial\vec{E}_x}{\partial t}\right)d\vec{s} = \frac{\Delta y\Delta z}{2}\cdot\varepsilon_0\cdot\frac{\partial E_x}{\partial t} + \frac{\Delta y\Delta z}{2}\cdot\left(\varepsilon_s\frac{\partial E_x}{\partial t} + \sigma_s E_x\right). \quad (1.28)$$

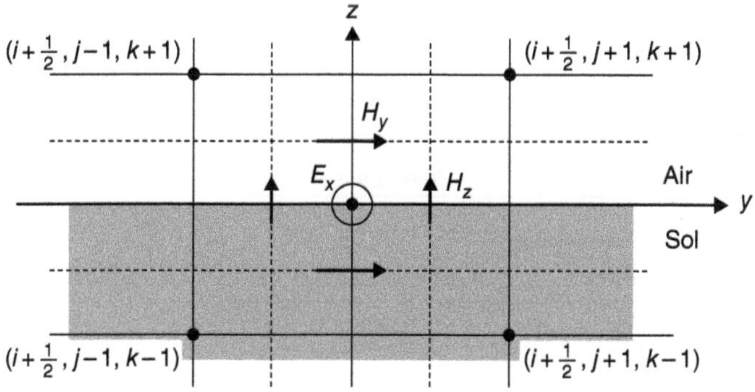


**Figure 1.4** Treatment of the air–ground interface by the "the contour integral approach" method.

Combining (1.27) and (1.28) yields

$$E_x^{n+1}(i+1/2,j,k) = \frac{K}{N} \cdot E_x^n(i+1/2,j,k) + \frac{1}{N \cdot \Delta y} \cdot \left[ H_z^{n+\frac{1}{2}}(i+1/2,j+1/2,k) - H_z^{n+\frac{1}{2}}(i+1/2,j-1/2,k) \right] - \frac{1}{N \cdot \Delta z} \cdot \left[ H_y^{n+\frac{1}{2}}(i+1/2,j,k+1/2) - H_y^{n+\frac{1}{2}}(i+1/2,j,k-1/2) \right], \tag{1.29}$$

where

$$K = \frac{1}{2\Delta t}(\varepsilon_0 + \varepsilon_s) - \frac{\sigma_s}{4}, \quad N = \frac{1}{2\Delta t}(\varepsilon_0 + \varepsilon_s) + \frac{\sigma_s}{4}. \tag{1.30}$$

The other field components can be derived in a similar manner. It is also necessary to truncate the calculation domain. Therefore, absorbing regions must be implemented at the domain's limits, simulating the wave propagation thus avoiding non-natural reflections in infinite domain. In this context, the absorbing conditions are used [12].

### 1.2.5 Computational Examples

Computational examples are related to the transient behavior of grounding systems in two-layer soil. FDTD solution of Maxwell's equations is performed by taking into account the variation in conductivity between the conductive layers of the soil. The obtained FDTD results are compared to the numerical results calculated via the TL approach [11].

The first example deals with a horizontal grounding electrode in two-layer stratified soil. Figure 1.5 shows the electrode buried horizontally at depth

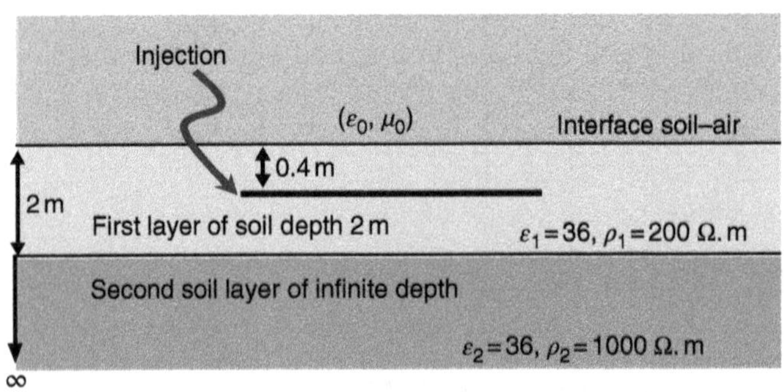


**Figure 1.5** Horizontal grounding electrode in two-layer soil.

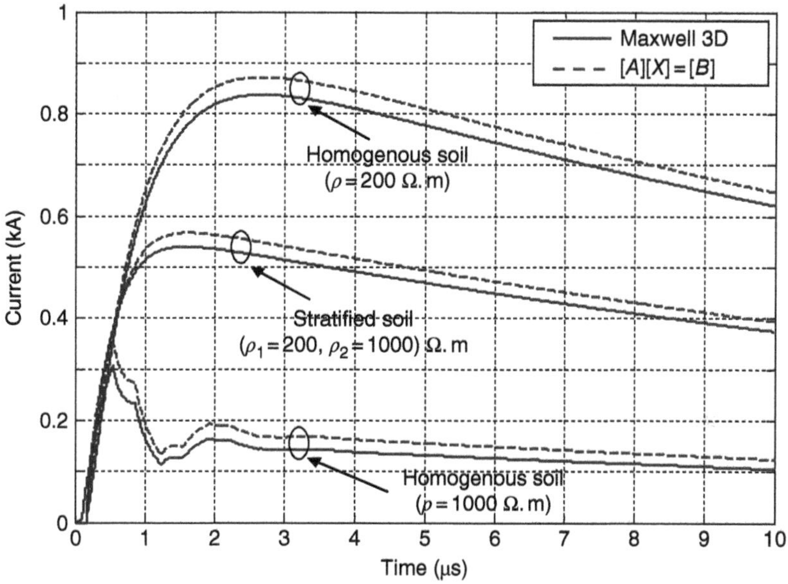


**Figure 1.6** Transinet current induced at the middle of the buried electrode.

$d = 0.4$ m from the soil–air interface. The radius of the conductor is $a = 7$ mm and its length is $L = 20$ m, while the height of the upper layer is $D = 2$ m.

The electrode is excited by a voltage source given by double exponential function [11]

$$V(t) = V_0(e^{-at} - e^{-bt}), \tag{1.31}$$

where $V_0 = 30$ kV, $a = 45\,099\ s^{-1}$, $b = 9\,022\,879\ s^{-1}$.

Physical constants $\varepsilon$, $\rho$ of the medium are depicted in Figure 1.5.

Figure 1.6 shows the transient current induced along the horizontal grounding electrode calculated by FDTD and the TL approach, respectively.

The results obtained via different approaches seem to agree satisfactorily. The next example is a grounding grid in two-layer stratified soil. Figure 1.7 shows the grid buried at $d = 0.8$ m depth in the stratified soil. The size of the grounding grid is 20 m × 20 m. The physical constants ($\varepsilon$, $\rho$) of the medium are shown in Figure 1.7. The radius of the conductors is $a = 7$ mm. The grounding grid system is excited by double exponential voltage impulse (1.31). Two cases are considered, with the injection point being on the corner and in the middle of the grounding grid system, respectively, as shown in Figure 1.7.

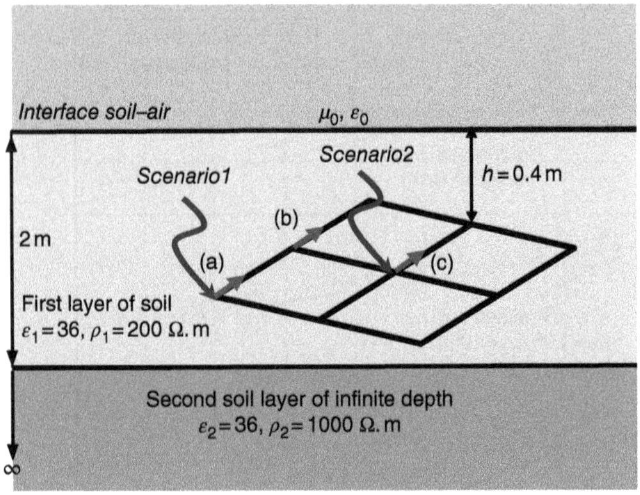


**Figure 1.7** Horizontal grounding grid in two-layer soil.

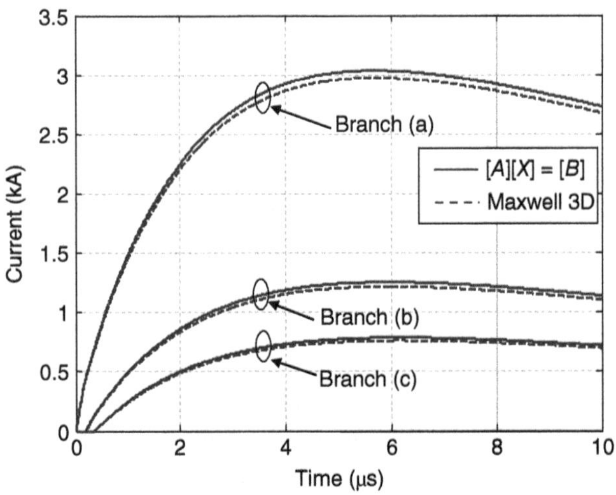


**Figure 1.8** Transient currents induced at different points for the horizontal grounding grid (Scenario 1).

Figures 1.8 and 1.9 show transient currents induced on the horizontal grounding grid obtained as FDTD solution of Maxwell's equations and using the TL approach, respectively.

Two different scenarios are studied (corner injection – scenario 1, and central point injection – scenario 2 of the current source).

The numerical results obtained by different approaches are in satisfactory agreement, again.

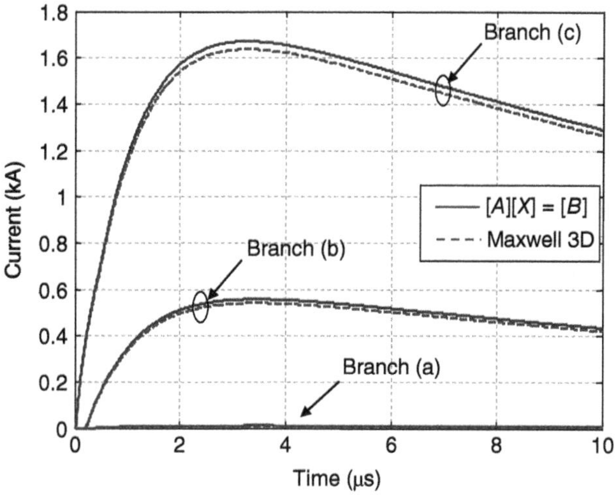


**Figure 1.9** Transient currents induced at different points for the horizontal grounding grid (Scenario 2).

## 1.3 The Electromagnetic Wave Equations

Maxwell's equations are coupled first order space–time partial differential equations that are very difficult to apply when solving boundary-value problems. One way to overcome the difficulty of solving coupling equations is to decouple these first order equations, thereby obtaining the second order electromagnetic wave equations.

The wave equations are readily derived from the Maxwell curl equations, by differentiation and substitution. Taking curl on both sides of equation (1.5) leads to

$$\nabla \times \nabla \times \vec{H} = \nabla \times \vec{J} + \frac{\partial}{\partial t}(\nabla \times \vec{D}). \tag{1.32}$$

Using constitutive equations (1.12) and (1.13) and assuming uniform scalar material properties yields

$$\nabla \times \nabla \times \vec{H} = \sigma \nabla \times \vec{E} + \varepsilon \frac{\partial}{\partial t}(\nabla \times \vec{E}). \tag{1.33}$$

According to the Maxwell equation (1.4) curl of $E$ is replaced by the rate of change of magnetic flux density, and using (1.14) it follows that

$$\nabla \times \nabla \times \vec{H} = -\mu\sigma \frac{\partial \vec{H}}{\partial t} - \mu\varepsilon \frac{\partial^2 \vec{H}}{\partial t^2}. \tag{1.34}$$

Performing some mathematical manipulations, the same equations can be derived for the electric field.

Using the standard vector identity valid for any vector $E$,

$$\nabla \times \nabla \times \vec{H} = \nabla \cdot (\nabla \vec{E}) - \nabla^2 \vec{H}. \tag{1.35}$$

Taking into account the solenoidal nature of the magnetic field (1.7) yields the final form of the wave equation:

$$\nabla^2 \vec{H} - \mu\sigma \frac{\partial \vec{H}}{\partial t} - \mu\varepsilon \frac{\partial^2 \vec{H}}{\partial t^2} = 0. \tag{1.36}$$

If a linear, isotropic, homogeneous, source-free medium is considered then the set of equations (1.36) simplifies into

$$\nabla^2 \vec{H} - \frac{1}{v^2} \frac{\partial^2 \vec{H}}{\partial t^2} = 0, \tag{1.37}$$

where $v$ denotes the wave propagation velocity in lossless homogeneous medium:

$$v = \frac{1}{\sqrt{\mu\varepsilon}}. \tag{1.38}$$

The velocity of wave propagation in free space is the velocity of light:

$$c = \frac{1}{\sqrt{\mu_0 \varepsilon_0}}, \tag{1.39}$$

where $c = 3 \times 10^8$ m s$^{-1}$, approximately.

## 1.4 Conservation Laws in the Electromagnetic Field

A general relationship for power and energy expressed in terms of electric and magnetic fields is given in the form of Poynting theorem.

The conservation law of electromagnetic energy can be obtained from curl Maxwell equations.

An equivalence of vector operators yields

$$\nabla \cdot (\vec{E} \times \vec{H}) = \vec{H} \cdot \nabla \times \vec{E} + \vec{E} \cdot \nabla \times \vec{H}. \tag{1.40}$$

Combining Equations (1.4), (1.5), and (1.40), one has

$$\vec{E} \cdot \frac{\partial \vec{D}}{\partial t} + \vec{H} \cdot \frac{\partial \vec{B}}{\partial t} = -\vec{E} \cdot \vec{J} + \vec{E} \cdot (\nabla \times \vec{H}) - \vec{H} \cdot (\nabla \times \vec{E}), \tag{1.41}$$

or in the alternative form,

$$\frac{\partial w}{\partial t} = -\vec{E} \cdot \vec{J} - \nabla \cdot (\vec{E} \times \vec{H}), \tag{1.42}$$

where the $w$ term represents the energy storage per unit volume for an electromagnetic field:

$$w = \frac{1}{2}(\vec{E}\vec{D} + \vec{H}\vec{B}). \tag{1.43}$$

Integrating (1.41) over some finite region in space one obtains the integral form of the electromagnetic energy conservation law:

$$\int_V w\mathrm{d}V = -\int_V \vec{E}\cdot\vec{J}\mathrm{d}V - \int_V \nabla\cdot(\vec{E}\times\vec{H})\mathrm{d}V. \tag{1.44}$$

The left-hand side term is the time rate of the stored energy in the electric and magnetic fields of the region.

The first term on the right-hand side represents the Joule heat (the ohmic power loss if $J$ is a conduction current density or the power required to accelerate charges if $J$ is a convection current arising from moving charges). If there is an energy source then the product $EJ$ is negative for that source and represents energy flow out of the region.

The other term on the right-hand side gives the flow into the domain boundary.

Applying the Gauss integral theorem to the last term of (1.44)

$$\int_V \nabla\cdot(\vec{E}\times\vec{H})\mathrm{d}V = \oint_S(\vec{E}\times\vec{H})\cdot\mathrm{d}\vec{S}, \tag{1.45}$$

the volume integral transforms to the surface integral over the boundary, where $\mathrm{d}\vec{S}$ is the outward drawn normal vector surface element.

Since all the energy changes must be supplied externally, this term represents the energy flow into the volume per unit time due to the minus sign of the surface integral. Changing sign, the rate of energy flow, or power flow, out through the enclosing surface is given by

$$P = \oint_S(\vec{E}\times\vec{H})\mathrm{d}\vec{S}, \tag{1.46}$$

where $\vec{E}\times\vec{H}$ is the Poynting vector representing power density flow – flow of energy per unit area per unit time at the surface (power density flow), known as Poynting vector:

$$\vec{P}_\mathrm{d} = \vec{E}\times\vec{H}. \tag{1.47}$$

The Poynting vector (1.47) gives the direction and magnitude of energy flow density at any point in space.

Power flow does not exist in the vicinity of a system of static charges having electric but no magnetic field. Also, in the vicinity of a perfect conductor there is a zero tangential component normal to the conductor and power flow into the perfect conductor is not possible.

Therefore, the final integral form of the conservation law in the electromagnetic field is then given by

$$\frac{\partial}{\partial t}\int_V \frac{1}{2}(\vec{E}\cdot\vec{D}+\vec{H}\cdot\vec{B})\mathrm{d}V = -\int_V \vec{E}\cdot\vec{J}\ \mathrm{d}V + \oint_S(\vec{E}\times\vec{H})\cdot \mathrm{d}\vec{S}. \tag{1.48}$$

Therefore, the rate of increase of electromagnetic energy in the domain equals the rate of flow of energy in through the domain surface less the Joule heat production in the domain.

For a battery with a nonelectrostatic field $E'$ pumping energy both into heat losses and into a magnetic field is considered; the corresponding current density can be written as

$$\vec{J} = \sigma(\vec{E}+\vec{E}') \tag{1.49}$$

and (1.48) becomes

$$\int_V \vec{E}'\vec{J}\ \mathrm{d}V = \frac{\partial}{\partial t}\int_V \frac{1}{2}(\vec{E}\cdot\vec{D}+\vec{H}\cdot\vec{B})\mathrm{d}V + \int_V \frac{|\vec{J}|}{\sigma}\mathrm{d}V + \oint_S(\vec{E}\times\vec{H})\cdot \mathrm{d}\vec{S}, \tag{1.50}$$

where the term on the left-hand side represents the sources within the volume of interest.

The first and second terms on the right-hand side of (1.50) are the total energy stored in the electric and magnetic fields, respectively.

## 1.5 Density of Quantity of Movement in the Electromagnetic Field

According to the laws of classical mechanics, force $\vec{F}$ is equal to the change in quantity of movement of matter:

$$\vec{F} = \frac{\mathrm{d}\vec{G}_{\mathrm{meh}}}{\mathrm{d}t} = \frac{\mathrm{d}}{\mathrm{d}t}(m\vec{v}) = \frac{\mathrm{d}}{\mathrm{d}t}\int_V \rho_m \vec{v}\,\mathrm{d}V, \tag{1.51}$$

where $\vec{G}_{\mathrm{meh}}$ is the momentum or quantity of movement, $\rho_{\mathrm{m}}$ g is the mass density, and $\vec{v}$ is the velocity.

If an electromagnetic system is subjected to an external force, where force density to charges and currents is given by

$$\vec{f} = \rho\vec{E}+\vec{J}\times\vec{B}. \tag{1.52}$$

The total force on the matter contained within volume $V$, i.e. to charges and currents, is defined by the expression

$$\vec{F} = \int_V \vec{f}\mathrm{d}V = \int_V (\rho\vec{E}+\vec{J}\times\vec{B})\mathrm{d}V. \tag{1.53}$$

Combining (1.51) and (1.53), Newton's second law yields

$$\frac{d\vec{G}_{meh}}{dt} = \int_V (\rho\vec{E} + \vec{J} \times \vec{B})dV. \tag{1.54}$$

Furthermore, for free space one has

$$\rho = \varepsilon_0 \nabla\vec{E} \tag{1.55}$$

and

$$\vec{J} = \frac{1}{\mu_0}\nabla \times \vec{B} - \varepsilon_0 \frac{\partial\vec{E}}{\partial t}. \tag{1.56}$$

So, it follows that

$$\vec{f} = \rho\vec{E} + \vec{J} \times \vec{B} = \varepsilon_0\vec{E}\nabla\vec{E} + \frac{1}{\mu_0}\vec{B} \times \nabla \times \vec{B} - \varepsilon_0 \frac{\partial\vec{E}}{\partial t} \times \vec{B}. \tag{1.57}$$

Now, the last term in (1.57) can be written in the following manner:

$$\varepsilon_0 \frac{\partial\vec{E}}{\partial t} \times \vec{B} = \frac{\partial}{\partial t}(\varepsilon_0\vec{E} \times \vec{B}) - \varepsilon_0\vec{E} \times \frac{\partial\vec{B}}{\partial t} = \frac{\partial}{\partial t}(\varepsilon_0\vec{E} \times \vec{B}) + \varepsilon_0\vec{E} \times \nabla \times \vec{E}. \tag{1.58}$$

and the force density (1.57) becomes

$$\vec{f} = \varepsilon_0\vec{E}\nabla\vec{E} - \frac{1}{\mu_0}\vec{B} \times \nabla \times \vec{B} - \frac{\partial}{\partial t}(\varepsilon_0\vec{E} \times \vec{B}) - \varepsilon_0\vec{E} \times \nabla \times \vec{E}. \tag{1.59}$$

As one has $\nabla\vec{H} = 0$, expression (1.59) can be written as

$$\vec{f} = \varepsilon_0\vec{E}\nabla\vec{E} - \varepsilon_0\vec{E} \times \nabla \times \vec{E} + \mu_0\vec{H}\nabla\vec{H} - \mu_0\vec{H} \times \nabla \times \vec{H} - \frac{\partial}{\partial t}(\varepsilon_0\vec{E} \times \vec{B}), \tag{1.60}$$

and the relation for the total force within the volume becomes

$$\vec{F} = \int_V \vec{f}dV = \varepsilon_0 \int_V (\vec{E}\nabla\vec{E} - \vec{E} \times \nabla \times \vec{E})dV + \frac{1}{\mu_0}\int_V (\vec{B}\nabla\vec{B} - \vec{B} \times \nabla \times \vec{B})dV - \frac{d}{dt}\int_V (\varepsilon_0\mu_0\vec{E} \times \vec{H})dV. \tag{1.61}$$

As for an arbitrary vector function $\vec{A}$ one has

$$\int_V (\vec{A}\nabla\vec{A} - \vec{A} \times \nabla \times \vec{A})dV = \oint_S \left[\vec{A}(\vec{n} \cdot \vec{A}) - \frac{1}{2}\vec{n}A^2\right] dS, \tag{1.62}$$

it follows that

$$\vec{F} = \oint_S \vec{T}\,dS - \frac{d}{dt}\int_V \frac{1}{c^2}\vec{E} \times \vec{H}\,dV, \tag{1.63}$$

where

$$\vec{T} = \varepsilon_0 \left[\vec{E}(\vec{n}\cdot\vec{E}) - \frac{1}{2}\vec{n}E^2\right] + \frac{1}{\mu_0}\left[\vec{B}(\vec{n}\cdot\vec{B}) - \frac{1}{2}\vec{n}B^2\right] \tag{1.64}$$

is the force over the surface unit and is referred as stress tensor.

According (9.4) and (9.13) it can be written that

$$\frac{d\vec{G}_{meh}}{dt} = \oint_S \vec{T}\, dS - \frac{d}{dt}\int_V \frac{1}{c^2}\vec{E}\times\vec{H}\, dV, \tag{1.65}$$

or

$$\frac{d}{dt}\left(\vec{G}_{meh} + \int_V \frac{1}{c^2}\vec{E}\times\vec{H} dV\right) = \oint_S \vec{T}\, dS. \tag{1.66}$$

If one deals with an isolated system the term from the right-hand side in (9.16.), which accounts for the surface stress, vanishes and from (9.1) it follows that

$$\frac{d}{dt}\int_V (\rho_m\vec{v} + \varepsilon_0\vec{E}\times\vec{B})dV = 0. \tag{1.67}$$

So the expression for the laws of the conservation of quantity of movement in an isolated mechanic-electromagnetic system must be modified, and the expression

$$\vec{g}_{EM} = \varepsilon_0\vec{E}\times\vec{B} = \frac{1}{c^2}\vec{E}\times\vec{H} \tag{1.68}$$

is considered to be the density of the quantity of movement of the electromagnetic field.

Electromagnetic momentum, in accordance to mechanics, is then given by

$$\vec{G}_{EM} = \int_V (\vec{D}\times\vec{B})dV = \int_V \left(\frac{1}{c^2}\vec{E}\times\vec{H}\right) dV. \tag{1.69}$$

Therefore, if an arbitrary isolated mechanical system of mass density $\rho_m$ and velocity $v$ within a small volume $V$ is considered the quantity of movement is conserved. The law of conservation of quantity of movement, which accounts for *electromagnetic quantity of movement*, could be written in the form

$$\frac{d}{dt}(\vec{G}_{meh} + \vec{G}_{EM}) = 0, \tag{1.70}$$

i.e. it follows that

$$\vec{G}_{meh.} + \vec{G}_{EM} = \text{Const}. \tag{1.71}$$

The existence of stress in the electromagnetic field and the expression of electromagnetic quantity of movement prove the existence of the electromagnetic field as a real physical entity, thus confirming the vision of Michael Faraday.

## 1.6 Electromagnetic Potentials

Instead of using fields the analysis of electric and magnetic fields can be simplified by using auxiliary potential functions, such as the electric scalar potential $\varphi$, or the magnetic vector potential $A$. The potential functions are readily derived from the Maxwell equations.

Thus, the Maxwell equation (1.7) is satisfied if the flux density $B$ can be expressed in terms of an auxiliary vector $A$, i.e.

$$\vec{B} = \nabla \times \vec{A}. \tag{1.72}$$

Maxwell curl equation (1.4) then becomes

$$\nabla \times \vec{E} = -\frac{\partial}{\partial t}(\nabla \times \vec{A}), \tag{1.73}$$

and by rearranging (1.73), it follows that

$$\nabla \times \left(\vec{E} + \frac{\partial \vec{A}}{\partial t}\right) = 0. \tag{1.74}$$

Furthermore, the quantity within brackets in (1.74) can be written as the gradient of the scalar potential function $\varphi$:

$$\vec{E} + \frac{\partial \vec{A}}{\partial t} = -\nabla \varphi \tag{1.75}$$

or

$$\vec{E} = -\frac{\partial \vec{A}}{\partial t} - \nabla \phi. \tag{1.76}$$

Thus, knowing the potential functions $A$ and $\varphi$, the magnetic and electric fields can be determined from Equations (1.72) and (1.76).

## 1.7 Solution of the Wave Equation and Radiation Arrow of Time

From the physical point of view natural laws expressed in the mathematical form are symmetric, i.e. they remain the same if one changes the direction of time [11, 13, 14].

Thus, Maxwell's equations appear to be time invariant (there is no preference regarding the time direction). However, the electromagnetic wave equations that are derived from Maxwell's equations have *two solutions* in the form of retarded and advanced potential, respectively. The *retarded* potential solution is considered to have physical meaning and is related to the electromagnetic waves detected at an observation point after they left the source, i.e. the time required for these waves to reach a receiver is delayed with respect to the time measured at the source. On the other hand, the advanced potential solutions pertaining to the waves that would propagate in such a way as to arrive at the detector before they leave the source are mathematically also possible. Such waves are never observed in nature and are eliminated by specifying certain set of boundary and initial conditions, respectively. Note that the physical laws formulated in terms of differential equations are not sufficient to fully describe the natural phenomena as the corresponding initial and boundary conditions, respectively, are to be specified [14].

Basically, this "wave asymmetry," i.e. nonexistence of convergent waves is due to the radiation. Thus, divergent radiated fields exist (related with accelerated charges) and their temporal inverses (convergent waves) are never observed in nature.

The radiation arrow of time could be analyzed by studying the solutions of wave equation for magnetic vector potential $A$:

$$\nabla^2\vec{A} - \mu\varepsilon\frac{\partial^2\vec{A}}{\partial t^2} = -\mu\vec{J}. \tag{1.77}$$

The wave equation (1.77) has *two solutions*, one in the form of *retarded potential*

$$\vec{A}_{\text{ret}}(r,t) = \frac{\mu}{4\pi}\int_{V'}\frac{\vec{J}(\vec{r}',t-R/c)}{R}\mathrm{d}V', \tag{1.78}$$

and the other one in the form of *advanced potential*

$$\vec{A}_{\text{adv}}(r,t) = \frac{\mu}{4\pi}\int_{V'}\frac{\vec{J}(\vec{r}',t+R/c)}{R}\mathrm{d}V', \tag{1.79}$$

where $R = |\vec{r} - \vec{r}'|$ is the distance from the source to the observation point, respectively.

Note that $R/c$ is the time necessary for the signal to arrive from the source point to the observation point (Figure 1.10).

In the case of divergent waves a signal at the source point at time $t' = t - R/c$ is advanced for time $R/c$ compared to the signal at the observation point, i.e. the signal at the observation point at time $t = t' + R/c$ is delayed for time $R/c$ compared to the signal at the source point. Similarly, in the case of convergent waves a signal at the source point at time $t' = t + R/c$ is delayed for $R/c$ compared to the signal at the observation point, or the signal at an observation point at time $t = t' - R/c$ is advanced for time $R/c$ at an observation point with respect to the source point. The source and observation point, respectively, are depicted in Figure 1.10.

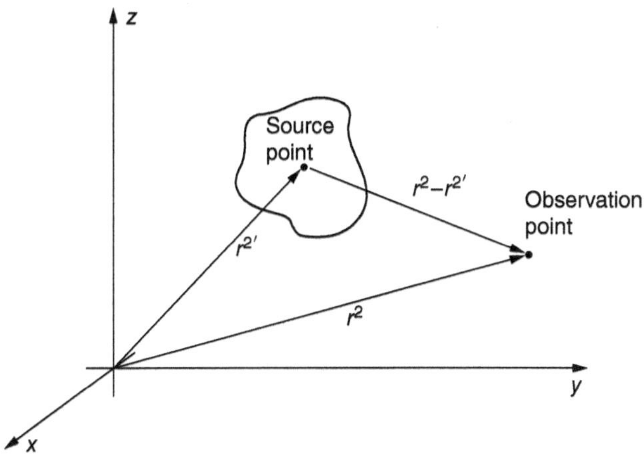


Figure 1.10 The source point and the observation point.

## 1.8 Complex Phasor Form of Equations in Electromagnetics

In many engineering scenarios, devices or systems are excited sinusoidally and a time-harmonic variation of electromagnetic fields is assumed. In such cases, it is convenient to represent the variables of interest in a complex phasor form. Also, if a transient response is of interest, inverse Fourier transform (IFT) is used to transform the frequency response in the time domain.

### 1.8.1 The Generalized Symmetric Form of Maxwell's Equations

For a simple medium the time-harmonic, *symmetric* form of Maxwell's equations, i.e. the form in which both electric and fictitious magnetic charges and currents are taken into account, is given by Poljak and Tham [15]

$$\nabla \times \vec{E} = -j\omega\mu\vec{H} - \vec{M}, \tag{1.80}$$

$$\nabla \times \vec{H} = j\omega\varepsilon\vec{E} + \vec{J}, \tag{1.81}$$

$$\nabla \cdot \vec{E} = \frac{1}{\varepsilon}\rho_e, \tag{1.82}$$

$$\nabla \cdot \vec{H} = \frac{1}{\mu}\rho_m, \tag{1.83}$$

in which the fictitious magnetic surface current $M$ and magnetic charge density $\rho_m$ are introduced. The time-harmonic factor $e^{j\omega t}$, which is implied, has been omitted in the equations.

Electric current $J$ and charge $\rho_e$ give rise to electric field $E$ and magnetic field $H$, which may be expressed in terms of the magnetic vector potential $A$ and electric scalar potential $\varphi$:

$$\vec{E} = -j\omega\vec{A} - \nabla\varphi, \tag{1.84}$$

$$\vec{H} = \frac{1}{\mu}\nabla \times \vec{A}. \tag{1.85}$$

Similarly, the effects of the fictitious magnetic current $M$ and charge $\rho_m$ may be expressed in terms of the electric vector potential **F** and the magnetic scalar potential $\Psi$:

$$\vec{H} = -j\omega\vec{F} - \nabla\Psi, \tag{1.86}$$

$$\vec{E} = -\frac{1}{\varepsilon}\nabla \times \vec{F}. \tag{1.87}$$

Two general field equations may be obtained by combining the effects of **J**, **M**, $\rho_{e,}$ and $\rho_m$ in Equations (1.84)–(1.87) as follows:

$$\vec{E} = -j\omega\vec{A} - \nabla\varphi - \frac{1}{\varepsilon}\nabla \times \vec{F}, \tag{1.88}$$

$$\vec{H} = -j\omega\vec{F} - \nabla\Psi - \frac{1}{\mu}\nabla \times \vec{A}. \tag{1.89}$$

To express the fields in Equations (1.88) and (1.89) in terms of the sources **J** and **M**, first the *non-homogeneous Helmholtz's equations* relating to **J** and $\rho_e$ are derived from Maxwell's equations as follows:

$$\nabla^2\vec{A} + k^2\vec{A} = -\mu\vec{J}, \tag{1.90}$$

$$\nabla^2\varphi + k^2\varphi = -\frac{\rho_e}{\varepsilon}, \tag{1.91}$$

where

$$k = \omega\sqrt{\mu\varepsilon} = \frac{\omega}{c} \tag{1.92}$$

is the *wave number* and c is the velocity of light.

The magnetic counterparts are

$$\nabla^2\vec{F} + k^2\vec{F} = -\varepsilon\vec{M}, \tag{1.93}$$

$$\nabla^2\Psi + k^2\Psi = -\frac{\rho_m}{\mu}. \tag{1.94}$$

Solutions of the Helmholtz's equations are given in terms of the four potentials as follows:

$$\vec{A}(\vec{r}) = \frac{\mu}{4\pi}\iint_S \vec{J}(\vec{r}')\ G(\vec{r},\vec{r}')\ \mathrm{d}\vec{S}(\vec{r}'), \tag{1.95}$$

$$\vec{F}(\vec{r}) = \frac{\varepsilon}{4\pi}\iint_S M(\vec{r}')\ G(\vec{r},\vec{r}')\ \mathrm{d}\vec{S}(\vec{r}'), \tag{1.96}$$

$$\varphi(\vec{r}) = -\frac{1}{j4\pi\omega\varepsilon} \iint_S \nabla_s' \cdot \vec{J}(\vec{r}')\ G(\vec{r},\vec{r}')\ \mathrm{d}\vec{S}(\vec{r}'), \tag{1.97}$$

$$\Psi(\vec{r}) = -\frac{1}{j4\pi\omega\mu} \iint_S \nabla_s' \cdot \vec{M}(\vec{r}')\ G(\vec{r},\vec{r}')\ \mathrm{d}\vec{S}(\vec{r}'), \tag{1.98}$$

where

$$G(\vec{r},\vec{r}') = \frac{e^{-jk|\vec{r}-\vec{r}'|}}{|\vec{r}-\vec{r}'|}, \tag{1.99}$$

is the *free space Green's function*, $\vec{r}$ being the position vector at the observation point and $\vec{r}'$ that at the source point. The equations of continuity expressing the conservation of charge in time-harmonic form given below are used in Equations (1.97) and (1.98).

$$\nabla' \cdot \vec{J}(\vec{r}') = -j\omega\rho_{\mathrm{e}}, \tag{1.100}$$

$$\nabla' \cdot \vec{M}(\vec{r}') = -j\omega\rho_{\mathrm{m}}. \tag{1.101}$$

Equations (1.95)–(1.98) may be substituted into the general field equations (1.88) and (1.89) to complete the expressions in which the field quantities $E$ and $H$ (the effects) are expressed in terms of the sources $J$ and $M$ (the causes).

### 1.8.2 Complex Phasor Form of Electromagnetic Wave Equations

The complex phasor representation of the wave equation (1.36) results in the following equation of the Helmholtz type:

$$\nabla^2\vec{H} - \gamma^2\vec{H} = 0, \tag{1.102}$$

where $\gamma$ is the complex propagation constant given by

$$\gamma = \sqrt{j\omega\mu\sigma - \omega^2\mu\varepsilon}. \tag{1.103}$$

For a linear, isotropic, homogeneous, source-free medium the Helmholtz equation (1.102) simplifies into

$$\nabla^2\vec{H} + k^2\vec{H} = 0, \tag{1.104}$$

where $k$ is the wave number of a lossless medium given by (1.92).

The complex form of potential wave equations could be derived similarly.

### 1.8.3 Poynting Theorem for Complex Phasors

Using the vector identity

$$\nabla(\vec{E} \times \vec{H}^*) = \vec{H}^*\nabla \times \vec{E} + \vec{E}\nabla \times \vec{H}^*, \tag{1.105}$$

where the asterisk stands for the complex conjugate, and taking into account the complex phasor form of curl Maxwell's equations (1.80) and (1.81) neglecting the fictitious magnetic sources,

$$\nabla(\vec{E}\times\vec{H}^*) = \vec{H}^*(-j\omega\vec{B}) - \vec{E}^*(\vec{J}^* - j\omega\vec{D}^*). \tag{1.106}$$

Integrating (1.106) through volume $V$ and applying the divergence theorem, one has

$$\begin{aligned}\int_V \nabla(\vec{E}\times\vec{H}^*)\mathrm{d}V &= \oint_S (\vec{E}\times\vec{H}^*)\mathrm{d}\vec{S}\\ &= -\int_V \left[\vec{E}\vec{J}^* + j\omega(\vec{H}^*\vec{B} - \vec{E}\vec{D}^*)\right]\mathrm{d}V.\end{aligned} \tag{1.107}$$

Equation (1.107) is the general Poynting theorem when dealing with complex phasors. In an isotropic medium (in which all losses occur through conduction currents $J = \sigma E$) (1.107) then becomes

$$\oint_S (\vec{E}\times\vec{H}^*)\mathrm{d}\vec{S} = -\int_V \sigma\vec{E}\vec{E}^*\,\mathrm{d}V - j\omega\int_V (\mu\vec{H}\vec{H}^* - \varepsilon\vec{E}\vec{E}^*)\mathrm{d}V. \tag{1.108}$$

Furthermore, the power density is given by Poljak [2]

$$\begin{aligned}\vec{P}_\mathrm{d}(\vec{r},t) &= \mathrm{Re}[\vec{E}(\vec{r})e^{j\omega t}]\times\mathrm{Re}[\vec{H}(\vec{r})e^{j\omega t}] = \frac{1}{2}[\vec{E}(\vec{r})\times\vec{H}^*(\vec{r})]\\ &\quad+\frac{1}{2}[\vec{E}(\vec{r})\times\vec{H}^*(\vec{r})e^{j2\omega t}],\end{aligned} \tag{1.109}$$

and the time average Poynting vector (the average power density) is then

$$\vec{P}_\mathrm{d,av} = \frac{1}{2\pi}\int_0^{2\pi}\vec{P}_\mathrm{d}(\vec{r},t)d(\omega t) = \frac{1}{2}\mathrm{Re}[\vec{E}\times\vec{H}^*]. \tag{1.110}$$

The ½ factor appears because $E$ and $H$ fields represent peak values, and it should be omitted for root mean square (*rms*) values.

The total average power is then given by the surface integral

$$P_\mathrm{av} = \oint_S \frac{1}{2}\mathrm{Re}[\vec{E}\times\vec{H}^*]\mathrm{d}\vec{S}. \tag{1.111}$$

For example, it could represent the power radiated by an antenna. The first volume integral in the right-hand side of (1.108) represents power loss in the conduction currents and it is just twice the average power loss, which is given by

$$P_\mathrm{L} = \frac{1}{2}\int_V \sigma|\vec{E}|^2\,\mathrm{d}V, \tag{1.112}$$

Finally, the second volume integral in the right-hand side of (1.108) is proportional to the difference between the average value of the energy stored in the magnetic field and the value of the energy stored in the electric field.

## References


1 d'Espagnat, B. (2006). *On Physics and Philosophy*. New York: Princeton University Press.

2 Poljak, D. (2007). *Advanced Modeling in Computational Electromagnetic Compatibility*. Hoboken, NJ: Wiley-Interscience.

3 Kuhn, T.S. (1962). *The Structure of Scientific Revolutions*. University of Chicago Press.

4 Howard, D. (2005). Albert Einstein as a philosopher of science. *Physics Today* **58**: 34–40.

5 Poljak, D., Sokolić, F., and Jakić, M. (2013). *Questioning of Philosophy-Science Relationship with Reference to Concepts of Thomas S. Kuhn (In Croatian)*, 167–184. Zagreb, Croatia: Institute of Philosophy.

6 Arthur, J.W. (2013). The evolution of Maxwell's equations from 1862 to the present day. *IEEE Antennas and Propagation Magazine* **55** (3): 61–81.

7 Maxwell, J.C. (1865). A dynamical theory of the electromagnetic field. *Philosophical Transactions of the Royal Society of London* **166**: 459–512.

8 Sarkar, T.K., Salazar-Palma, M., and Sengupta, D.L. (2009). Who was James Clerk Maxwell and what was and is his electromagnetic theory? *IEEE Antennas and Propagation Magazine* **51** (4): 97–116.

9 Poljak, D., Nekhoul, B., and El Khamlichi Drissi, K. (2015). Some remarks related to first 150 years of Maxwell's equations. Proceedings of the 23rd International Conference of Software, Telecommunications and Computer Networks, SoftCOM 2015, Brac Island (16–18 September 2015). Split.

10 Sekki, D., Nekhoul, B., Kerroum, K. et al. (2014). Transient behaviour of grounding system in a two-layer soil using the transmission line theory. *Automatika* **55**: 48–55.

11 Yee, K.S. (1966). Numerical solution of initial boundary value problems involving Maxwell's Equations in isotropic media. *IEEE Transactions on Antennas and Propagation* **14**: 302–307.

12 Taflove, A. and Brodwin, M.E. (1975). Numerical solution of steady-state electromagnetic scattering problems using time-difference-dependent Maxwell's Equations. *IEEE Transactions on Microwave Theory and Techniques* **MTT-23**: 623–630.

13 Dyag, W.M.G., Sarkar, T.K., Garcia-Lamperez, A. et al. (2013). A critical look at the principles of electromagnetic time reversal and its consequences. *IEEE Antennas and Propagation Magazine* **55** (5): 28–62.

14 Albert, D.Z. and Albert, D.Z. (2009). *Time and Chance*. Harvard University Press.

15 Poljak, D. and Tham, C.Y. (2003). *Integral Equation Techniques in Transient Electromagnetics*. Southampton-Boston: WIT Press.

# 2

# Antenna Theory versus Transmission Line Approximation – General Considerations

The chapter deals with simple antenna models applied to EMC problems and related numerical solution methods. The models are based on the corresponding electric field integral equations (EFIE). Also, a correlation between the rigorous antenna theory (AT) approach and transmission line (TL) approximation is presented.

## 2.1 A Note on EMC Computational Models

It is the rapid progress in the development of digital computers that has provided advances in EMC computational models in the last few decades.

Electromagnetic modeling provides the simulation of the electromagnetic behavior of an electrical system for a rather wide variety of parameters including different initial and boundary conditions, excitation types, and different configuration of the system itself. The important fact is that modeling can be undertaken within a significantly shorter time than it would be necessary for building and testing the appropriate prototype via experimental procedures. A basic purpose of an EMC computational model is to predict a victim response to the external excitation generated by a certain EMI source.

In principle, all EMC models arise from the rigorous electromagnetic theory concepts and foundations based on Maxwell's equations. The governing equations of a particular problem in differential, integral, or integro-differential form can be readily derived from the four Maxwell's equations.

EMC models are analyzed using either analytical or numerical methods. Although both approaches can be used in the design of the electrical systems, analytical models are not useful for accurate simulation of electric systems, or their use is restricted to the solution of rather simplified geometries with a high degree of symmetry (canonical problems).

On the other hand, a more accurate simulation of various practical engineering problems is possible by the use of numerical methods. In this case, errors

*Computational Methods in Electromagnetic Compatibility: Antenna Theory Approach versus Transmission line Models*, First Edition. Dragan Poljak and Khalil El Khamlichi Drissi.

are primarily related to the limitations of the mathematical model itself and the applied numerical method of solution, respectively.

EMC computational models can be validated via experimental measurements or by theoretically comparing the results to already well-established numerical models.

It is also possible to a test a new model on some standard benchmark problems, or on some canonical problems for which the closed-form solution is available.

### 2.1.1 Classification of EMC Models

There are many possible classifications of EMC computational models used in research and practical purposes [1–4].

Regarding the underlying theoretical background, EMC models can be classified as

- circuit theory models featuring the concentrated electrical parameters;
- TL models using distributed parameters in which low frequency electromagnetic field coupling is taken into account;
- models based on the full-wave approach taking into account radiation effects for the treatment of electromagnetic wave propagation problems.

It is worth emphasizing that this book mostly deals with the full-wave EMC models based on the thin wire antenna (scattering) theory.

Furthermore, taking into account different character of EMI sources, EMC problems [1, 2] can be classified as

- continuous wave (CW) problems;
- transient phenomena.

The most general classification of EMI sources is the one related to natural and artificial (manmade) sources, respectively. The natural EMI source most commonly being analyzed is lightning. Regarding coupling path EMI can be divided into two groups:

- Conducted disturbances (induced overvoltages, voltage dips, switching, harmonics).
- Radiated disturbances (lightning-induced voltages, antenna radiation, crosstalk).

### 2.1.2 Summary Remarks on EMC Modeling

To reliably separate EMI source from its victim, regulations and standards have become necessary. Standards set limits defining the acceptable and plausible (reasonable) level of susceptibility and provide individual testing of equipment.

EMI measurements and calculations carried out are related to

- radiated and conducted emission;
- radiated and conducted susceptibility.

Physical phenomena that represent EMI source, EMI victim, and coupling path between an EMI source and susceptible device can be modeled to a certain degree. The most important question is the level of accuracy achieved within a given model. The main limits to EMC modeling arise from the physical complexity of the considered electric system. Sometimes, even the electrical properties of the system are too difficult to determine, or the number of independent parameters necessary for building a valid EMC model is too large for a practical computer code to handle.

The EMC modeling approach presented in this book is based on integral equation formulations in the frequency and time domain and related boundary element method (BEM) of solution featuring the direct and indirect approaches, respectively.

This approach is preferred over a partial differential equation formulation and related numerical methods of solution, as the integral equation approach is based on the corresponding fundamental solution of the linear operator and, therefore, provides more accurate results. This higher accuracy level is paid with more complex formulation than is required within the framework of the partial differential equation approach, and related computational cost.

## 2.2 Generalized Telegrapher's Equations for the Field Coupling to Finite Length Wires

The electromagnetic field coupling to wire configurations can be analyzed via a rigorous AT approach or an approximate TL model [1]. The TL theory does not provide a complete solution if the wavelength of the line is comparable to or less than the transverse electrical dimensions of the line. The TL approximation is usually considered as a compromise between a quasi-static approximation and the receiving antenna model [1]. The TL approach, although a sufficient approximation if long lines with electrically small cross sections are considered, fails for the case of finite length lines and high frequency excitations. Thus, the TL model fails to predict resonances, fails to take into account properly the presence of a lossy ground, and the effects at the line ends also cannot be taken into account utilizing this approach [1–7]. So, when lines of finite length are of interest the AT should be used.

Thus, the wavelike behavior at higher frequencies requires a more general approach that is based on integral equations arising from wire AT. On the other hand, the restrictions of the wire antenna model applied to overhead lines are often related to the long computational time required for the calculations pertaining to long lines.

An extension of the standard TL approach to the combined electromagnetic field-to-TL coupling equations valid for wires above PEC ground has been presented in [3] and [4]. Further improvement of this approach has been carried out aiming to

1) include the effect of a lossy ground in Tkatchenko et al. corrected form of telegrapher's equations;
2) provide a clear relationship between TL equations and AT.

Namely, a generalized type of the telegrapher's equations, including the presence of a lossy ground and conductor loss, are derived in both frequency and time domain [8–11], thus providing a clear correlation between the rigorous AT approach to the analysis of finite length lines based on the standard space frequency and space–time Pocklington's integral equation formulation, respectively, and the classic telegrapher's equation approach arising from the TL theory in both frequency and time domain. The influence of a lossy ground is taken into account via the corresponding reflection coefficient (RC) appearing within the Green function, while conductor losses are taken into account via the surface impedance per unit length.

In some simple cases, the standard set of telegrapher's equations is deduced from the generalized telegrapher's equations.

### 2.2.1 Frequency Domain Analysis for Straight Wires above a Lossy Ground

The straight wire of a finite length $L$ and radius $a$ located at height $h$ above a lossy ground and excited by an incident electromagnetic field, see Figure 2.1, is considered.

The wire is subjected to electromagnetic fields arriving from a distant source and inducing current to flow along the line. The key to understanding the behavior of induced fields is the knowledge that the currents induced in victim equipment by the incident fields also create scattered fields propagating away from the equipment and generate the scattered voltage along the wire [1, 2].

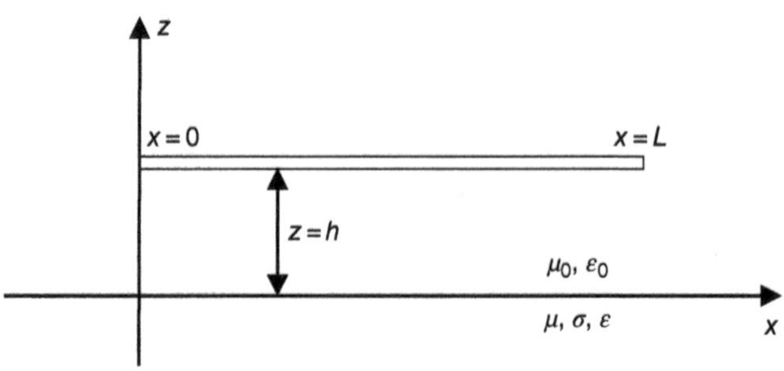


**Figure 2.1** Finite length line above a lossy ground.

#### 2.2.1.1 Integral Equation for PEC Wire of Finite Length above a Lossy Ground

The governing equations for the induced current are derived by enforcing the interface conditions for the tangential components of the electric field:

$$\vec{e}_x \cdot \vec{E}^{\text{tot}} = 0 \text{ on the surface of the wire,} \tag{2.1}$$

where the total field $\vec{E}^{\text{tot}}$ is composed from the excitation $\vec{E}^{\text{exc}}$ and scattered $\vec{E}^{\text{sct}}$ field components, respectively:

$$\vec{E}^{\text{tot}} = \vec{E}^{\text{exc}} + \vec{E}^{\text{sct}}. \tag{2.2}$$

Assuming the wire to be perfectly conducting (PEC), the tangential electric field vanishes, i.e.

$$\vec{e}_x \cdot (\vec{E}^{\text{exc}} + \vec{E}^{\text{sct}}) = 0 \text{ on the surface of the wire,} \tag{2.3}$$

where the excitation field represents the sum of the incident field $\vec{E}^{\text{inc}}$ and field reflected from the lossy ground $\vec{E}^{\text{ref}}$:

$$\vec{E}^{\text{exc}} = \vec{E}^{\text{inc}} + \vec{E}^{\text{ref}}. \tag{2.4}$$

The scattered field is given by

$$\vec{E}^{\text{sct}} = -j\omega\vec{A} - \nabla\phi, \tag{2.5}$$

where $\vec{A}$ is the magnetic vector potential and $\phi$ is the electric scalar potential.

According to the thin wire approximation (TWA) [2] only the axial component of the magnetic potential differs from zero and Equation (2.5) simplifies into

$$E_x^{\text{sct}} = -j\omega A_x - \frac{\partial\phi}{\partial x}, \tag{2.6}$$

where the magnetic vector potential and electric scalar potential are respectively defined as

$$A_x = \frac{\mu}{4\pi}\int_0^L I(x')g(x,x')\mathrm{d}x', \tag{2.7}$$

$$\varphi(x) = \frac{1}{4\pi\varepsilon}\int_0^L q(x')g(x,x')\mathrm{d}x', \tag{2.8}$$

where $q(x)$ is the induced charge along the wire, $I(x')$ is the induced current along the line, and $g(x, x')$ stands for the Green function given by

$$g(x,x') = g_0(x,x') - R_{\text{TM}}g_i(x,x'), \tag{2.9}$$

where $g_0(x, x')$ is the free space Green function:

$$g_0(x,x') = \frac{\mathrm{e}^{-jk_oR_o}}{R_o}, \tag{2.10}$$

while $g_i(x, x')$ arises from the image theory and is given by

$$g_i(x, x') = \frac{e^{-jk_o R_i}}{R_i}, \tag{2.11}$$

and $R_0$ and $R_i$ denote the corresponding distance from the source to the observation point, respectively.

$R_{TM}$ is the RC for the transverse polarization:

$$R_{TM} = \frac{n \cos\Theta - \sqrt{n - \sin^2\Theta}}{n \cos\Theta + \sqrt{n - \sin^2\Theta}}, \tag{2.12}$$

which accounts for the presence of a lossy lower half-space.

The refraction index $n$ is given by

$$n = \varepsilon_r - j\frac{\sigma}{\omega\varepsilon_0}, \tag{2.13}$$

and argument $\theta$ is defined as

$$\Theta = \text{arctg}\frac{|x - x'|}{2h}. \tag{2.14}$$

The linear charge density and the current distribution along the line wire are related through the equation of continuity [8]:

$$q = -\frac{1}{j\omega}\frac{dI}{dx}. \tag{2.15}$$

Substituting Equation (2.15) into (2.8) yields

$$\varphi(x) = -\frac{1}{j4\pi\omega\varepsilon}\int_0^L \frac{\partial I(x')}{\partial x'} g(x, x')dx'. \tag{2.16}$$

Combining Equations (2.6), (2.7), and (2.16) results in the integral relationship for the scattered field:

$$E_x^{sct} = -j\omega\frac{\mu}{4\pi}\int_0^L I(x')g(x, x')dx' + \frac{1}{j4\pi\omega\varepsilon}\frac{\partial}{\partial x}\int_0^L \frac{\partial I(x')}{\partial x'} g(x, x')dx'. \tag{2.17}$$

Finally, Equations (2.3) and (2.17) result in the integral equation for the unknown current distribution induced along the wire:

$$E_x^{exc} = j\omega\frac{\mu}{4\pi}\int_0^L I(x')g(x, x')dx' - \frac{1}{j4\pi\omega\varepsilon}\frac{\partial}{\partial x}\int_0^L \frac{\partial I(x')}{\partial x'} g(x, x')dx'. \tag{2.18}$$

Integral equation (2.18) is well known in AT and is one of the most commonly used variants of the Pocklington's integral equation.

#### 2.2.1.2 Integral Equation for a Lossy Conductor above a Lossy Ground

If an overhead wire is not PEC, the tangential component of the electric field along the conductor differs from zero and the interface condition (2.3) has to be modified. The finite wire conductivity can be taken into account by introducing the concept of the surface impedance $Z_s(x)$, and the tangential component of the total electric field at the conductor surface is equal to the product of the wire current $I(x)$ and surface internal impedance $Z_s(x)$ per unit length of the wire, i.e. (2.3) becomes

$$\vec{e}_x \cdot (\vec{E}^{\text{exc}} + \vec{E}^{\text{sct}}) = Z_s(x)I(x) \text{ on the wire surface,} \tag{2.19}$$

while the surface internal impedance $Z_s(x)$ is given by [1]

$$Z_s(x) = \frac{Z_{cw}}{2\pi a}\frac{I_0(\gamma_{wa})}{I_1(\gamma_{wa})}, \tag{2.20}$$

where $I_0(\gamma_w)$and $I_1(\gamma_w)$ are modified Bessel functions of the zero and first order respectively, and $Z_{cw}$ and $\gamma_w$ are given by [1]

$$Z_{cw} = \sqrt{\frac{j\omega\mu_w}{\sigma_w + j\omega\varepsilon_w}}, \tag{2.21}$$

$$\gamma_w = \sqrt{j\omega\mu(\sigma_w + j\omega\varepsilon_w)}. \tag{2.22}$$

Combining Equations (2.17), (2.19)–(2.22) yields the following integral equation:

$$E_x^{\text{exc}} = j\omega\frac{\mu}{4\pi}\int_0^L I(x')g(x,x')\mathrm{d}x' - \frac{1}{j4\pi\omega\varepsilon}\frac{\partial}{\partial x}\int_0^L \frac{\partial I(x')}{\partial x'}g(x,x')\mathrm{d}x' + Z_s(x)I(x). \tag{2.23}$$

The expression for $Z_s(x)$ can be simplified if low or high frequencies (HF) are of interest.

Thus, at low frequencies where $|\gamma_w a| \ll 1$ (2.20) simplifies into

$$Z_s(x) = \frac{1}{\pi a^2\sigma} + j\omega\frac{\mu_w}{4\pi}. \tag{2.24}$$

On the other hand, at HF where $|\gamma_w a| \gg 1$ (2.20) becomes

$$Z_s(x) = \frac{1+j}{2\pi a}\sqrt{\frac{\omega\mu_w}{2\sigma_w}}. \tag{2.25}$$

For the case of very good conductors, the surface impedance $Z_s(x)$ can be neglected.

#### 2.2.1.3 Generalized Telegraphers Equations for PEC Wires

To derive the telegrapher's type equations for the currents induced along the line, the so-called scattered voltage has to be included in the formulation.

A conceptual difficulty in handling the half-space problem arises in the definition of the line voltage [1]. In the case of PEC the line voltage is defined as follows:

$$V^{\mathrm{sct}}(x) = -\int_0^h E_z^{\mathrm{sct}}(x,z)\mathrm{d}z. \tag{2.26}$$

For the case of an imperfectly conducting ground the electric field at $z = 0$ is not zero, while the non-zero voltage reference point is placed somewhere within a lossy ground. Assumption of the zero voltage at $z \to -\infty$ discussed in [1] is adopted in this work as well.

Therefore, scattered voltage along the line above a lossy ground is defined as an integral of a scattered vertical field component from the point in the remote soil to the line surface:

$$V^{\mathrm{sct}}(x) = -\int_{-\infty}^h E_z^{\mathrm{sct}}(x,z)\mathrm{d}z. \tag{2.27}$$

The vertical field component can be expressed in terms of the scalar potential gradient, i.e.

$$E_z^{\mathrm{sct}} = -\frac{\partial \phi}{\partial z}, \tag{2.28}$$

and the scattered voltage along the wire can be written as

$$V^{\mathrm{sct}}(x) = \int_{-\infty}^h \frac{\partial \varphi}{\partial z}\mathrm{d}z = \frac{\mathrm{d}}{\mathrm{d}z}\int_{-\infty}^h \varphi(x,z)\mathrm{d}z. \tag{2.29}$$

Note that integrals

$$\int_{-\infty}^h E_z^{\mathrm{sct}}(x,z)\mathrm{d}z \tag{2.30}$$

and

$$\int_{-\infty}^h \varphi(x,z)\mathrm{d}z \tag{2.31}$$

should be rigorously written as follows:

$$\int_{-\infty}^h E_z^{\mathrm{sct}}(x,z)\mathrm{d}z = \int_{-\infty}^0 E_{z,\mathrm{gnd}}^{\mathrm{sct}}(x,z)\mathrm{d}z + \int_0^h E_{z,\mathrm{air}}^{\mathrm{sct}}(x,z)\mathrm{d}z, \tag{2.32}$$

$$\int_{-\infty}^h \varphi(x,z)\mathrm{d}z = \int_{-\infty}^0 \varphi_{\mathrm{gnd}}(x,z)\mathrm{d}z + \int_0^h \varphi_{\mathrm{air}}(x,z)\mathrm{d}z, \tag{2.33}$$

where ground and air quantities are separated.

Note that potential in the ground, $\varphi_{\mathrm{gnd}}$ is not defined by (2.8) or (2.16), respectively, which is related to the potential in the air.

Performing the integration from the infinite soil to the electrode surface, it simply follows that

$$V^{\mathrm{sct}}(x) = \varphi(x,z)|_{z=-\infty}^{z=h} = \varphi_{\mathrm{air}}(x,z)\big|_{z=h} - \varphi_{\mathrm{gnd}}(x,z)\Big|_{z=-\infty}. \tag{2.34}$$

Assuming the scalar potential in the remote soil to be zero yields

$$V^{\mathrm{sct}}(x) = \varphi(x,h), \tag{2.35}$$

where $\varphi(x,h)$ is determined by relation (2.16).

Combining Equations (2.16) and (2.23), the final relationship for the scattered voltage along the TL of finite length above a lossy ground is obtained:

$$V^{\mathrm{sct}}(x) = -\frac{1}{j4\pi\omega\varepsilon}\int_0^L \frac{\partial I(x')}{\partial x'} g(x,x')\mathrm{d}x'. \tag{2.36}$$

Using expression (2.36), Equation (2.18) can be written in the form:

$$E_x^{\mathrm{exc}} = j\omega\frac{\mu}{4\pi}\int_0^L I(x')g(x,x')\mathrm{d}x' + \frac{\partial V^{\mathrm{sct}}(x)}{\partial x}, \tag{2.37}$$

and finally, by slightly rearranging Equations (2.36) and (2.37) one obtains the general form of the telegrapher's equations, valid for PEC wires of finite length:

$$\frac{\mathrm{d}V^{\mathrm{sct}}(x)}{\mathrm{d}x} + j\omega\frac{\mu}{4\pi}\int_0^L I(x')g(x,x')\mathrm{d}x' = E_x^{\mathrm{exc}}, \tag{2.38}$$

$$\int_0^L \frac{\partial I(x')}{\partial x'} g(x,x')\mathrm{d}x' + j4\pi\omega\varepsilon V^{\mathrm{sct}}(x) = 0. \tag{2.39}$$

The set of Equations (2.38) and (2.39) can also be written in the following form:

$$\frac{\mathrm{d}V^{\mathrm{sct}}(x)}{\mathrm{d}x} + j\omega L''\int_0^L I(x')g(x,x')\mathrm{d}x' = E_x^{\mathrm{exc}}, \tag{2.40}$$

$$\int_0^L \frac{\partial I(x')}{\partial x'} g(x,x')\mathrm{d}x' + j\omega C'' V^{\mathrm{sct}}(x) = 0, \tag{2.41}$$

where the corresponding equivalent inductance and capacitance per unit length of the line are given by the relations:

$$L'' = \frac{\mu}{4\pi} \tag{2.42}$$

and

$$C'' = 4\pi\varepsilon. \tag{2.43}$$

The correlation of the generalized telegrapher's equations (2.40) and (2.41) with the classic telegrapher's equation for a lossless conductor above a PEC ground [1] can be readily carried out.

Assuming the perfect ground PEC case with the following conditions satisfied

$$k \cdot h \ll 1, \tag{2.44a}$$

$$L \gg 2h, \tag{2.44b}$$

and by taking into account the fact that the effective length for the integration along the line is approximately $2h$ [3] the quantities $I(x')$ and $\frac{\partial I(x')}{\partial x'}$ can be removed from the integral operator, and the characteristic integral term over Green function is

$$\int_0^L g(x,x')\mathrm{d}x' = 2\ln\frac{2h}{a}, \tag{2.45}$$

which simply results in the following form of (4.38) and (4.39):

$$V^{\mathrm{sct}}(x) + \frac{1}{j4\pi\omega\varepsilon} \cdot 2\ln\frac{2h}{a} \cdot \frac{\partial I(x)}{\partial x} = 0, \tag{2.46}$$

$$E_x^{\mathrm{exc}}(x) = j\omega\frac{\mu}{4\pi} \cdot 2\ln\frac{2h}{a} \cdot I(x) + \frac{\partial V^{\mathrm{sct}}(x)}{\partial x}, \tag{2.47}$$

and can be rewritten in the standard well-known form of the Agrawal et al. [12] field-to-TL coupling equations:

$$\frac{\mathrm{d}V^{\mathrm{sct}}(x)}{\mathrm{d}x} + j\omega L' I(x) = E_x^{\mathrm{exc}}(x), \tag{2.48}$$

$$\frac{\mathrm{d}I(x)}{\mathrm{d}x} + j\omega C' V^{\mathrm{sct}}(x) = 0, \tag{2.49}$$

where the corresponding equivalent inductance and capacitance per unit length of the aboveground line are

$$L' = \frac{\mu}{4\pi}\ln\frac{2h}{a}, \tag{2.50}$$

$$C' = \frac{2\pi\varepsilon}{\ln\frac{2h}{a}}. \tag{2.51}$$

The induced current and voltage along the line are obtained by solving Equations (2.18) and (2.36), respectively via the Galerkin–Bubnov indirect boundary element method (GB-IBEM) [2].

#### 2.2.1.4 Generalized Telegraphers Equations for Lossy Conductors

For the case of lossy wires, expression (2.23) combined with relation (2.36) yields

$$E_x^{\mathrm{exc}} = j\omega\frac{\mu}{4\pi}\int_0^L I(x')g(x,x')\mathrm{d}x' + \frac{\partial V^{\mathrm{sct}}(x)}{\partial x} + Z_s(x)I(x), \tag{2.52}$$

and by rearranging equation (2.52), the general form of the first telegrapher's equation valid for lossy conductors is obtained:

$$\frac{\mathrm{d}V^{\mathrm{sct}}(x)}{\mathrm{d}x} + j\omega\frac{\mu}{4\pi}\int_0^L I(x')g(x,x')\mathrm{d}x' + Z_s(x)I(x) = E_x^{\mathrm{exc}}. \tag{2.53}$$

Equation (2.53) can be written as follows:

$$\frac{\mathrm{d}V^{\mathrm{sct}}(x)}{\mathrm{d}x} + j\omega L'' \int_0^L I(x')g(x,x')\mathrm{d}x' + Z_s(x)I(x) = E_x^{\mathrm{exc}}, \tag{2.54}$$

where the corresponding equivalent inductance per unit length of the line is given by relation (2.42).

The correlation of the generalized first telegrapher's equation (2.54) with the classic first telegrapher's equation [8] can be carried out in a rather straightforward manner.

Assuming the PEC ground case with the condition (44) satisfied and by taking into account the fact that the effective length for the integration along the line is approximately $2h$ [2], the quantities $I(x')$ and $\frac{\partial I(x')}{\partial x'}$ can be removed from the integral operator and the characteristic integral term over Green function is given by Equation (2.45).

This simply leads to the following form of Equation (2.52):

$$E_x^{\mathrm{exc}}(x) = j\omega\frac{\mu}{4\pi}\cdot 2\ln\frac{2h}{a}\cdot I(x) + Z_s(x)I(x) + \frac{\partial V^{\mathrm{sct}}(x)}{\partial x}, \tag{2.55}$$

which can be rewritten in the standard well-known form of the Agrawal et al. [1, 12] field-to-TL coupling equations:

$$\frac{\mathrm{d}V^{\mathrm{sct}}(x)}{\mathrm{d}x} + Z'I(x) = E_x^{\mathrm{exc}}(x), \tag{2.56}$$

where the corresponding total per-unit-length impedance $Z'$ is given by

$$Z' = j\omega L' + Z_s(x), \tag{2.57}$$

while equivalent inductance and capacitance per unit length of the above-ground line are given by (2.42) and (2.43), respectively. The induced current and voltage along the line are obtained by solving Equations (2.23) and (2.36), respectively via the GB-IBEM [2].

#### 2.2.1.5 Numerical Solution of Integral Equations

An operator form of the Pocklington's integro-differential equation (2.18) or (2.23), respectively, can be, for convenience, symbolically written as

$$KI = E, \tag{2.58}$$

where $K$ is a linear operator and $I$ is the unknown function to be found for a given excitation $E$.

The current is expanded into a finite sum of linearly independent basis functions $\{f_i\}$ with unknown complex coefficients $\alpha_i$:

$$I \cong I_n = \sum_{i=1}^{n} \alpha_i f_i. \tag{2.59}$$

Substituting (2.59) into (2.58) yields

$$KI \cong KI_n = \sum_{i=1}^{n} \alpha_i K f_i = E_n = P_n(E), \tag{2.60}$$

where $P_n(E)$ is called a projection operator [2].

Now the residual $R_n$ is formed as follows:

$$R_n = KI_n - E = P_n(E) - E. \tag{2.61}$$

In accordance to the definition of the scalar product of functions in Hilbert function space, the error $R_n$ is weighted to zero with respect to certain weighting functions $\{W_j\}$, i.e.

$$\langle R_n, W_j \rangle = 0; \quad j = 1, 2, \dots, n, \tag{2.62}$$

where the expression in brackets stands for a scalar product of functions given by

$$\langle R_n, W_j \rangle = \int_\Omega R_n W_j^* \, d\Omega, \tag{2.63}$$

where $\Omega$ denotes the actual calculation domain.

Since the operator $K$ is linear, a system of linear equations is obtained by choosing $W_j = f_j$, which implies the Galerkin–Bubnov procedure. Thus, it can be written as

$$\sum_{i=1}^{n} \alpha_i \langle Kf_i, f_j \rangle = \langle E, f_j \rangle \quad j = 1, 2, \dots, n. \tag{2.64}$$

Equation (2.64) is the strong Galerkin–Bubnov formulation of the Pocklington's integral equation of (2.18). Utilizing the integral equation kernel symmetry and taking into account the Dirichlet boundary conditions for the current at the free ends of the cylinder, after integration by parts Equation (2.64) becomes

$$\begin{aligned}
&\sum_{i=1}^{n} \alpha_i \frac{1}{j4\pi\omega\varepsilon} \left\{ \int_0^L \frac{\mathrm{d}f_j(x)}{\mathrm{d}x} \int_0^L \frac{\mathrm{d}f_i(x)}{\mathrm{d}x} g(x, x') \mathrm{d}x \, \mathrm{d}x' \right. \\
&\qquad \left. + k^2 \int_0^L f_j(x) \int_0^L f_i(x) g(x, x') \mathrm{d}x \, \mathrm{d}x' + \int_0^L f_j(x) Z_s(x) \mathrm{d}x \right\} \\
&= \int_0^L E_x^{\mathrm{exc}}(x) f_j(x) \mathrm{d}x, j = 1, 2, \dots, n.
\end{aligned} \tag{2.65}$$

For the case of lossless conductors $Z_s = 0$.

Equation (2.65) represents the weak Galerkin–Bubnov formulation of the integral equation of (2.61).

The resulting system of algebraic equations arising from the boundary element discretization of (2.65) is given by [2]

$$\sum_{j=1}^{M} [Z]_{ji} \{I\}_i = \{V\}_j, \quad \text{and} \quad j = 1, 2, \dots, M, \tag{2.66}$$

where $[Z]_{ji}$ is the local matrix representing the interaction of the $i$th source boundary element with the $j$th observation wire segment:

$$[Z]_{ji} = -\frac{1}{4j\pi\omega\varepsilon}\left(\int_{\Delta l_j}\{D\}_j\int_{\Delta l_i}\{D'\}_i^T g(x,x')\mathrm{d}x'\,\mathrm{d}x\right.$$
$$\left.+\,k^2\int_{\Delta l_j}\{f\}_j\int_{\Delta l_i}\{f\}_i^T g(x,x')\mathrm{d}x'\,\mathrm{d}x\right)$$
$$+\int_{\Delta l_j} Z_L(x)\{f\}_j\{f\}_i^T\mathrm{d}x. \tag{2.67}$$

The vector $\{I\}$ contains the unknown coefficients of the solution, and it represents the local voltage vector. Matrices $\{f\}$ and $\{f'\}$ contain the shape functions while $\{D\}$ and $\{D'\}$ contain their derivatives, $M$ is the total number of line segments, and $\Delta l_i$, $\Delta l_j$ are the widths of the $i$th and $j$th segment.

Functions $f_k(z)$ are the Lagrange's polynomials and $\{V\}_j$ is the local right-side vector for the $j$th observation segment,

$$\{V\}_j = \int_{\Delta l_j} E_x^{\mathrm{exc}}\{f\}_j\mathrm{d}z, \tag{2.68}$$

representing the local voltage vector.

Since the functions $f(x)$ are required to be of class $C^1$ (once differentiable), a convenient choice for the shape functions over the finite elements is the family of Lagrange's polynomials given by

$$L_i(x) = \prod_{j=1}^{m}\frac{x-x_j}{x_i-x_j},\quad j\neq i. \tag{2.69}$$

Linear approximation over a segment is used as it has been shown that this choice provides accurate and stable results [2].

Once the current distribution is determined the scattered voltage defined as

$$V^{\mathrm{sct}}(x) = -\frac{1}{j4\pi\omega\varepsilon}\int_0^L\frac{\partial I(x')}{\partial x'}g(x,x')\mathrm{d}x' \tag{2.70}$$

can be readily computed using the boundary element formalism.

A linear approximation over a boundary element is used as this choice provides efficient calculation of integral (2.70).

The local approximation functions are given by

$$f_i = \frac{x_{i+1}-x'}{\Delta x}\quad f_{i+1} = \frac{x'-x_i}{\Delta x}. \tag{2.71}$$

The differentiation of the current distribution variation along the segment

$$\frac{\partial I(x')}{\partial x'} = I_i\frac{\partial f_i}{\mathrm{d}x'} + I_{i+1}\frac{\partial f_{i+1}}{\mathrm{d}x'} \tag{2.72}$$

is simply

$$\frac{\partial I(x')}{\partial x'} = \frac{I_{i+1} - I_i}{\Delta x}. \tag{2.73}$$

The scattered voltage then can be written as follows:

$$V^{\text{sct}}(x) = -\frac{1}{j4\pi\omega\varepsilon} \sum_{i=1}^{M} \int_{\Delta l_i} \{D\}^T g(x, x')\mathrm{d}x' \cdot I_i. \tag{2.74}$$

Therefore, it follows that

$$V^{\text{sct}}(x) = -\frac{1}{j4\pi\omega\varepsilon} \sum_{i=1}^{M} \int_{x_i}^{x_{i+1}} \frac{I_{i+1} - I_i}{\Delta x} g(x, x')\mathrm{d}x', \tag{2.75}$$

where $M$ is the total number of elements along the entire line.

#### 2.2.1.6 Simulation Results

The excitation field at the conductor surface is the time-harmonic plane wave. The TL of interest is characterized by a variable length, with wire radius $a = 1$ cm, while the height above ground is $h = 2.5$ m. The wire is assumed to be PEC. The excitation field at the conductor surface is the time-harmonic plane wave for the case of normal incidence:

$$E^{\text{exc}} = E_0(1 - R_{\text{TM}}\,\mathrm{e}^{-j2kh}). \tag{2.76}$$

The relative dielectric constant of the ground is $\varepsilon_r = 10$ and the ground conductivity is $\sigma = 0.01\ \text{S m}^{-1}$. The magnitude of the plane wave is $E_0 = 1\ \text{V m}^{-1}$ and the frequency is $f = 50$ MHz.

The proposed approach is tested using as reference the Numerical Electromagnetics Code NEC-2 [13]. Figure 2.2 presents examples of comparison for a line length of 5 m and a frequency of 50 MHz. Figure 2.3 corresponds to a 20 m long line excited by a 300 MHz plane wave. It can be seen that for both cases, the results are in excellent agreement with those obtained using NEC-2.

#### 2.2.1.7 Simulation Results and Comparison with TL Theory

Figures 2.4–2.6 present the real and imaginary parts of the induced current along the line, for different line lengths, namely, $L = 5$, 10, and 20 m. The same figures present the results obtained by applying the TL theory (for a PEC ground and a lossy ground).

It can be seen that for relatively short line lengths ($L/h < 4$, Figures 2.4 and 2.5), the TL approximation fails in reproducing accurately the current distribution along the line. For longer lines ($L/h = 8$, Figure 2.6), the results obtained using the TL approximation become acceptable.

Note that the simulations presented in Figures 2.4–2.6 correspond to a 6 m wavelength incident field, about 2.5 times the height of the line.

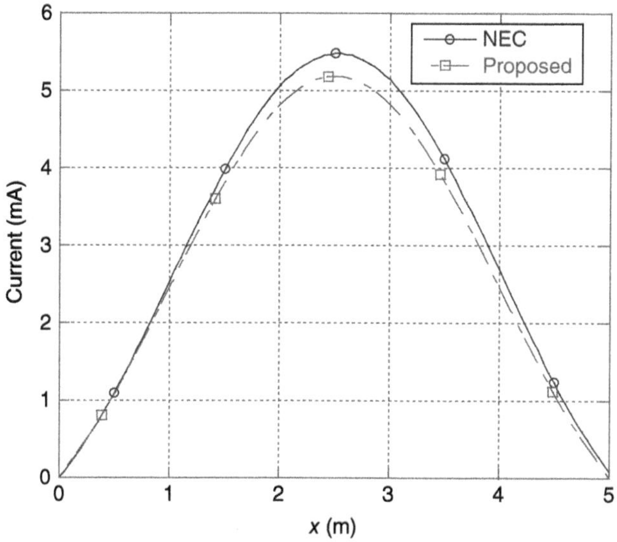


**Figure 2.2** Magnitude of the induced current distribution along the line. $L = 5$ m, $f = 50$ MHz. The calculations have been performed using (i) the derived generalized telegrapher's equations and (ii) using NEC-2.

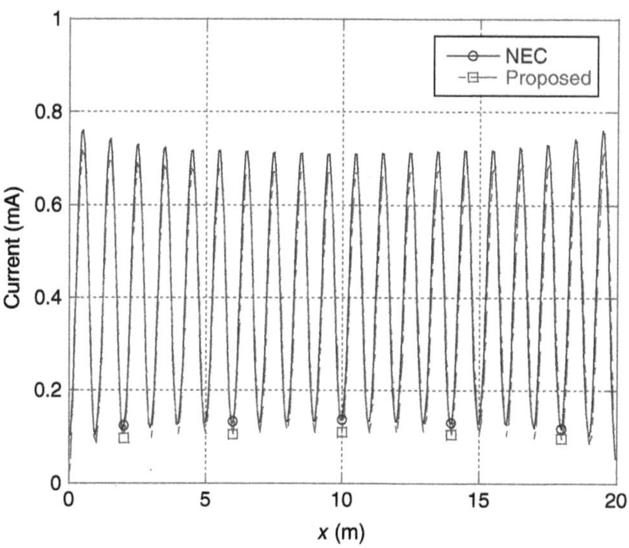


**Figure 2.3** Magnitude of the induced current distribution along the line. $L = 20$ m, $f = 300$ MHz. The calculations have been performed using (i) the derived generalized telegrapher's equations and (ii) using NEC-2.

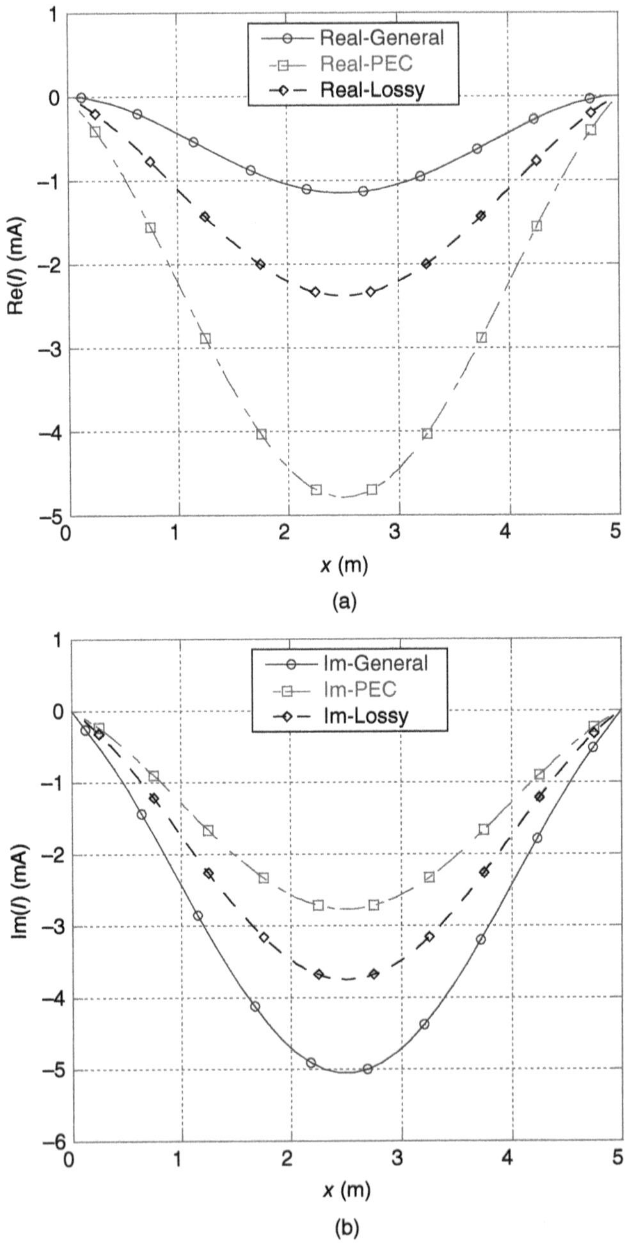


**Figure 2.4** Induced current distribution along the line. $L = 5$ m. (a) Real part and (b) imaginary part. The calculations have been performed using (i) the derived generalized telegrapher's equations, (ii) the classical transmission line theory for a perfectly conducting ground (PEC), and (iii) the classical transmission line theory for a lossy ground.

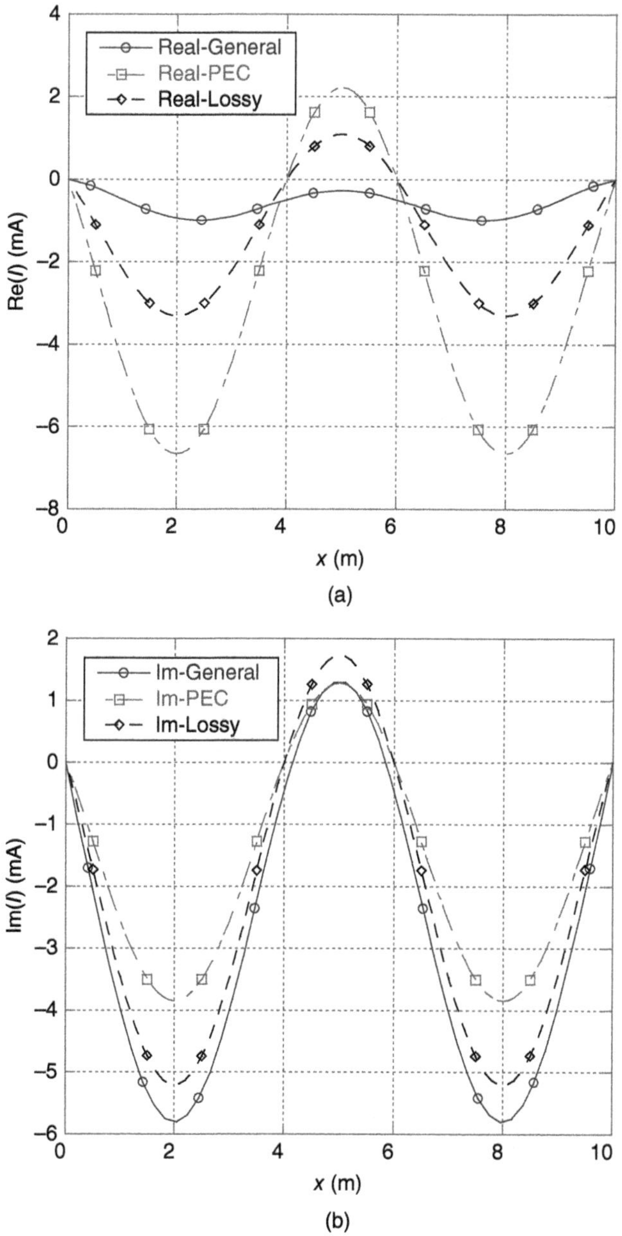


**Figure 2.5** Induced current distribution along the line. $L = 10$ m. (a) Real part and (b) imaginary part. The calculations have been performed using (i) the derived generalized telegrapher's equations, (ii) the classical transmission line theory for a perfectly conducting ground (PEC), and (iii) the classical transmission line theory for a lossy ground.

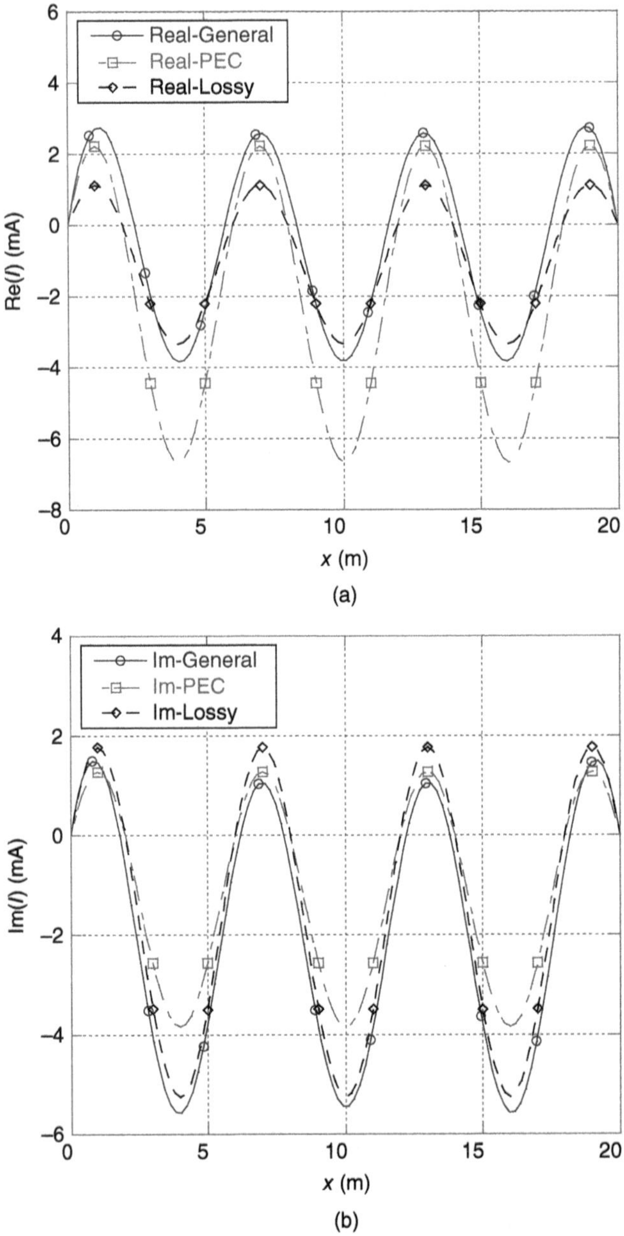


**Figure 2.6** Induced current distribution along the line. $L = 20$ m. (a) Real part and (b) imaginary part. The calculations have been performed using (i) the derived generalized telegrapher's equations, (ii) the classical transmission line theory for a perfectly conducting ground (PEC), and (iii) the classical transmission line theory for a lossy ground.

Figure 2.7 presents similar results for a 20 m long line, but considering a frequency of the incident field equal to 300 MHz. This frequency is well beyond the validity of the TL approximation (corresponding to a wavelength of $h/2.5$).

Indeed, the results show a clear discrepancy (up to one order of magnitude) between the TL approximation and the generalized solutions.

### 2.2.2 Frequency Domain Analysis for Straight Wires Buried in a Lossy Ground

A finitely conducting horizontal wire of length $L$ and radius $a$, buried in a lossy ground at depth $d$, illuminated by a transmitted $E$-field s shown in Figure 2.8.

The buried wire is subjected to electromagnetic fields arriving from a distant source, thus inducing current to flow along the wire.

Knowledge of the current distribution induced along the wire enables analysis of induced current and voltages. In particular, the induced currents in victim equipment generate scattered fields and consequently the scattered voltage along the buried wire [9].

#### 2.2.2.1 Integral Equation for Lossy Conductor Buried in a Lossy Ground

The governing equations for the current induced along the buried wire can be derived by enforcing the interface conditions for the tangential components of the electric field.

The total field composed of the transmitted excitation field $\vec{E}^{\text{exc}}$ and scattered field $\vec{E}^{\text{sct}}$ is equal to the product of the current along the wire $I(x)$ and surface internal impedance $Z_s(x)$ per unit length of the wire as expressed by (2.19).

The scattered electric field component can be expressed in terms of the magnetic vector potential $\vec{A}$ and the electric scalar potential $\phi$ (2.6). The vector potential is given by (2.7) while the scalar potential is

$$\phi(x) = \frac{1}{4\pi\varepsilon_{\text{eff}}}\int_0^L q(x')g(x,x')\mathrm{d}x'. \tag{2.77}$$

The complex permittivity of the lossy ground $\varepsilon_{\text{eff}}$ is

$$\varepsilon_{\text{eff}} = \varepsilon_r\varepsilon_0 - j\frac{\sigma}{\omega}, \tag{2.78}$$

where $\varepsilon_{rg}$ and $\sigma$ are the corresponding permittivity and conductivity, respectively, of the ground, while $q(x)$ denotes the charge distribution along the line, $I(x')$ is the induced current along the line, and $g(x, x')$ is the Green function given by

$$g(x,x') = g_0(x,x') - \Gamma_{\text{ref}}g_i(x,x'), \tag{2.79}$$

where $g_0(x, x')$ denotes the lossy medium Green function

$$g_0(x,x') = \frac{\mathrm{e}^{-\gamma R_1}}{R_1}, \tag{2.80}$$

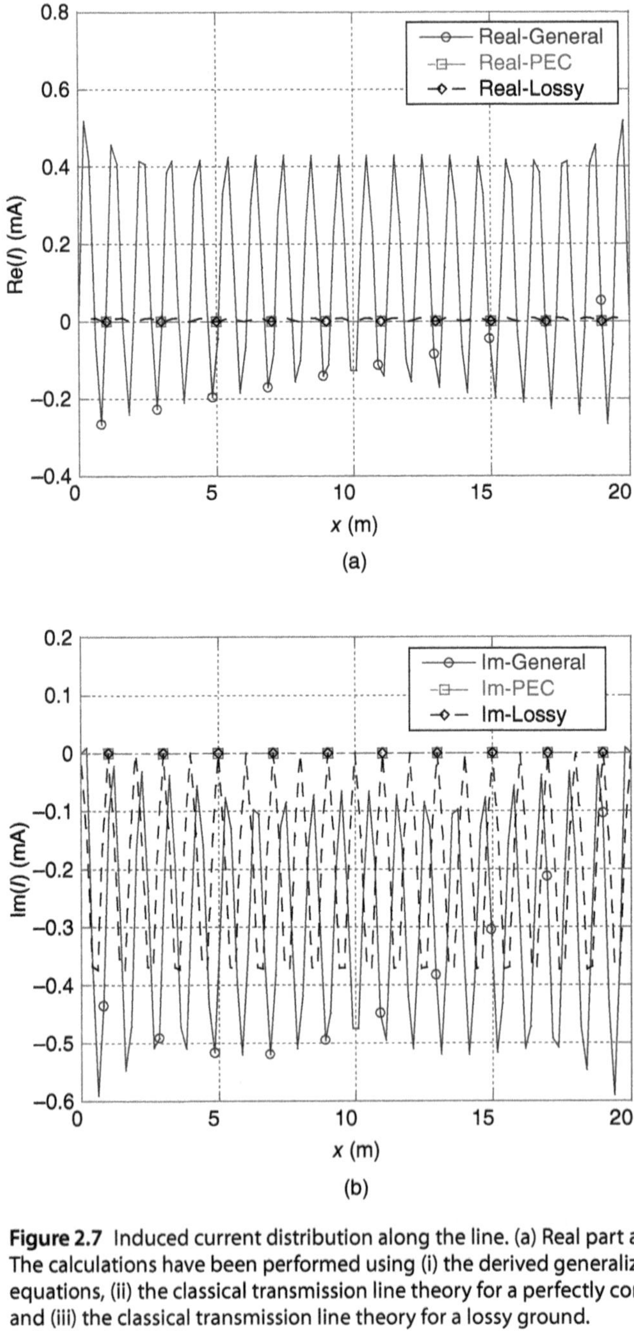


**Figure 2.7** Induced current distribution along the line. (a) Real part and (b) imaginary part. The calculations have been performed using (i) the derived generalized telegrapher's equations, (ii) the classical transmission line theory for a perfectly conducting ground (PEC), and (iii) the classical transmission line theory for a lossy ground.

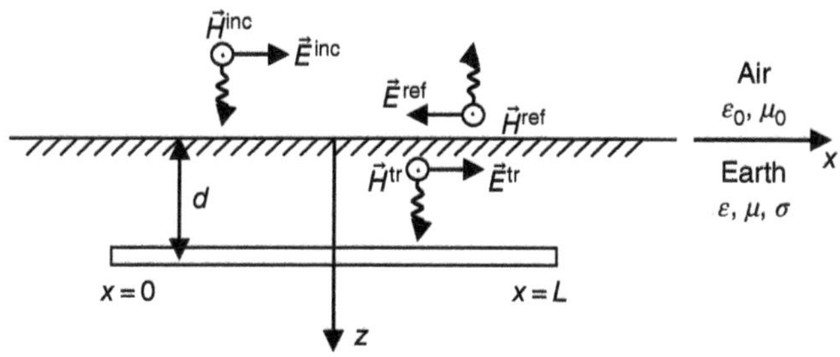


**Figure 2.8** A straight thin wire buried in lossy earth.

and $g_i(x, x')$ is, according to the image theory, given by

$$g_i(x, x') = \frac{e^{-\gamma R_2}}{R_2}, \tag{2.81}$$

while the propagation constant of the lower medium is defined as

$$\gamma = \sqrt{j\omega\mu\sigma - \omega^2\mu\varepsilon}, \tag{2.82}$$

and $R_1$ and $R_2$ are given by

$$R_1 = \sqrt{(x - x')^2 + a^2}, \quad R_2 = \sqrt{(x - x')^2 + 4d^2}. \tag{2.83}$$

The influence of a nearby ground interface is taken into account via the plane wave RC [9]:

$$\Gamma_{ref} = \frac{\frac{1}{\underline{n}}\cos\theta - \sqrt{\frac{1}{\underline{n}} - \sin^2\theta}}{\frac{1}{\underline{n}}\cos\theta + \sqrt{\frac{1}{\underline{n}} - \sin^2\theta}}; \quad \theta = \text{arctg}\frac{|x - x'|}{2d}; \quad \underline{n} = \frac{\varepsilon_{eff}}{\varepsilon_0}. \tag{2.84}$$

The linear charge density and the current distribution along the line are related through the widely used equation of continuity (2.15).

Substituting continuity equation (2.15) into (2.78) yields

$$\phi(x) = -\frac{1}{j4\pi\omega\varepsilon_{eff}}\int_0^L \frac{\partial I(x')}{\partial x'} g(x, x')dx', \tag{2.85}$$

while substituting Equations (2.77) and (2.85) into (2.6) gives the following integral expression for the scattered electric field:

$$E_x^{sct} = -j\omega\frac{\mu}{4\pi}\int_0^L I(x')g(x, x')dx' + \frac{1}{j4\pi\omega\varepsilon_{eff}}\frac{\partial}{\partial x}\int_0^L \frac{\partial I(x')}{\partial x'} g(x, x')dx'. \tag{2.86}$$

Finally, combining Equations (2.19) and (2.86) result in the following integral equation for the unknown current distribution induced along the buried wire:

$$E_x^{\mathrm{tr}} = j\omega\frac{\mu}{4\pi}\int_0^L I(x')g(x,x')\mathrm{d}x' - \frac{1}{j4\pi\omega\varepsilon_{\mathrm{eff}}}\frac{\partial}{\partial x}\int_0^L \frac{\partial I(x')}{\partial x'}g(x,x')\mathrm{d}x' + Z_s(x)I(x). \tag{2.87}$$

Integral equation (2.87) represents a single straight wire buried in a lossy medium (Pocklington's integral equation).

The excitation function given in the form of the electric field transmitted into the lossy ground (normal incidence) and illuminating the buried wire can be written as follows [9]:

$$E_x^{\mathrm{exc}} = E_x^{\mathrm{tr}} = E_0\Gamma_{\mathrm{TM}}\mathrm{e}^{-\gamma z}, \tag{2.88}$$

where $\Gamma_{\mathrm{TM}}$ is the corresponding Fresnel transmission coefficient, respectively, at the air–earth interface given by [1, 12]:

$$\Gamma_{\mathrm{TM}} = \frac{2}{1+\sqrt{\underline{n}}}. \tag{2.89}$$

Solving the Pocklington's integro-differential equation (2.87) the current distribution at the given frequency is obtained.

Knowledge of the current along the wire enables analysis of induced scattered field and voltage, respectively.

#### 2.2.2.2 Generalized Telegraphers Equations for Buried Lossy Wires

To derive the telegrapher's type equations for the currents induced along the buried wire, the so-called scattered voltage has to be incorporated in the formulation.

A conceptual difficulty in handling the half-space problem is the definition of the wire voltage [9]. The scattered voltage along the wire buried in a lossy ground is determined by an integral of a scattered vertical electric field component from the point in the remote soil to the wire surface, i.e. by the following line integral:

$$V^{\mathrm{sct}}(x) = -\int_\infty^d E_z^{\mathrm{sct}}(x,z)\mathrm{d}z, \tag{2.90}$$

where the non-zero voltage reference point is placed somewhere within a lossy ground.

Assumption of the zero voltage at $z \to \infty$ discussed in [9] is adopted, as well. Expressing the vertical field component in terms of the scalar potential gradient (2.28) the scattered voltage along the wire can be written as follows:

$$V^{\mathrm{sct}}(x) = \int_\infty^d \frac{\partial\varphi}{\partial z}\mathrm{d}z = \frac{\mathrm{d}}{\mathrm{d}z}\int_\infty^d \varphi(x,z)\mathrm{d}z. \tag{2.91}$$

Taking the integral from the infinite soil to the conductor surface it simply follows that

$$V^{\text{sct}}(x) = \varphi(x,z)|_{z=\infty}^{z=d} = \varphi(x,z)|_{z=d} - \varphi(x,z)|_{z=\infty}. \tag{2.92}$$

Assuming the scalar potential in the remote soil to be zero yields

$$V^{\text{sct}}(x) = \varphi(x,d), \tag{2.93}$$

where $\varphi(x,d)$ is determined by relation (2.85), i.e. the scattered voltage along the buried wire is given by

$$V^{\text{sct}}(x) = -\frac{1}{j4\pi\omega\varepsilon_{\text{eff}}}\int_0^L \frac{\partial I(x')}{\partial x'}g(x,x')\text{d}x'. \tag{2.94}$$

Combining Equations (2.94) and (2.87) yields

$$E_x^{\text{tr}} = j\omega\frac{\mu}{4\pi}\int_0^L I(x')g(x,x')\text{d}x' + \frac{\partial V^{\text{sct}}(x)}{\partial x} + Z_s(x)I(x). \tag{2.95}$$

Now, it is convenient to rewrite Equations (2.94) and (2.95) in the following form:

$$\frac{\text{d}V^{\text{sct}}(x)}{\text{d}x} + j\omega\frac{\mu}{4\pi}\int_0^L I(x')g(x,x')\text{d}x' + Z_s(x)I(x) = E_x^{\text{exc}}, \tag{2.96}$$

$$\int_0^L \frac{\partial I(x')}{\partial x'}g(x,x')\text{d}x' + j4\pi\omega\varepsilon_{\text{eff}}V^{\text{sct}}(x) = 0. \tag{2.97}$$

The set of integro-differential relationships for lossy buried wires (2.96) and (2.97) can also be written as follows:

$$\frac{\text{d}V^{\text{sct}}(x)}{\text{d}x} + j\omega L'\int_0^L I(x')g(x,x')\text{d}x' + Z_s(x)I(x) = E_x^{\text{tr}}, \tag{2.98}$$

$$\int_0^L \frac{\partial I(x')}{\partial x'}g(x,x')\text{d}x' + Y'V^{\text{sct}}(x) = 0, \tag{2.99}$$

where the corresponding equivalent inductance and admittance per unit length of the wire are given by

$$L' = \frac{\mu}{4\pi}, \tag{2.100}$$

$$Y' = j4\pi\omega\varepsilon_{\text{eff}} = G' + j\omega C', \tag{2.101}$$

and the equivalent capacitance and conductance are

$$C' = 4\pi\varepsilon, \tag{2.102}$$

$$G' = 4\pi\sigma. \tag{2.103}$$

If the case of grounding electrodes is of interest the electric field excitation on the right-hand side vanishes [9] and Equation (2.98) becomes

$$\frac{\text{d}V^{\text{sct}}(x)}{\text{d}x} + j\omega L'\int_0^L I(x')g(x,x')\text{d}x' + Z_s(x)I(x) = 0. \tag{2.104}$$

The current source excitation is taken into account as a boundary condition [9, 14]. The correlation of the generalized telegrapher's equations (2.98) and (2.99), or Equations (2.98) and (2.104) with the classic telegrapher's equation for lossy conductor immersed in an infinite earth can be readily undertaken.

Assuming the infinite ground case with the following condition satisfied

$$\gamma \cdot R \ll 1, \tag{2.105}$$

and assuming that the quantities $I(x')$ and $\frac{\partial I(x')}{\partial x'}$ can be removed from the integral operator [9], the characteristic integral term over Green function is

$$\int_0^L g(x,x')\mathrm{d}x' \doteq 2\ln\frac{L}{a}, \tag{2.106}$$

which simply gives the following form of Equations (2.96) and (2.97):

$$j\omega\frac{\mu}{4\pi}\cdot 2\ln\frac{L}{a}\cdot I(x) + Z_s(x)I(x) + \frac{\partial V^{\mathrm{sct}}(x)}{\partial x} = E_x^{\mathrm{tr}}(x), \tag{2.107}$$

$$j4\pi\omega\varepsilon_{\mathrm{eff}}V^{\mathrm{sct}}(x) + \cdot 2\ln\frac{L}{a}\cdot\frac{\partial I(x)}{\partial x} = 0, \tag{2.108}$$

and can be rewritten in the standard form of field-to-TL coupling equations:

$$\frac{\mathrm{d}V^{\mathrm{sct}}(x)}{\mathrm{d}x} + Z''I(x) = E_x^{\mathrm{exc}}(x), \tag{2.109}$$

$$\frac{\mathrm{d}I(x)}{\mathrm{d}x} + Y''V^{\mathrm{sct}}(x) = 0, \tag{2.110}$$

where the corresponding equivalent inductance, capacitance, and impedance per unit length of the buried wire are given by

$$L'' = \frac{\mu}{2\pi}\ln\frac{L}{a}, \tag{2.111}$$

$$Y'' = \frac{Y'}{2\ln\frac{L}{a}}, \tag{2.112}$$

and

$$Z'' = j\omega L'' + Z_s(x). \tag{2.113}$$

For the case of grounding wire, Equation (2.109) simplifies into

$$\frac{\mathrm{d}V^{\mathrm{sct}}(x)}{\mathrm{d}x} + Z''I(x) = 0. \tag{2.114}$$

The induced current and voltage along the wire are obtained by solving Equations (2.98), (2.99), or (2.104), respectively via the GB-IBEM. An outline of the implemented numerical procedures can be found elsewhere, e.g. in [2].

#### 2.2.2.3 Computational Examples

The excitation field at the earth surface is the time-harmonic plane wave. The TL of interest is characterized by a variable length, with conductor radius

$a = 1$ cm, while the burial depth is $d = 2.5$ m. The wire is assumed to be PEC. The excitation field at the conductor surface is the time-harmonic plane wave for the case of the normal incidence:

$$E^{exc} = E^{tr} = \Gamma_{TM} E_0 e^{-\gamma d}. \tag{2.115}$$

The relative dielectric constant of the ground is $\varepsilon_r = 10$ and the ground conductivity is $\sigma = 0.01$ S m$^{-1}$. The magnitude of the plane wave is $E_0 = 1$ V m$^{-1}$ and the frequency is $f = 50$ MHz.

The approach proposed in [9] is tested using the NEC-2 as reference [13]. Figure 2.9 shows the real and imaginary parts of the current distribution

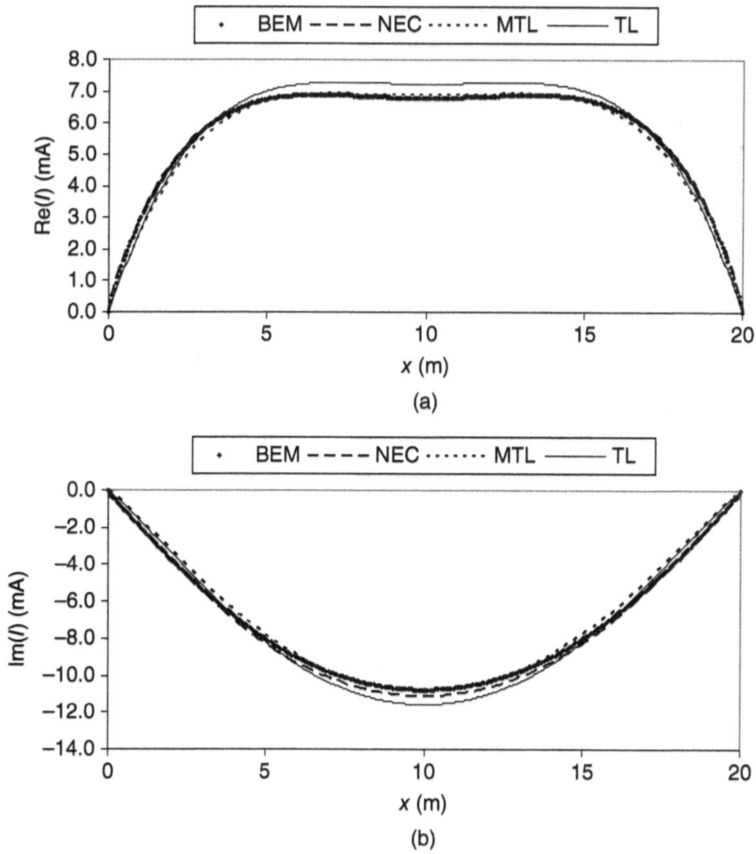


**Figure 2.9** Induced current distribution along the wire ($f = 1$ MHz, $L = 20$ m, $a = 0.01$ m, $d = 2.5$ m, $\sigma = 0.01$ S m$^{-1}$, $\varepsilon_r = 10$). (a) Real part and (b) imaginary part. The calculations have been performed using (i) the derived generalized telegrapher's equations, (ii) the NEC-2, (iii) the modified transmission line theory for a lossy ground, and (iv) the classical transmission line theory for a lossy ground.

induced along a 20 m long wire excited by a 1 MHz plane wave. The results obtained via BEM, NEC, TL, and MTL seem to be in very good agreement.

Figures 2.10–2.12 present the real and imaginary parts of the induced current along the wire, for different line lengths, namely, $L = 5$, 10, and 20 m. The excitation is the same as in the previous example.

It can be seen that for relatively shorter wire lengths, the TL and MTL approximation fail in reproducing accurately the current distribution along the wire. For longer wires ($L = 20$ m), the results obtained using the TL and MTL approximation become acceptable.

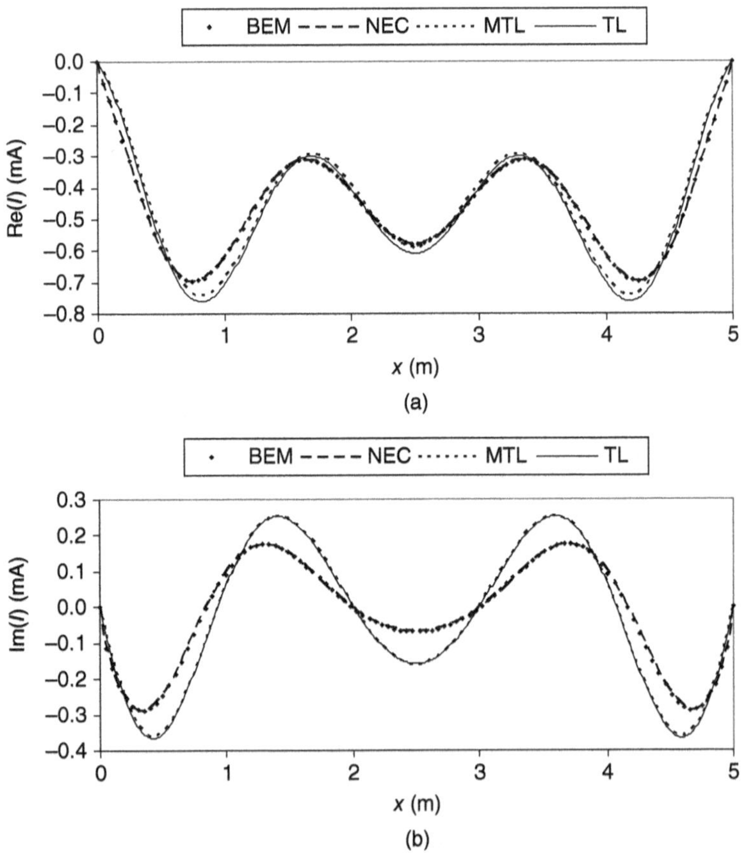


**Figure 2.10** Induced current distribution along the wire ($f = 50$ MHz, $L = 5$ m, $a = 0.01$ m, $d = 2.5$ m, $\sigma = 0.01$ S m$^{-1}$, $\varepsilon_r = 10$). (a) Real part and (b) imaginary part. The calculations have been performed using (i) the derived generalized telegrapher's equations, (ii) the NEC-2, (iii) the modified transmission line theory for a lossy ground, and (iv) the classical transmission line theory for a lossy ground.

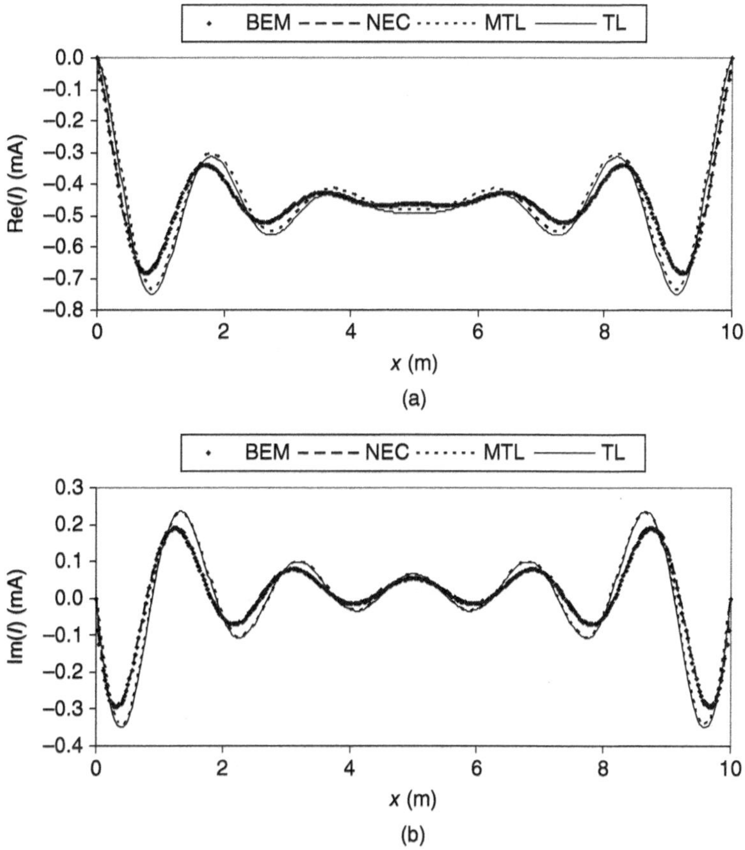


**Figure 2.11** Induced current distribution along the wire ($f = 50$ MHz, $L = 10$ m, $a = 0.01$ m, $d = 2.5$ m, $\sigma = 0.01$ S m$^{-1}$, $\varepsilon_r = 10$). (a) Real part and (b) imaginary part. The calculations have been performed using (i) the derived generalized telegrapher's equations, (ii) the NEC-2, (iii) the modified transmission line theory for a lossy ground, and (iv) the classical transmission line theory for a lossy ground.

Therefore, the use of TL and MTL approach, respectively, may result in significant errors for shorter wires at HF. Furthermore, observing results for longer wires leads to the conclusion that the MTL method is almost equivalent to the classical TL approximation.

It is also necessary to make a trade-off between the rigorous Sommerfeld integral approach and the approximation RC approach, respectively, to account for the effect of the earth–air interface reflected field upon the current distribution along the buried wire. The rigorous Sommerfeld integral approach has been found to be numerically stable and thus reliable for buried wire located to within $10^{-6}$ wavelengths of the interface [9].

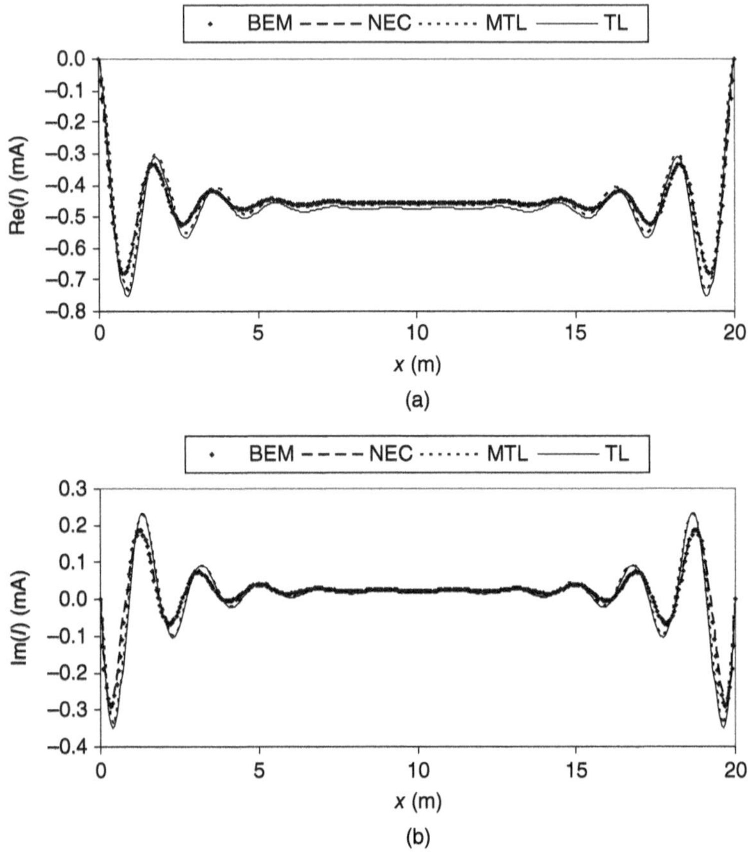


**Figure 2.12** Induced current distribution along the wire ($f = 50$ MHz, $L = 20$ m, $a = 0.01$ m, $d = 2.5$ m, $\sigma = 0.01$ S m$^{-1}$, $\varepsilon_r = 10$). (a) Real part and (b) imaginary part and. The calculations have been performed using (i) the derived generalized telegrapher's equations, (ii) the NEC-2, (iii) the modified transmission line theory for a lossy ground, and (iv) the classical transmission line theory for a lossy ground.

In addition, the results arising from the use of Sommerfeld integral approach and the RC approximation approach are found to be in a very good agreement for depths greater, or at least equal to [2]

$$d \geq \frac{\lambda}{4\sqrt{\varepsilon_{\text{eff}}}}. \tag{2.116}$$

As a rough guideline, the RC approach has been found to produce 10% of those obtained via the computationally very expensive Sommerfeld integral approach.

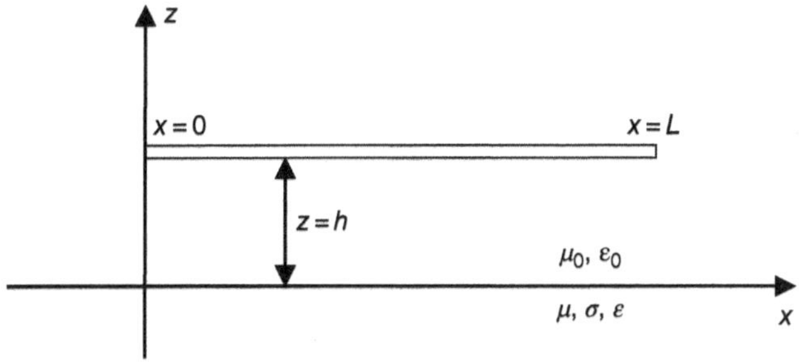


**Figure 2.13** Finite length line above a lossy ground.

### 2.2.3 Time Domain Analysis for Straight Wires above a Lossy Ground

A TL of finite length $L$ and radius $a$ located at height $h$ above an imperfect ground (see Figure 2.13) is considered.

The aboveground wire is subjected to a transient electromagnetic field generated by a distant source that induces currents and charges along the line [1].

#### 2.2.3.1 Space–Time Integro-Differential Equation for PEC Wire above a Lossy Ground

The governing equations for the transient current induced along the wire are derived by enforcing the interface conditions for the tangential field components

$$\vec{e}_x \cdot \vec{E}^{tot}(\vec{r}, t) = 0 \text{ on the surface of the wire,} \tag{2.117}$$

where the total field $\vec{E}^{tot}$ can be decomposed into its excitation component $\vec{E}^{exc}$ and scattered component $\vec{E}^{sct}$, respectively:

$$\vec{E}^{tot}(\vec{r}, t) = \vec{E}^{exc}(\vec{r}, t) + \vec{E}^{sct}(\vec{r}, t). \tag{2.118}$$

For the case of PEC wire the tangential electric field vanishes:

$$\vec{e}_x \cdot \left[\vec{E}^{exc}(\vec{r}, t) + \vec{E}^{sct}(\vec{r}, t)\right] = 0 \text{ on the surface of the wire,} \tag{2.119}$$

where the excitation field represents the sum of the incident field $\vec{E}^{inc}$ and field reflected from the lossy ground $\vec{E}^{ref}$, both determined in absence of the wire:

$$\vec{E}^{exc}(\vec{r}, t) = \vec{E}^{inc}(\vec{r}, t) + \vec{E}^{ref}(\vec{r}, t). \tag{2.120}$$

Furthermore, the scattered field component is

$$\vec{E}^{sct} = -\frac{\partial \vec{A}}{\partial t} - \nabla \varphi, \tag{2.121}$$

where $\vec{A}$ and $\varphi$ are the magnetic vector potential and electric scalar potential respectively.

According to the TWA only the axial component of the magnetic potential differs from zero and equation (2.121) becomes [10]

$$E_x^{\mathrm{sct}}(x,t) = -\frac{\partial A_x(x,t)}{\partial t} - \frac{\partial \varphi(x,t)}{\partial x}. \tag{2.122}$$

In the homogeneous dielectric medium the magnetic vector potential is given by

$$A_x(x,t) = \frac{\mu}{4\pi}\int_0^L \frac{i(x',t-R/v)}{4\pi R}\mathrm{d}x', \tag{2.123}$$

where $i(x', t - R/v)$ is the unknown space–time varying current along the wire, and $v$ is the related propagation velocity in the lossless dielectric, and for the case of free space it is equal to the speed of light.

The vector and scalar potentials are related through the Lorentz gauge

$$\nabla\vec{A} + \mu\varepsilon\frac{\partial\varphi}{\partial t} = 0, \tag{2.124}$$

which, for the case of thin wire geometry, simplifies into

$$\frac{\partial A_x(x,t)}{\partial x} = -\mu\varepsilon\frac{\partial\varphi(x,t)}{\partial t}. \tag{2.125}$$

The scalar potential can be written as follows:

$$\begin{aligned}\varphi(x,t) &= -\frac{1}{4\pi\varepsilon}\frac{\partial}{\partial x}\int_0^{t'} A(x,t)\mathrm{d}t \\ &= -\frac{1}{4\pi\varepsilon}\frac{\partial}{\partial x}\int_0^L\int_0^{t'} i(x,t-R/v)\frac{1}{R}\mathrm{d}t\,\mathrm{d}x'.\end{aligned} \tag{2.126}$$

Taking into account the kernel property

$$\frac{\partial}{\partial x}\left(\frac{1}{R}\right) = -\frac{\partial}{\partial x'}\left(\frac{1}{R}\right), \tag{2.127}$$

and performing some mathematical manipulations, it can be written that

$$\varphi(x,t) = -\frac{1}{4\pi\varepsilon}\int_0^L\int_0^{t'}\frac{\partial i(x,t-R/v)}{\partial x'}\frac{1}{R}\mathrm{d}t\,\mathrm{d}x'. \tag{2.128}$$

For the case of a wire located above a dissipative half-space, the magnetic vector potential (2.123) has to be extended by an additional term arising from the image theory to account for the ground effects.

In the frequency domain an additional term due to the image wire is included by multiplying the Green function by a corresponding RC.

The Laplace transform of (2.123) is

$$A_x(x,s) = \frac{\mu}{4\pi}\int_0^L I(x',s)\frac{\mathrm{e}^{-s\frac{R}{v}}}{R}\mathrm{d}x'. \tag{2.129}$$

Now the extended magnetic vector potential taking into account the presence of a lossy half soil can be written as

$$A_x(x,s) = \frac{\mu}{4\pi}\int_0^L I(x',s)\frac{e^{-s\frac{R}{v}}}{R}dx' - \frac{\mu}{4\pi}\int_0^L R_{TM}(\theta,s)I(x',s)\frac{e^{-s\frac{R^*}{v}}}{R^*}dx', \tag{2.130}$$

where, $I(x, s)$ is the unknown current distribution along the wire, $s$ is the Laplace domain variable, $R$ is the distance from the source point located at the wire in the air to the observation point in the air, and $R^*$ is the distance from the image wire to the observation point in the air. The presence of the air–earth interface is taken into account in terms of the Fresnel RC for the transverse magnetic polarization.

The RC $R_{TM}$ is given by [16]

$$R_{TM} = \frac{n\cos\Theta - \sqrt{n-\sin^2\Theta}}{n\cos\Theta + \sqrt{n-\sin^2\Theta}}. \tag{2.131}$$

The refraction index $n$ is given by

$$n = \varepsilon_r + \frac{\sigma}{\varepsilon_0 s}, \tag{2.132}$$

where $\varepsilon_r$ and $\sigma$ are the ground permittivity and conductivity, respectively, and argument $\theta$ is defined as

$$\Theta = \text{arctg}\frac{|x-x'|}{2h}. \tag{2.133}$$

As the multiplication in the frequency domain requires the convolution in the time domain, an additional term due to the image wire involves convolution of the space–time dependent current with the corresponding air–earth RC. Consequently, the time domain counterpart of (2.130) is of the form

$$A_x(x,t) = \frac{\mu}{4\pi}\left[\int_0^L \frac{i(x',t-R/v)}{4\pi R}dx' - \int_{-\infty}^t \int_0^L r_{TM}(\theta,\tau)\frac{i(x',t-R^*/v-\tau)}{4\pi R^*}dx'\,d\tau\right], \tag{2.134}$$

where $r_{TM}(\theta, \tau)$ is the space–time RC obtained by the inverse Laplace transform of (2.131), i.e.

$$r_{TM}(\theta,t) = L^{-1}[R(\theta,s)]. \tag{2.135}$$

Furthermore, the Laplace transform of (2.128) is given by

$$\varphi(x,s) = -\frac{1}{4\pi\varepsilon s}\int_0^L \frac{\partial I(x',s)}{\partial x'}\frac{e^{-s\frac{R}{v}}}{R}dx', \tag{2.136}$$

and the extended scalar potential with an additional term representing the effect of a lossy soil in the Laplace domain is of the form

$$\varphi(x,s) = -\frac{1}{4\pi\varepsilon s}\int_0^L \frac{\partial I(x',s)}{\partial x'}\frac{e^{-s\frac{R}{v}}}{R}dx' + \frac{1}{4\pi\varepsilon s}\int_0^L R_{TM}(\theta,s)\frac{\partial I(x',s)}{\partial x'}\frac{e^{-s\frac{R^*}{v}}}{R^*}dx'. \tag{2.137}$$

Applying the convolution operator yields the time domain counterpart of expression (2.137):

$$\varphi(x,t) = -\frac{1}{4\pi\varepsilon}\int_0^L\int_0^{t'} \frac{\partial i(x,t-R/v)}{\partial x'}\frac{1}{R}dt\,dx' + \frac{1}{4\pi\varepsilon}\int_0^L\int_0^{t'}\int_0^t r_{TM}(\theta,\tau)\frac{\partial i(x,t-R^*/v-\tau)}{\partial x'}\frac{1}{R^*}dt'\,dt\,dx'. \tag{2.138}$$

Combining Equations (2.122), (2.134), and (2.138) leads to the following integral relationship for the scattered field:

$$\begin{aligned} E_x^{sct}(x,t) = &-\frac{\mu}{4\pi}\left[\int_0^L \frac{\partial I(x',t-R/v)}{\partial t}\frac{1}{R}dx' - \int_{-\infty}^t\int_0^L r_{TM}(\theta,\tau)\frac{\partial I(x',t-R^*/v-\tau)}{\partial t}\frac{1}{R^*}dx'\,d\tau\right] \\ &+ \frac{1}{4\pi\varepsilon}\frac{\partial}{\partial x}\int_0^L\int_0^{t'} \frac{\partial i(x,t-R/v)}{\partial x'}\frac{1}{R}dt\,dx' \\ &- \frac{1}{4\pi\varepsilon}\frac{\partial}{\partial x}\int_0^L\int_0^{t'}\int_0^t r_{TM}(\theta,\tau)\frac{\partial i(x,t-R^*/v-\tau)}{\partial x'}\frac{1}{R^*}dt'\,dt\,dx'. \end{aligned} \tag{2.139}$$

Now condition (2.119) at the PEC thin wire surface can be written as

$$E_x^{exc}(x,t) + E_x^{sct}(x,t) = 0. \tag{2.140}$$

Inserting (2.139) into (2.140) gives the time domain Pocklington's integral equation for a straight thin wire above a lossy ground:

$$\begin{aligned} E_x^{exc}(x,t) = &\frac{\mu}{4\pi}\left[\int_0^L \frac{\partial I(x',t-R/v)}{\partial t}\frac{1}{R}dx' - \int_{-\infty}^t\int_0^L r_{TM}(\theta,\tau)\frac{\partial I(x',t-R^*/v-\tau)}{\partial t}\frac{1}{R^*}dx'\,d\tau\right] \\ &- \frac{1}{4\pi\varepsilon}\frac{\partial}{\partial x}\int_0^L\int_0^{t'} \frac{\partial i(x,t-R/v)}{\partial x'}\frac{1}{R}dt\,dx' \end{aligned}$$

$$+\frac{1}{4\pi\varepsilon}\frac{\partial}{\partial x}\int_0^L\int_0^{t'}\int_0^t r_{\mathrm{TM}}(\theta,\tau)\frac{\partial i(x,t-R^*/v-\tau)}{\partial x'}\frac{1}{R^*}\mathrm{d}t'\,\mathrm{d}t\,\mathrm{d}x'. \tag{2.141}$$

By differentiating Equation (2.141) over time one obtains the following form of the Pocklington's integro-differential equation:

$$-\varepsilon\frac{\partial E_x^{\mathrm{exc}}}{\partial t}=\left[\frac{\partial^2}{\partial x^2}-\frac{1}{v^2}\frac{\partial^2}{\partial t^2}\right]$$
$$\times\left[\int_0^L\frac{I(x',t-R/c)}{4\pi R}\mathrm{d}x'-\int_{-\infty}^t\int_0^L r_{\mathrm{TM}}(\theta,\tau)\frac{I(x',t-R^*/c-\tau)}{4\pi R^*}\mathrm{d}x'\,\mathrm{d}\tau\right]. \tag{2.142}$$

The integral equation (2.142) represents one of the commonly used variants of the Pocklington's integral equation in the wire AT.

#### 2.2.3.2 Space–Time Integro-Differential Equation for Lossy Conductors

For the case of lossy conductors, the tangential component of the electric field along the wire differs from zero and the condition (2.119) has to be modified. Namely, in the Laplace domain the tangential component of the total electric field at the wire surface is equal to the product of the wire current $I(x, s)$ and the internal impedance $Z_s(x, s)$ per unit length of the wire, i.e. an extended counterpart of Equation (2.119) in the Laplace domain becomes [10]

$$\vec{e}_x\cdot[\vec{E}^{\mathrm{exc}}(\vec{r},s)+\vec{E}^{\mathrm{sct}}(\vec{r},s)]=Z_s(\vec{r},s)I(\vec{r},s)\ \text{ on the wire surface.} \tag{2.143}$$

For the case of a straight thin wire along the $x$ axis, (2.143) reduces to

$$\vec{e}_x\cdot[\vec{E}^{\mathrm{exc}}(xs)+\vec{E}^{\mathrm{sct}}(x,s)]=Z_s(x,s)I(x,s)\ \text{ on the wire surface,} \tag{2.144}$$

and the surface internal impedance $Z_s(x)$ is given by [10]

$$Z_s(x,s)=\frac{Z_{cw}}{2\pi a}\frac{I_0[\gamma_w(s)a]}{I_1[\gamma_w(s)a]}, \tag{2.145}$$

where $I_0(\gamma_w)$ and $I_1(\gamma_w)$ are modified Bessel functions of the zero and first order respectively, while $Z_{cw}$ and $\gamma_w$ are determined by expressions [10]

$$Z_{cw}(s)=\sqrt{\frac{s\mu_w}{\sigma_w+s\varepsilon_w}}, \tag{2.146}$$

$$\gamma_w=\sqrt{s\mu(\sigma_w+s\varepsilon_w)}. \tag{2.147}$$

The time domain counterpart of (2.142) is

$$\vec{e}_x\cdot\left[\vec{E}^{\mathrm{exc}}(x,t)+\vec{E}^{\mathrm{sct}}(x,t)\right]=\int_0^t z_s(x,\tau)i(x,t-\tau)\mathrm{d}\tau\ \text{ on the wire surface,} \tag{2.148}$$

where $z_s(x,t)$ is the inverse Laplace transform of (2.145)

$$z_s(x,t) = L^{-1} Z_s(x,s). \tag{2.149}$$

Combining Equations (2.139) and (2.148) yields the following integro-differential equation:

$$\begin{aligned}
E_x^{\text{exc}}(x,t) = \frac{\mu}{4\pi}\Bigg[&\int_0^L \frac{\partial I(x',t-R/v)}{\partial t}\frac{1}{R}\mathrm{d}x' \\
&- \int_{-\infty}^{t}\int_0^L r_{\text{TM}}(\theta,\tau)\frac{\partial I(x',t-R^*/v-\tau)}{\partial t}\frac{1}{R^*}\mathrm{d}x'\,\mathrm{d}\tau\Bigg] \\
&- \frac{1}{4\pi\varepsilon}\frac{\partial}{\partial x}\int_0^L\int_0^{t'} \frac{\partial i(x,t-R/v)}{\partial x'}\frac{1}{R}\mathrm{d}t\,\mathrm{d}x' \\
&+ \frac{1}{4\pi\varepsilon}\frac{\partial}{\partial x}\int_0^L\int_0^{t'}\int_0^t r_{\text{TM}}(\theta,\tau)\frac{\partial i(x,t-R^*/v-\tau)}{\partial x'}\frac{1}{R^*}\mathrm{d}t'\,\mathrm{d}t\,\mathrm{d}x' \\
&+ \int_0^t z_s(x,\tau)i(x,t-\tau)\mathrm{d}\tau.
\end{aligned} \tag{2.150}$$

Differentiating Equation (2.150) over time gives

$$\begin{aligned}
-\varepsilon\frac{\partial E_x^{\text{exc}}}{\partial t} = \left[\frac{\partial^2}{\partial x^2} - \frac{1}{v^2}\frac{\partial^2}{\partial t^2}\right]\cdot\Bigg[&\int_0^L \frac{I(x',t-R/c)}{4\pi R}\mathrm{d}x' \\
&- \int_{-\infty}^{t}\int_0^L r_{\text{TM}}(\theta,\tau)\frac{I(x',t-R^*/c-\tau)}{4\pi R^*}\mathrm{d}x'\,\mathrm{d}\tau\Bigg] \\
&+ \int_0^t z_s(x,\tau)i(x,t-\tau)\mathrm{d}\tau.
\end{aligned} \tag{2.151}$$

For the case of very good conductors with conductivity around order of $10^6$ S m$^{-1}$ the surface internal impedance can be neglected.

#### 2.2.3.3 Generalized Telegraphers Equations for PEC Wires

To derive telegrapher-like equations for the currents induced along the line, the concept of the scattered voltage has to be used [10].

A conceptual difficulty in handling the half-space problem arising from the definition of the line voltage [1] for the case of frequency domain analysis has been discussed in details in [8, 9]. In the time domain for the case of PEC ground the space–time dependent line voltage is defined by the following line integral:

$$u^{\text{sct}}(x,t) = -\int_0^h E_z^{\text{sct}}(x,z,t)\mathrm{d}z. \tag{2.152}$$

On the other hand, for the case of a finitely conducting ground, the electric field at $z = 0$ is not zero, while the non-zero voltage reference point is placed somewhere within the lossy ground. The assumption of a zero voltage at $z \to -\infty$ [10] is adopted.

Therefore, the scattered voltage along the horizontal straight wire above a lossy ground is defined by the integral of a scattered vertical field component from the point in the remote soil to the conductor surface:

$$u^{\text{sct}}(x,t) = -\int_{-\infty}^{h} E_z^{\text{sct}}(x,z,t)\text{d}z. \tag{2.153}$$

As the magnetic vector potential has only the axial component, the vertical field component can be expressed in terms of the scalar potential gradient,

$$E_z^{\text{sct}} = -\frac{\partial \phi}{\partial z}, \tag{2.154}$$

and the scattered voltage along the wire can be written as follows:

$$u^{\text{sct}}(x,t) = \int_{-\infty}^{h} \frac{\partial \varphi(x,z,t)}{\partial z}\text{d}z = \frac{\text{d}}{\text{d}z}\int_{-\infty}^{h} \varphi(x,z,t)\text{d}z. \tag{2.155}$$

Now, the integration in the air and ground, respectively, can be taken separately, i.e. it can be written as

$$\int_{-\infty}^{h} \varphi(x,z)\text{d}z = \int_{-\infty}^{0} \varphi_{\text{gnd}}(x,z,t)dz + \int_{0}^{h} \varphi_{\text{air}}(x,z,t)\text{d}z. \tag{2.156}$$

Integrating the scalar potential from the infinite soil to the conductor surface it simply follows that

$$u^{\text{sct}}(x,t) = \varphi(x,z,t)|_{z=-\infty}^{z=h} = \varphi_{\text{air}}(x,z,t)\big|_{z=h} - \varphi_{\text{gnd}}(x,z,t)\Big|_{z=-\infty}. \tag{2.157}$$

Finally, assuming the scalar potential in the remote soil to be zero yields

$$u^{\text{sct}}(x,t) = \varphi(x,h,t), \tag{2.158}$$

where $\varphi(x,h,t)$ is determined by (2.138).

Combining Equations (2.138) and (2.158), the expression for the scattered voltage along the TL of finite length above a lossy ground is given by

$$\begin{aligned} u^{\text{sct}}(x,t) = &-\frac{1}{4\pi\varepsilon}\int_0^L\int_0^t \frac{\partial i(x',t)}{\partial x'}\frac{1}{R}\text{d}x'\,\text{d}t \\ &+\frac{1}{4\pi\varepsilon}\int_0^L\int_0^t\int_0^{t'} \frac{\partial i(x',t-\tau)}{\partial x'}\frac{1}{R^*}r(\theta,\tau)\text{d}x'\,\text{d}\tau\,\text{d}t. \end{aligned} \tag{2.159}$$

Using expression (2.159), Equation (2.150) can be written in the form

$$\begin{aligned} E_x^{\text{exc}} = \frac{\mu}{4\pi}\Bigg[&\int_0^L \frac{\partial i(x',t-R/v)}{\partial t}\frac{1}{R}\text{d}x' \\ &-\int_0^L\int_0^t r(\theta,\tau)\frac{\partial i(x',t-\tau-R^*/v)}{\partial t}\frac{1}{R}\text{d}x'\,\text{d}\tau\Bigg] + \frac{\partial u^{\text{sct}}(x,t)}{\partial x}, \end{aligned} \tag{2.160}$$

and finally, by slightly rearranging Equations (2.159) and (2.160) the general form of the time domain telegrapher's equations, valid for PEC wires of finite length, is obtained:

$$\frac{\partial u^{\mathrm{sct}}(x,t)}{\partial x} + L'' \left[ \int_0^L \frac{\partial i(x',t-R/v)}{\partial t} \frac{1}{R} \mathrm{d}x' - \int_0^L \int_0^t r(\theta,\tau) \frac{\partial i(x',t-\tau-R^*/v)}{\partial t} \frac{1}{R^*} \mathrm{d}x' \, \mathrm{d}\tau \right] = E_x^{\mathrm{exc}}, \quad (2.161)$$

$$C'' \frac{\partial u^{\mathrm{sct}}(x,t)}{\partial t} + \left( \int_0^L \frac{\partial i(x',t-R/v)}{\partial x'} \frac{1}{R} \mathrm{d}x' - \int_0^L \int_0^t r(\theta,\tau) \frac{\partial i(x',t-\tau-R^*/v)}{\partial x'} \frac{1}{R^*} \mathrm{d}x' \, \mathrm{d}\tau \right) = 0, \quad (2.162)$$

where the corresponding equivalent inductance and capacitance per unit length of the line are given by the relations:

$$L'' = \frac{\mu}{4\pi}, \quad (2.163)$$

$$C'' = 4\pi\varepsilon. \quad (2.164)$$

The correlation of the generalized telegrapher's equation (2.161) and (2.162) with the classic telegrapher's equation for lossless conductor above a lossy half-space [17] and PEC ground [1] can be performed in a rather straightforward manner.

For convenience, it can be written that

$$\int_0^L \frac{\partial i(x',t-R/v)}{\partial t} \frac{1}{R} \mathrm{d}x' - \int_0^L \int_0^t r(\theta,\tau) \frac{\partial i(x',t-\tau-R^*/v)}{\partial t} \frac{1}{R^*} \mathrm{d}x' \, \mathrm{d}\tau = \frac{\partial}{\partial t} \left( \int_0^L \frac{i(x',t-R/v)}{R} \mathrm{d}x' - \int_0^L \int_0^t r(\theta,\tau) \frac{i(x',t-\tau-R^*/v)}{R^*} \mathrm{d}x' \, \mathrm{d}\tau \right) \quad (2.165)$$

and

$$\int_0^L \frac{\partial i(x',t-R/v)}{\partial x'} \frac{1}{R} \mathrm{d}x' - \int_0^L \int_0^t r(\theta,\tau) \frac{\partial i(x',t-\tau-R^*/v)}{\partial x'} \frac{1}{R^*} \mathrm{d}x' \, \mathrm{d}\tau = \frac{\partial}{\partial x} \left( \int_0^L \frac{i(x',t-R/v)}{R} \mathrm{d}x' - \int_0^L \int_0^t r(\theta,\tau) \frac{i(x',t-\tau-R^*/v)}{R^*} \mathrm{d}x' \, \mathrm{d}\tau \right). \quad (2.166)$$

Now, using the addition–subtraction technique [10], the characteristic integrals for the source and image wires, respectively, yield

$$\int_0^L \frac{i(x',t-R/v) + i(x,t) - i(x,t)}{R} \mathrm{d}x' = i(x,t) \int_0^L \frac{1}{R} \mathrm{d}x' + \int_0^L \frac{i(x',t-R/v) - i(x,t)}{R} \mathrm{d}x', \quad (2.167)$$

$$\frac{\partial}{\partial t}\int_0^L\int_0^t r(\theta,\tau)\frac{i(x',t-\tau-R^*/v)+i(x,t-\tau)-i(x,t-\tau)}{R^*}\mathrm{d}x'\,\mathrm{d}\tau$$
$$=\int_0^t \frac{\partial i(x,t-\tau)}{\partial t}\int_0^L r(\theta,\tau)\frac{1}{R^*}\mathrm{d}x'\,\mathrm{d}\tau$$
$$+\frac{\partial}{\partial t}\int_0^L\int_0^t r(\theta,\tau)\frac{i(x',t-\tau-R^*/v)-i(x,t-\tau)}{R^*}\mathrm{d}x'\,\mathrm{d}\tau, \quad (2.168)$$

$$\frac{\partial}{\partial x}\int_0^L\int_0^t r(\theta,\tau)\frac{i(x',t-\tau-R^*/v)+i(x,t-\tau)-i(x,t-\tau)}{R^*}\mathrm{d}x'\,\mathrm{d}\tau$$
$$=\int_0^L \frac{r(\theta,\tau)}{R^*}\int_0^t \frac{\partial i(x,t-\tau)}{\partial x}\mathrm{d}\tau\,\mathrm{d}x'$$
$$+\frac{\partial}{\partial x}\int_0^L\int_0^t r(\theta,\tau)\frac{i(x',t-\tau-R^*/v)-i(x,t-\tau)}{R^*}\mathrm{d}x'\,\mathrm{d}\tau. \quad (2.169)$$

For the case of PEC ground the RC vanishes:

$$\int_0^L \frac{i(x',t-R^*/v)+i(x,t)-i(x,t)}{R^*}\mathrm{d}x'$$
$$=i(x,t)\int_0^L \frac{1}{R^*}\mathrm{d}x'+\int_0^L \frac{i(x',t-R^*/v)-i(x,t)}{R^*}\mathrm{d}x'. \quad (2.170)$$

Neglecting the second integrals at the right-hand side of (2.170) the solutions of characteristic integrals over Green function are straightforward:

$$\int_0^L \frac{1}{R}\mathrm{d}x' = 2\ln\frac{L}{a}, \quad (2.171)$$
$$\int_0^L \frac{1}{R^*}\mathrm{d}x' = 2\ln\frac{L}{2h}. \quad (2.172)$$

For the case of a lossy ground the first generalized telegrapher's equation (2.161) becomes

$$\frac{\partial u^{\mathrm{sct}}(x,t)}{\partial x}+L'_{lg}\frac{\partial i(x,t)}{\partial t}+\int_0^t z_g(x,\tau)\frac{\partial i(x,t-\tau)}{\partial t}d\tau = E_x^{\mathrm{exc}}(x,t), \quad (2.173)$$

where

$$L'_{lg}=\frac{\mu}{2\pi}\ln\frac{L}{a}, \quad (2.174)$$

and $z_g(x,t)$ is the corresponding transient ground impedance given by

$$z_g(x,t)=-L''\int_0^L r(\theta,t)\frac{1}{R^*}\mathrm{d}x' = -\frac{\mu}{4\pi}\int_0^L r(\theta,t)\frac{1}{R^*}\mathrm{d}x'. \quad (2.175)$$

Basically, expression (2.173) corresponds to the first telegrapher's equation presented in [18].

Furthermore, for the case of a lossy ground the second generalized telegrapher's equation (2.162) can be written in the form

$$C'_{lg}\frac{\partial u^{\text{sct}}(x,t)}{\partial t}+\frac{\partial i(x,t)}{\partial x}+\int_0^t \frac{\partial i(x,t-\tau)}{\partial x}h(x,t)\mathrm{d}\tau=0, \tag{2.176}$$

where

$$C'_{lg}=\frac{2\pi\varepsilon}{\ln\frac{L}{a}}, \tag{2.177}$$

and $z_g(x,t)$ is the corresponding transient ground impedance given by

$$h(x,t)=\frac{1}{2\pi\ln\frac{L}{a}}\int_0^L r(\theta,t)\frac{1}{R^*}\mathrm{d}x'. \tag{2.178}$$

Relation (2.176) corresponds to the second telegrapher's equation given in [18].

For the case of a PEC ground the set of generalized telegrapher's equations (2.161) and (2.162) simplifies into

$$\frac{\partial u^{\text{sct}}(x,t)}{\partial x}+L'_{pg}\frac{\partial i(x,t)}{\partial t}=E_x^{\text{exc}}, \tag{2.179}$$

$$\frac{\partial u^{\text{sct}}(x,t)}{\partial t}+C'_{pg}\frac{\partial i(x,t)}{\partial x}=0, \tag{2.180}$$

where the corresponding equivalent inductance and capacitance per unit length of the overhead wire are now given by

$$L'_{pg}=\frac{\mu}{2\pi}\ln\frac{2h}{a}, \tag{2.181}$$

$$C'_{pg}=\frac{2\pi\varepsilon}{\ln\frac{2h}{a}}. \tag{2.182}$$

Note that the set of Equations (2.179) and (2.180) represents the standard well-known form of the field-to-TL coupling equations in the time domain [1, 18].

The mathematical details of the solution method can be found elsewhere, e.g. in [18].

#### 2.2.3.4 Generalized Telegrapher's Equations for Lossy Conductors

For the case of finitely conducting wires, taking into account (2.148) the following generalized telegrapher's equation is obtained:

$$\frac{\partial u^{\text{sct}}(x,t)}{\partial x}+L''\left[\int_0^L \frac{\partial i(x',t-R/v)}{\partial t}\frac{1}{R}\mathrm{d}x' - \int_0^L\int_0^t r(\theta,\tau)\frac{\partial i(x',t-\tau-R^*/v)}{\partial t}\frac{1}{R^*}\mathrm{d}x'\,\mathrm{d}\tau\right]$$

$$+\int_0^t z_s(x,\tau)i(x',t-\tau)\mathrm{d}\tau = E_x^{\mathrm{exc}}. \tag{2.183}$$

The correlation of the generalized telegrapher's equation (2.183) with its classic counterpart is undertaken in the same manner as it was undertaken for the case of lossless conductors.

For the case of a lossy medium it simply follows that

$$\frac{\partial u^{\mathrm{sct}}(x,t)}{\partial x} + L'_{lg}\frac{\partial i(x,t)}{\partial t} + \int_0^t z_g(x,\tau)\frac{\partial i(x,t-\tau)}{\partial t}\mathrm{d}\tau$$
$$+\int_0^t z_s(x,\tau)i(x,t-\tau)\mathrm{d}\tau = E_x^{\mathrm{exc}}(x,t), \tag{2.184}$$

and for the case of a PEC ground expression (2.184) simplifies into

$$\frac{\partial u^{\mathrm{sct}}(x,t)}{\partial x} + L'_{pg}\cdot\frac{\partial i(x,t)}{\partial t} + \int_0^t z_s(x,\tau)i(x,t-\tau)\mathrm{d}\tau = E_x^{\mathrm{exc}}(x,t). \tag{2.185}$$

For the case of constant, frequency-independent per-unit-length resistance $R'$, Equation (2.184) reduces to

$$\frac{\partial u^{\mathrm{sct}}(x,t)}{\partial x} + L'_{lg}\frac{\partial i(x,t)}{\partial t} + \int_0^t z_g(x,\tau)\frac{\partial i(x,t-\tau)}{\partial t}\mathrm{d}\tau + R'\cdot i(x,t)$$
$$= E_x^{\mathrm{exc}}(x,t), \tag{2.186}$$

while relation (2.185) becomes

$$\frac{\partial u^{\mathrm{sct}}(x,t)}{\partial x} + R'\cdot i(x,t) + L'\frac{\partial i(x,t)}{\partial t} = E_x^{\mathrm{exc}}, \tag{2.187}$$

which represents the standard form of the classical time domain field-to-TL telegrapher's equation [10].

Computational examples

The line shown in Figure 2.13 is excited by a plane wave with $\psi = 90°$ and $\varphi = 0°$. Its time dependence is a unit amplitude double exponential waveform defined as [1]

$$E^{\mathrm{inc}}(t) = E_0\left(\mathrm{e}^{-\alpha t} - \mathrm{e}^{-\beta t}\right), \tag{2.188}$$

where $E_0 = 1.3\ \mathrm{V\,m^{-1}}$, $a = 4\times10^7$, and $b = 6\times10^8$.

This gives a waveform with unit amplitude, a rise time of 4 ns, and a decay time to half peak of 24 ns. The approximate maximum frequency content of this waveform is about 80 MHz corresponding to a minimum wavelength of 3.8 m. In all cases, the line is assumed to be open at both ends. Also, in all simulations, the wire radius is at $a = 2$ mm.

Finitely conducting ground with a conductivity of $\sigma = 0.001\ \mathrm{S\,m^{-1}}$ and relative permittivity of $\varepsilon_r = 10$ is assumed. Simulation results of the induced transient current at the wire center point aimed at the validation of the proposed

GB-IBEM solution of the presented formulation are shown in Figure 2.14 for a wire of length $L = 1$ m located at three different heights above the ground, namely, $h = 0.25$ m, $h = 0.5$ m, and $h = 1$ m. The results are compared with the method of moments (MoM) solution of the thin wire EFIE implemented in the NEC-4 [19] and also the classical TL theory [1], both implemented in the frequency domain. Time to frequency and frequency to time conversions are carried out through appropriate fast Fourier transform (FFT) and inverse fast Fourier transform (IFFT) routines using $N = 4096$ frequency samples starting from $f_0 = 1.22$ MHz corresponding to a sampling interval of $\Delta t = 0.1$ ns up to $T_{max} = 819.2$ ns. In the classical TL solution, the effect of ground is taken into account via the approximate grounding impedance expression proposed by Sunde [15]:

$$Z'_g = \frac{j\omega\mu_0}{\pi} \ln\left[\frac{1 + \gamma_g h}{\gamma_g h}\right], \tag{2.189}$$

where

$$\gamma_g = \sqrt{j\omega\mu_0(\sigma + j\omega\varepsilon_0\varepsilon_r)} \tag{2.190}$$

is the complex propagation constant inside the lossy ground.

The Sunde approximation has been shown to be accurate within the limits of the TL theory [20].

As can be seen from the three plots in Figure 2.14a–c, the GB-IBEM solution of the presented generalized telegrapher's equations is in excellent agreement with the MoM solution of the governing thin wire EFIE implemented in NEC-4. The agreement is indeed well established for the three heights considered, especially for the third case of $h = 1$ m, which is clearly beyond the validity range of the classical TL theory. In fact, the classical TL theory is known to give accurate results if (1) the line length to height $L/h$ ratio is large, (2) the line cross section $h/\lambda$ is electrically small, and (3) the ground is highly conducting giving a TEM response.

While for the curves in Figure 2.14c none of the above conditions is met, for the curves in Figure 2.14b conditions (1) and (3) are not satisfied and for those in Figure 2.14a the first two conditions are marginally respected. This is why the agreement between the predictions of the classical TL theory with

**Figure 2.14** Transient current induced at the center of the line with $L = 1$ m length and $a = 2$ mm radius above a lossy ground with conductivity of $\sigma = 0.001$ S m$^{-1}$ and relative permittivity of $\varepsilon_r = 10$ for three different heights above the ground: (a) $h = 0.25$ m, (b) $h = 0.5$ m, and (c) $h = 1$ m. The calculations have been performed (i) using the GB-IBEM solution of the derived generalized telegrapher's equations, (ii) NEC-4, and (iii) classical TL solution based on the grounding impedance formula proposed by Sunde [15]. *Source*: From Ref. [15]. Reproduced with permission of John Wiley and Sons.

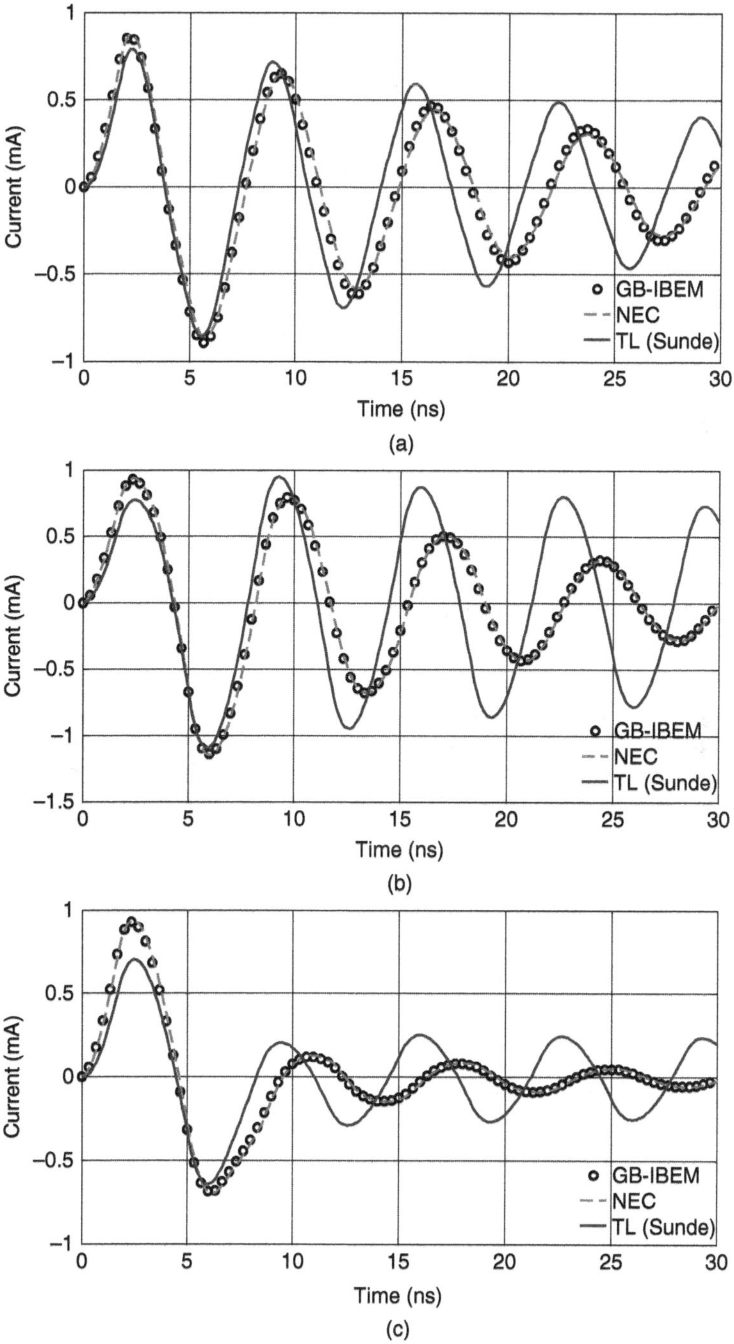
Current (mA)
Time (ns)
GB-IBEM
NEC
TL (Sunde)
(a)
Current (mA)
Time (ns)
GB-IBEM
NEC
TL (Sunde)
(b)
Current (mA)
Time (ns)
GB-IBEM
NEC
TL (Sunde)
(c)

those of the accurate solutions obtained using either the proposed method or NEC-4 improves from Figure 2.14c to Figure 2.14b and to Figure 2.14a. This is clearly observed in the first peaks predicted by the three techniques in the three plots. Further, the late time disagreement is again more pronounced in Figure 2.14c than in Figure 2.14b, and in Figure 2.14b than in Figure 2.14a. Such a difference is clearly visible for both current amplitudes and its zero crossing times. Accordingly, one expects that when the three necessary conditions are well satisfied the differences between the classical TL theory and the full-wave techniques would be minimal. To check this, the case in Figure 2.15a is considered. The height is assumed to be $h = 0.1$ m, while the ground conductivity is to a value of $\sigma = 10\,\mathrm{S\,m^{-1}}$. The simulation results are shown in Figure 2.15a and b using the three different methods.

It can be seen that, compared to the curves in Figure 2.14a, when the wire is brought closer to the ground surface, the differences between the predictions of the traditional TL theory and the two full-wave techniques are, as expected, reduced.

However, no improvement is seen in the predictions of the classical TL theory from Figure 2.15a to Figure 2.15b. This could be attributed to the fact that for the case in Figure 2.15b where the ground becomes highly conducting, the resonance nature of the TL becomes important and it starts to govern the solution.

One can also observe that, for both cases in Figure 2.15, the predictions of the GB-IBEM solution of the proposed generalized telegrapher's equations appear to be closer to the classical TL solution. Such deviations from the full-wave solution implemented in NEC-4, although minima, can be clearly observed in Figure 2.15b. This deviation is due to the fact that for a wire very close to the ground plane, the RC approximation starts to fail in modeling ground effects. For such a case when the ground is highly conducting one should indeed rely on the image theory in order to account for the contribution of the ground.

### 2.2.4 Time Domain Analysis for Straight Wires Buried in a Lossy Ground

A single wire TL of a finite length $L$ and radius $a$ buried at depth $d$ inside a lossy medium, see Figure 2.16, is considered.

The buried straight wire is subjected to transient electromagnetic fields from a distant source transmitted through a lossy medium, inducing a current to flow along the line.

#### 2.2.4.1 Space–Time Integro-Differential Equation for PEC Wire below a Lossy Ground

The integro-differential expressions for the transient current induced along the buried wire are derived starting from the interface condition for the tangential

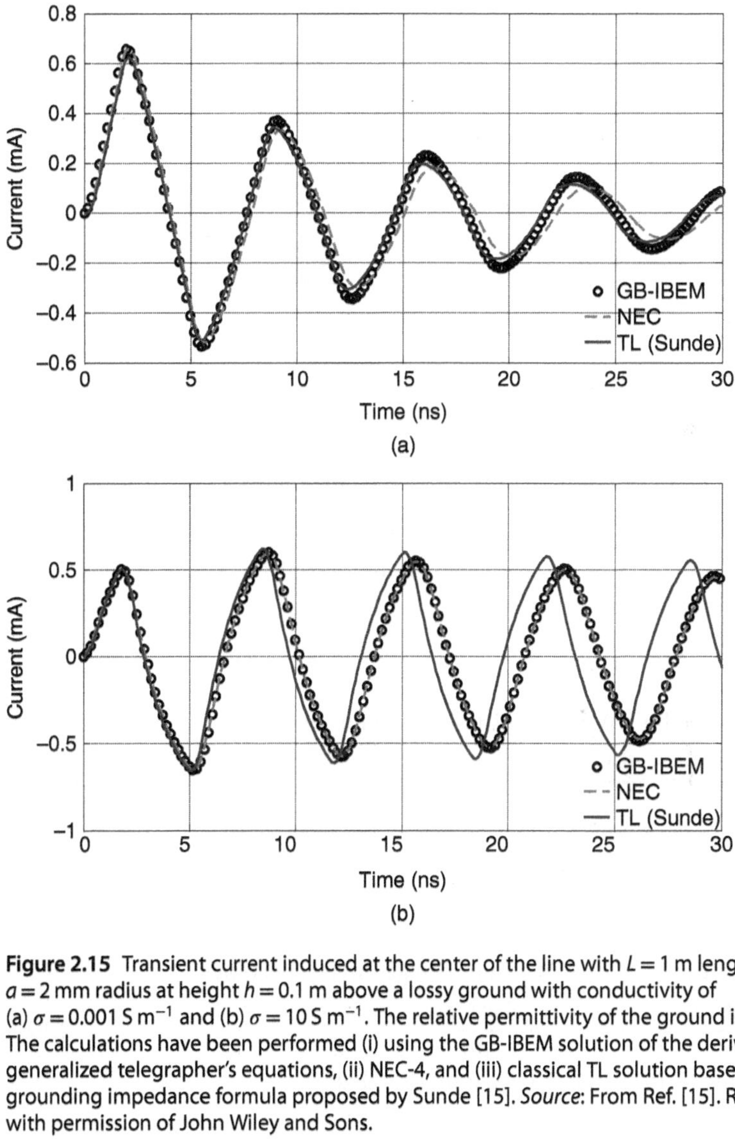


**Figure 2.15** Transient current induced at the center of the line with $L = 1$ m length and $a = 2$ mm radius at height $h = 0.1$ m above a lossy ground with conductivity of (a) $\sigma = 0.001\ \mathrm{S\,m^{-1}}$ and (b) $\sigma = 10\ \mathrm{S\,m^{-1}}$. The relative permittivity of the ground is $\varepsilon_r = 10$. The calculations have been performed (i) using the GB-IBEM solution of the derived generalized telegrapher's equations, (ii) NEC-4, and (iii) classical TL solution based on the grounding impedance formula proposed by Sunde [15]. *Source*: From Ref. [15]. Reproduced with permission of John Wiley and Sons.

components of the electric field given by

$$\vec{e}_x \cdot \vec{E}^{\mathrm{tot}}(\vec{r}, t) = 0 \text{ on the surface of the wire,} \tag{2.191}$$

where the total field $\vec{E}^{\mathrm{tot}}$ is the sum of excitation $\vec{E}^{\mathrm{exc}}$ and scattered $\vec{E}^{\mathrm{sct}}$ field component, respectively:

$$\vec{E}^{\mathrm{tot}}(\vec{r}, t) = \vec{E}^{\mathrm{exc}}(\vec{r}, t) + \vec{E}^{\mathrm{sct}}(\vec{r}, t). \tag{2.192}$$

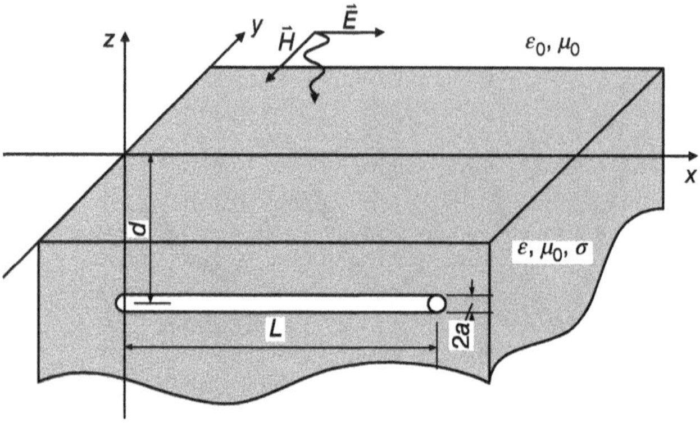


**Figure 2.16** Finite length wire below a lossy ground.

If the conductor is assumed to be PEC, the tangential electric field vanishes:

$$\vec{e}_x \cdot [\vec{E}^{\text{exc}}(\vec{r}, t) + \vec{E}^{\text{sct}}(\vec{r}, t)] = 0 \text{ on the surface of the wire.} \tag{2.193}$$

The scattered field component can be expressed by means of the vector potential $\vec{A}$ and the scalar potential $\varphi$:

$$\vec{E}^{\text{sct}} = -\frac{\partial \vec{A}}{\partial t} - \nabla\varphi. \tag{2.194}$$

According to the TWA [11], (2.194) simplifies into

$$E_x^{\text{sct}}(x, t) = -\frac{\partial A_x(x, t)}{\partial t} - \frac{\partial \varphi(x, t)}{\partial x}. \tag{2.195}$$

For the case of finitely conducting media, the scalar and vector potentials are related through the Lorentz gauge

$$\nabla\vec{A} + \mu\sigma\phi + \mu\varepsilon\frac{\partial \phi}{\partial t} = 0, \tag{2.196}$$

which, for the special case of a thin wire geometry, becomes

$$\frac{\partial A_x(x, t)}{\partial x} = -\mu\sigma\varphi - \mu\varepsilon\frac{\partial \varphi(x, t)}{\partial t}. \tag{2.197}$$

Combining (2.195) and (2.197), it follows that

$$\left(\frac{1}{v^2}\frac{\partial}{\partial t} + \mu\sigma\right) E_x = \frac{\partial^2 A_x}{\partial x^2} - \mu\sigma\frac{\partial A_x}{\partial t} - \mu\varepsilon\frac{\partial^2 A_x}{\partial t^2}. \tag{2.198}$$

Also, the wave equation for the vector potential in a lossy media due to a volume current source is [21, 22]

$$\nabla^2\vec{A} - \mu\sigma\frac{\partial \vec{A}}{\partial t} - \mu\varepsilon\frac{\partial^2 \vec{A}}{\partial t^2} = -\mu\vec{J}_i, \tag{2.199}$$

where $\vec{J}_i$ is the volume current density.

The solution of the inhomogeneous wave equation (2.199) for the case of a straight thin wire of length $L$ in an unbounded lossy medium is given in the form of the particular integral [11, 21]:

$$A_x(x,t) = \frac{\mu}{4\pi}\int_0^L I(x', t - R/v)\frac{\mathrm{e}^{-\frac{t}{\tau_g}\frac{R}{v}}}{R}\mathrm{d}x', \tag{2.200}$$

where, $I(x, t)$ is the unknown space–time current along the buried conductor.

For the case of a dissipative half-space, the magnetic vector potential (2.200) is extended by an additional term due to the image wire in the air. In the frequency domain, the term due to the image wire is added by multiplying the Green function with a corresponding RC.

Thus, the Laplace transform of (2.200) is given by

$$A_x(x,s) = \frac{\mu}{4\pi}\int_0^L I(x', s)\frac{\mathrm{e}^{-s\frac{R}{v}}\mathrm{e}^{-\frac{1}{\tau_g}\frac{R}{v}}}{R}\mathrm{d}x', \tag{2.201}$$

where, $I(x, s)$ is the unknown space–time current distribution along the wire, $s$ is the Laplace domain variable, and $R$ is the distance from the source point located on the wire to the observation point in a lossy medium.

The corresponding time constant $\tau_g$ and propagation velocity in the lossy medium $v$ are given by

$$\tau_g = \frac{2\varepsilon}{\sigma}, \tag{2.202}$$

$$v = \frac{1}{\sqrt{\mu\varepsilon}}. \tag{2.203}$$

The magnetic vector potential due to a straight wire buried in a lossy ground extended with an additional term to account for the image wire in the air can be written as follows:

$$A_x(x,s) = \frac{\mu}{4\pi}\int_0^L I(x', s)\frac{\mathrm{e}^{-s\frac{R}{v}}\mathrm{e}^{-\frac{1}{\tau_g}\frac{R}{v}}}{R}\mathrm{d}x' - \frac{\mu}{4\pi}\int_0^L \Gamma_{\mathrm{ref}}(s) I(x', s)\frac{\mathrm{e}^{-s\frac{R^*}{v}}\mathrm{e}^{-\frac{1}{\tau_g}\frac{R^*}{v}}}{R}\mathrm{d}x', \tag{2.204}$$

where $R^*$ is the distance from the image wire to the observation point in the lossy medium and $\Gamma_{\mathrm{ref}}(s)$ is the RC to account for the effects due to the presence of the earth–air interface arising from the modified image theory (MIT).

$$\Gamma_{\mathrm{ref}}^{\mathrm{MIT}}(s) = -\frac{\tau_1 s + 1}{\tau_2 s + 1}. \tag{2.205}$$

The corresponding time constants are

$$\tau_1 = \frac{\varepsilon_0(\varepsilon_r - 1)}{\sigma}, \tag{2.206}$$

$$\tau_2 = \frac{\varepsilon_0(\varepsilon_r + 1)}{\sigma}. \tag{2.207}$$

It is worth noting that the RC (2.205) represents a rather simple characterization of the earth–air interface, taking into account only the medium properties. The accuracy of (2.205) has been discussed elsewhere, e.g. in [11] and [21].

The time domain counterpart of (2.204) is given by

$$A_x(x,t) = \frac{\mu}{4\pi}\int_0^L i(x', t - R/v)\frac{e^{-\frac{t}{\tau_g}\frac{R}{v}}}{R}\mathrm{d}x' - \frac{\mu}{4\pi}\int_{-\infty}^{t}\int_0^L \Gamma_{\text{ref}}(\theta,\tau) i(x', t - R^*/v - \tau)\frac{e^{-\frac{t}{\tau_g}\frac{R^*}{v}}}{R^*}\mathrm{d}x'\,\mathrm{d}\tau, \tag{2.208}$$

where the time domain version of (2.205) is [11]

$$\Gamma_{\text{ref}}^{\text{MIT}}(t) = -\left[\frac{\tau_1}{\tau_2}\delta(t) + \frac{1}{\tau_2}\left(1 - \frac{\tau_1}{\tau_2}\right)e^{-\frac{t}{\tau_2}}\right]. \tag{2.209}$$

For the considered geometry of a PEC thin wire, we have

$$E_x^{\text{exc}}(x,t) + E_x^{\text{sct}}(x,t) = 0. \tag{2.210}$$

Now combining (2.195), (2.208), and (2.210) yields

$$\left[\frac{\sigma}{\varepsilon} + \frac{\partial}{\partial t}\right]E_x^{\text{tr}} = \frac{\mu}{4\pi}\cdot\left[-v^2\frac{\partial^2}{\partial x^2} + \frac{\partial^2}{\partial t^2} + \frac{\sigma}{\varepsilon}\frac{\partial}{\partial t}\right] \times\left[\int_0^L i(x', t - R_a/v)\frac{e^{-\frac{t}{\tau_g}\frac{R_a}{v}}}{R_a}\mathrm{d}x' - \int_{-\infty}^{t}\int_0^L \Gamma_{\text{ref}}(\theta,\tau) i(x', t - R_d^*/v - \tau)\frac{e^{-\frac{t}{\tau_g}\frac{R_d^*}{v}}}{R_d^*}\mathrm{d}x'\,\mathrm{d}\tau\right], \tag{2.211}$$

where $R_a$ is the distance from the source point to the observation point, both located at the buried wire,

$$R_a = \sqrt{(x - x')^2 + a^2}, \tag{2.212}$$

while $R_d^*$ is the distance from the source point located at the image wire in the air to the observation point located at the wire immersed in a lossy medium,

$$R_d^* = \sqrt{(x - x')^2 + 4d^2}. \tag{2.213}$$

Expression (2.211) represents a variant of the Pocklington's integro-differential equation for the straight wire scatterer buried in the lossy half-space.

For the case of horizontal grounding electrodes, the excitation function in the form of tangential electric field does not exist and the electrode is driven by an equivalent current generator as shown in Figure 2.17.

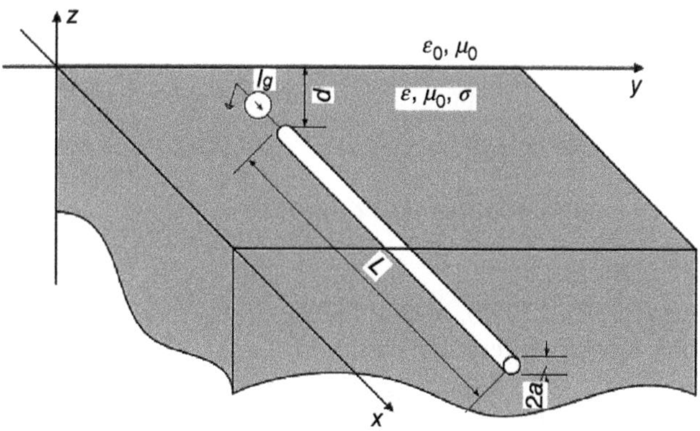


**Figure 2.17** Horizontal grounding electrode buried in a lossy medium.

In such a case, (2.211) becomes homogeneous:

$$\frac{\mu}{4\pi}\cdot\left[-v^2\frac{\partial^2}{\partial x^2}+\frac{\partial^2}{\partial t^2}+\frac{\sigma}{\varepsilon}\frac{\partial}{\partial t}\right]\left[\int_0^L i(x',t-R_a/v)\frac{\mathrm{e}^{-\frac{t}{\tau_g}\frac{R_a}{v}}}{R_a}\mathrm{d}x'\right.$$
$$\left.-\int_{-\infty}^{t}\int_0^L \Gamma_{\mathrm{ref}}(\theta,\tau)i(x',t-R_d^*/v-\tau)\frac{\mathrm{e}^{-\frac{t}{\tau_g}\frac{R_d^*}{v}}}{R_d^*}\mathrm{d}x'\,\mathrm{d}\tau\right]=0. \quad (2.214)$$

The current source excitation is included in the formulation through boundary and initial conditions, respectively [11].

#### 2.2.4.2 Space–Time Integro-Differential Equation for Lossy Conductors

If lossy conductors are considered, the tangential component of the electric field along the wire differs from zero and it is necessary to modify the condition (2.193). Thus, in the Laplace domain the tangential component of the total field at the conductor surface is equal to the product of the wire current $I(x, s)$ and per-unit-length internal impedance $Z_s(x, s)$ of the wire.

$$\vec{e}_x\cdot[\vec{E}^{\mathrm{exc}}(\vec{r},s)+\vec{E}^{\mathrm{sct}}(\vec{r},s)]=Z_s(\vec{r},s)I(\vec{r},s)\ \text{on the wire surface,} \quad (2.215)$$

and, for the case of a straight thin wire, (2.215) simplifies into

$$\vec{e}_x\cdot[\vec{E}^{\mathrm{exc}}(x,s)+\vec{E}^{\mathrm{sct}}(x,s)]=Z_s(x,s)I(x,s)\ \text{along the wire,} \quad (2.216)$$

where the surface internal impedance $Z_s(x, s)$ of a cylindrical wire is given by (2.145).

Furthermore, the convolution operator applied to (2.218) yields

$$\vec{e}_x \cdot [\vec{E}^{\mathrm{exc}}(x,t) + \vec{E}^{\mathrm{sct}}(x,t)] = \int_0^t z_s(x,\tau) i(x, t-\tau) \mathrm{d}\tau \text{ along the wire,} \tag{2.217}$$

where $z_s(x, t)$ is the inverse Laplace transform of $Z_s(x, s)$,

$$z_s(x,t) = L^{-1} Z_s(x,s). \tag{2.218}$$

Combining Equations (2.190), (2.205), and (2.217) results in the following form of the Pocklington's integro-differential equation:

$$\begin{aligned}\left[\frac{\sigma}{\varepsilon} + \frac{\partial}{\partial t}\right] E_x^{\mathrm{exc}}(x,t) = \frac{\mu}{4\pi} \cdot \left[-v^2 \frac{\partial^2}{\partial x^2} + \frac{\partial^2}{\partial t^2} + \frac{\sigma}{\varepsilon}\frac{\partial}{\partial t}\right] \\ \times \left[\int_0^L i(x', t - R_a/v) \frac{\mathrm{e}^{-\frac{t}{\tau_g}\frac{R_a}{v}}}{R_a} \mathrm{d}x' \right. \\ \left. - \int_{-\infty}^t \int_0^L \Gamma_{\mathrm{ref}}(\theta,\tau) i(x', t - R_d^*/v - \tau) \frac{\mathrm{e}^{-\frac{t}{\tau_g}\frac{R_d^*}{v}}}{R_d^*} \mathrm{d}x' \, \mathrm{d}\tau \right] \\ + \left[\frac{\sigma}{\varepsilon} + \frac{\partial}{\partial t}\right] \int_0^t z_s(x,\tau) i(x, t-\tau) \mathrm{d}\tau.\end{aligned} \tag{2.219}$$

Furthermore, for the case of horizontal grounding electrodes, (2.219) simplifies into

$$\begin{aligned}\frac{\mu}{4\pi} \cdot \left[-v^2 \frac{\partial^2}{\partial x^2} + \frac{\partial^2}{\partial t^2} + \frac{\sigma}{\varepsilon}\frac{\partial}{\partial t}\right] \\ \times \left[\int_0^L i(x', t - R/v) \frac{\mathrm{e}^{-\frac{t}{\tau_g}\frac{R}{v}}}{R} \mathrm{d}x' \right. \\ \left. - \int_{-\infty}^t \int_0^L \Gamma_{\mathrm{ref}}(\theta,\tau) i(x', t - R_d^*/v - \tau) \frac{\mathrm{e}^{-\frac{t}{\tau_g}\frac{R_d^*}{v}}}{R^*} \mathrm{d}x' \, \mathrm{d}\tau \right] \\ + \left[\frac{\sigma}{\varepsilon} + \frac{\partial}{\partial t}\right] \int_0^t z_s(x,\tau) i(x, t-\tau) \mathrm{d}\tau = 0.\end{aligned} \tag{2.220}$$

If one deals with very good conductors the surface internal impedance can be neglected and the last term on the left-hand side of (2.219) and (2.220) vanishes.

#### 2.2.4.3 Generalized Telegrapher's Equations for Buried Wires

There are several advantages in using generalized telegrapher's equations to analyze straight buried wires instead of solving the corresponding Pocklington's equation, such as a clear correlation of the AT model with TL formulation and a direct inclusion of scattered voltage into the formulation.

The fact that derivation of telegrapher's type equations for the currents induced along the buried wire requires the scattered voltage to be included in the formulation raises a conceptual difficulty in handling the half-space problem arising from the definition of the line voltage [1]. For the case of a frequency domain analysis, this matter has been discussed in detail elsewhere, e.g. in [8, 9].

The scattered voltage along the horizontal straight wire buried in a lossy ground is defined by the integral of a scattered vertical field component from the point in the remote soil to the conductor surface:

$$u^{\text{sct}}(x,t) = -\int_{-\infty}^{-d} E_z^{\text{sct}}(x,z,t)\text{d}z. \tag{2.221}$$

Assuming the zero voltage at $z \to -\infty$ [11] and expressing the scattered vertical field component in terms of the scalar potential gradient,

$$E_z^{\text{sct}} = -\frac{\partial \phi}{\partial z}, \tag{2.222}$$

the scattered voltage along the wire can be written as follows:

$$u^{\text{sct}}(x,t) = \int_{-\infty}^{-d} \frac{\partial \varphi(x,z,t)}{\partial z}\text{d}z = \frac{\text{d}}{\text{d}z}\int_{-\infty}^{-d} \varphi(x,z,t)\text{d}z. \tag{2.223}$$

Now, integrating the scalar potential from the infinite soil to the wire surface and assuming the scalar potential in the remote soil to be zero yields

$$u^{\text{sct}}(x,t) = \varphi(x,z,t)|_{z=-\infty}^{z=-d} = \varphi(x,-d,t), \tag{2.224}$$

where $\varphi(x,-d,t)$ can be determined from (2.197) and (2.208), respectively.

Combining (2.182) and (2.195), (2.207), and (2.224) one deals with the following expression:

$$\begin{aligned} E_x^{\text{exc}} = \frac{\mu}{4\pi}\Bigg[&\int_0^L i(x',t-R/v)\frac{\text{e}^{-\frac{t}{\tau_g}\frac{R}{v}}}{R}\text{d}x' \\ &-\int_{-\infty}^{t}\int_0^L \Gamma_{\text{ref}}(\theta,\tau) i(x',t-R^*/v-\tau)\frac{\text{e}^{-\frac{t}{\tau_g}\frac{R^*}{v}}}{R^*}\text{d}x'\,\text{d}\tau\Bigg] \\ &+\frac{\partial u^{\text{sct}}(x,t)}{\partial x}. \end{aligned} \tag{2.225}$$

Furthermore, from (2.182) and (2.195) it follows that

$$\begin{aligned} \mu\varepsilon\frac{\partial u^{\text{sct}}(x,t)}{\partial t} + \mu\sigma u^{\text{sct}}(x,t) + \frac{\mu}{4\pi}\Bigg(&\int_0^L \frac{\partial i(x',t-R/v)}{\partial x'}\frac{\text{e}^{-\frac{t}{\tau_g}\frac{R}{v}}}{R}\text{d}x' \\ &+\int_0^L\int_0^t r(\theta,\tau)\frac{\partial i(x',t-\tau-R^*/v)}{\partial x'}\frac{\text{e}^{-\frac{t}{\tau_g}\frac{R^*}{v}}}{R^*}\text{d}x'\,\text{d}\tau\Bigg) = 0, \end{aligned} \tag{2.226}$$

and finally, by slightly rearranging Equations (2.225) and (2.226) the general form of the time domain telegrapher's equations is obtained, valid for PEC wires of finite length:

$$\frac{\partial u^{\mathrm{sct}}(x,t)}{\partial x} + L''\left[\int_0^L \frac{\partial i(x',t-R/v)}{\partial t}\frac{\mathrm{e}^{-\frac{t}{\tau_g}\frac{R}{v}}}{R}\mathrm{d}x' - \int_0^L\int_0^t \Gamma(\theta,\tau)\frac{\partial i(x',t-\tau-R^*/v)}{\partial t}\frac{\mathrm{e}^{-\frac{t}{\tau_g}\frac{R^*}{v}}}{R^*}\mathrm{d}x'\,\mathrm{d}\tau\right] = E_x^{\mathrm{exc}}, \tag{2.227}$$

$$C''\frac{\partial u^{\mathrm{sct}}(x,t)}{\partial t} + G''u^{\mathrm{sct}}(x,t) + \left(\int_0^L \frac{\partial i(x',t-R/v)}{\partial x'}\frac{\mathrm{e}^{-\frac{t}{\tau_g}\frac{R}{v}}}{R}\mathrm{d}x' + \int_0^L\int_0^t \Gamma(\theta,\tau)\frac{\partial i(x',t-\tau-R^*/v)}{\partial x'}\frac{\mathrm{e}^{-\frac{t}{\tau_g}\frac{R^*}{v}}}{R^*}\mathrm{d}x'\,\mathrm{d}\tau\right) = 0. \tag{2.228}$$

Note that the corresponding equivalent inductance, capacitance, and conductance per unit length of the line are given by the relations

$$L'' = \frac{\mu}{4\pi}, \tag{2.229}$$

$$C'' = 4\pi\varepsilon, \tag{2.230}$$

$$G'' = 4\pi\sigma. \tag{2.231}$$

For the case of a finitely conducting wire, expressions (2.197), (2.208), (2.220), and (2.224) lead to the following generalized telegrapher's equation:

$$\frac{\partial u^{\mathrm{sct}}(x,t)}{\partial x} + L''\left[\int_0^L \frac{\partial i(x',t-R/v)}{\partial t}\frac{\mathrm{e}^{-\frac{t}{\tau_g}\frac{R}{v}}}{R}\mathrm{d}x' - \int_0^L\int_0^t \Gamma(\theta,\tau)\frac{\partial i(x',t-\tau-R^*/v)}{\partial t}\frac{\mathrm{e}^{-\frac{t}{\tau_g}\frac{R^*}{v}}}{R^*}\mathrm{d}x'\,\mathrm{d}\tau\right] + \int_0^t z_s(x,\tau)i(x',t-\tau)\mathrm{d}\tau = E_x^{\mathrm{exc}}. \tag{2.232}$$

If a horizontal grounding electrode is of interest, (2.227) and (2.232) become homogeneous.

The governing equations given in this section are solved by means of the analytical method presented in [22].

#### 2.2.4.4 Computational Results: Buried Wire Scatterer

The buried wire scatterer of interest is specified by a variable length $L$ and conductor radius $a$, and the burial depth is $d$. The wire is assumed to be PEC. The excitation field at the air–earth interface is the double exponential electromagnetic pulse (EMP) tangential to the interface (2.188), with $E_0 = 1\,\mathrm{V\,m^{-1}}$, $\alpha = 4\times10^6\,\mathrm{s^{-1}}$, $\beta = 4.78\times10^8\,\mathrm{s^{-1}}$.

The transmitted electric field exciting the buried wire surface in the Laplace domain is given by

$$E_x^{\text{tr}}(s) = \Gamma_{\text{tr}}(s)E_x(s)\mathrm{e}^{-\gamma d}, \tag{2.233}$$

where $\Gamma_{\text{tr}}(s)$ is the Fresnel transmission coefficient [11]

$$\Gamma_{\text{tr}}(s) = \frac{2\sqrt{s\varepsilon_0}}{\sqrt{s\varepsilon + \sigma} + \sqrt{s\varepsilon_0}}. \tag{2.234}$$

As the analytical convolution, i.e. the time domain counterpart of (2.233), would be too complex to perform, numerical convolution is carried out, as reported in [11].

Some illustrative computational examples related to the transient response of a buried wire excited by a transient plane wave (normal incidence) transmitted into the ground, as shown in Figure 2.16, are presented in Figures 2.18 and 2.19.

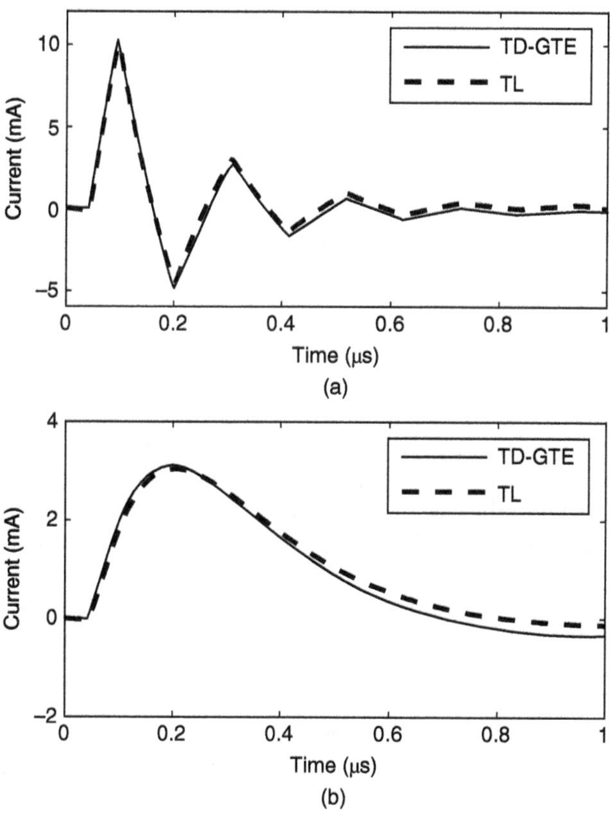


**Figure 2.18** Transient current at the center of the straight wire, $L = 10$ m, $d = 4$ m. (a) $\sigma = 1$ mS m$^{-1}$. (b) $\sigma = 10$ mS m$^{-1}$.

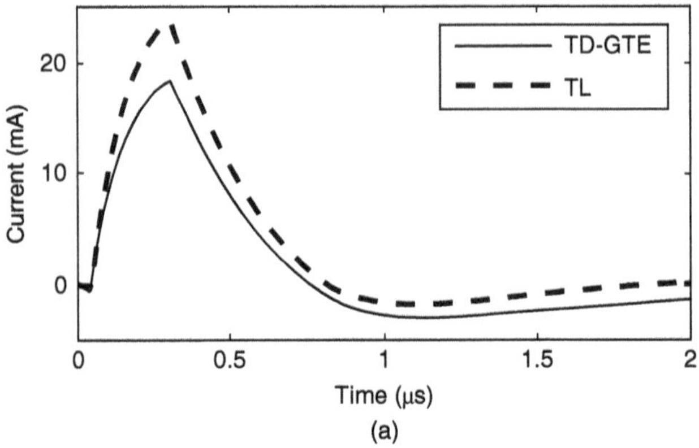


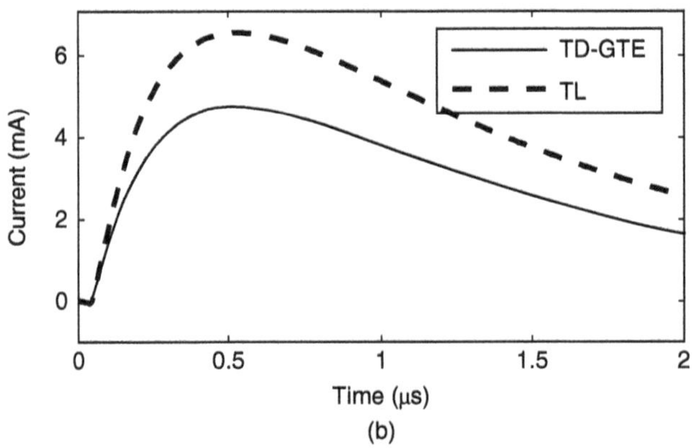


**Figure 2.19** Transient current at the center of the straight wire, $L = 50$ m, $d = 4$ m. (a) $\sigma = 1$ mS m$^{-1}$. (b) $\sigma = 10$ mS m$^{-1}$.

The results obtained using the TD-GTE approach are compared to the results obtained by the TL approach [22]. Figure 2.18 shows the transient current induced at the center of the wire ($L = 10$ m, $d = 4$ m, $a = 5$ mm, $\varepsilon_r = 10$) for different values of ground conductivity. Figure 2.19 shows the transient current induced at the center of a longer wire ($L = 50$ m).

The results calculated via different approaches agree satisfactorily in the case of a 10 m long wire, while bigger discrepancies between different approaches appear for a 50 m long wire scatterer.

#### 2.2.4.5 Computational Results: Horizontal Grounding Electrode

The horizontal electrode of radius $a = 5$ mm, buried in a lossy half-space with relative permittivity $\varepsilon_r = 10$ at depth $d = 0.5$ m is considered. The grounding electrode is excited with the double exponential current source:

$$I(t) = I_0(e^{-at} - e^{-bt}), \tag{2.235}$$

with $I_0 = 1.1043$ A, $a = 0.07924 \times 10^7\ s^{-1}$, $b = 4.0011 \times 10^7\ s^{-1}$.

The analytical results obtained by solving TD-GTE are compared to the numerical results obtained by the TL approach.

Figure 2.20 shows the transient response at the center of the 10 m long electrode buried in a lossy medium for different values of soil conductivity and excited by the 0.1/1 μs pulse.

Figure 2.21 shows the transient response at the center of the 50 m long electrode and excited by the 0.1/1 μs pulse.

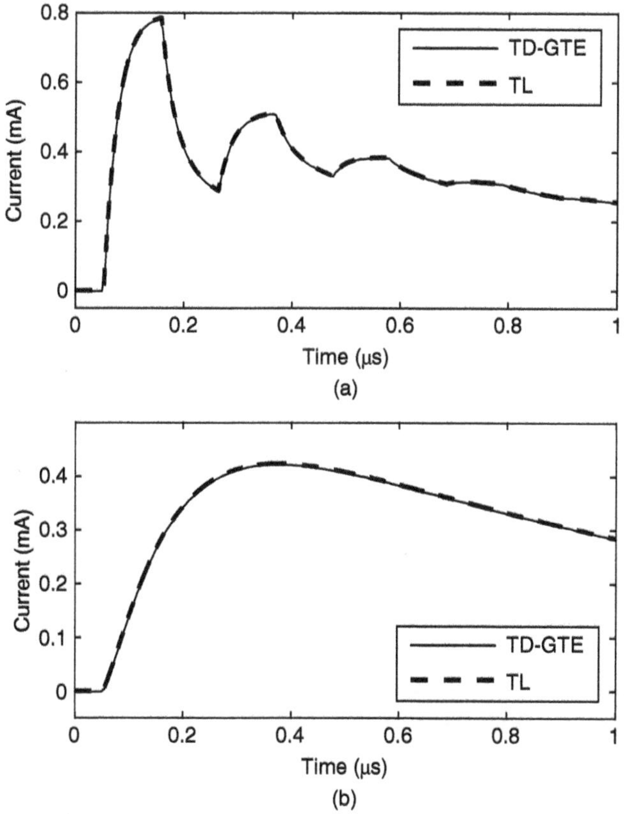


**Figure 2.20** Transient current at the center of the electrode, $L = 10$ m, 0.1/1 μs pulse. $\sigma = 1\ mS\ m^{-1}$. (b) $\sigma = 10\ mS\ m^{-1}$.

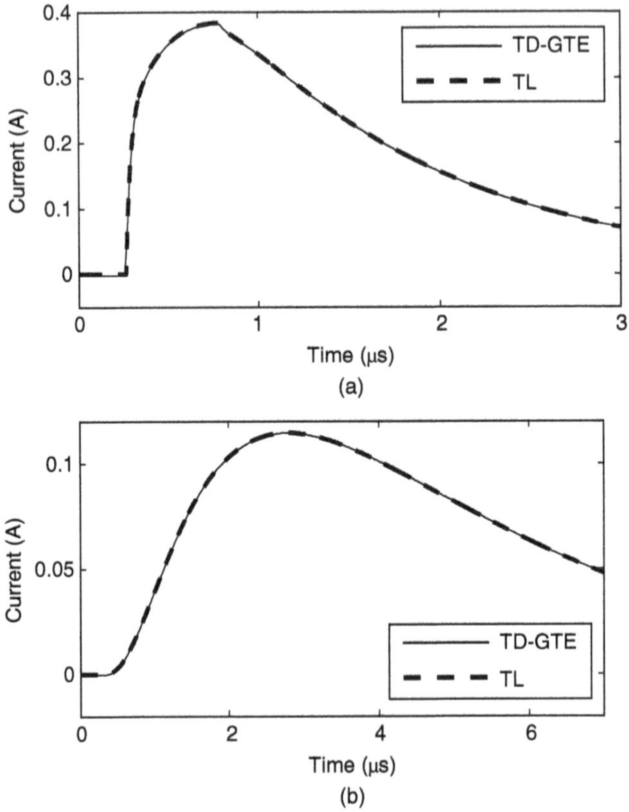


**Figure 2.21** Transient current at the center of the electrode, $L = 50$ m, 0.1/1 μs pulse. $\sigma = 1$ mS m$^{-1}$. (b) $\sigma = 10$ mS m$^{-1}$.

It can be observed that the results obtained by means of the AT and TL approach, respectively, agree rather satisfactorily.

## 2.3 Single Horizontal Wire in the Presence of a Lossy Half-Space: Comparison of Analytical Solution, Numerical Solution, and Transmission Line Approximation

As already mentioned, the analysis of electromagnetic field coupling to overhead and buried wires of finite length can be carried out by using the thin wire AT, or TL model [1–4, 8, 23, 24]. The TL approach does not provide a complete solution if the wavelength of the field coupling to either aboveground or belowground wires is comparable to, or less than, the transverse electrical dimensions of the line. Thus, wires of finite length have to be analyzed by means of the wire AT. The principal disadvantage of the wire antenna approach, with a related numerical method applied, is the relatively high computational cost if longer lines are analyzed [1]. An extension of the TL approach to the combined electromagnetic field-to-TL coupling equations valid for finite length lines above PEC ground has been reported in [3] and [4]. A comparison between the TL model and AT approach to the analysis of finite length lines has been discussed elsewhere, e.g. in [8, 23].

Consequently, it is of certain practical interest to consider an analytical solution of the Pocklington's equation under some plausible set of approximation in order to get some results useful in an average sense for engineering applications and to compare it with numerical treatment of the AT and TL model, respectively. One of the applications of particular interest is related to power line communications (PLC) [25]. Such solutions could serve to rapidly estimate the phenomena of interest. Of particular interest would be an analytical treatment of arbitrary configuration of multiconductor lines in the presence of an imperfectly conducting ground.

Different methods for the analysis of electromagnetic field coupling to a single wire of finite length above or below a PEC ground are considered in [26]. The formulation arising from the wire AT is based on the Pocklington's integro-differential equation, while the TL model is based on the related telegrapher's equations. The Pocklington's equation is solved both numerically and analytically. The numerical solution is carried out using the GB-IBEM. The obtained results are compared to the results computed via NEC and to the results obtained via TL approximation featuring the corresponding telegrapher's equations.

The analysis presented in [26] is restricted to PEC wires. An extension to lossy conductors is straightforward [8]. A PEC wire of length $L$ and radius $a$, located at height $h$ above, or buried at depth $d$ in the lossy ground, as shown in Figure 2.22a and b, respectively, is considered.

The wire is excited by a corresponding tangential electric field in both cases. The essential parameter to study the behavior of scattered fields and voltages is the knowledge of the induced current distribution along the wire. This axial current generates scattered fields propagating away from the wire and the so-called scattered voltage along the line [1, 8].

The spatial current distribution along the single wire above or below a finitely conducting ground is governed by the general form of the Pocklington integro-differential equation [8, 26]:

$$\left[\frac{\partial^2}{\partial x^2} + k^2\right] \int_0^L I(x')g(x,x')\mathrm{d}x' = -j4\pi \frac{k}{Z} E_x^{\mathrm{exc}}, \tag{2.236}$$

where $I(x')$ is the induced current along the line, $g(x, x')$ is the corresponding total Green function, and $E_x^{\mathrm{exc}}$ is the electric field excitation. For the sake of simplicity, only the normal incidence is analyzed.

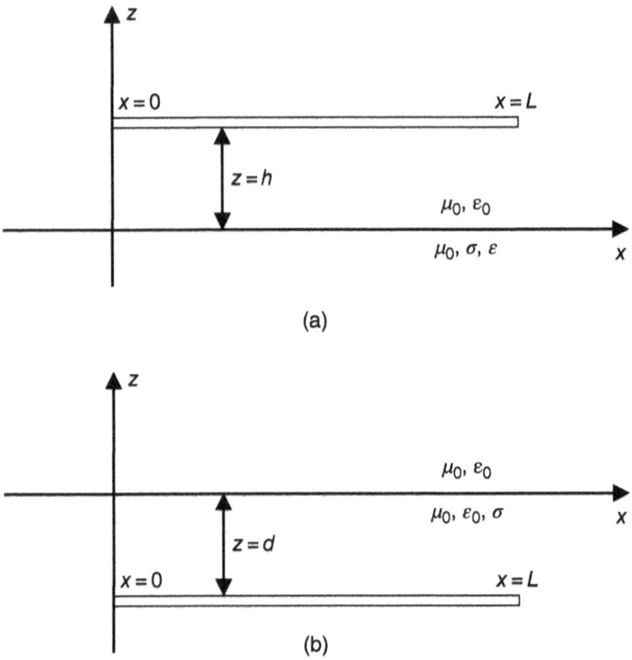


**Figure 2.22** Finite length line in the presence of a lossy ground. (a) Aboveground line. (b) Belowground line.

It is convenient to separately analyze the case of wire above PEC ground, above lossy ground, and below lossy ground.

### 2.3.1 Wire above a Perfect Ground

The Green function for a horizontal wire above a PEC ground is given by

$$g(x, x') = g_0(x, x') - g_i(x, x'), \tag{2.237}$$

where $g_0(x, x')$ is the free space Green function:

$$g_0(x, x') = \frac{e^{-jkR_o}}{R}, \tag{2.238}$$

while $g_i(x, x')$ arises from the image theory and is given by

$$g_i(x, x') = \frac{e^{-jkR_i}}{R_i}, \tag{2.239}$$

where $R_o$ and $R_i$ denote the corresponding distance from the source and image to the observation point, respectively, while the propagation constant

$k$ is given by

$$k = k_0 = \omega\sqrt{\mu_0 \varepsilon_0}, \tag{2.240}$$

and $Z$ is the free space impedance defined as follows:

$$Z = Z_0 = \sqrt{\frac{\mu_0}{\varepsilon_0}}, \tag{2.241}$$

where index "0" is related to free space.

For the case of normal incidence the electric field is given by

$$E_x^{\text{exc}} = E_0 \left(1 - e^{-j2kh}\right), \tag{2.242}$$

where $E_0$ is the electric field at the wire surface.

### 2.3.2 Wire above an Imperfect Ground

The Green function for a wire over an imperfect ground is given by

$$g(x, x') = g_0(x, x') - \Gamma_0^{\text{ref}} g_i(x, x'). \tag{2.243}$$

The RC due to an air–earth interface arises from the MIT and is given by [27]

$$\Gamma_0^{\text{ref}} = \frac{\varepsilon_{\text{eff}} - \varepsilon_0}{\varepsilon_{\text{eff}} + \varepsilon_0}, \tag{2.244}$$

where $\varepsilon_{\text{eff}}$ is the complex permittivity of the earth:

$$\varepsilon_{\text{eff}} = \varepsilon_r \varepsilon_0 + \frac{\sigma}{j\omega}. \tag{2.245}$$

The electric field excitation is of the form:

$$E_x^{\text{exc}} = E_0 \left(1 - \Gamma_0^{\text{ref}} e^{-j2kh}\right), \tag{2.246}$$

where $E_0$ is the field value at the wire surface.

### 2.3.3 Wire Buried in a Lossy Ground

The Green function for wire buried in a lossy half-space is given by

$$g(x, x') = g_E(x, x') - \Gamma_E^{\text{ref}} g_{Ei}(x, x'), \tag{2.247}$$

where $g_E(x, x')$ is the Green function for infinite lossy earth:

$$g_E(x, x') = \frac{e^{-jk_E R_E}}{R_E}, \tag{2.248}$$

while $g_{Ei}(x, x')$ arises from the image theory and is given by

$$g_{Ei}(x, x') = \frac{e^{-jk_E R_{Ei}}}{R_{Ei}}, \tag{2.249}$$

and $R_E$ and $R_{Ei}$ denote the corresponding distance from the source and image to the observation point, respectively.

The RC due to an earth–air interface is given by [27]

$$\Gamma_E^{\text{ref}} = -\frac{\varepsilon_{\text{eff}} - \varepsilon_0}{\varepsilon_{\text{eff}} + \varepsilon_0}. \tag{2.250}$$

The corresponding propagation constant $k$ is

$$k = k_E = \omega\sqrt{\mu_0 \varepsilon_{\text{eff}}}, \tag{2.251}$$

and the impedance $Z$ is defined as

$$Z = Z_E = \sqrt{\frac{\mu_0}{\varepsilon_{\text{eff}}}}, \tag{2.252}$$

where index $E$ is related to lossy earth.

The electric field excitation (normal incidence) can be expressed as follows:

$$E_x^{\text{exc}} = E_0 \Gamma_E^{\text{tr}} \mathrm{e}^{-jkd}, \tag{2.253}$$

where the transmission coefficient due to an earth–air interface is given by [27]

$$\Gamma_E^{\text{tr}} = -\frac{\varepsilon_{\text{eff}} - \varepsilon_0}{\varepsilon_{\text{eff}} + \varepsilon_0}, \tag{2.254}$$

and $E_0$ in (2.253) is related to the field value at the interface between two media.

### 2.3.4 Analytical Solution

To solve Equation (2.236) analytically it is convenient to write the integral on the left-hand side of (2.236) as follows:

$$\int_0^L I(x')g(x,x')\mathrm{d}x' = I(x)\int_0^L g(x,x')\mathrm{d}x' + \int_0^L [I(x') - I(x)]g(x,x')\mathrm{d}x'. \tag{2.255}$$

The integral on the left-hand side can be approximated by the first term on the right-hand side of (2.255) under conditions documented in [24], thus evaluating the characteristic integral term over the Green function analytically. The effective length for the integration along the line is approximately equal to $2h$ and the integration simplifies significantly, if the observation points are sufficiently far from the wire ends [3].

For case of the wire above a PEC ground, this integration has been discussed elsewhere, e.g. in [3, 4], i.e. it simply follows that

$$\int_0^L g(x,x')\mathrm{d}x' = \psi = 2\ln\frac{2h}{a}. \tag{2.256}$$

For the case of finitely conducting ground, the appropriate analytical integration of the first integral on the right-hand side of (2.255) has been carried out in this work as well.

Thus, for the case of the wire above a lossy ground, it follows that

$$\int_0^L g(x,x')\mathrm{d}x' = \psi = 2\left(\ln\frac{L}{a} - \Gamma_0^{\mathrm{ref}}\ln\frac{L}{2h}\right), \tag{2.257}$$

while when dealing with wire buried in a lossy ground, one has

$$\int_0^L g(x,x')\mathrm{d}x' = \psi = 2\left(\ln\frac{L}{a} - \Gamma_E^{\mathrm{ref}}\ln\frac{L}{2d}\right). \tag{2.258}$$

Consequently, the Pocklington's equation (2.236) significantly simplifies, thus taking the following partial differential equation form:

$$\left[\frac{\partial^2}{\partial x^2} + k^2\right] I(x) = -j4\pi\frac{k}{\psi Z}E_x^{\mathrm{exc}}. \tag{2.259}$$

The analytical solution of Equation (2.259) is straightforward and is given by

$$I(x) = -j\frac{4\pi}{\psi Z}\frac{E_x^{\mathrm{exc}}}{k}\left[1 - \frac{\cos k\left(\frac{L}{2} - x\right)}{\cos k\frac{L}{2}}\right]. \tag{2.260}$$

Alternatively, Equation (2.259), for the case of overhead wire, can be also obtained from the generalized form of telegrapher's equations [8]:

$$\frac{\mathrm{d}V^{\mathrm{sct}}(x)}{\mathrm{d}x} + j\omega\frac{\mu}{4\pi}\int_0^L I(x')g(x,x')\mathrm{d}x' = E_x^{\mathrm{exc}}, \tag{2.261}$$

$$\int_0^L \frac{\partial I(x')}{\partial x'}g(x,x')\mathrm{d}x' + j4\pi\omega\varepsilon V^{\mathrm{sct}}(x) = 0, \tag{2.262}$$

as shown in [26].

Starting from the generalized telegrapher's equations (2.261) and (2.262) and taking into account approximation (2.255) it follows that

$$V^{\mathrm{sct}}(x) + \frac{1}{j4\pi\omega\varepsilon}\psi\frac{\partial I(x)}{\partial x} = 0, \tag{2.263}$$

$$E_x^{\mathrm{exc}} = j\omega\frac{\mu}{4\pi}\psi I(x) + \frac{\mathrm{d}V^{\mathrm{sct}}(x)}{\mathrm{d}x}. \tag{2.264}$$

Differentiating Equation (2.263) yields

$$\frac{\mathrm{d}V^{\mathrm{sct}}(x)}{\mathrm{d}x} + \frac{\psi}{j4\pi\omega\varepsilon}\frac{\partial^2 I(x)}{\partial x^2} = 0. \tag{2.265}$$

Finally, combining (2.265) with (2.264) gives

$$-j4\pi\frac{k}{\psi Z_0}E_x^{\mathrm{exc}} = \left[\frac{\partial^2}{\partial x^2} + k^2\right] I(x), \tag{2.266}$$

i.e. Equation (2.259) is obtained.

Therefore, there is a clear correlation between the analytical solution of Pocklington integro-differential equation and the corresponding telegrapher's equation. The case of the wire buried in a lossy ground is very similar and it is based on the corresponding set of generalized telegrapher's equations [9]. The presented approach could be also extended to wires of arbitrary shape.

### 2.3.5 Boundary Element Procedure

The integro-differential equation (2.236) is numerically solved via the GB-IBEM whose basic steps are outlined in this section. More mathematical details can be found elsewhere, e.g. in [2].

The unknown current $I^e(x)$ along the electrode segment can be written as follows:

$$I^e(x') = \{f\}^T\{I\}. \tag{2.267}$$

Assembling the contributions from each segment the integro-differential equation (2.236) is transferred into the matrix equation

$$\sum_{j=1}^{M} [Z]_{ji}\{I\}_i = \{V\}_j, \quad \text{and} \quad j = 1, 2, \ldots, M, \tag{2.268}$$

where $M$ is the total number of segments and $[\mathbf{Z}]_{ji}$ is the mutual impedance matrix representing the interaction of the $i$th source with the $j$th observation segment, respectively:

$$[Z]_{ji} = -\frac{1}{4j\pi\omega\varepsilon_{\text{eff}}} \times \left( \int_{\Delta l_j} \{D\}_j \int_{\Delta l_i} \{D'\}_i^T g(x,x')\mathrm{d}x'\,\mathrm{d}x + k^2 \int_{\Delta l_j} \{f\}_j \int_{\Delta l_i} \{f\}_i^T g(x,x')\mathrm{d}x'\,\mathrm{d}x \right). \tag{2.269}$$

Matrices $\{\mathbf{f}\}$ and $\{\mathbf{f}'\}$ contain the shape functions while $\{\mathbf{D}\}$ and $\{\mathbf{D}'\}$ contain their derivatives, and $\Delta l_i$, $\Delta l_j$ are the widths of $i$th and $j$th boundary elements. Note that in the case of wires above ground $\varepsilon_{\text{eff}} = \varepsilon_0$.

A linear approximation over a wire segment requires the shape functions to be of the form

$$f_i = \frac{x_{i+1} - x'}{\Delta x} \quad f_{i+1} = \frac{x' - x_i}{\Delta x}, \tag{2.270}$$

and the related derivatives are simply given by

$$\frac{\partial I(x')}{\partial x'} = \frac{I_{i+1} - I_i}{\Delta x}, \tag{2.271}$$

as this choice was proved to be the optimal one in modeling various wire structures [2].

The voltage vector containing the excitation in the form of the plane wave electric field is

$$\{V\}_j = -\int_{x_i}^{x_{i+1}} E_x^{\text{exc}}\{f\}\mathrm{d}x. \tag{2.272}$$

For the case of plane wave normal incidence integral (2.272) becomes rather straightforward and can be evaluated analytically [2].

### 2.3.6 The Transmission Line Model

Voltages and currents along the line shown in Figure 2.22a induced by an external field excitation can be obtained using the field-to-TL equations [23]:

$$\frac{\mathrm{d}V}{\mathrm{d}x} + ZI(x) = \hat{V}_F(x), \tag{2.273}$$

$$\frac{\mathrm{d}I(x)}{\mathrm{d}x} + YV(x) = \hat{I}_F(x), \tag{2.274}$$

where the longitudinal impedance is

$$\hat{Z} = j\omega L + \hat{Z}_\mathrm{w} + \hat{Z}_\mathrm{g} \tag{2.275}$$

and the transversal admittance is

$$\hat{Y} = j\omega C + G, \tag{2.276}$$

where $L$ is the per-unit-length longitudinal inductance for a perfect soil; $C$ and $G$ are the per-unit-length transverse capacitance and conductance matrixes of the line, respectively, $\hat{Z}_\mathrm{w}$ is the per-unit-length internal impedance of the conductor, $\hat{Z}_\mathrm{g}$ is the per-unit-length ground impedance [28, 29], while $V_F(x)$ and $\hat{I}_F(x)$ are sources vectors expressed in terms of the incident magnetic and electric field [23]. The set of Equations (2.273) and (2.274) have been solved by means of the chain matrix method and modal equation to derive per unit length parameters [23, 28].

Furthermore, with reference to Figure 2.22b, the TL equations describing the interaction of an external electromagnetic field and a bare buried wire are given by [9]

$$\frac{\mathrm{d}V(x)}{\mathrm{d}x} + Z'I(x) = E_x^e(x, z = -d), \tag{2.277}$$

$$\frac{\mathrm{d}I(x)}{\mathrm{d}x} + Y'V(x) = 0, \tag{2.278}$$

where the longitudinal impedance is

$$Z' = Z'_\mathrm{w} + Z'_\mathrm{g}, \tag{2.279}$$

and the transverse admittance is

$$Y' = Y'_\mathrm{g}. \tag{2.280}$$

In this case, $Z'_\mathrm{w}$ is the wire impedance, and $Z'_\mathrm{g}$ and $Y'_\mathrm{g}$ are, respectively, the ground impedance and admittance.

Several expressions are available in the literature to describe the ground impedance. In this section the relation proposed in [9] is used

$$Z'_\mathrm{g} = \frac{j\omega\mu_0}{2\pi}\left\{\ln\left(\frac{1+\gamma_\mathrm{g}a}{\gamma_\mathrm{g}a}\right)+\left[\frac{2e^{-2d|\gamma_\mathrm{g}|}}{4+\gamma_\mathrm{g}^2a^2}\right]\right\}, \tag{2.281}$$

where $a$ is the radius of the buried wire, $d$ is its depth below the ground surface, and $\gamma_\mathrm{g}$ is the propagation constant inside the ground, which can be expressed as

$$\gamma_\mathrm{g} = \sqrt{j\omega\mu_0(\sigma_\mathrm{g}+j\omega\varepsilon_0\varepsilon_\mathrm{rg})}, \tag{2.282}$$

while the ground admittance is given by [1]

$$Y'_\mathrm{g} = \frac{\gamma_\mathrm{g}^2}{Z'_\mathrm{g}}. \tag{2.283}$$

### 2.3.7 Modified Transmission Line Model

The TL equations for a horizontal grounding wire excited by an external field are given by (2.277) and (2.278).

In the modified transmission line model (MTL) the per-unit-length parameters $Z'(\omega)$ and $Y'(\omega)$ of buried conductor are computed using modal equation available in [26] and are frequency dependent:

$$Z'[\Gamma(\omega)]\cdot Y'[\Gamma(\omega)] = [\Gamma(\omega)]^2, \tag{2.284}$$

where

$$Z'(\Gamma) = \frac{j\omega\mu_0}{2\pi}[K_0(\gamma_1 a)+K_0(\gamma_1(2d-a))+I_1(\Gamma)] \tag{2.285}$$

and

$$Y'_\mathrm{g}(\Gamma) = \frac{j2\pi\omega\varepsilon_\mathrm{eff}}{K_0(\gamma_1 a)+K_0(\gamma_1(2d-a))+k_1^2 I_2(\Gamma)}, \tag{2.286}$$

where expressions $I_1$ and $I_2$ are given by

$$I_1(\Gamma) = \int_{-\infty}^{+\infty}\frac{\exp(-2u_1d)}{u_1+u_2}\mathrm{d}\lambda, \tag{2.287}$$

$$I_2(\Gamma) = \int_{-\infty}^{+\infty}\frac{\exp(-2u_1d)}{k_2^2u_1+k_1^2u_2}\mathrm{d}\lambda, \tag{2.288}$$

where $K_0$ is the zero order Bessel function of the second kind, while $u_1$ and $u_2$ are given by

$$u_1 = (\lambda^2+\gamma_1^2)^{\frac{1}{2}} \quad \text{with} \quad \gamma_1^2 = \Gamma^2+k_1^2, \tag{2.289}$$

$$u_2 = (\lambda^2+\gamma_2^2)^{\frac{1}{2}} \quad \text{with} \quad \gamma_2^2 = \Gamma^2+k_2^2. \tag{2.290}$$

$k_1, k_2$ are the propagation constants in a lossy ground and air, respectively:

$$k_1^2 = \omega^2 \mu_0 \varepsilon_{\text{eff}}, \tag{2.291}$$

$$k_2^2 = \omega^2 \mu_0 \varepsilon_0. \tag{2.292}$$

In order to obtain numerical values of $\Gamma$, an optimization for each frequency is carried out by means of the Muller's method generalizing the secant method in a complex plane using a quadratic interpolation among three points instead of a linear interpolation between two points. Solving for the zeros of the quadratic form enables the method to find complex pairs of roots. As there are two possible values for $\Gamma$ at each frequency, $+\Gamma$ and $-\Gamma$, to respect the uniqueness of the solution, the real parts of $u_1$ and $u_2$ are chosen to retain a positive value.

### 2.3.8 Computational Examples

The numerical results are presented in three sections: single wire above a PEC ground, single wire above an imperfectly conducting ground, and single wire buried in a lossy ground. The results obtained via the analytical and numerical solution of the Pocklington's equation (2.236) are compared to the results computed via NEC [12] and to the results obtained by the TL approach [23, 28].

#### 2.3.8.1 Wire above a PEC Ground

A computational example deals with an overhead wire of length $L$, radius $a = 0.01$ m located at height $h = 2.5$ m above PEC ground and illuminated by the plane wave of normal incidence with amplitude $E_0 = 1\ \text{V m}^{-1}$. Figures 2.23–2.25 show the related real, imaginary, and absolute value of spatial current distribution for line $L = 5$ m, $L = 10$ m and $L = 20$ m, and operating frequency $f = 50$ MHz.

The results obtained via different methods all agree in the waveform. The agreement between the analytical results and the TL results stress out the fact that the analytical solution of Pocklington's integro-differential equation essentially implies the shift to the TL approximation. On the other hand, the results obtained via BEM agree satisfactorily with results computed with NEC. Figure 2.26 shows the real, imaginary, and absolute value of the current distribution along the 20 m long wire for the operating frequency $f = 1$ MHz.

The difference between the results obtained via the different approaches is found to be less satisfactory than in previous cases.

#### 2.3.8.2 Wire above a Lossy Ground

The next example deals with a horizontal wire of length $L$, radius $a = 0.01$ m located at height $h = 2.5$ m above an imperfect ground and excited by the plane wave of normal incidence with amplitude $E_0 = 1\ \text{V m}^{-1}$ at the wire surface. Figures 2.27–2.29 present the real, imaginary, and absolute value of spatial current distribution for line $L = 5$ m, $L = 10$ m, and $L = 20$ m. The operating

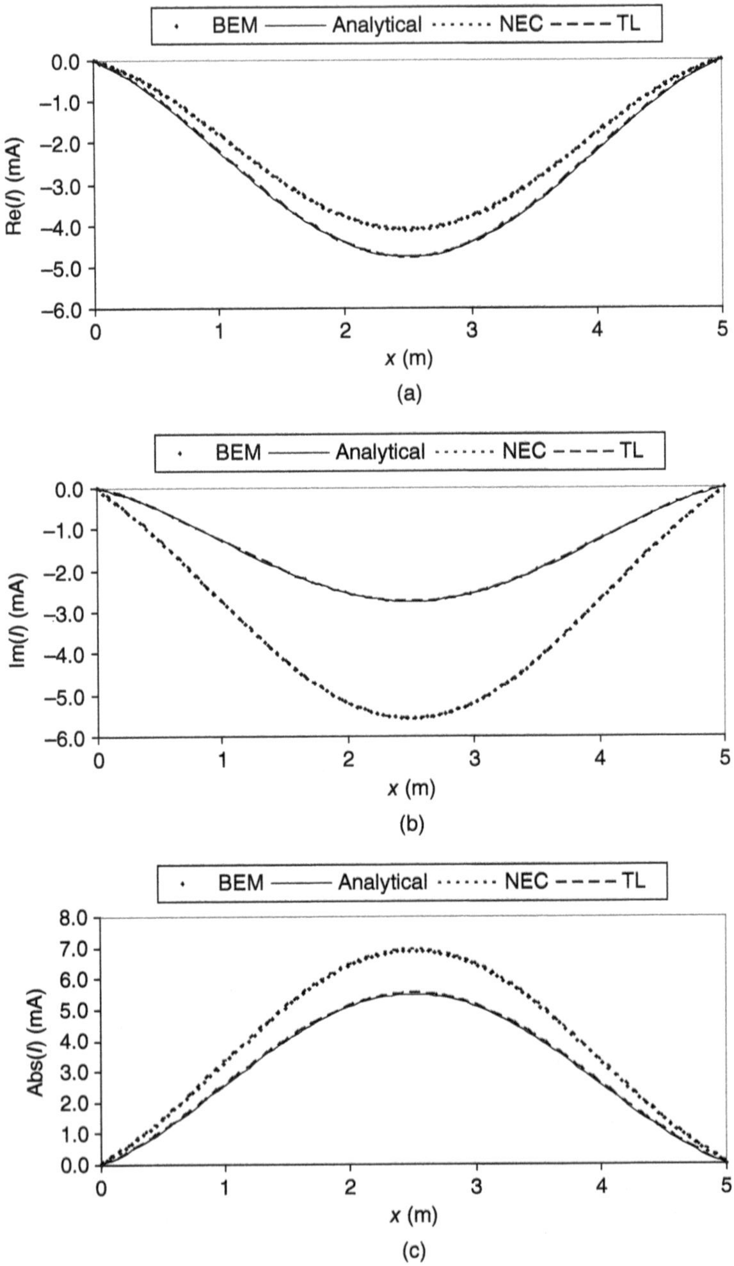


**Figure 2.23** (a) Real, (b) imaginary, and (c) absolute value of the current distribution along the single wire above a PEC ground: $f = 50$ MHz, $L = 5$ m, $a = 0.01$ m, $h = 2.5$ m.

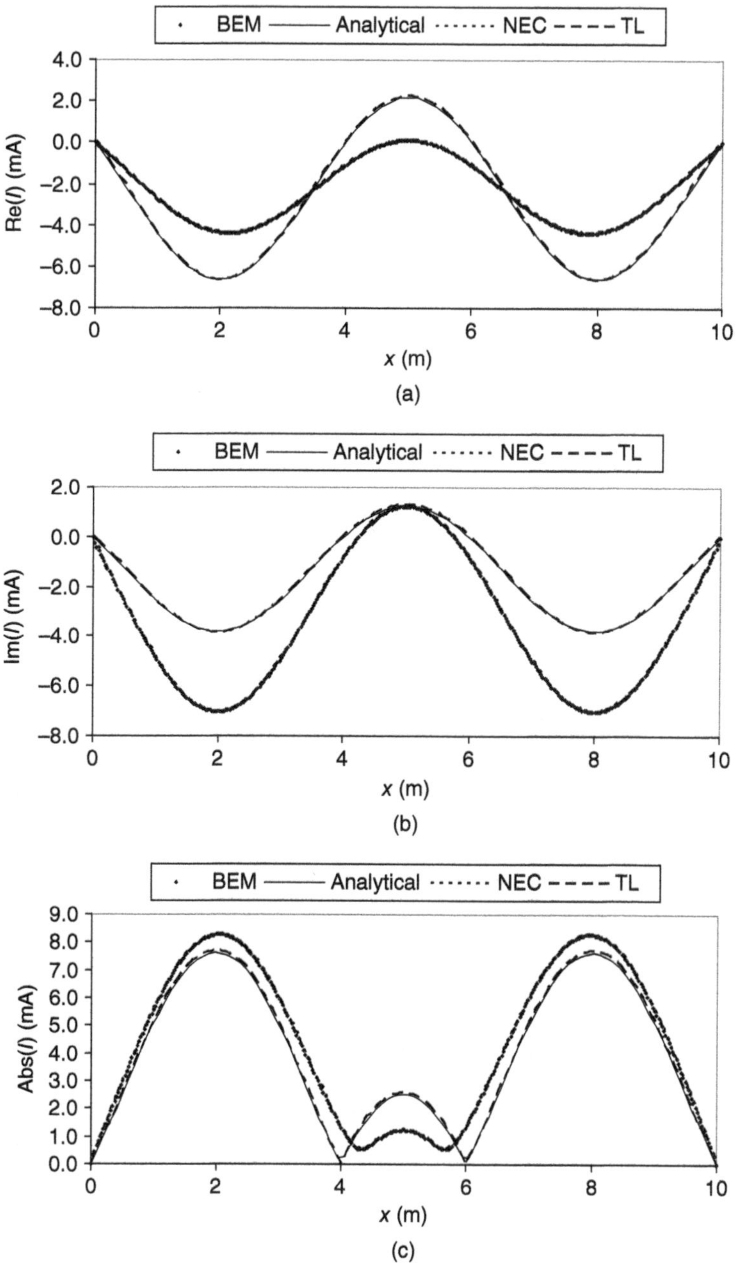


**Figure 2.24** a) Real, b) imaginary, and c) absolute value of current distribution along the single wire above a PEC ground: $f = 50$ MHz, $L = 10$ m, $a = 0.01$ m, $h = 2.5$ m.

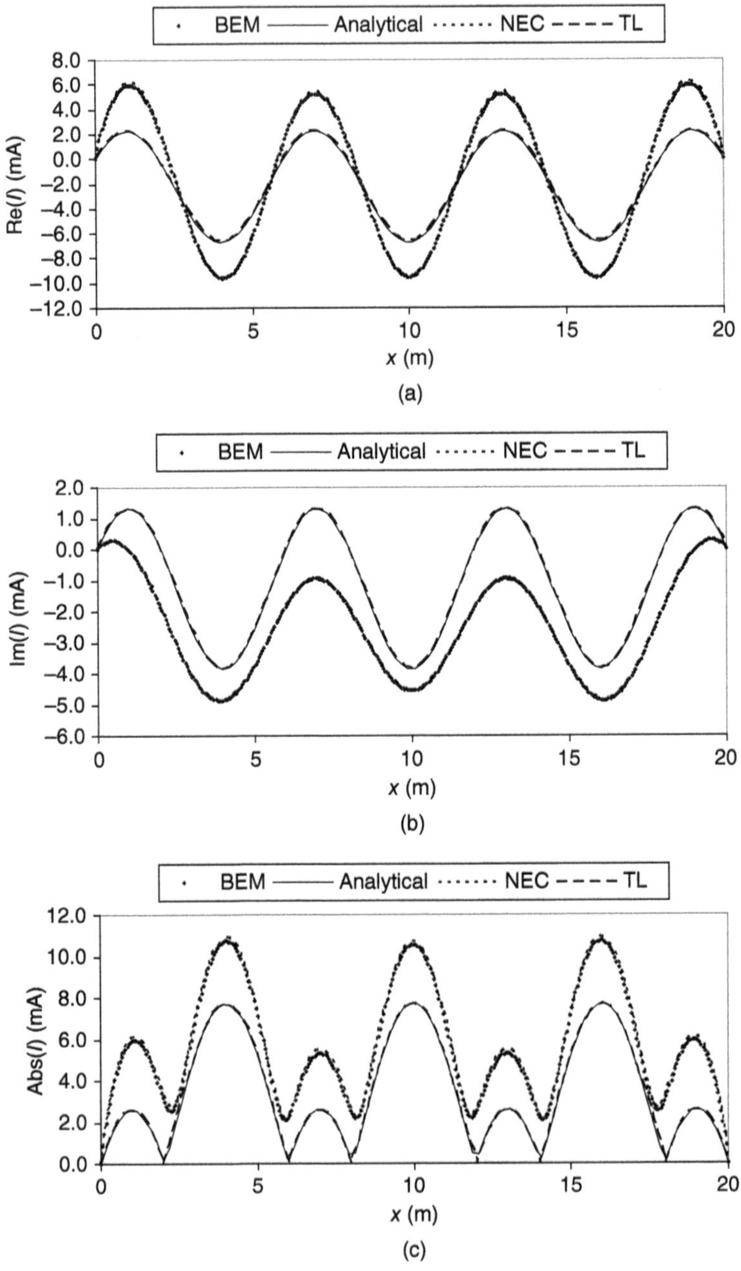


**Figure 2.25** (a) Real, (b) imaginary, and (c) absolute value of current distribution along the single wire above a PEC ground: $f = 50$ MHz, $L = 20$ m, $a = 0.01$ m, $h = 2.5$ m.

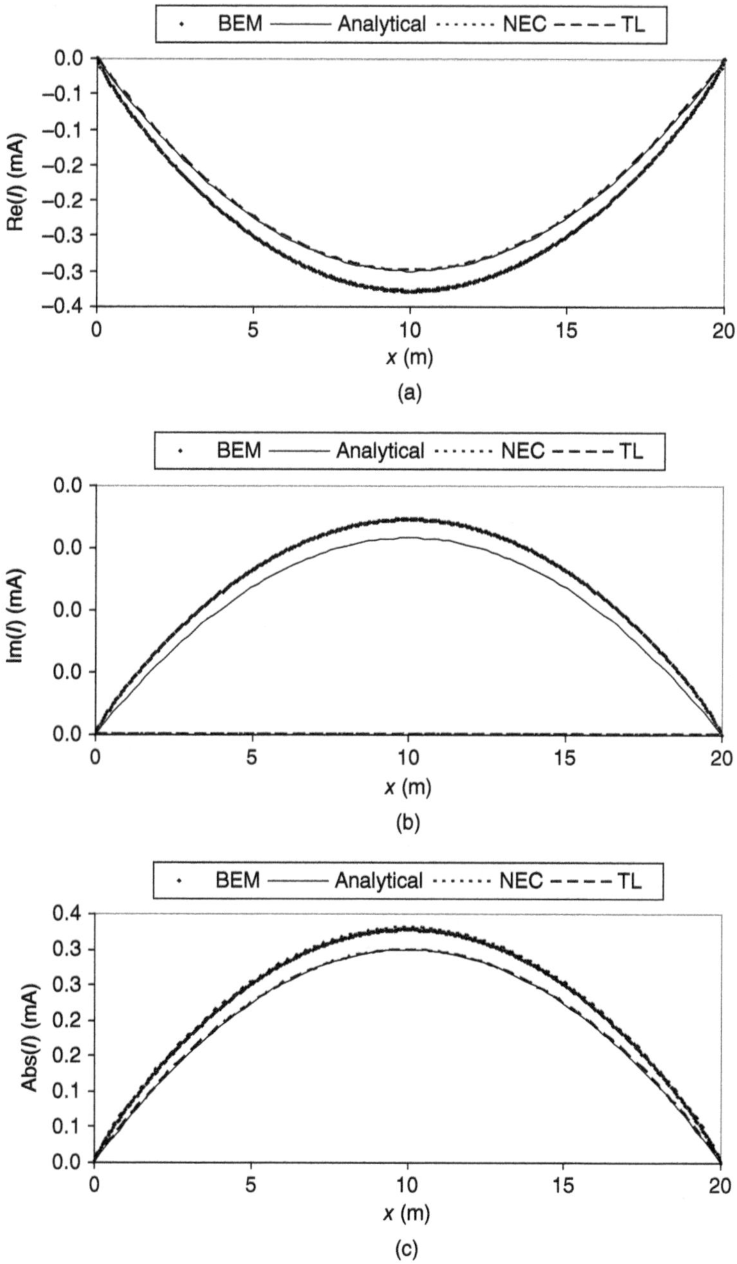


**Figure 2.26** (a) Real, (b) imaginary, and (c) absolute value of current distribution along the single wire above a PEC ground: $f = 1$ MHz, $L = 20$ m, $a = 0.01$ m, $h = 2.5$ m.

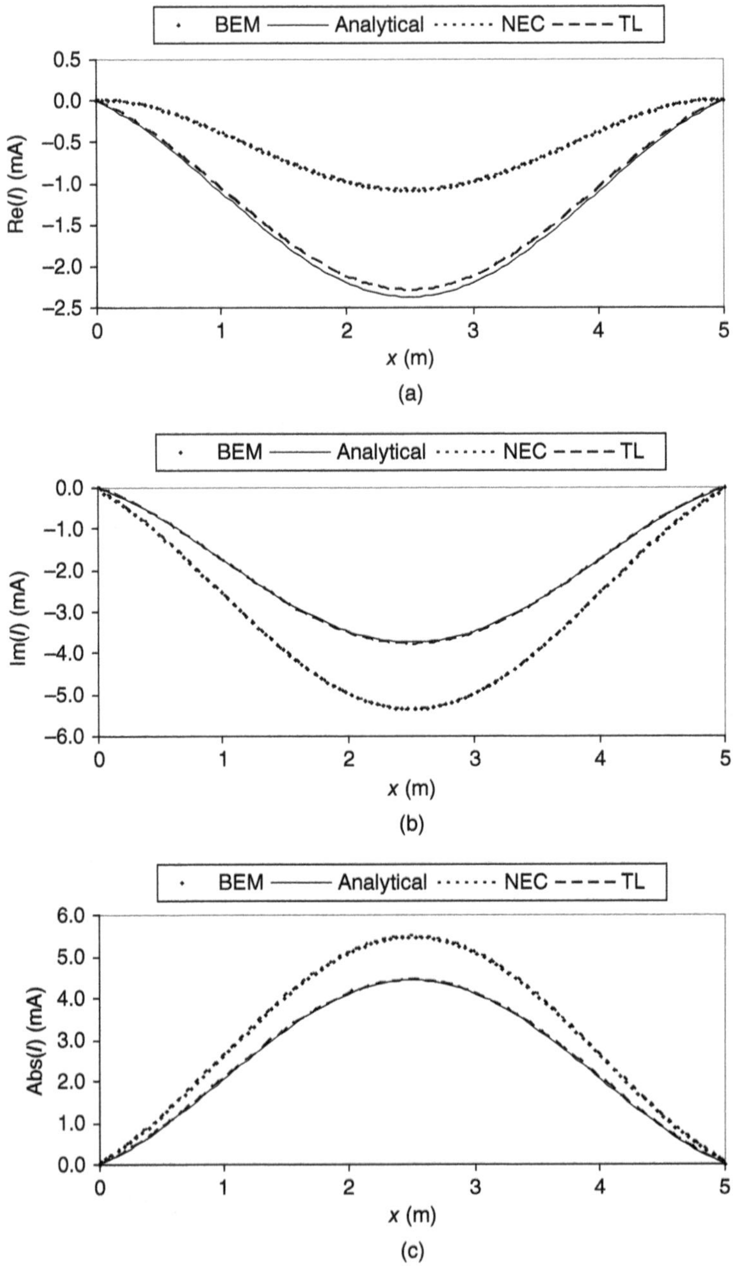


**Figure 2.27** a) Real, b) imaginary, and c) absolute value of current distribution along the single wire above an imperfect ground: $f = 50$ MHz, $L = 5$ m, $a = 0.01$ m, $h = 2.5$ m, $\sigma = 0.01$ S m$^{-1}$, $\varepsilon_r = 10$.

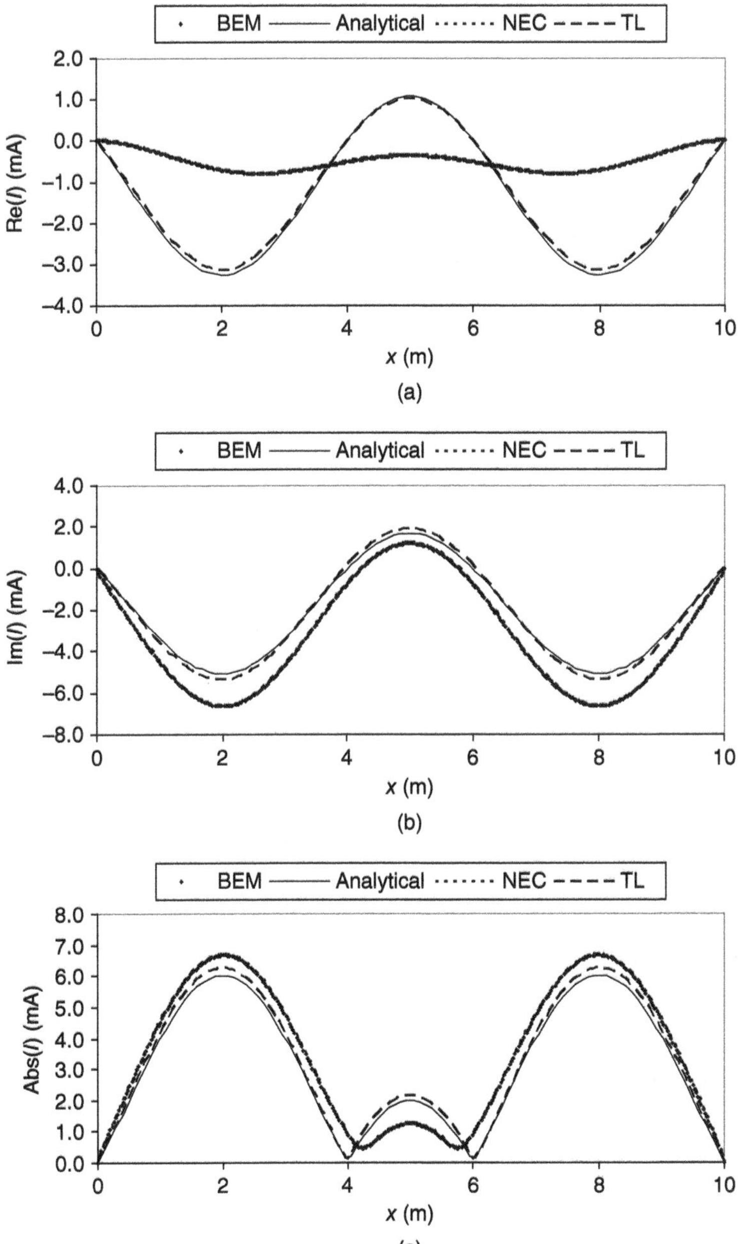


**Figure 2.28** (a) Real, (b) imaginary, and (c) absolute value of current distribution along the single wire above an imperfect ground: $f = 50$ MHz, $L = 10$ m, $a = 0.01$ m, $h = 2.5$ m, $\sigma = 0.01\ \mathrm{S\,m^{-1}}$, $\varepsilon_r = 10$.

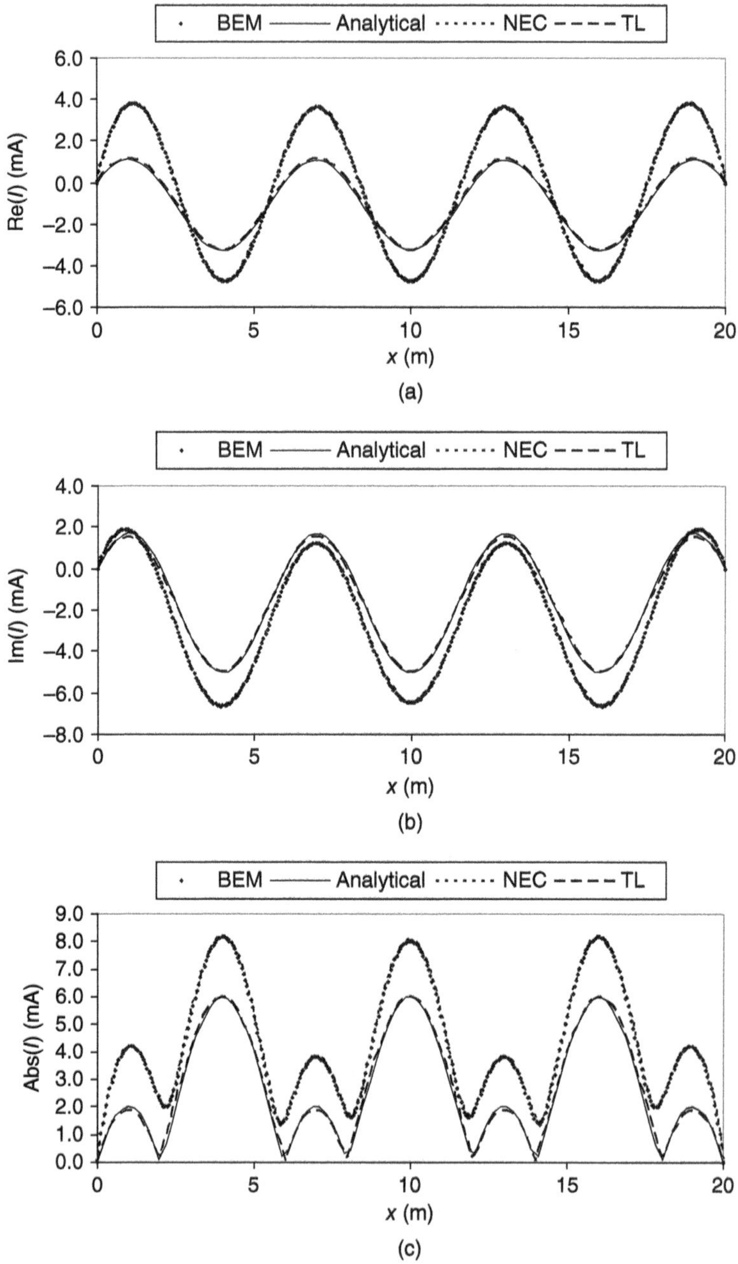


**Figure 2.29** (a) Real, (b) imaginary, and (c) absolute value of current distribution along the single wire above an imperfect ground: $f = 50$ MHz, $L = 20$ m, $a = 0.01$ m, $h = 2.5$ m, $\sigma = 0.01$ S m$^{-1}$, $\varepsilon_r = 10$.

frequency is $f = 50$ MHz. The conductivity of the ground is $\sigma = 0.01$ S m$^{-1}$ and permittivity is $\varepsilon_r = 10$.

The presented curves show the similar waveform behavior and the differences between the results are slightly less than in the PEC ground case. Figure 2.30 shows the real, imaginary, and absolute value of the current distribution along the 20 m long wire for the operating frequency $f = 1$ MHz. Again, a satisfactory agreement can be noticed between analytical and TL results and between BEM and NEC results.

The agreement between the results obtained via the different approaches is found to be more satisfactory than in previous cases of wires above a dissipative half-space related to $f = 50$ MHz.

#### 2.3.8.3 Wire Buried in a Lossy Ground

The last set of figures is related to a horizontal wire of length $L$, radius $a = 0.01$ m buried at depth $d = 2.5$ m in a lossy ground and illuminated by the plane wave of normal incidence transmitted into the ground with amplitude $E_0 = 1$ V m$^{-1}$ at the interface between two media. Figures 2.31–2.33 show the real, imaginary, and absolute value of spatial current distribution for line $L = 5$ m, $L = 10$ m, and $L = 20$ m, and the operating frequency is $f = 50$ MHz. The operating frequency is $f = 50$ MHz. The conductivity of the ground is $\sigma = 0.01$ S m$^{-1}$ and permittivity is $\varepsilon_r = 10$.

The behavior of all waveforms is similar, but the difference between analytical and the numerical results obtained by other methods is significant. Figure 2.34 shows the real, imaginary, and absolute value of the current distribution along the 20 m long wire for the operating frequency $f = 1$ MHz. The agreement between the results obtained via the different approaches is found to be satisfactory than in previous cases.

Observing Figure 2.34, the difference is slightly less in this case compared to the results obtained for buried wires at $f = 50$ MHz.

### 2.3.9 Field Transmitted in a Lower Lossy Half-Space

Knowledge of the energy transmitted into the dissipative half-space due to radiation of dipole antenna is important in many areas of computational electromagnetics, such as GPR analysis, and can be carried out in the frequency or time domain, respectively, [2, 30, 31]. Many efficient GPR antenna models based on the finite difference time domain (FDTD) method of solution have been reported, e.g. [32, 33].

Contrary to the widely used FDTD approach the assessment of transmitted electric field in the ground due to the GPR dipole antenna by means of the BEM is undertaken in [34]. The formulation is based on the space–frequency integro-differential equation of the Pocklington's type and the corresponding filed formulas. Once the current distribution along the dipole is obtained by solving the space–frequency Pocklington's equation via the GB-IBEM the

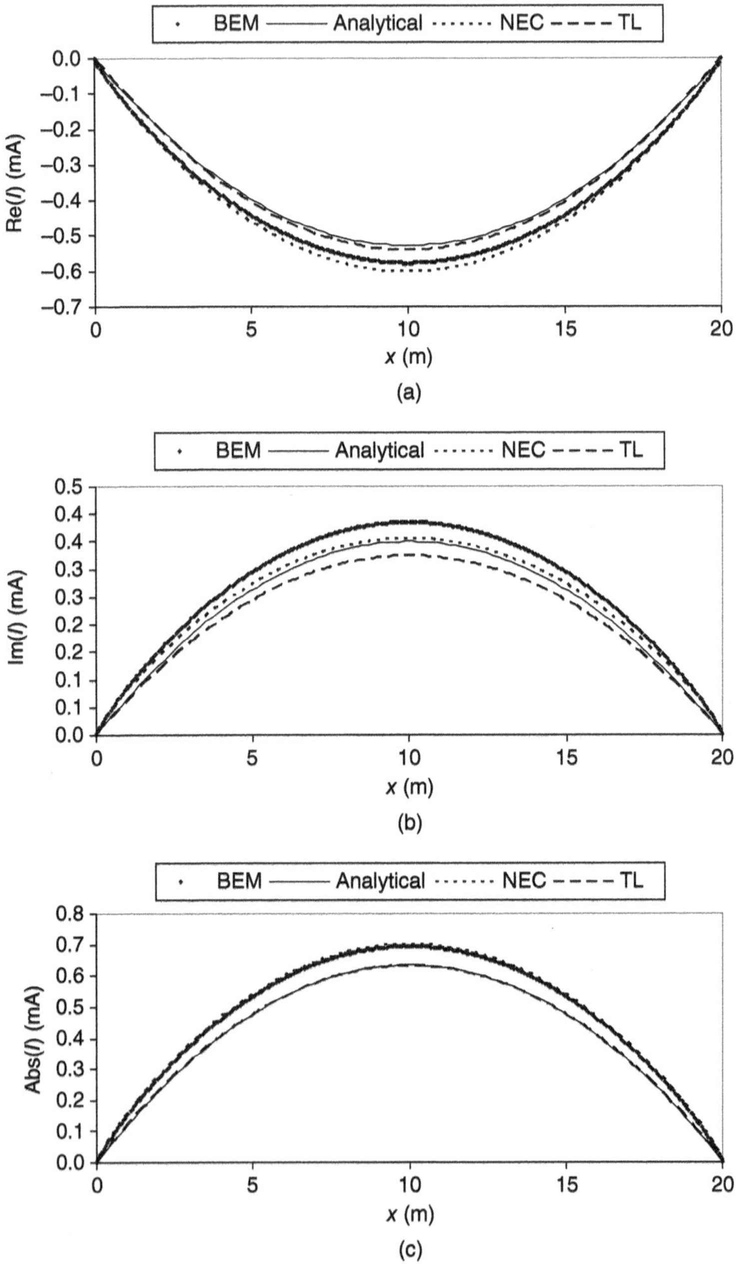


**Figure 2.30** (a) Real, (b) imaginary, and (c) absolute value of current distribution along the single wire above an imperfect ground: $f = 1$ MHz, $L = 20$ m, $a = 0.01$ m, $h = 2.5$ m, $\sigma = 0.01$ S m$^{-1}$, $\varepsilon_r = 10$.

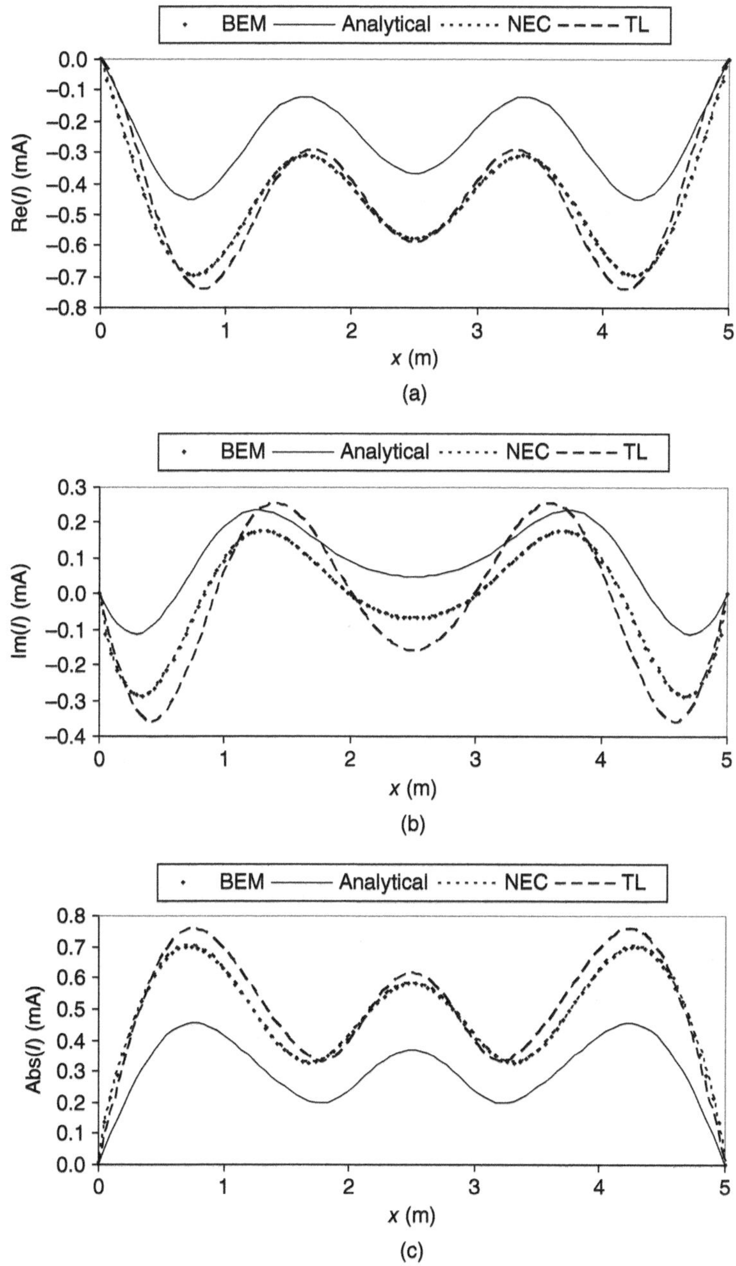


**Figure 2.31** (a) Real, (b) imaginary, and (c) absolute value of current distribution along the single wire buried in a lossy ground: $f = 50$ MHz, $L = 5$ m, $a = 0.01$ m, $d = 2.5$ m, $\sigma = 0.01$ S m$^{-1}$, $\varepsilon_r = 10$.

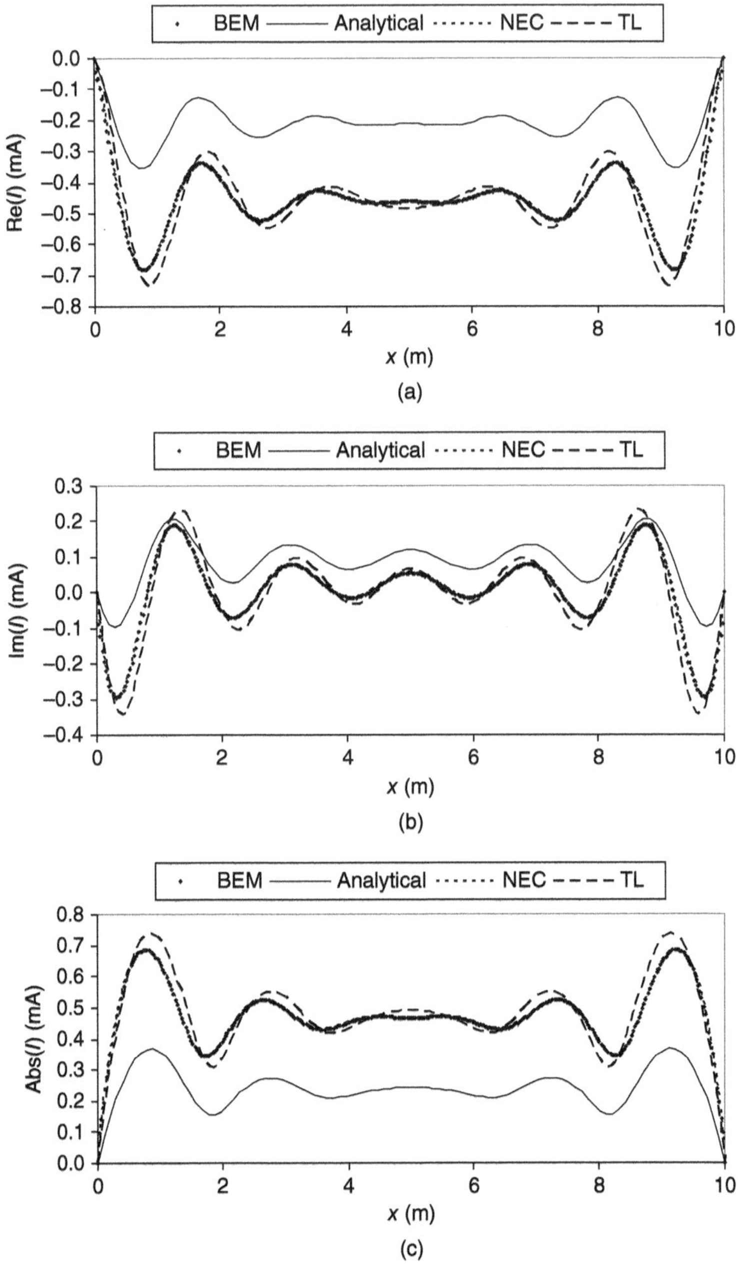


**Figure 2.32** (a) Real, (b) imaginary, and (c) absolute value of current distribution along the single wire buried in a lossy ground: $f = 50$ MHz, $L = 10$ m, a = 0.01 m, $d = 2.5$ m, $\sigma = 0.01$ S m$^{-1}$, $\varepsilon_r = 10$.

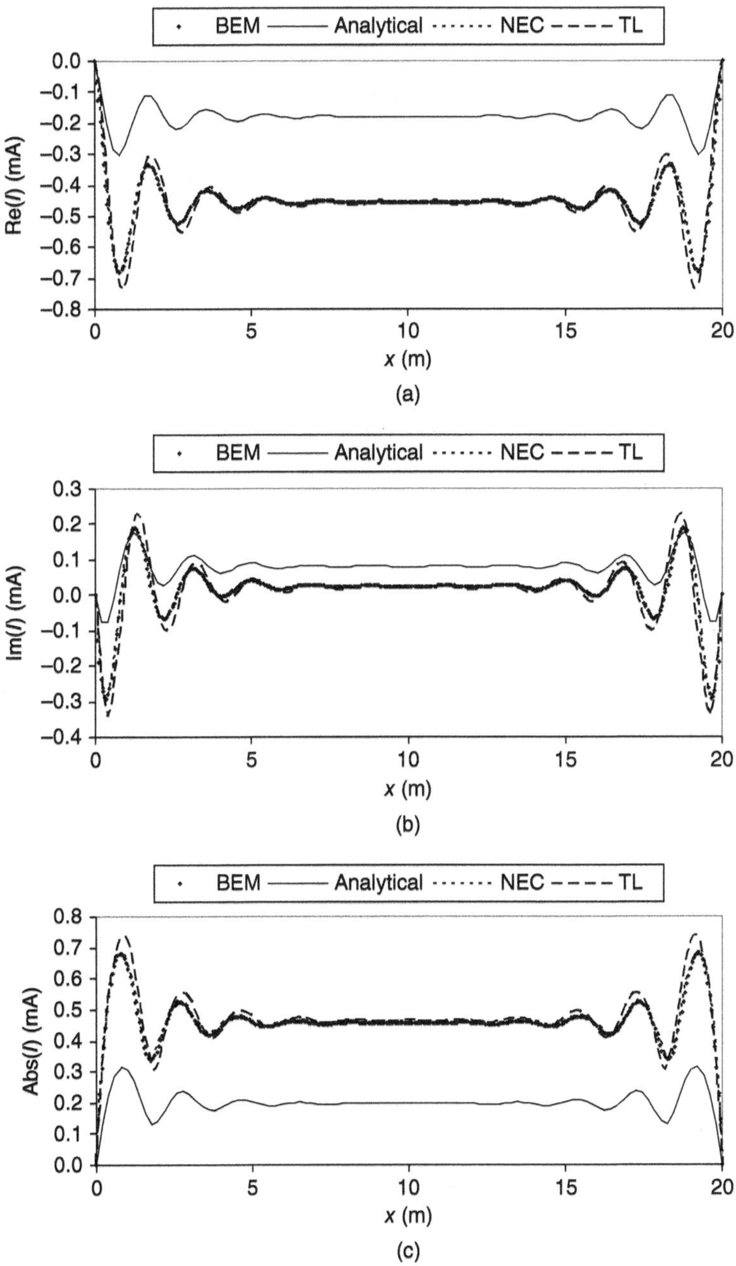


**Figure 2.33** a) Real, b) imaginary, and c) absolute value of current distribution along the single wire buried in a lossy ground: $f = 50$ MHz, $L = 20$ m, $a = 0.01$ m, $d = 2.5$ m, $\sigma = 0.01$ S m$^{-1}$, $\varepsilon_r = 10$.

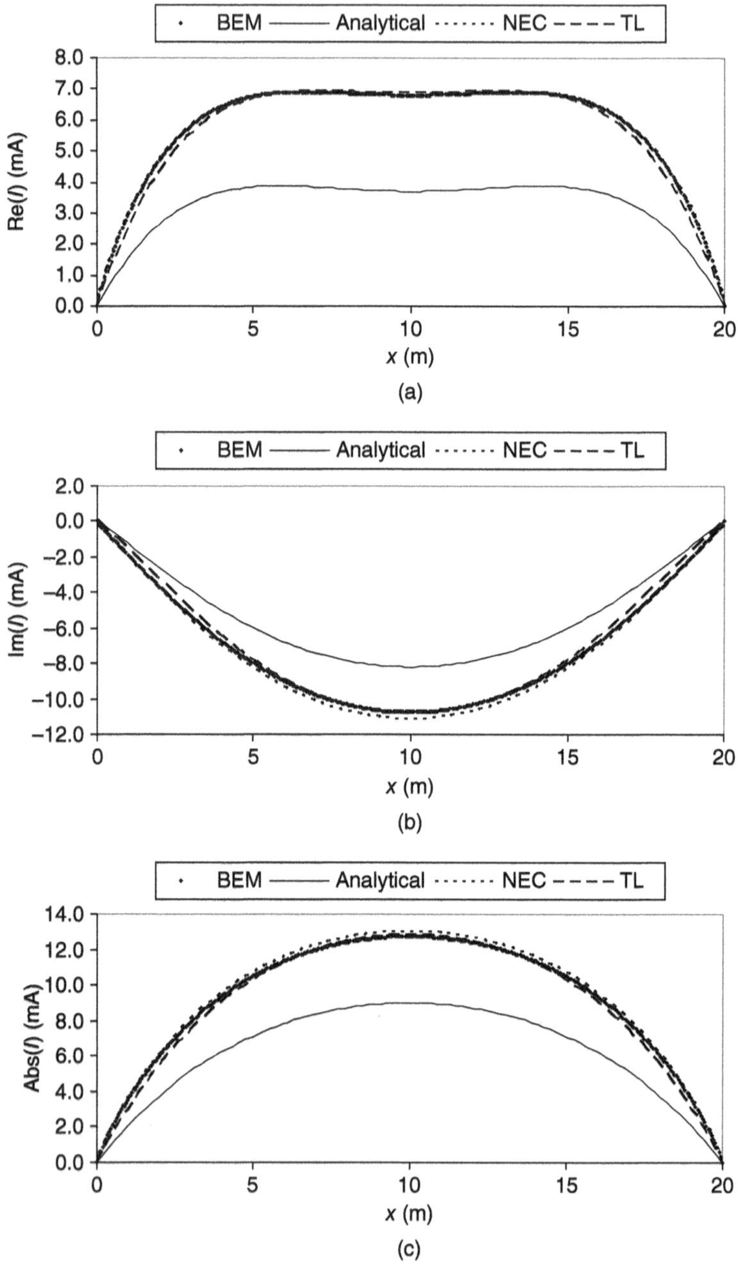


**Figure 2.34** (a) Real, (b) imaginary, and (c) absolute value of current distribution along the single wire buried in a lossy ground: $f = 1$ MHz, $L = 20$ m, $a = 0.01$ m, $d = 2.5$ m, $\sigma = 0.01$ S m$^{-1}$, $\varepsilon_r = 10$.

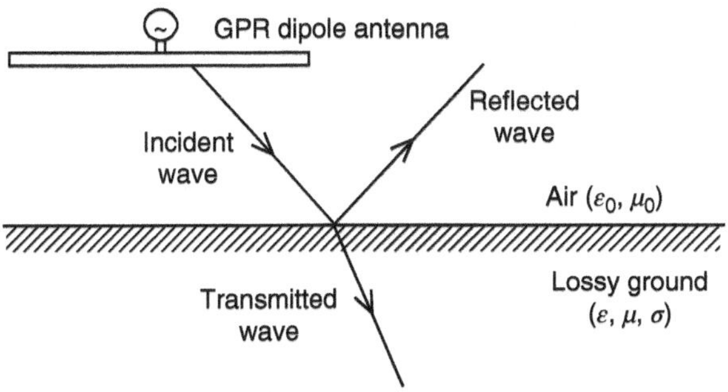


**Figure 2.35** GPR dipole antenna above a lossy half-space.

corresponding transmitted field is obtained by numerically computing the related field integrals.

The geometry of interest is related to the dipole radiating above a lossy medium, as it is shown in Figure 2.35.

Generally, a directive transmission of signal into a material half-space can be analyzed by using the rigorous Sommerfeld integral formulation, or the approximate Fresnel reflection/transmission coefficient approach [2]. The validity of each approach has been discussed elsewhere and some general remarks can be, for example, found in [16], or [26]. The use of the MIT [27] in the transmitted field formula has been proposed in [34].

Once the current distribution is obtained by solving the Pocklington's equation (2.236) [34],

$$E_x = \frac{1}{j4\pi\omega\varepsilon_0}\left[-\int_{-L/2}^{L/2}\frac{\partial I(x')}{\partial x'}\frac{\partial G(x,x',z)}{\partial x'}\mathrm{d}x' + k^2\int_{-L/2}^{L/2} G(x,x',z)I(x')\mathrm{d}x'\right], \tag{2.293}$$

$$E_z = \frac{1}{j4\pi\omega\varepsilon_0}\int_{-L/2}^{L/2}\frac{\partial I(x')}{\partial x'}\frac{\partial G(x,x',z)}{\partial z}\mathrm{d}x', \tag{2.294}$$

where $I(x')$ is the induced current along the antenna and the total Green function $G(x, x')$ is given by

$$G(x,x') = \Gamma_{\mathrm{tr}}^{\mathrm{MIT}} g_E(x,x',z), \tag{2.295}$$

with

$$g(x,x') = g_0(x,x') - R_{\mathrm{TM}} g_i(x,x'), \tag{2.296}$$

where $g_0(x, x')$ is the free space Green function:

$$g_o(x,x') = \frac{\mathrm{e}^{-jk_oR_o}}{R_o}, \tag{2.297}$$

while $g_i(x, x')$ arises from the image theory:

$$g_i(x,x') = \frac{\mathrm{e}^{-jk_oR_i}}{R_i}, \tag{2.298}$$

where $R_o$ and $R_i$ denote the corresponding distance from the source to the observation point, respectively.

The transmission coefficient $\Gamma_{tr}^{MIT}$ arising from the MIT is given by

$$\Gamma_{tr}^{MIT} = \frac{2n}{n+1}, \tag{2.299}$$

where the refraction index $n$ is

$$n = \varepsilon_r - j\frac{\sigma}{\omega\varepsilon_0}. \tag{2.300}$$

Provided the induced current is known, the related transmitted electric field is obtained by numerically computing field integrals (2.293) and (2.294).

Thus, the current and its first derivative at the $i$th wire segment are given by

$$I(x') = I_{1i}\frac{x_{2i} - x'}{\Delta x} + I_{2i}\frac{x' - x_{1i}}{\Delta x}, \tag{2.301}$$

$$\frac{\partial I(x')}{\partial x'} = \frac{I_{2i} - I_{1i}}{\Delta x}, \tag{2.302}$$

where $I_{1i}$ and $I_{2i}$ are the values of current at the local nodes of the $i$th boundary element, with coordinates $x_{1i}$ and $x_{2i}$, and $\Delta x = x_{2i} - x_{1i}$ denotes the element length.

Discretizing (2.293) and (2.294) and substituting (2.301) and (2.302) into (2.293) and (2.294) yields

$$E_x = \frac{1}{j4\pi\omega\varepsilon_0}\sum_{i=1}^{N_j}\left[-\frac{I_{2i} - I_{1i}}{\Delta x_j}\int_{x_{1ij}}^{x_{2i}}\frac{\partial G(x,x',z)}{\partial x'}\mathrm{d}x' - \gamma^2\int_{x_{1ij}}^{x_{2ij}}\left[I_{1i}\frac{x_{2i} - x'}{\Delta x} + I_{2i}\frac{x' - x_{1i}}{\Delta x}\right]G(x,x',z)I(x')\mathrm{d}x'\right], \tag{2.303}$$

$$E_z = \frac{1}{j4\pi\omega\varepsilon_0}\sum_{j=1}^{M}\sum_{i=1}^{N_j}\frac{I_{2ij} - I_{1ij}}{\Delta x_j}\int_{x_{1ij}}^{x_{2ij}}\frac{\partial G(x,x',z)}{\partial z}\mathrm{d}x', \tag{2.304}$$

where $N_j$ denotes the total number of boundary elements along the wire.

Integrals in (2.303) and (2.304) are numerically evaluated using the Gaussian quadrature. The quasi-singularity of the Green function is avoided by approximating the first order differential operator with finite differences [3].

### 2.3.10 Numerical Results

The computational example is related to the dipole antenna of length $L = 1$ m and radius $a = 2$ mm horizontally placed at height $h = 0.25$ m above a real ground with permittivity $\varepsilon_{rg} = 10$ and conductivity $\sigma = 10$ mS m$^{-1}$. Terminal voltage is $V_T = 1$ V. The operating frequency is varied from 1 to 300 MHz.

Figures 2.36–2.38 show the related field components for $f = 1$ MHz, $f = 10$ MHz, and $f = 100$ MHz.

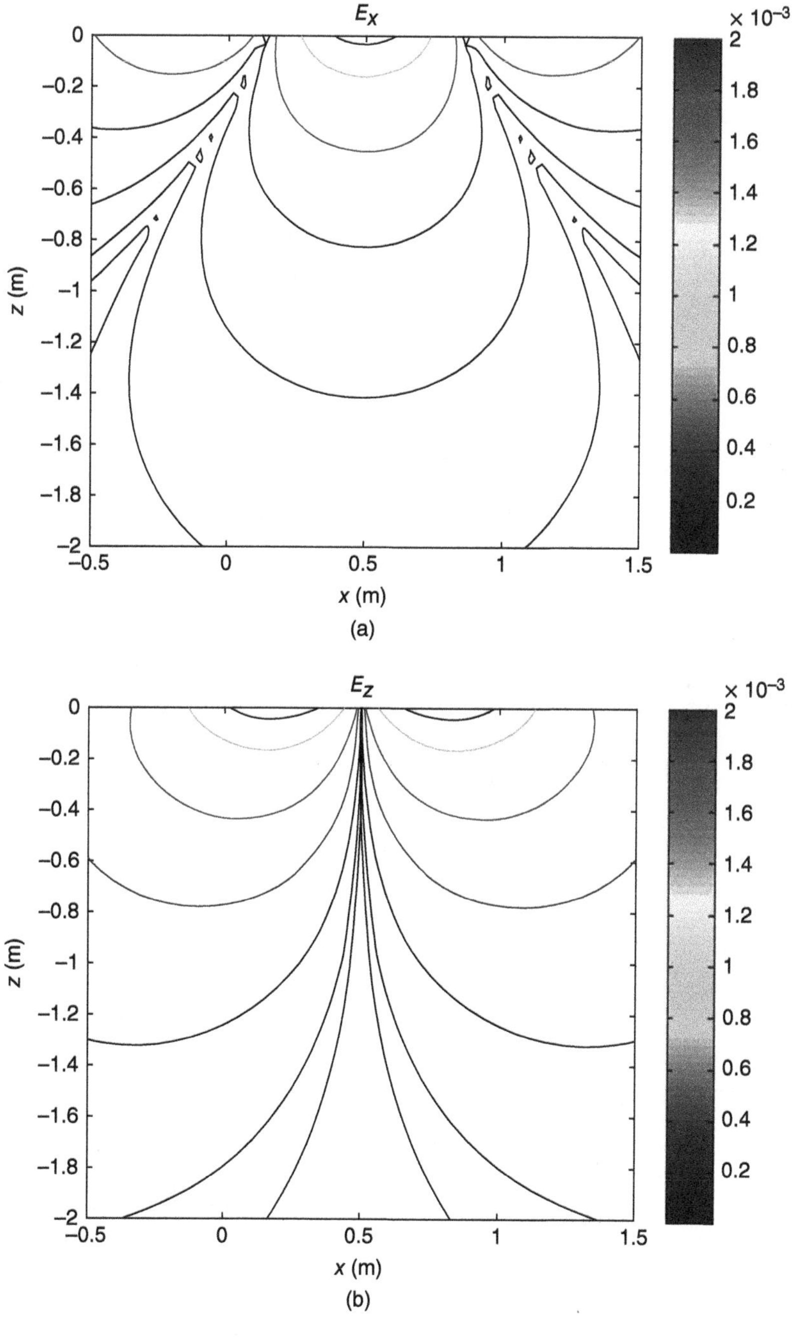


**Figure 2.36** Transmitted field (V m$^{-1}$) into the ground at $f = 1$ MHz. (a) $E_x$-component. (b) $E_z$-component.

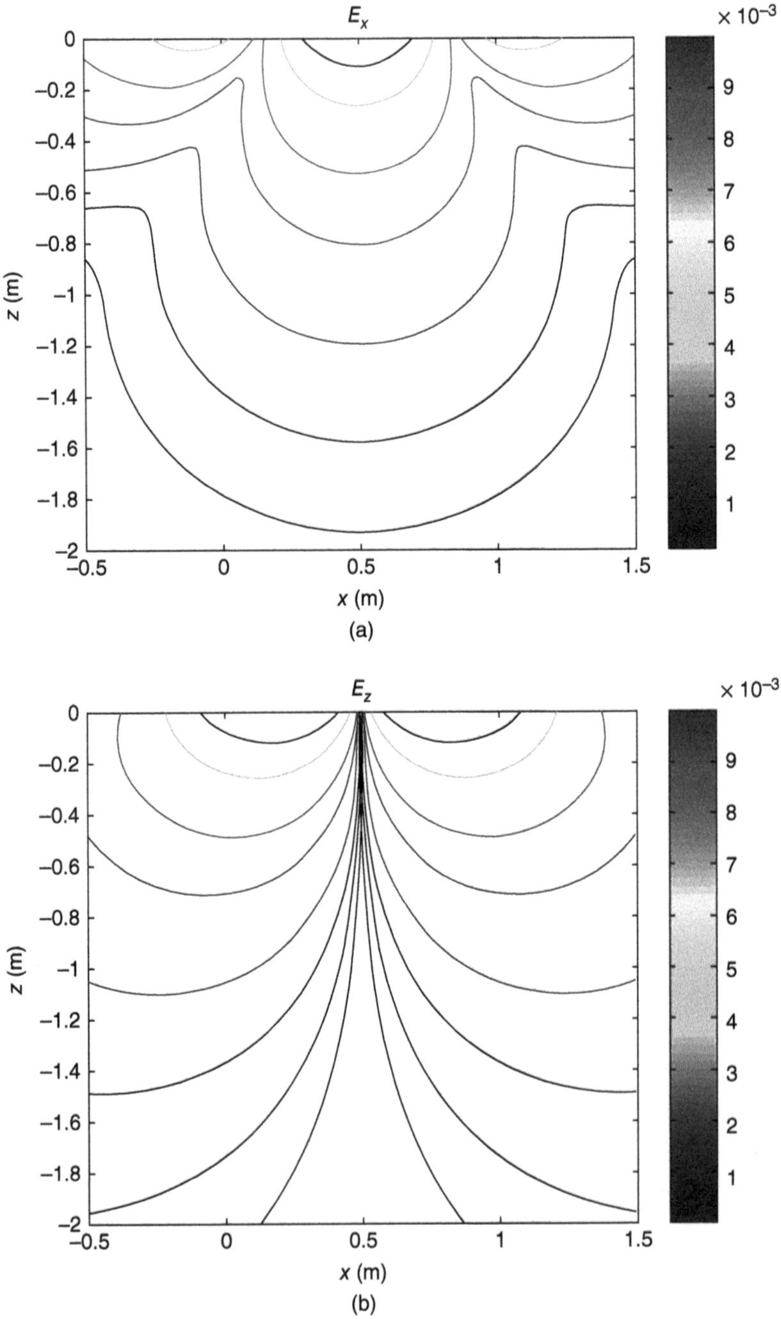


**Figure 2.37** Transmitted field (V m$^{-1}$) into the ground at $f = 10$ MHz. (a) $E_x$-component. (b) $E_z$-component.

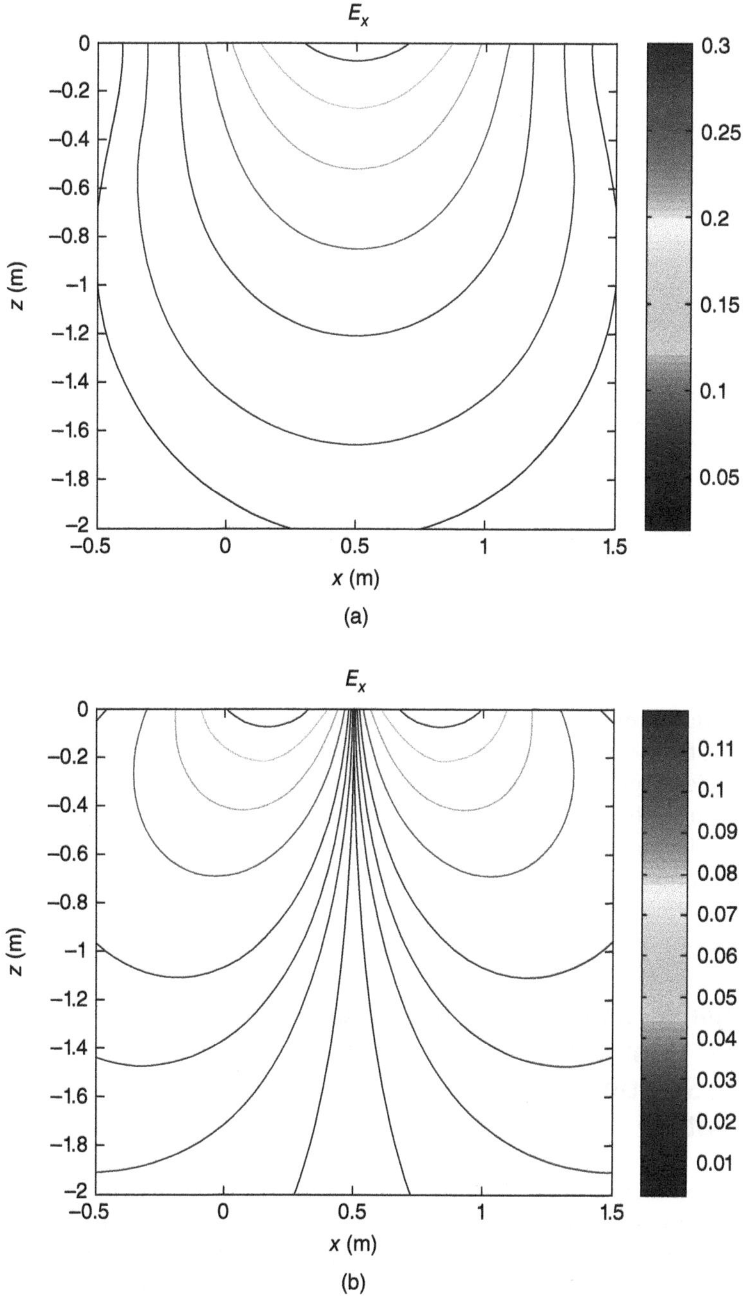


**Figure 2.38** Transmitted field (V m$^{-1}$) into the ground at $f = 100$ MHz. (a) $E_x$-component. (b) $E_z$-component.

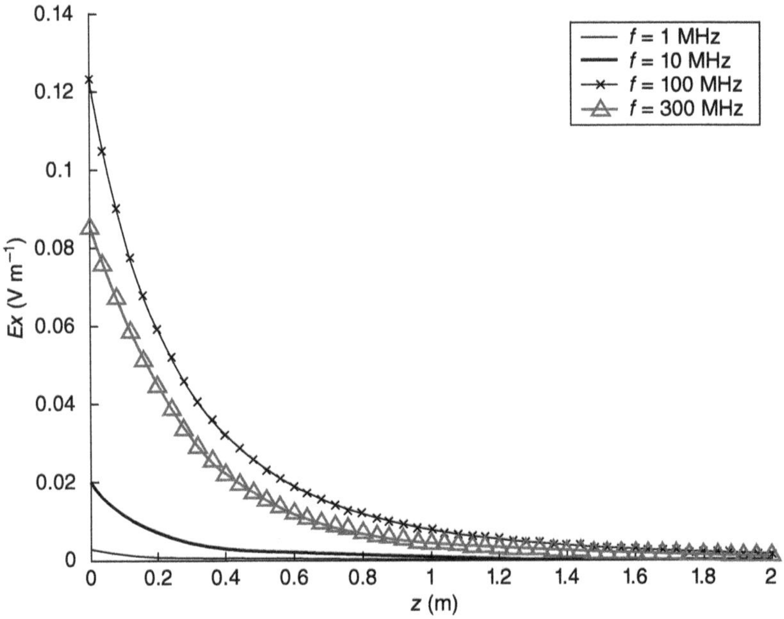


**Figure 2.39** Broadside transmitted field (V m$^{-1}$) into the ground for different frequencies.

Analyzing the numerical results presented in Figures 2.36–2.38 it can be observed that the field distribution remains relatively stable over the considered frequencies. Figure 2.39 shows the $E_x$ component of the transmitted field versus depth in the broadside direction for different operating frequencies.

Note that, due to the symmetry of the problem, $E_z$ component in the broadside direction is zero, which is readily visible from Figures 2.36b, 2.37b, and 2.38b.

## 2.4 Single Vertical Wire in the Presence of a Lossy Half-Space: Comparison of Analytical Solution, Numerical Solution, and Transmission Line Approximation

The problem of vertical wire radiating above a lossy ground has numerous applications in radio communications and EMC [1]. The analysis is based on the solution of the corresponding Pocklington's integro-differential equation using either analytical or numerical methods. It is of practical interest to seek an analytical solution of the Pocklington's equation. Namely, under some plausible set of approximation, a rapid estimation of a phenomenon being useful in an average sense for engineering applications can be obtained [26]. One of the applications of particular interest is related to shipboard antennas [35], or a human being represented by an equivalent parasitic thick wire antenna [36].

In this section, the Pocklington's equation is solved both numerically and analytically. The numerical solution is carried out using the GB-IBEM [2]. The obtained numerical results are compared to the results computed via NEC [13].

The configuration of interest is related to a straight thin vertical wire positioned at certain height $h$ above a real ground, characterized by permittivity $\varepsilon$ and conductivity $\sigma$, as shown in Figure 2.40. The corresponding geometry of the wire and its image are shown in Figure 2.41.

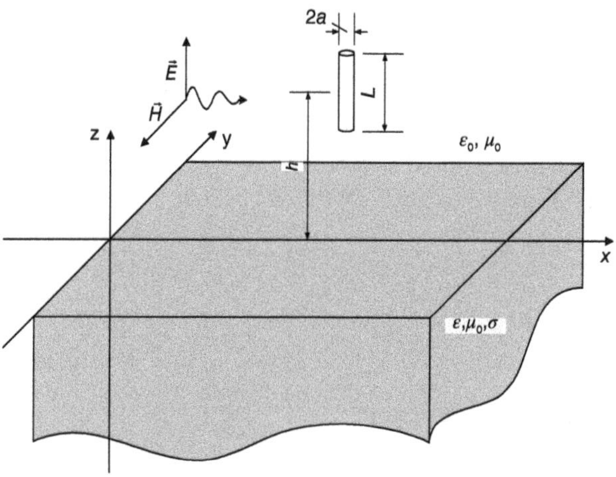


**Figure 2.40** Vertical straight thin wire above a lossy ground.

**Figure 2.41** The geometry of a straight wire and its image.

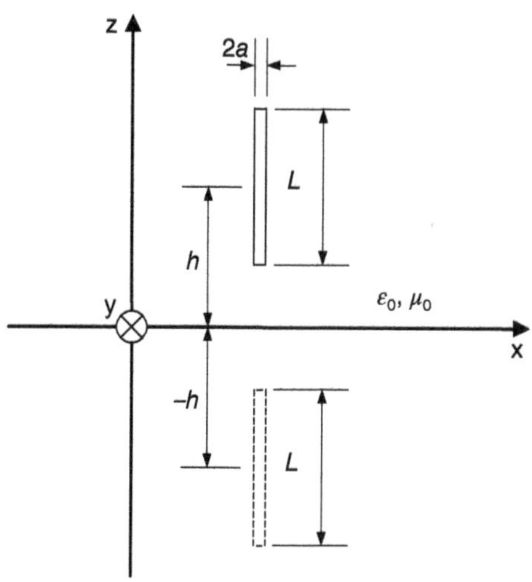

The key parameter in the analysis of the vertical wire excited by either a voltage source or a plane wave is the induced current along the wire, which is governed by an integro-differential equation of the Pocklington's type. This integral equation can be readily derived from Maxwell's equations and is given by [26]

$$\left[\frac{\partial^2}{\partial z^2}+k^2\right]\int_{h-\frac{L}{2}}^{h+\frac{L}{2}} I(z')g(z,z')\mathrm{d}z' = -j4\pi\frac{k}{Z_0}E_z^{\mathrm{exc}}, \tag{2.305}$$

where $I(z')$ is the induced current along the line, $g(z, z')$ denotes the corresponding total Green function, and $E_z^{\mathrm{exc}}$ stands for the electric field excitation.

For the sake of simplicity, only the case of normal incidence is considered.

The propagation constant $k$ is given by

$$k = \omega\sqrt{\mu_0\varepsilon_0}, \tag{2.306}$$

and $Z_0$ is the free space impedance defined as follows:

$$Z_0 = \sqrt{\frac{\mu_0}{\varepsilon_0}}. \tag{2.307}$$

The Green function is given by

$$g(z,z') = g_0(z,z') - \Gamma_{\mathrm{Fr}}^{\mathrm{ref}} g_i(z,z'), \tag{2.308}$$

where $g_0(z, z')$ is the free space Green function:

$$g_o(z,z') = \frac{\mathrm{e}^{-jkR}}{R}, \tag{2.309}$$

while $g_i(z, z')$ arises from the image theory and is given by

$$g_i(z,z') = \frac{\mathrm{e}^{-jkR_i}}{R_i}, \tag{2.310}$$

where $R_o$ and $R_i$ are related to corresponding distances from the source and image to the observation point, respectively:

$$R = \sqrt{(z-z')^2+a^2}, \quad R_i = \sqrt{(z+z')^2+a^2}. \tag{2.311}$$

The Fresnel RC by which the air–earth interface is taken into account is given by [16]

$$\Gamma_{\mathrm{Fr}}^{\mathrm{ref}} = \frac{\sqrt{n}-1}{\sqrt{n}+1}, \quad n = \frac{\varepsilon_{\mathrm{eff}}}{\varepsilon_0}, \tag{2.312}$$

where $\varepsilon_{\mathrm{eff}}$ is the complex permittivity of the earth:

$$\varepsilon_{\mathrm{eff}} = \varepsilon_{\mathrm{r}}\varepsilon_0 + \frac{\sigma}{j\omega}. \tag{2.313}$$

The Fresnel RC approach is a satisfactory approximation replacing the rigorous Sommerfeld integral formulation. A rather comprehensive study on the subject has been carried out in [16]. Thus, the RC approximation has been shown to be rather accurate if the wire is appreciably away from the interface [16, 37, 38].

Once the current is determined the input impedance spectrum can be evaluated using the simple formula:

$$Z_{\text{in}} = \frac{V_g}{I_0}, \tag{2.314}$$

where $V_g$ stands for source voltage and the $I_0$ is the current at the wire center.

The electromagnetic field radiated by vertical dipole above a PEC ground is given by [1]

$$\begin{aligned} E_\rho(\rho, z) &= \frac{1}{j4\pi\omega\varepsilon} \int_{h-\frac{L}{2}}^{h+\frac{L}{2}} I(z') \frac{\partial^2 g(\rho, z, z')}{\partial\rho\partial z} \mathrm{d}z' \\ &= \frac{1}{j4\pi\omega\varepsilon} \int_{h-\frac{L}{2}}^{h+\frac{L}{2}} \frac{\partial I(z')}{\partial z'} \frac{\partial g(\rho, z, z')}{\partial\rho} \mathrm{d}z', \end{aligned} \tag{2.315}$$

$$\begin{aligned} E_z(\rho, z) &= \frac{1}{j4\pi\omega\varepsilon} \left[ \frac{\partial^2}{\partial z^2} + k^2 \right] \int_{h-\frac{L}{2}}^{h+\frac{L}{2}} I(z') g(\rho, z, z') \mathrm{d}z' \\ &= -\frac{1}{j4\pi\omega\varepsilon} \left[ \int_{h-\frac{L}{2}}^{h+\frac{L}{2}} \frac{\partial I(z')}{\partial z'} \frac{\partial g(\rho, z, z')}{\partial z} \mathrm{d}z' + k^2 \int_{-L}^{L} I(z') \rho(\rho, z, z') \mathrm{d}z' \right], \end{aligned} \tag{2.316}$$

$$H_\phi(\rho, z) = -\frac{1}{4\pi} \int_{h-\frac{L}{2}}^{h+\frac{L}{2}} I(z') \frac{\partial g(\rho, z, z')}{\partial\rho} \mathrm{d}z'. \tag{2.317}$$

Therefore, knowing the induced current on the vertical wire provides the assessment of the radiated field.

### 2.4.1 Numerical Solution

The integro-differential equation (2.305) is solved via the GB-IBEM.

The unknown current $I^e(z')$ along the straight wire segment can be written as follows:

$$I^e(z') = \{f\}^T \{I\}. \tag{2.318}$$

Assembling the contributions from each segment the integro-differential equation (2.305) is transferred into the matrix equation:

$$\sum_{j=1}^{M} [Z]_{ji} \{I\}_i = \{V\}_j, \quad \text{and} \quad j = 1, 2, \ldots, M, \tag{2.319}$$

where $M$ is the total number of segments and $[Z]_{ji}$ is the mutual impedance matrix representing the interaction of the $i$th source with the $j$th observation segment, respectively:

$$[Z]_{ji} = -\int_{\Delta l_j} \{D\}_j \int_{\Delta l_i} \{D'\}_i^T g(z,z')\mathrm{d}z' + \int_{\Delta l_j} \{D\}_j \int_{\Delta l_i} \{D'\}_i^T \Gamma_{\mathrm{ref}}^{\mathrm{Fr}} g_i(z,z')\mathrm{d}z' + k^2 \int_{\Delta l_j} \{f\}_j \int_{\Delta l_i} \{f\}_i^T g(z,z')\mathrm{d}z'\,\mathrm{d}z. \tag{2.320}$$

Matrices $\{f\}$ and $\{f'\}$ contain the shape functions while $\{D\}$ and $\{D'\}$ contain their derivatives, and $\Delta l_i$, $\Delta l_j$ are the widths of $i$th and $j$th boundary elements.

For a linear approximation over a wire segment the shape functions are of the form:

$$f_i(z) = \frac{z_{i+1} - z'}{\Delta z} \quad f_{i+1}(z) = \frac{z' - z_i}{\Delta z}, \tag{2.321}$$

and the related derivatives are simply given by

$$\frac{\partial f_i(z')}{\partial z'} = \mp \frac{1}{\Delta z}, \tag{2.322}$$

as this choice was proved to be the optimal one in modeling various wire configurations [9].

The voltage vector containing the excitation term is given by

$$\{V\}_j = -j4\pi \frac{k}{Z_0} \int_{z_i}^{z_{i+1}} E_z^{\mathrm{exc}} \{f\}\mathrm{d}z. \tag{2.323}$$

For the case of transmitting antenna and plane wave normal incidence integral (2.323) is evaluated analytically [2].

If the delta-function voltage generator is used (antenna mode) the right-side vector differs from zero only in the feed-gap area, i.e. the incident electric field is given by

$$E_z^{\mathrm{exc}}(z) = \frac{V_g}{\Delta l_g}, \tag{2.324}$$

where $V_g$ is the source voltage and $\Delta l_g$ is the feed-gap width.

Using the linear shape functions it follows that

$$V_1 = V_2 = -j4\pi \frac{k}{Z_0} \int_{-\Delta l_g/2}^{\Delta l_g/2} \frac{V_g}{\Delta l_g} \frac{z_2 - z}{\Delta l_g} \mathrm{d}z = -j4\pi \frac{k}{Z_0} \frac{V_g}{2}. \tag{2.325}$$

In the scattering mode for the simple case of normal incidence the wire is excited by a plane wave, i.e. the incident field is

$$E_z^{\mathrm{exc}}(z) = E_0, \tag{2.326}$$

and the right-hand side vector differs from zero on each segment and the local voltage vector is given by

$$V_j = -j4\pi\frac{k}{Z_0}\int_{-\Delta l/2}^{\Delta l/2} E_0\frac{z_2 - z}{\Delta l}\mathrm{d}z = -j4\pi\frac{k}{Z_0}E_0\frac{\Delta l}{2}. \tag{2.327}$$

The full description on the GB-IBEM procedures can be found in [2].

The field components can be calculated using the boundary element formalism, as well, and they are, as follows [39]:

$$E_\rho(\rho, z) = \frac{1}{j4\pi\omega\varepsilon}\sum_{i=1}^{M}\int_{\Delta l_i}\{D'\}_i^T\frac{\partial g(\rho, z, z')}{\partial \rho}\mathrm{d}z', \tag{2.328}$$

$$E_z(\rho, z) = -\frac{1}{j4\pi\omega\varepsilon} \times \sum_{i=1}^{M}\left\{\int_{\Delta l_i}\{D'\}_i^T\frac{\partial g(\rho, z, z')}{\partial z}\mathrm{d}z' + k^2\int_{\Delta l_i}\{f'\}_i^T g(\rho, z, z')\mathrm{d}z'\right\}, \tag{2.329}$$

$$H_\phi(\rho, z) = -\frac{1}{4\pi}\sum_{i=1}^{M}\int_{\Delta l_i}\{f'\}_i^T\frac{\partial g(\rho, z, z')}{\partial \rho}\mathrm{d}z'. \tag{2.330}$$

The full description of GB-IBEM can be found in [2].

### 2.4.2 Analytical Solution

The first step in the analytical solution of (2.305) is as follows:

$$\int_{h-\frac{L}{2}}^{h+\frac{L}{2}} I(z')g(z, z')\mathrm{d}z' = I(z)\int_{h-\frac{L}{2}}^{h+\frac{L}{2}} g(z, z')\mathrm{d}z' + \int_{h-\frac{L}{2}}^{h+\frac{L}{2}} [I(z') - I(z)]g(z, z')\mathrm{d}z'. \tag{2.331}$$

The integral on the left-hand side of (2.331) is now approximated by the first term on the right-hand side of (2.331), thus neglecting the second integral on the right-hand side. Such an approach has been used in a number of papers, e.g. in [39].

Furthermore, the characteristic integral from (2.331) can be written as follows:

$$\psi(z) = \int_{h-\frac{L}{2}}^{h+\frac{L}{2}} g(z, z')\mathrm{d}z' = \int_{h-\frac{L}{2}}^{h+\frac{L}{2}} \frac{\mathrm{e}^{-jkR}}{R}\mathrm{d}z' + \Gamma_{\mathrm{ref}}^{\mathrm{Fr}}\int_{h-\frac{L}{2}}^{h+\frac{L}{2}} \frac{\mathrm{e}^{-jkR_i}}{R_i}\mathrm{d}z'. \tag{2.332}$$

Using the well-known approximation [39, 40]

$$\frac{\mathrm{e}^{-jkR}}{R} \doteq \frac{1}{R} - jk \tag{2.333}$$

yields

$$\psi(z) = \ln\left(\frac{z-h+\frac{L}{2}+\sqrt{\left(z-h+\frac{L}{2}\right)^2+a^2}}{z-h-\frac{L}{2}+\sqrt{\left(z-h-\frac{L}{2}\right)^2+a^2}}\right) + \Gamma_{\mathrm{ref}}^{\mathrm{Fr}} \cdot \ln\left(\frac{z+h+\frac{L}{2}+\sqrt{\left(z+h+\frac{L}{2}\right)^2+a^2}}{z+h-\frac{L}{2}+\sqrt{\left(z+h-\frac{L}{2}\right)^2+a^2}}\right) - jkL\left(1+\Gamma_{\mathrm{ref}}^{\mathrm{Fr}}\right). \tag{2.334}$$

Consequently, the Pocklington's equation (2.305) simplifies into the partial differential equation of the form:

$$\left[\frac{\partial^2}{\partial z^2}+k^2\right] I(z) = -j4\pi \frac{k}{\psi(z)Z_0} E_z^{\mathrm{exc}}. \tag{2.335}$$

The analytical solution of Equation (2.335) is [39]

$$I(z) = -j\frac{4\pi}{\psi(z)Z_0}\frac{E_z^{\mathrm{exc}}}{k}\left[1-\frac{\cos k(h-z)}{\cos k\frac{L}{2}}\right]. \tag{2.336}$$

Note that $E_z^{\mathrm{exc}}$ stands for the excitation given in the form of an equivalent voltage source or plane wave, respectively.

On the other hand, at higher frequencies the sinusoidal approximate current distribution along center-fed vertical dipole above a PEC ground is commonly assumed [41]. First, the case of a thin wire antenna insulated in free space is considered, Figure 2.42. Then a vertical dipole above a PEC ground is of interest.

The radiated field is determined by following integral expressions [41]:

$$E_\rho = -\frac{Z_0}{j8\pi k}\int_{-\frac{L}{2}}^{\frac{L}{2}} \sin 2\theta_0 \left[\frac{\partial^2 G_0}{\partial R_0^2} - \frac{1}{R_0}\frac{\partial G_0}{\partial R_0}\right] I(z')\mathrm{d}z', \tag{2.337}$$

$$E_z = \frac{Z_0}{j4\pi k}\int_{-\frac{L}{2}}^{\frac{L}{2}} \left[\cos^2\theta_0 \frac{\partial^2 G_0}{\partial R_0^2} + \frac{\sin^2\theta_0}{R_0}\frac{\partial^2 G_0}{\partial R_0^2} + k^2 G_0\right] I(z')\mathrm{d}z', \tag{2.338}$$

$$H_\phi = -\frac{1}{4\pi}\int_{-\frac{L}{2}}^{\frac{L}{2}} \sin\theta_0 \frac{\partial G_0}{\partial R_0} I(z')\mathrm{d}z', \tag{2.339}$$

where

$$G_o = \frac{\mathrm{e}^{-jkR_0}}{R_0}. \tag{2.340}$$

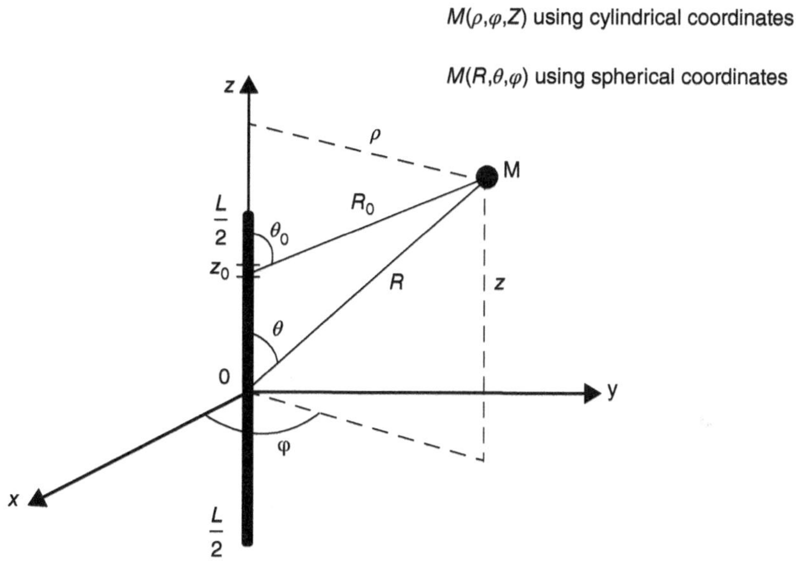


**Figure 2.42** Straight thin wire in free space.

Using the notation depicted in Figure 2.43 the current along the dipole is given by [41]

$$I_r(z') = \begin{cases} I_m \sin[k(h - |z'|)], h \le z' \le \left(\frac{l}{2} + h\right)^+ \\ -I_m \sin[k(l + h - |z'|)], \left(\frac{l}{2} + h\right)^- \le z' \le l + h \end{cases}, \tag{2.341}$$

while the current along the image wire is then [41]

$$I_{im}(z') = \begin{cases} I_m \sin[k(l + h - |z'|)], -l - h \le z' \le \left(-\frac{l}{2} - h\right)^- \\ -I_m \sin[k(h - |z'|)], \left(-\frac{l}{2} - h\right)^+ \le z' \le -h \end{cases}. \tag{2.342}$$

Adapting the formulas (2.337)–(2.339) to the geometry of vertical dipole presented in Figure 2.43 and utilizing the approximate current distribution (2.341) and (2.342) the closed form expressions for the electric and magnetic field components can be obtained [41].

### 2.4.3 Computational Examples

The first set of numerical results is related to following two sections: vertical straight wire in a transmitting (antenna) mode and a vertical straight wire in a receiving (scattering) mode.

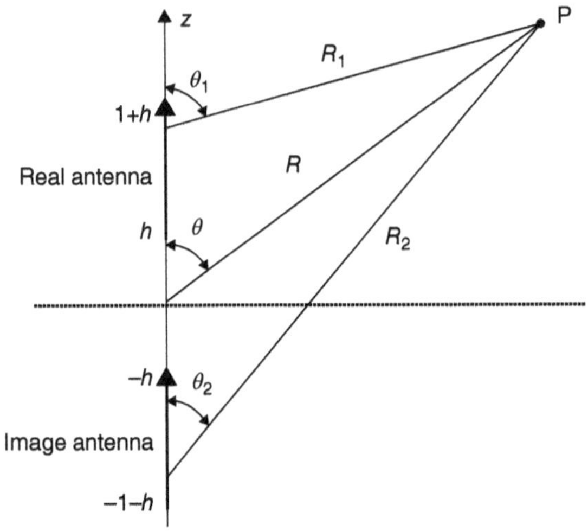


**Figure 2.43** Vertical dipole above a PEC ground and its image.

#### 2.4.3.1 Transmitting Antenna

A straight wire of length $L = 1$ m and radius $a = 5$ mm, located at height $h$ above a lossy ground ($\sigma = 1$ mS m$^{-1}$, $\varepsilon_r = 10$) is considered. The wire is excited by a unit voltage source, while the operating frequency is $f = 168.2$ MHz. This frequency is of interest in certain air-traffic control applications and some EMI problems occur at this frequency as well. Figures 2.44 and 2.45 show the real and imaginary parts of the current distribution, respectively.

The results calculated via the GB-IBEM solution of the Pocklington's equation (2.305) are compared to the results computed via NEC [13]. The obtained results agree satisfactorily.

Figures 2.46 and 2.47 show the real and imaginary parts of the input impedance spectrum respectively of the half wave vertical dipole versus height $h$ at $f = 3$ MHz. The wire radius expressed in wavelengths is $a/\lambda = 0.0005$.

The results calculated via the GB-IBEM (both RC approximation and rigorous Sommerfeld solution presented in [19]) agree satisfactorily with the NEC results and are in accordance to the results obtained in [42]. Also, Figures 2.46 and 2.47 clearly show that RC approximation solution provides accurate results compared to the rigorous Sommerfeld approach, especially when the height of the wire gets close to the wavelength.

#### 2.4.3.2 Receiving Antenna

The vertical straight wire illuminated by a plane wave ($L = 1$ m, $a = 0.005$ m, $f = 168.2$ MHz), whose magnitude is $E_0 = 1$ V m$^{-1}$, is of interest.

Figures 2.48 and 2.49 show the real and imaginary parts of the current distribution, respectively, for the case of $h = 2$ m above a PEC ground,

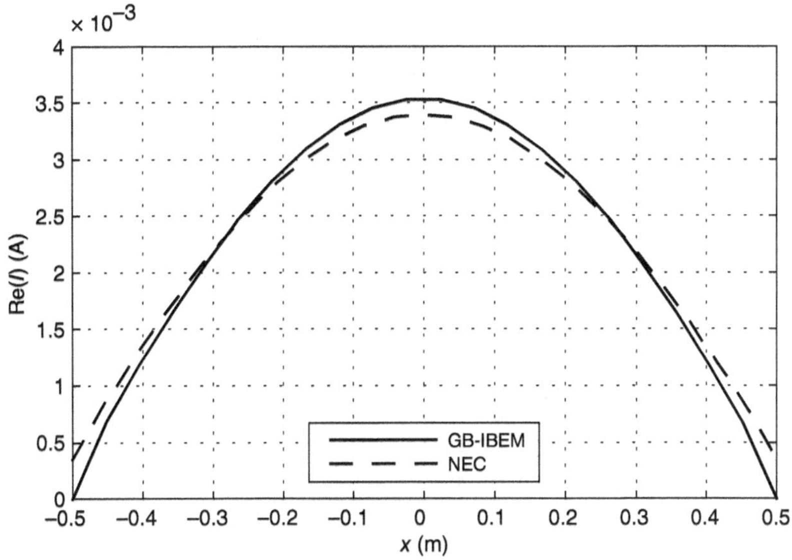


**Figure 2.44** The real part of the current distribution ($L = 1$ m, $a = 0.005$ m, $h = 2$ m, $\sigma = 1$ mS m$^{-1}$, $\varepsilon_r = 10$, $V_0 = 1$ V, $f = 168.2$ MHz).

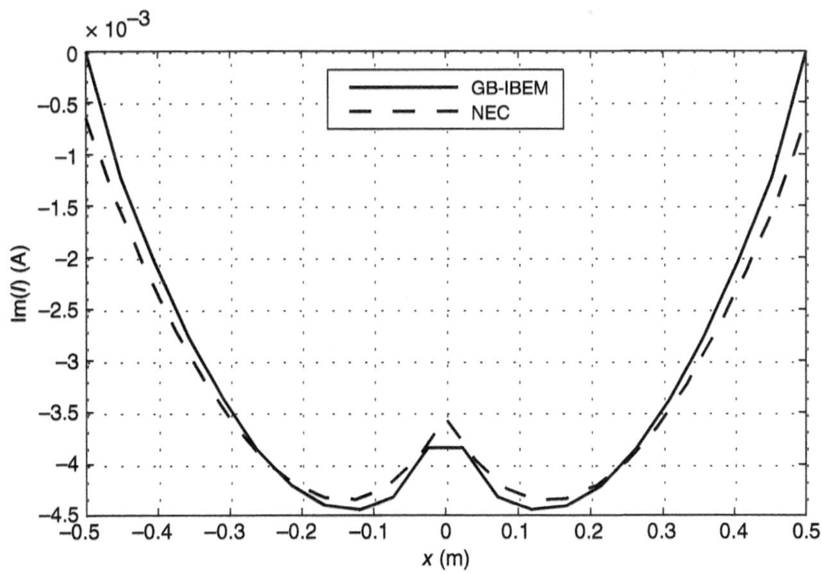


**Figure 2.45** The imaginary part of the current distribution ($L = 1$ m, $a = 0.005$ m, $h = 2$ m, $\sigma = 1$ mS m$^{-1}$, $\varepsilon_r = 10$, $V_0 = 1$ V, $f = 168.2$ MHz).

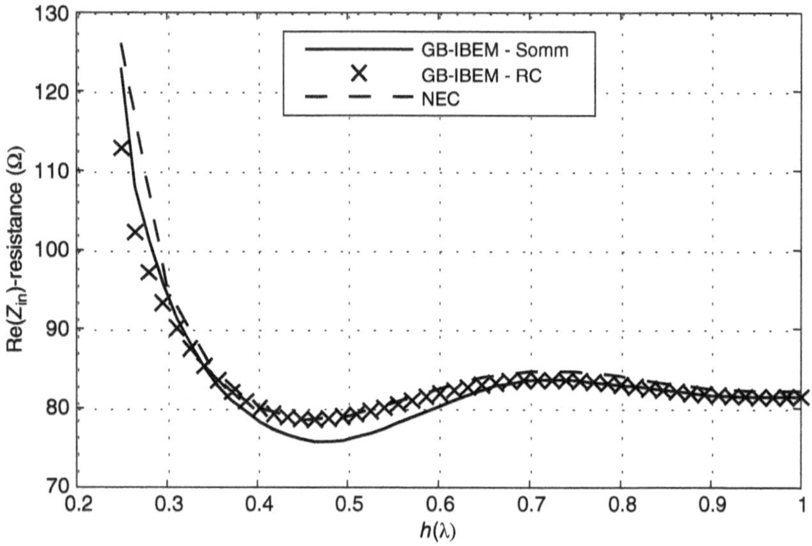


**Figure 2.46** Resistance of vertical dipole above a real ground ($L = 1$ m, $a/\lambda = 0.0005$ m, $\sigma = 3$ mS m$^{-1}$, $\varepsilon_r = 10$, $f = 3$ MHz).

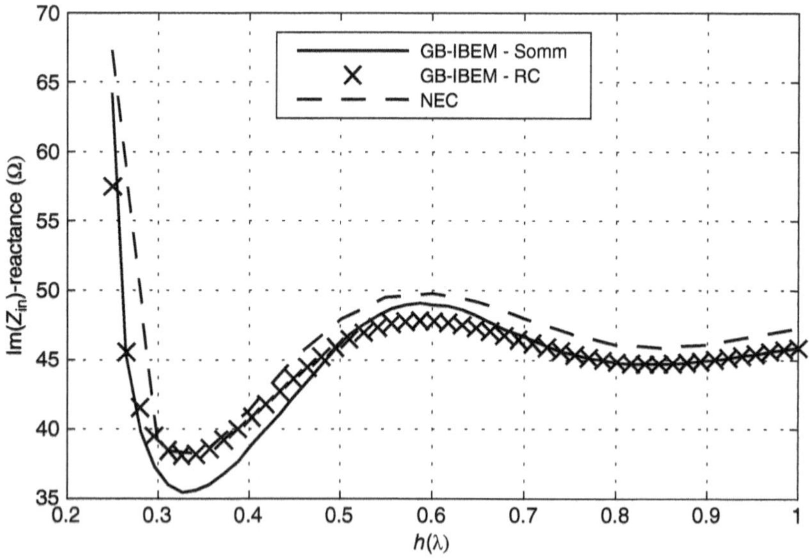


**Figure 2.47** Reactance of vertical dipole above a real ground ($L = 1$ m, $a/\lambda = 0.0005$ m, $\sigma = 3$ mS m$^{-1}$, $\varepsilon_r = 10$, $f = 3$ MHz).

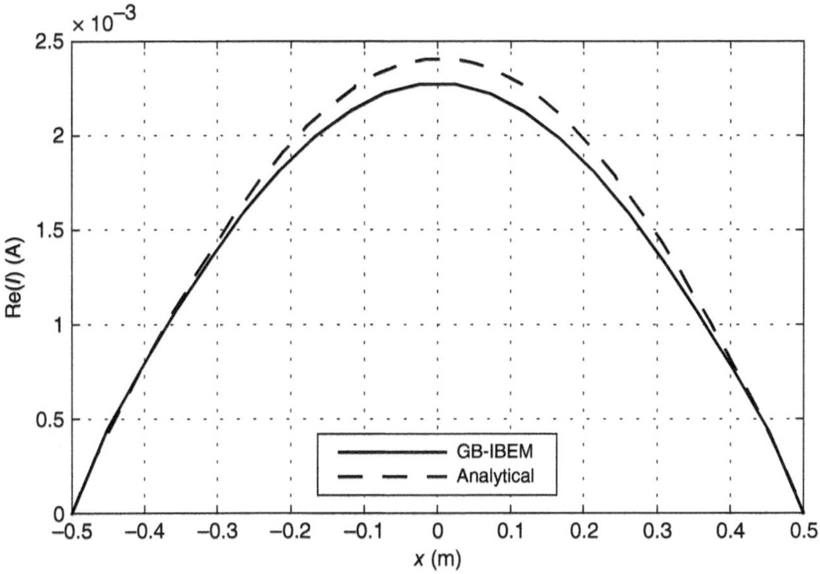


**Figure 2.48** The real part of the current distribution ($L = 1$ m, $a = 0.005$ m, $h = 2$ m, $E_0 = 1$ V m$^{-1}$, $f = 168.2$ MHz).

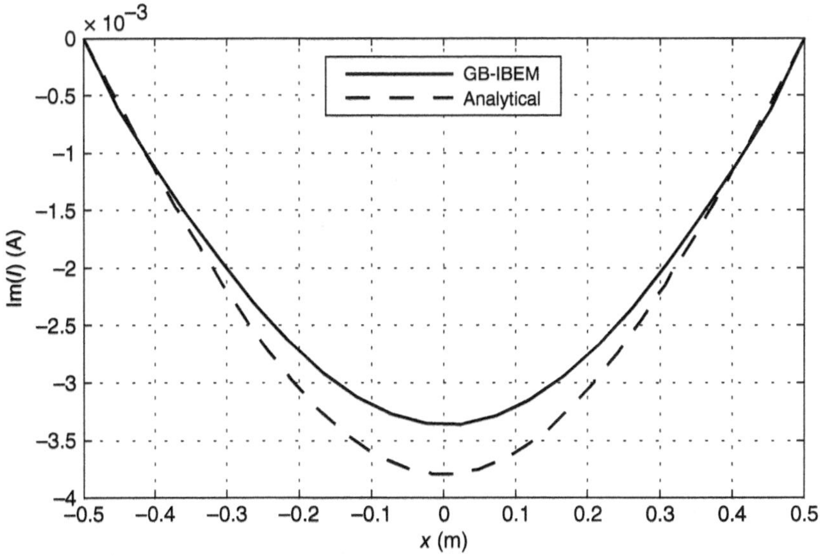


**Figure 2.49** The imaginary part of the current distribution ($L = 1$ m, $a = 0.005$ m, $h = 2$ m, $E_0 = 1$ V m$^{-1}$, $f = 168.2$ MHz).

while Figures 2.50 and 2.51 are related to the vertical wire above a lossy ground ($\sigma = 1\,\mathrm{mS\,m^{-1}}$, $\varepsilon_r = 10$, $h = 2\,\mathrm{m}$).

The results obtained via the analytical solution method and GB-IBEM agree relatively satisfactorily.

Figure 2.52 shows the spectrum of the absolute value of the current induced at the center of a straight wire for the frequency range from order Hz to 200 MHz computed via analytical and numerical approach, respectively.

As can be seen from Figure 2.52 the results seem to agree satisfactorily for frequencies less than 100 MHz. There are some discrepancies around the resonant frequency $f = 150\,\mathrm{MHz}$, i.e. there is a problem with the cosine function in the denominator of (2.336), which becomes singular. This difficulty arises due to the approximations used to obtain the analytical expression. These approximations give rather satisfactory results when the receiving antenna is embedded in a lossy medium. Additional corrections have to be made in order to achieve better agreement of the results for the antenna located above a lossy ground.

The results for the radiated electric field are related to the case of a straight thin wire radiating in free space and a vertical dipole above a PEC

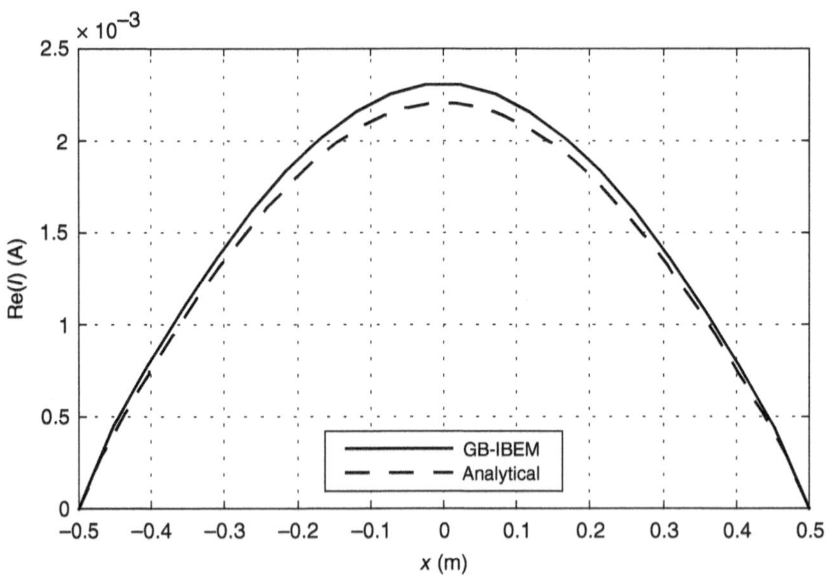


**Figure 2.50** The real part of the current distribution ($L = 1$ m, $a = 0.005$ m, $h$=2 m, $\sigma = 1\,\mathrm{mS\,m^{-1}}$, $\varepsilon_r = 10$, $E_0 = 1\,\mathrm{V\,m^{-1}}$, $f = 168.2$ MHz).

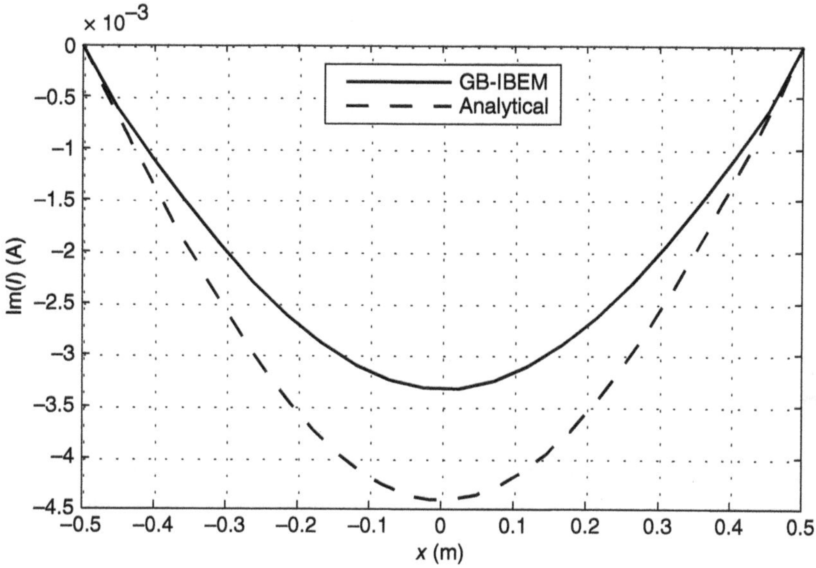


**Figure 2.51** The imaginary part of the current distribution ($L = 1$ m, $a = 0.005$ m, $h = 2$ m, $\sigma = 1$ mS m$^{-1}$, $\varepsilon_r = 10$, $E_0 = 1$ V m$^{-1}$, $f = 168.2$ MHz).

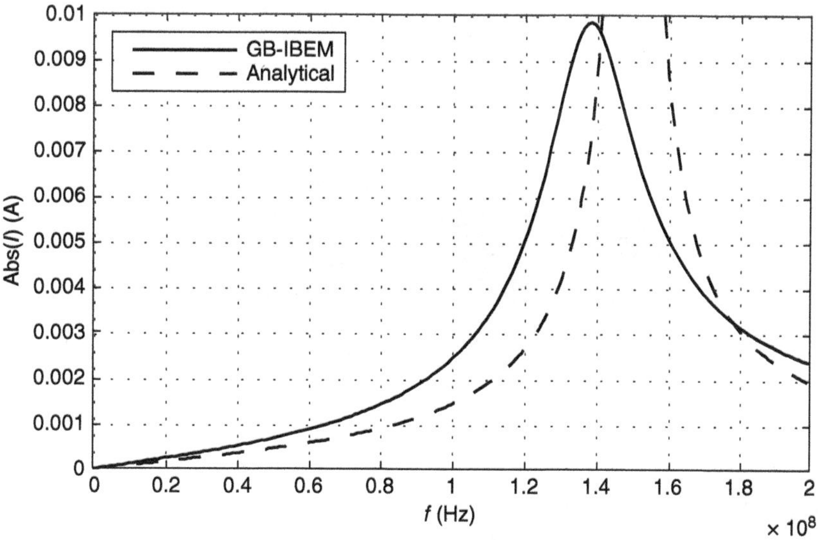


**Figure 2.52** Spectrum of the induced current at the center of the straight wire scatterer ($L = 1$ m, $a = 0.005$ m, $h = 1$ m, $\sigma = 1$ mS m$^{-1}$, $\varepsilon_r = 10$, $E_0 = 1$ V m$^{-1}$).

ground, respectively. The numerical results obtained by using GB-IBEM [2] is compared to the results calculated analytically assuming the current distribution waveform. The geometry of interest is a straight wire; length $L = 3$ m long, radius $a = 5$ mm, insulated in free space, and above a PEC ground plane. The wire is center fed by a voltage source at $f = 3$ MHz. The first set of figures is related to the case of straight wire in free space. In Figures 2.53–2.56, the tangential electric field computed via GB-IBEM and analytical model, respectively, is presented.

Observing the results obtained for the tangential electric field component it is evident that the waveforms obtained by different approaches agree rather satisfactorily. Some discrepancies appear at the peaks.

The next set of results is related to vertical dipole radiating over a perfect ground. Figures 2.57–2.60 show the tangential electric field computed via GB-IBEM and analytical model, respectively.

Analyzing the results for the tangential electric field component calculated via different approaches it is visible that the waveforms obtained by different approaches agree satisfactorily again. As in the case of free space, some discrepancies appear at the peaks.

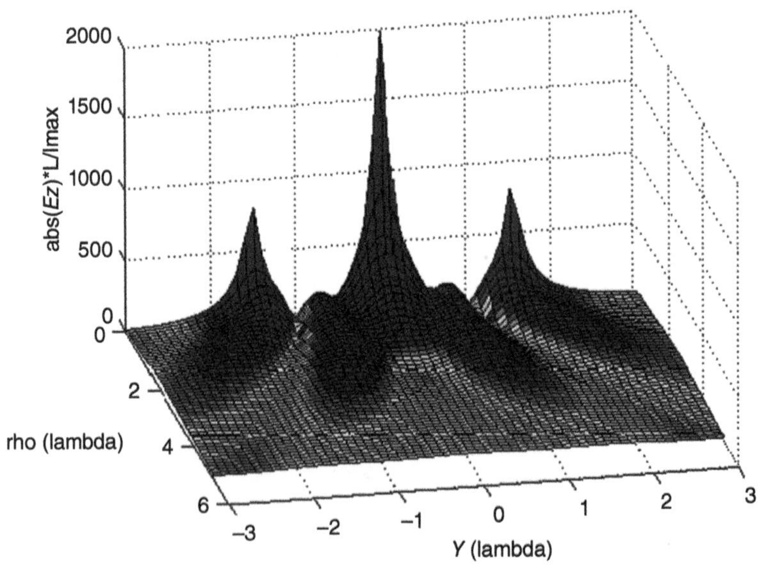


**Figure 2.53** Tangential field component radiated by the dipole in free space – numerical model.

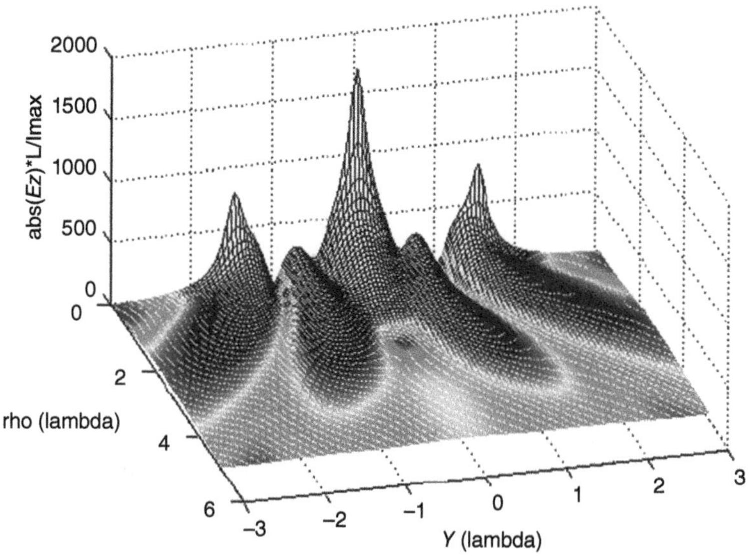


**Figure 2.54** Tangential field component radiated by the dipole in free space – analytical model.

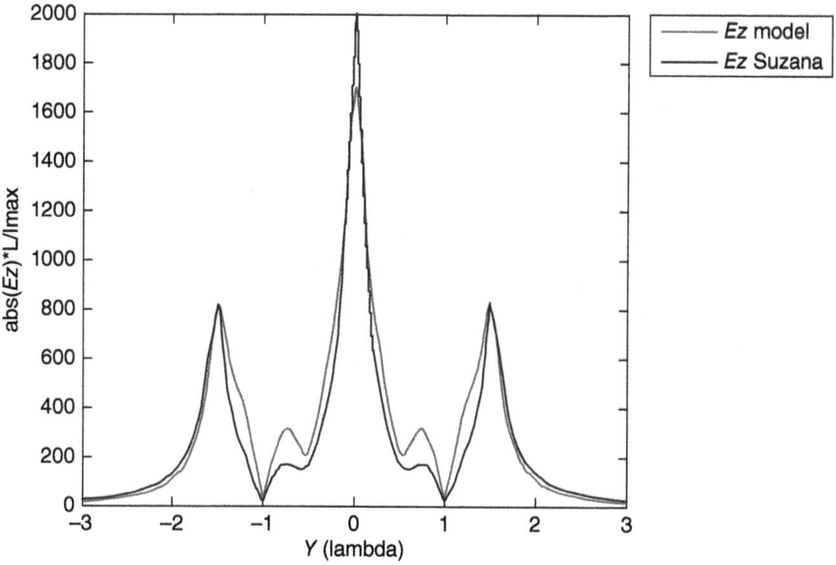


**Figure 2.55** Tangential field component radiated by the dipole in free space – comparison of analytical and numerical model at a distance $\rho = 0.1$ m from the wire.

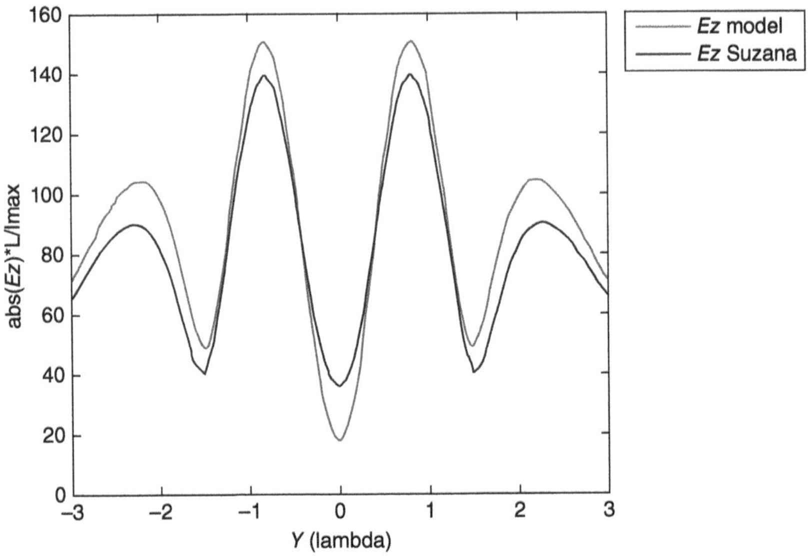


**Figure 2.56** Tangential field component radiated by the dipole in free space – comparison of analytical and numerical model at a distance $\rho = 2$ m from the wire.

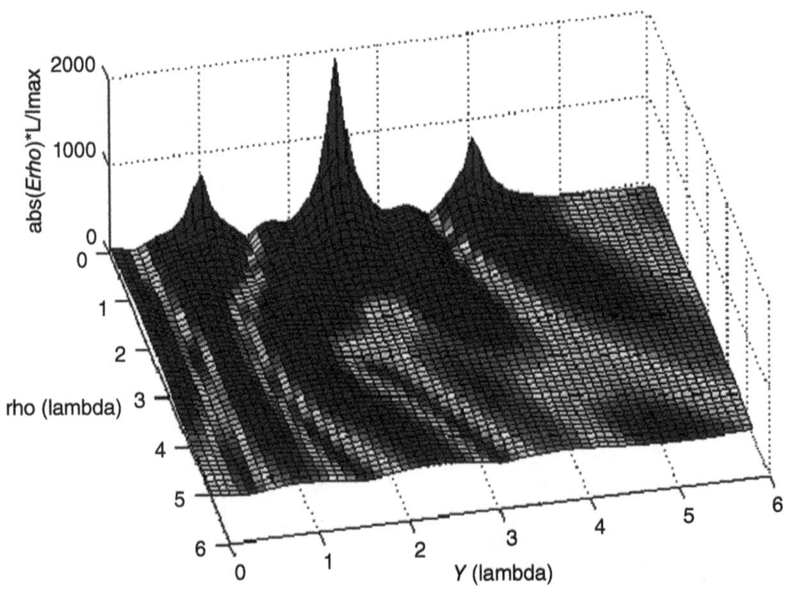


**Figure 2.57** Tangential field component radiated by the dipole above a PEC ground – numerical model.

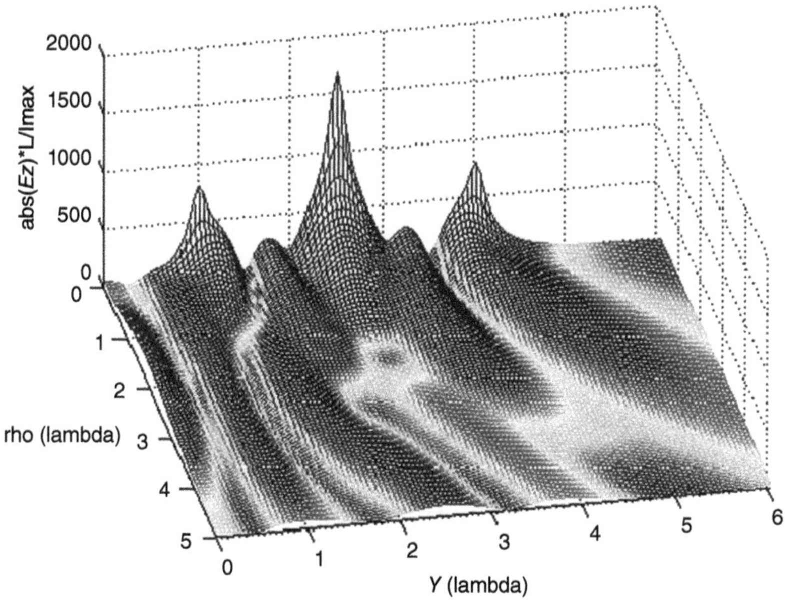


**Figure 2.58** Tangential field component radiated by the dipole in free space – analytical model.

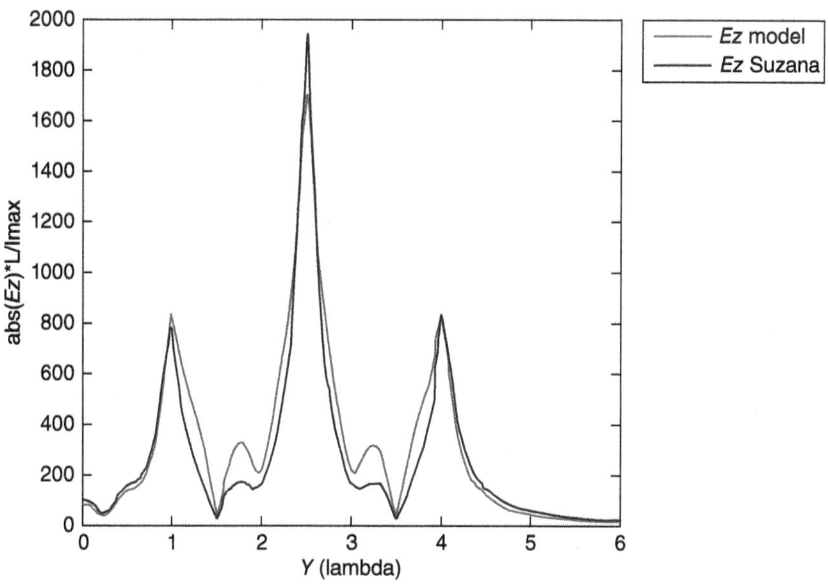


**Figure 2.59** Tangential field component radiated by the dipole above a PEC ground – comparison of analytical and numerical model at a distance $\rho = 0.1$ m from the wire.

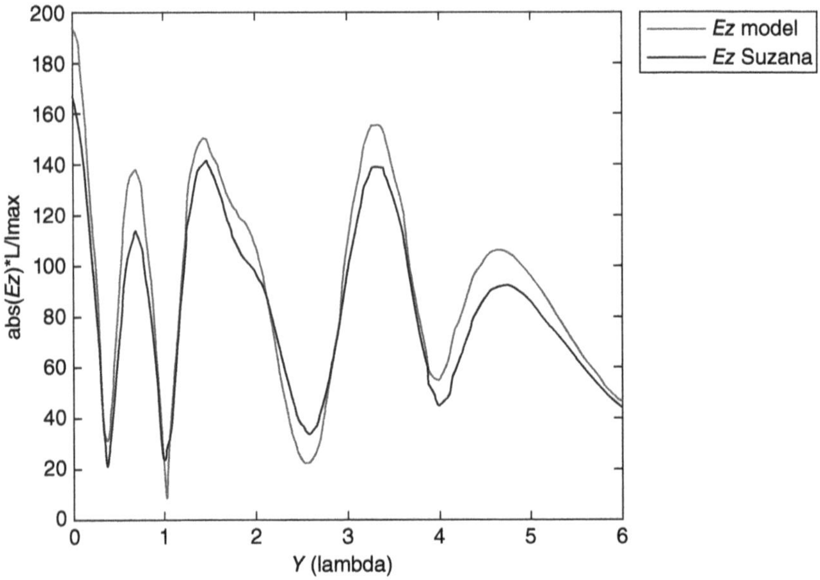


**Figure 2.60** Tangential field component radiated by the dipole above a PEC ground – comparison of analytical and numerical model at a distance $\rho = 2$ m from the wire.

## 2.5 Magnetic Current Loop Excitation of Thin Wires

The simplest configuration of thin wire antennas is a cylindrical dipole whose behavior is determined by the corresponding integral equation of the Pocklington [43] or the Hallen type [44]. Furthermore, there is a choice of two kernel types: the exact and reduced (approximated) kernel [2]. In the case of the finite length wires these integral equations are treated using numerical methods, usually with the MoM [40], or GB-IBEM [2].

The transmitting mode of an antenna requires a certain model of the excitation at input terminals. In engineering practice, antennas are driven by an open-wire TL or by a coaxial feed through the ground plane. Almost without exceptions, the distance between conductors of a feeding line in practice is appreciably small compared to the wavelength of the impressed field. Despite the fact that excitation area is electrically small, its geometry has a huge influence on the antenna admittance (impedance). The influence of the excitation model to the antenna radiation pattern is less pronounced.

Traditionally, there are two methods used to model the excitation: delta gap (DG) and magnetic frill (MF) [45, 46]. The DG source assumes the electric field impressed by the feeding line existing only in the gap between the antenna terminals, thus being zero outside. The DG source approach has been used extensively, primarily due to its simplicity. However, a serious drawback of this excitation model is the dependence of the current at the excitation element on the element length, thus making the input admittance also dependent on the length of wire segments. The MF models the excitation as a coaxial line that terminates into a monopole over the ground plane. The use of MF typically results in more accurate input admittance values at the expense of the increased computations complexity in the numerical solution.

Apart from these models there are other methods for excitation modeling reported in the literature, but they are rarely used. In [47] a source model in which the incident field has Gaussian distribution is presented, while in [48] a constant gap source model is used. In [47] and [48] the exact kernel is used, while in [49] a RGF DG source is introduced to improve the convergence and the accuracy. Furthermore, the so-called belt generator is presented in [50] while a biconical source is used in NEC [10], which generally should provide more accurate results.

A magnetic current loop (MCL) source model promoted in [51] is used to excite the cylindrical thin wire antennas. In [51] the analysis of the magnetic loop source model in terms of accuracy and numerical stability, with the respect to the commonly used DG and MF, is carried out. It should be noted that some similar ideas exist in the literature, e.g. see [52, 53]. The use of MCL to excite the finite length cylindrical antenna is reported in [51]. The similarities and differences with the ideas presented in [52] and [53] are discussed in [51], as well.

The analysis presented in [51] used the frequency domain formulation based on the Pocklington's integro-differential equation, numerically handled via different numerical methods: GB-IBEM [2] and the MoM (both point-matching and Galerkin version).

The problem of interest is the single dipole antenna of length $L$ and radius $a$, located in the unbounded medium and excited by a voltage source. The wire dimensions satisfy the TWA [2]. The current distribution along a dipole is governed by the Pocklington's integro-differential equation:

$$\left[\frac{\partial^2}{\partial z^2} + k^2\right] \int_{-\frac{L}{2}}^{\frac{L}{2}} I(z')g(z,z')\mathrm{d}z' = -j4\pi \frac{k}{Z_0} E_z^{\mathrm{exc}}, \tag{2.343}$$

where $I(z')$ is the induced current along the line and $E_z^{\mathrm{exc}}$ is the electric field excitation. The reduced kernel of the integral equation is given by

$$g(z,z') = \frac{\mathrm{e}^{-jkR}}{R}, \tag{2.344}$$

where $R$ is the corresponding distance from the source to the observation point.

The dipole is excited by a voltage source expressed in terms of electric field $E_z^{\mathrm{exc}}$. Traditionally, there are two methods for modeling excitation $E_z^{\mathrm{exc}}$: the *DG* excitation and the *MF* generator. Other models of excitation are not commonly used.

### 2.5.1 Delta Gap and Magnetic Frill

The DG source model (Figure 2.61.a) assumes the excitation field $E_z^{exc}$ to exist only over the gap between the antenna input terminals and is zero outside. Impressed electric field in the thin finite gap between terminals is expressed as [45]

$$E_z^{exc} = \frac{V_g}{\Delta z}, \tag{2.345}$$

where $\Delta z$ is the feed-gap width and $V_g$ is the feed voltage. Mostly, in numerical simulation it is assumed that this field exists only on one wire segment.

The DG model is the simplest and most widely used, but, at the same time the least accurate model, particularly for input impedance assessment. The smaller the gap, the more accurate results are [45, 46].

In MF source model, the fed gap is replaced with a circularly directed magnetic current density existing over the annular aperture with an inner radius $a$ (usually radius of the wire) and an outer radius $b$ (Figure 2.61b). As the antenna is often fed by a TL, the other radius $b$ is found using the expression for characteristic impedance of the TL. Therefore, radius $b$ is not always known and defined.

Originally, the model is introduced in [54] and successfully implemented in [55] and [56] on the cylindrical dipoles. Although the model is presented and

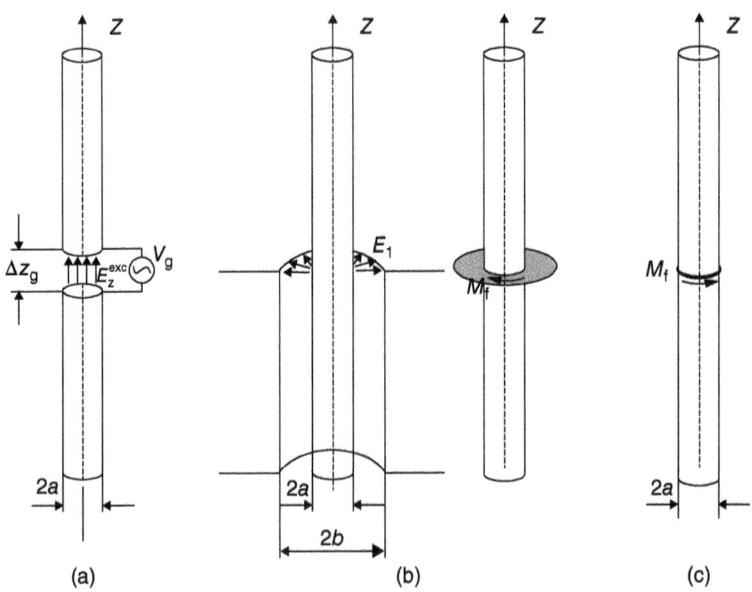


**Figure 2.61** Antenna excitation: (a) delta gap; (b) magnetic frill; (c) magnetic current loop.

explained in many books (e.g. [45, 46, 52, 57, 58]), there are not many papers analyzing MF in detail (e.g. [53, 59–65]).

Generally, in the MF models the excitation is considered as a coaxial line that terminates into a monopole over a ground plane. Assuming the field distribution over the aperture to be purely TEM, one can use the image theory and replace the ground plane and the aperture with MF as shown in Figure 2.61b.

This MF generates the electric filed on the axis of the wire exciting the antenna [66]:

$$E_z^{\text{exc}} = \frac{1}{2}\frac{V_g}{\ln\left(\frac{b}{a}\right)}\left[\frac{e^{-jk\sqrt{a^2+(z-z')^2}}}{\sqrt{a^2+(z-z')^2}} - \frac{e^{-jk\sqrt{b^2+(z-z')^2}}}{\sqrt{b^2+(z-z')^2}}\right]. \tag{2.346}$$

The use of magnetic fill generally gives more accurate input impedance results at the cost of increased computational complexity in computing the excitation vector.

### 2.5.2 Magnetic Current Loop

On the basis of DG and MF a MCL source model can be used for excitation modeling. Hence, similarly to MF, on the antenna feeding point a simple MCL, with radius $a$ (same as a wire) and magnetic current density $\vec{M}_{f,}$ is placed as shown in Figure 2.61c.

The simple magnetic current loop, with unity magnetic current density, generates axial electric field that feeds the antenna:

$$E_z(0,z) = \frac{a^2}{2}\left[jk + \frac{1}{\sqrt{a^2+(z-z')^2}}\right]\frac{e^{-jk\sqrt{a^2+(z-z')^2}}}{a^2+(z-z')^2}. \tag{2.347}$$

The axial electric field from the magnetic current loop (2.347) can be readily derived using a similar procedure as for MF explained in [66].

This type of source is connected with and can be derived from both DG and MF.

In the DG source, the electric field can be replaced with the narrow strip of equivalent magnetic current density $\vec{M}_g$:

$$\vec{M}_g = -\vec{n}\times\vec{E}_z^{\text{exc}} = -\vec{e}_\rho\times\vec{e}_z\frac{V_g}{\Delta z} = \vec{e}_\phi\frac{V_g}{\Delta z}; -\frac{\Delta z}{2} \le z' \le \frac{\Delta z}{2}, \tag{2.348}$$

as shown in Figure 2.62.

In the case when $\Delta z \to 0$ and the unit voltage is $V_g$, the axial electric field of narrow magnetic strip $\vec{M}_g$ acquires the form (2.347). For this reason the magnetic current loop is considered in [52], just as a special form of DG source (with infinitesimally small gap), despite the fact that it gives different results (particularly in the case of input admittance). Namely, some theoretical and

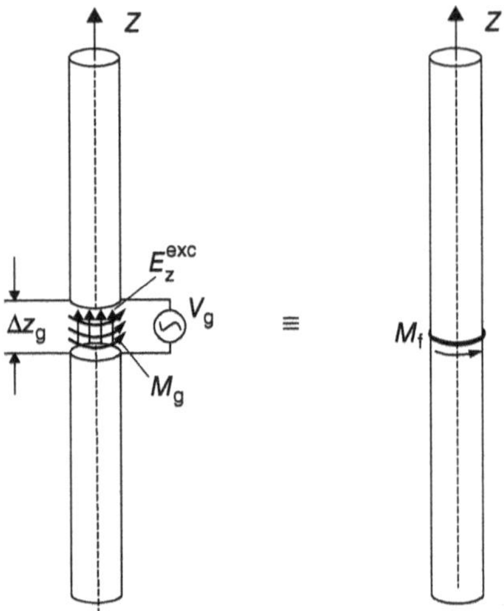


**Figure 2.62** Delta gap source and simple magnetic current loop.

mathematical aspects of DG source and magnetic current loop source, respectively, are given in [52], but without any analysis on the difference between the two approaches.

An alternative way to derive the magnetic current loop source model is to start from the MF. Namely, by letting the radius $b$ tends to $a$ in (2.346),

$$E_z(0,z) = \lim_{b\to a}\left\{\frac{1}{2}\frac{V_0}{\ln\left(\dfrac{b}{a}\right)}\left[\frac{e^{-jk\sqrt{b^2+(z-z')^2}}}{\sqrt{a^2+(z-z')^2}} - \frac{e^{-jk\sqrt{b^2+(z-z')^2}}}{\sqrt{b^2+(z-z')^2}}\right]\right\}, \tag{2.349}$$

the expression for axial electric field of magnetic current antenna (2.346) is obtained. Consequently, in [53] the magnetic current loop source as a special case of magnetic fill, to be referred as "small-frill limit," is considered. The DG and magnetic current loop source are proved to be different, particularly when approximate kernel (2.344) is used. The only case when DG and magnetic current loop give the same result is when $a \to 0$. Note that the analysis undertaken in [53] is related only to infinite-length dipole and analytical approach.

### 2.5.3 Numerical Solution

Thin wire integral equations are often treated by using some scheme of the MoM [40, 46]. Choice of basis and weight functions determines the form of MoM used. There are two classes of basis functions: subdomain and entire domain basis function. The former involves the use of basis that are applied in a repetitive manner over the segments of wire (piecewise constant –"pulse," piecewise linear (triangle), piecewise sinusoid, three-term sinusoidal) [46, 67]. Entire domain basis functions are defined over the entire length of the structure (usually Tschebyscheff, Maclaurin, Legendre, or Hermite polynomials) [46, 67]. Generally, the subdomain base functions provide more flexibility and can easily cope with the arbitrary and complex wire configurations. On the other hand, unlike the entire domain basis function, subdomain bases may suffer from some problems with discontinuity occurring at domain boundaries. The simplest choice of weight functions is related to the set of Dirac impulses leading to commonly used point-matching or collocation techniques. Although simple, the point-matching approach has a drawback due to relatively poor convergence rate and kernel quasi-singularities [2, 46, 67]. On the other hand, the Galerkin approach where the weight functions are the same as the basis functions provides more accurate and stable results [46, 67].

In contrast to the usual approach featuring MoM, the approach proposed in [2, 51] numerically handles the Pocklington's integro-differential equation (2.343) via GB-IBEM approach [2], being a combination of classical boundary methods for solving integral equations and some numerical techniques originated from finite element method (FEM). In particular, the approach reported in [31] deals with isoparametric elements [2].

The operator form of Equation (2.343) can be symbolically written as

$$K(I) = E, \tag{2.350}$$

where $K$ is the linear operator, and $I$ is the unknown current to be found for a given excitation $E$.

The unknown current is expressed by the sum of a finite number of linearly independent basis functions $\{f_n\}$, with unknown complex coefficients $I_n$:

$$I(z') = \sum_{n=1}^{N_g} I_n f_n(z'). \tag{2.351}$$

Applying the weighted residual approach with the Galekin–Bubnov procedure (same test and basis functions) the operator equation (2.350) transforms into a system of algebraic equations. After performing certain mathematical manipulations, including integration by parts follows the weak formulation of Pocklington's integro-differential equation (2.343):

$$\sum_{n=1}^{N_g} I_n \left[ -\int_{-\frac{L}{2}}^{\frac{L}{2}} \int_{-\frac{L}{2}}^{\frac{L}{2}} \frac{\partial f_n(z')}{\partial z'} \cdot \frac{\partial f_m(z)}{\partial z} g_0(z,z')\mathrm{d}z'\,\mathrm{d}z \right.$$
$$\left. + k^2 \int_{-\frac{L}{2}}^{\frac{L}{2}} \int_{-\frac{L}{2}}^{\frac{L}{2}} f_n(z') \cdot f_m(z) \cdot g_0(z,z')\mathrm{d}z'\,\mathrm{d}z \right]$$
$$= -j4\pi \frac{k}{Z_0} \int_{-\frac{L}{2}}^{\frac{L}{2}} E_z^{\mathrm{exc}}(z) f_m(z)\mathrm{d}z \quad j = 1, 2 \ldots, N_g. \tag{2.352}$$

Discretizing the antenna into wire segments results in the following matrix equation:

$$\sum_{i=1}^{N_e} [Z]_{ji}\{I\}_i = \{V\}_j, \quad j = 1, 2, \dots, N_e, \tag{2.353}$$

where the vector $\{I\}$ contains the unknown coefficients of the solution, i.e. unknown current, and $N_e$ is total number of elements. Mutual impedance $[Z]_{ji}$ represents the interaction of the $i$th source boundary element with $j$th observation boundary element. Choosing isoparametric boundary elements, the mutual impedance matrix acquires the following form:

$$\begin{aligned}[Z]_{ji} = &-\int_{-1}^{1}\int_{-1}^{1} \{D\}_j\{D'\}_i^T g_0(z,z')\frac{dz'}{d\zeta'}d\zeta'\frac{dz}{d\zeta}d\zeta \\ &+ k^2\int_{-1}^{1}\int_{-1}^{1} \{f\}_j\{f'\}_i^T g_0(z,z')\frac{dz'}{d\zeta'}d\zeta'\frac{dz}{d\zeta}d\zeta.\end{aligned} \tag{2.354}$$

Matrices $\{f\}$ and $\{f'\}$ contain the shape functions while $\{D\}$ and $\{D'\}$ contain their derivatives. The excitation vector $\{V\}_j$ represents the voltage along the segments and is given as follows:

$$\{V\}_j = -j4\pi\frac{k}{Z_0}\int_{-1}^{1} E_z^{exc}(\zeta)\{f\}_j\frac{dz}{d\zeta}d\zeta. \tag{2.355}$$

The local approximation for current at the segment is given by $f(\zeta)$

$$I^e(\zeta) = \sum_{i=1}^{nl} I_i f_i(\zeta) = \{f\}_n^T\{I\}_n, \tag{2.356}$$

where $n_l$ is the number of local nodes on the segment.

A transformation function $z(\zeta)$ defined over the element can be expressed in terms of the element approximating function:

$$z(\zeta) = \sum_{i=1}^{nl} z_i f_i(\zeta). \tag{2.357}$$

Thus, such a transformation in which the same families of approximation functions is used for the unknown quantity and also for the element transformation itself, is referred to be isoparametric. In this section, two families of shape functions over the elements are considered: linear and quadratic, respectively:

$$f_1(\zeta) = \frac{1}{2}(1-\zeta) \quad f_2(\zeta) = \frac{1}{2}(1+\zeta), \tag{2.358}$$

$$f_1(\zeta) = \frac{1}{2}\zeta(\zeta-1); \quad f_2(\zeta) = 1-\zeta^2; \quad f_3(\zeta) = \frac{1}{2}\zeta(\zeta+1). \tag{2.359}$$

In the case of the DG source the excitation vector (2.355) is readily obtained in the close form, as the excitation function (2.345) is constant across the gap, while in the case of MF and magnetic current loop a careful numerical integration should be performed, especially because of the very steep characteristic of the fields generated by the MF (2.346) and magnetic loop (2.347).

### 2.5.4 Numerical Results

Three different antenna lengths are analyzed, $L = 0.5\lambda$, $L = 0.75\lambda$, and $L = \lambda$, respectively, with radius $a = 0.007022\lambda$ resulting in values of the parameter $\Omega$ ($\Omega = 2\log(L/a)$) of 8.53, 9.34, and 9.92, respectively. The antennas are excited at the center via the DG, MF, and magnetic current loop (MA), respectively, with unit voltage. Different numerical techniques are used: GB-IBEM featuring linear (GB-IBEM ISO2) and quadratic (GB-IBEM ISO3) isoparametric approximation, widely used Galerkin version of MoM featuring piecewise triangle functions. Also, some of the results are obtained via NEC, which uses point-matching technique with tree-term sinusoidal basis functions. For the purpose of comparison, the input admittance of the center-fed dipoles is calculated using different numerical methods and with different excitations. As shown in [51], the input admittance is the most sensitive parameter in the antenna analysis, particularly regarding the impressed excitation and numerical solution. Special attention is dedicated to the numerical stability, efficiency, and accuracy of the solution in terms of number of elements and whether odd or even number of elements is used. Namely, the use of DG excitation (for center-fed) is only possible with odd number of elements (in both GB-IBEM and Galerkin MM) since it is placed on the whole central element. On the other hand, the use of the MF and magnetic loop does not require such a restriction as these source models can be placed either at the center of the central element (odd number of elements) or between two elements (on the central global node) in the case of even number of elements, as illustrated in Figure 2.63.

Figures 2.64–2.66 show the input admittance of the half wave dipole calculated via the GB-IBEM ISO2, GB-IBEM ISO3, and MoM, respectively. The obtained results are compared with the measurement data reported in [31]. All three numerical methods produce satisfactory results for all range of element numbers. The GB-IBEM (ISO2 and ISO3) seem to give slightly better results compared to MoM, with the proviso that the difference between ISO2 and ISO3 does not justify the use of ISO3, which is noticeably more demanding in both the mathematical and computational sense.

Concerning the use of different excitation models, it is clear that the DG gives the worst results with lowest rate of convergence, while MF and MA ensure better results. It is the case particularly when GB-IBEM is used with even number of elements. Namely, then the current at the dipole center

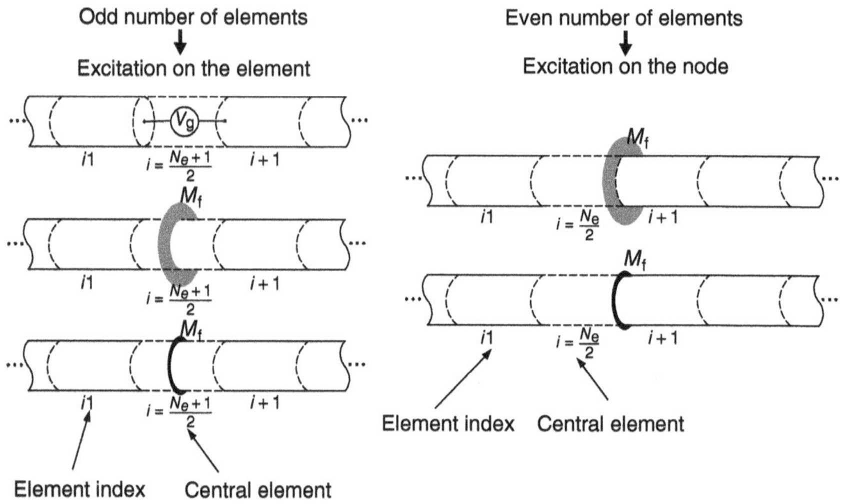


**Figure 2.63** Excitation placement.

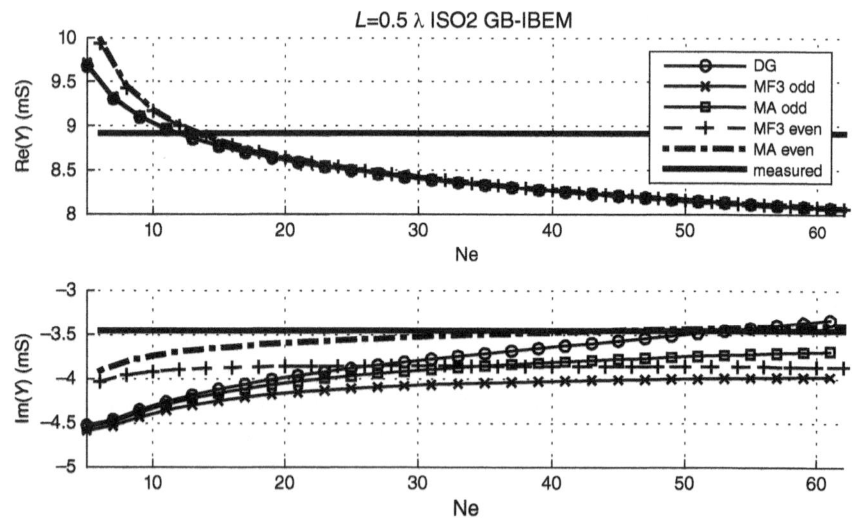


**Figure 2.64** Antenna input admittance for various number of elements (GB-IBEM ISO2; $L = 0.5\lambda$; $a = 0.007022\lambda$).

is much more independent of the elements number. Also, the difference between the methods (numerical technique, excitation and parity of elements) is more expressed in the imaginary part (susceptance) than in the real part (conductance) of admittance.

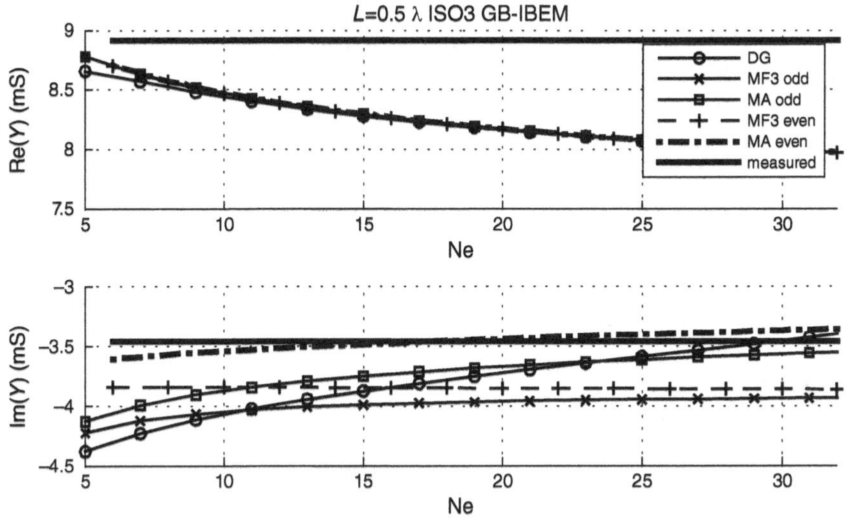


**Figure 2.65** Antenna input admittance for various number of elements (GB-IBEM ISO3; $L = 0.5\lambda$; $a = 0.007022\lambda$).

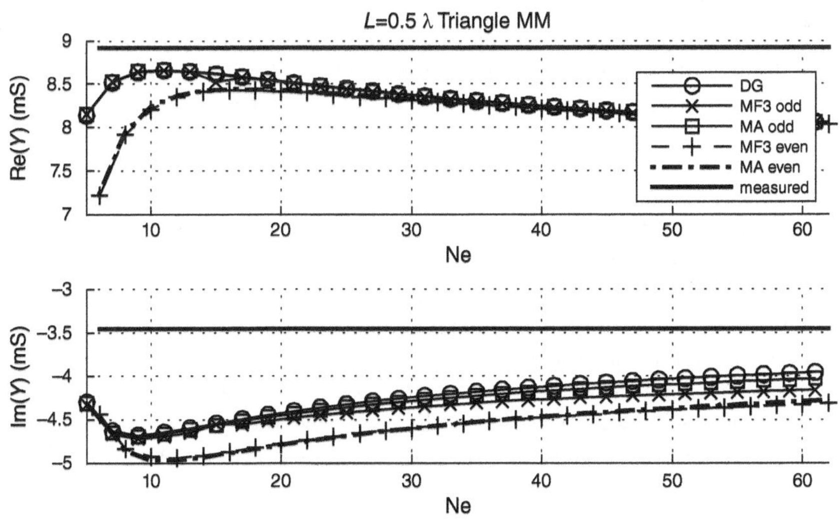


**Figure 2.66** Antenna input admittance for various number of elements (MM; $L = 0.5\lambda$; $a = 0.007022\lambda$).

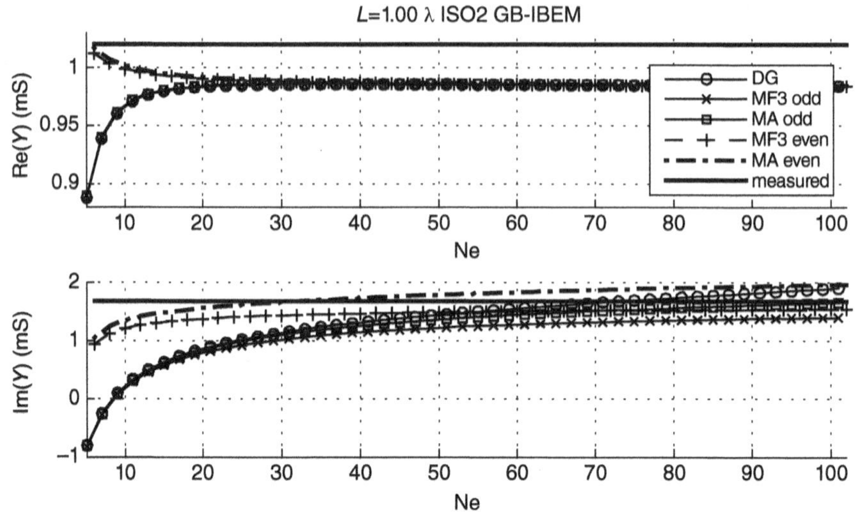


**Figure 2.67** Antenna input admittance for various number of elements (GB-IBEM ISO2; $L = 1.00\lambda$; $a = 0.007022\lambda$).

Similar conclusion can be drown out from the results in Figures 2.67–2.69 presenting the input admittance for full wave dipole calculated with GB-IBEM ISO2, GB-IBEM ISO3, and MoM, respectively.

Table 2.1 presents numerical and experimental results for the admittance of three dipole antennas of different lengths ($L = 0.5\lambda$, $L = 0.75\lambda$, and $L = \lambda$) and fixed radius. The number of elements used in all numerical calculations is set to be the maximal allowed with $\Delta z < 5a$. This condition aims to avoid the so-called "pancake element" effect that can occur if the element becomes comparable with the wire radius [51]. On the other hand, it is sufficiently large for gaining convergence.

The results from Table 2.1 show that the best method for dipole admittance assessment is the GB-IBEM ISO2 method with the magnetic current loop excitation and even number of elements. Furthermore, the worst results are obtained using NEC as it features the point-matching version of MoM. Also, the choice of excitation and parity of elements affects the antenna susceptance largely then the conductance. The results also prove the MA excitation model to numerically behave in a rather different way than traditional DG.

Table 2.2 presents the values of admittances of several monopole antennas calculated using the GB-IBEM ISO2 numerical technique as it provides the best solution. Two types of excitations are examined: the MF ($b/a = 3$) and magnetic current loop, respectively, and the results are compared with those obtained by measurements [68]. Measurement setup included isolated monopole antennas driven by coaxial line ($b/a = 3$). In this case, the model

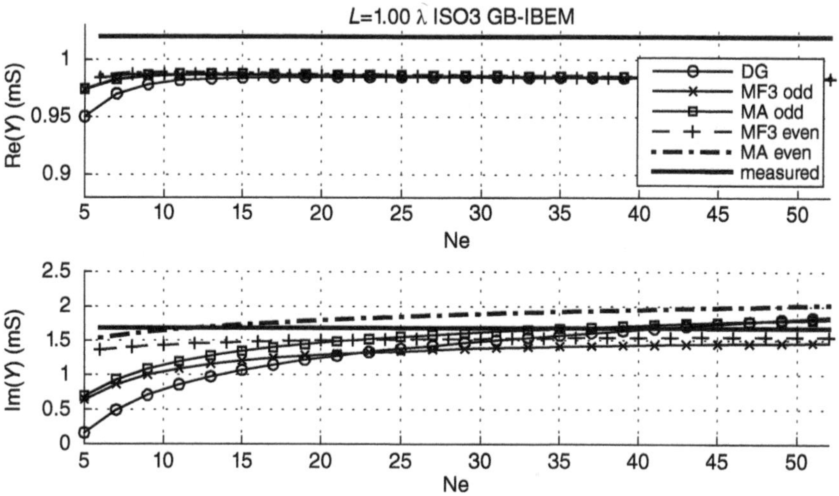


**Figure 2.68** Antenna input admittance for various number of elements (GB-IBEM ISO3; $L = 1.00\lambda$; $a = 0.007022\lambda$).

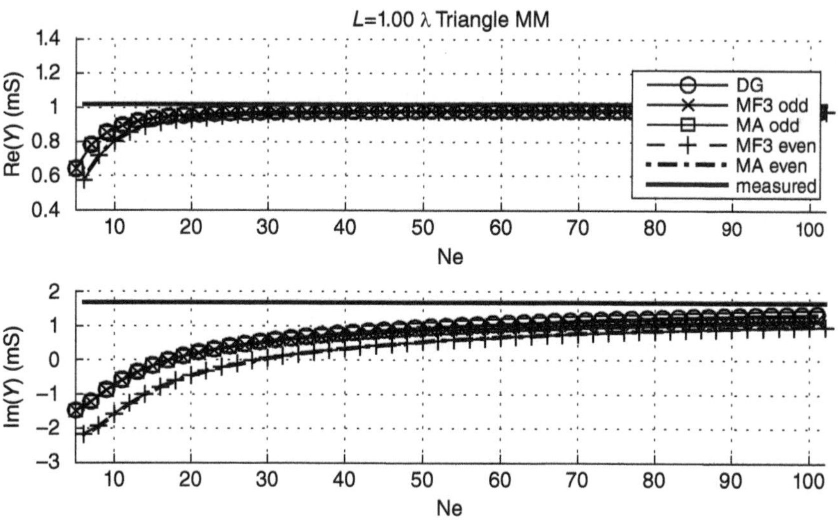


**Figure 2.69** Antenna input admittance for various number of elements (MM; $L = 1.00\lambda$; $a = 0.007022\lambda$).

**Table 2.1** Admittance of dipoles calculated by different numerical techniques.

| $a = 0.007022\lambda$ | | $L = 0.5\lambda$ | $L = 0.75\lambda$ | $L = \lambda$ |
|---|---|---|---|---|
| Measured [31][a)] | | 8.92−$i$3.46 | 1.58−$i$0.18 | 1.02+$i$1.68 |
| GB-IBEM ISO2 | DG | 8.86−$i$4.18 | 1.54−$i$0.77 | 0.99+$i$1.07 |
| | MF odd | 8.87−$i$4.27 | 1.54−$i$0.85 | 0.99+$i$0.99 |
| | MA odd | 8.86−$i$4.22 | 1.54−$i$0.80 | 0.99+$i$1.04 |
| | MF even | 8.88−$i$3.81 | 1.54−$i$0.40 | 0.99+$i$1.49 |
| | MA even | 8.89−$i$3.66 | 1.55−$i$0.25 | 0.99+$i$1.65 |
| GB-IBEM ISO3 | DG | 8.34−$i$3.95 | 1.52−$i$0.41 | 0.98+$i$1.43 |
| | MF odd | 8.35−$i$3.94 | 1.52−$i$0.43 | 0.99+$i$1.43 |
| | MA odd | 8.36−$i$3.79 | 1.52−$i$0.26 | 0.99+$i$1.58 |
| | MF even | 8.32−$i$3.75 | 1.52−$i$0.26 | 0.98+$i$1.61 |
| | MA even | 8.32−$i$3.50 | 1.52−$i10^{-5}$ | 0.99+$i$1.87 |
| MM | DG | 8.64−$i$4.60 | 1.51−$i$1.30 | 0.97+$i$0.47 |
| | MF odd | 8.62−$i$4.63 | 1.51−$i$1.34 | 0.97+$i$0.44 |
| | MA odd | 8.65−$i$4.62 | 1.51−$i$1.32 | 0.97+$i$0.46 |
| | MF even | 8.41−$i$4.92 | 1.49−$i$1.83 | 0.96−$i$0.02 |
| | MA even | 8.42−$i$4.93 | 1.49−$i$1.84 | 0.96−$i$0.03 |
| NEC | DG | 8.25−$i$4.36 | 1.51−$i$6.99 | 0.98+$i$1.14 |
| | Biconical | 7.12−$i$3.02 | 1.24−$i$1.44 | 0.81+$i$1.68 |

a) Measurement results are obtained from the measured results for admittance of isolated monopole antennas driven by coaxial line by correcting them in order to approximately take into account the influence on the antenna susceptance of the specific coaxial line used in measurements in contrast to the delta gap excitation [31].

**Table 2.2** Admittance of monopoles calculated using different excitation (GB-IBEM ISO2).

| $a = 3.175$ mm; $f = 663.5$ MHz | $L = 0.25$ | $L = 0.375\lambda$ | $L = 0.5\lambda$ | $L = 0.625\lambda$ |
|---|---|---|---|---|
| Measured [31][a)] | 17.84−$i$7.5 | 3.16−$i$0.93 | 2.05+$i$2.78 | 2.96+$i$7.86 |
| MF $b/a = 3$ | 17.76−$i$7.76 | 3.09−$i$0.93 | 1.98+$i$2.85 | 2.89+$i$8.01 |
| MA | 17.77−$i$7.32 | 3.09−$i$0.49 | 1.98+$i$3.29 | 2.89+$i$8.45 |

a) Measurement results are obtained from the measured results for admittance of isolated monopole antennas driven by coaxial line $b/a = 3$.

with the MF excitation clearly gives better results as it fully describes the coaxial line excitation compared to the magnetic current loop.

The next set of results show input admittance for half wave monopole with radius $a = 0.0245\lambda$ and $a = 0.0509\lambda$, respectively. The monopole in this example is modeled as a center-fed dipole. The input admittance is calculated for different excitation models using GB-IBEM ISO2 with various number of elements, as it was proved to be optimal. The results are compared with the measured results from [47] and [69]. As it can be seen from Figures 2.70 and 2.71 all examined excitation models provide satisfactory results for the antenna input conductance, despite the antenna radius. On the other hand, it is also noticed that the magnetic current loop source gives the best results for susceptance, despite the radius of antenna. Namely, DG source, for thinner radius, provides satisfactory results only for lower number of elements, while for the higher radius, it becomes completely off the chart. MF provides reasonably good results although slightly worse when compared to the magnetic current loop approach. All the results presented show that the magnetic current loop excitation model ensures more accurate results than the classical DG source. Also, it is the best choice for GB-IBEM excitation simulations of antennas that are not driven by a known coaxial line. On the other hand, monopole antennas driven by a known coaxial line are best simulated using MF, despite the fact that the magnetic current loop excitation also gives very good results.

The presented analysis also shows numerical stability of used GB-IBEM. Namely, despite the fact that the $L/a$ (length/radius) ratio of the last two

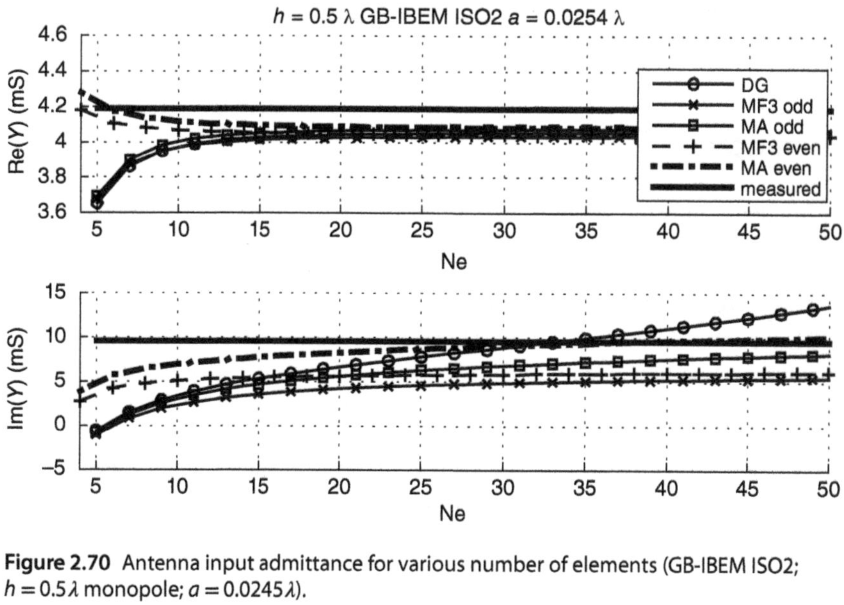


**Figure 2.70** Antenna input admittance for various number of elements (GB-IBEM ISO2; $h = 0.5\lambda$ monopole; $a = 0.0245\lambda$).

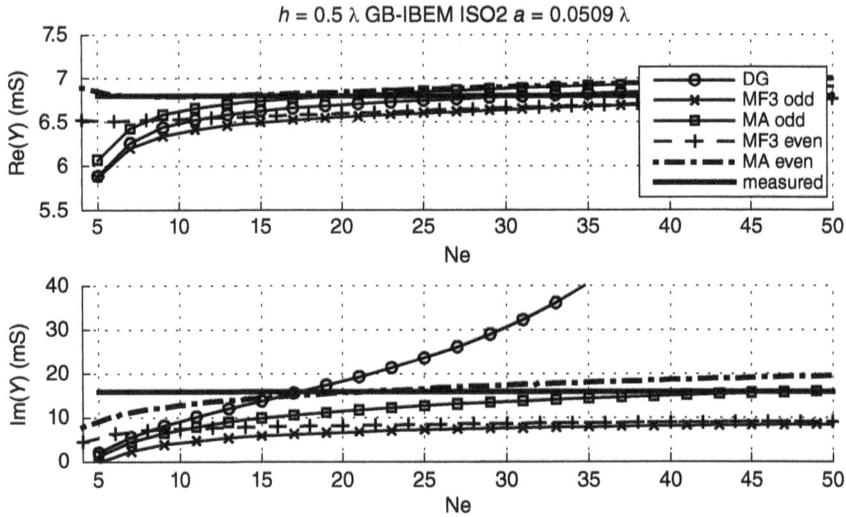


**Figure 2.71** Antenna input admittance for various number of elements (GB-IBEM ISO2; $h = 0.5\lambda$ monopole; $a = 0.0509\lambda$).

examples (especially the second one) is not large (19.68 and 9.82 respectively) and the use of reduced kernel, the GB-IBEM provides stable and accurate results for the input admittance for rather wide range of elements when DG is not used. Problems with "pancake element" effect start to emerge only for rather high number of elements (e.g. 130 elements for the case of monopole with $h = 0.5\lambda$ and $a = 0.0509\lambda$).

## References


1 Tesche, F., Ianoz, M., and Karlsson, T. (1997). *EMC Analysis Methods and Computational Models.* New York: Wiley.

2 Poljak, D. (2007). *Advanced Modeling in Computational Electromagnetic Compatibility.* New York: Wiley.

3 Tkatchenko, S., Rachidi, F., and Ianoz, M. (1995). Electromagnetic field coupling to a line of a finite length: theory and fast iterative solutions in frequency and time domains. *IEEE Transactions on Electromagnetic Compatibility* **37** (4): 509–518.

4 Tkatchenko, S., Rachidi, F., and Ianoz, M. (2001). High frequency electromagnetic field coupling to long terminated lines. *IEEE Transactions on Electromagnetic Compatibility* **43** (2): 117–129.

5 Ianoz, M. (1999). Electromagnetic field coupling to lines, cables and networks, review of problems and solutions. Proceedings of the International

Conference on Electromagnetics in Advanced Applications, ICEAA'95, Turin, Italy (12–15 September 1999), 75–80.

6 Arreghini, F., Ianoz, M., and Rachidi, F. (1999). Frequency and time domain applications in EMP coupling; A comparison between different methods of calculations, electromagnetic field coupling to lines, cables and networks, a review of problems and solutions. Proceedings of the International Conference on Electromagnetics in Aerospace Applications, ICEAA'95, Turin, Italy (17–20 September 1991), 209–214.

7 Degauque, P. and Zeddam, A. (1998). Remarks on the transmission approach to determining the current induced on above-ground cables. *IEEE Transactions on Electromagnetic Compatibility* **30** (1): 77–80.

8 Poljak, D., Rachidi, F., and Tkachenko, S. (2007). Generalized form of telegrapher's equations for the electromagnetic field coupling to finite length lines above a lossy ground. *IEEE Transactions on Electromagnetic Compatibility* **49** (3): 689–697.

9 Poljak, D., Doric, V., Rachidi, F. et al. (2009). Generalized form of telegrapher's equations for the electromagnetic field coupling to buried wires of finite length. *IEEE Transactions on Electromagnetic Compatibility* **51** (2): 331–337.

10 Poljak, D., Shoory, A., Rachidi, F. et al. (2012). Time-domain generalized form of telegrapher's equations for the electromagnetic field coupling to finite length lines above a Lossy ground. *IEEE Transactions on Electromagnetic Compatibility* **54** (1): 218–224.

11 Poljak, D., Sesnic, S., El Khamlichi Drisii, K., et al. (2017). Submitted to EPSR.

12 Agrawal, A.K., Price, H.J., and Gurbaxani, S.H. (1980). Transient response of a multiconductor transmission line excited by a nonuniform electromagnetic field. *IEEE Transactions on Electromagnetic Compatibility* **EMC-22** (2): 119–129.

13 Burke, G.J., Poggio, A.J., Logan, I.C., and Rockway, J.W. (1979). Numerical electromagnetics code – a program for antenna system analysis. Proceedings of the EMC '79 Symposium, Rotterdam, 89–94.

14 Grcev, L.D. and Menter, F.E. (1996). Transient electromagnetic fields near large earthing systems. *IEEE Transactions on Magnetics* **32**: 1525–1528.

15 Sunde, E.D. (1968). *Earth Conduction Effects in Transmission Systems*. New York: Dover Publication.

16 Miller, E.K., Poggio, A.J., Burke, G.J., and Selden, E.S. (1972). Analysis of wire antennas in the presence of a conducting half-space, Part II: the horizontal antenna in free space. *Canadian Journal of Physics* **50**: 2614–2627.

17 Poljak, D., Antonijevic, S., El Khamlichi Drissi, K., and Kerroum, K. (2010). Transient response of straight thin wires located at different heights above a ground plane using antenna theory and transmission line approach. *IEEE Transactions on Electromagnetic Compatibility* **52**: 108–116.

18 Rachidi, F. and Tkatchenko, S. ed. (2008). *Electromagnetic Field Interaction with Transmission Lines: From Classical Theory to HF Radiation Effects*. Southampton: WIT Press.

19 Burke, G.J. (1992). Numerical Electromagnetics Code NEC-4 Method of Moments. Lawrence Livermore National Laboratory. *Tech. Rep. UCRL-MA-109338*.

20 Rachidi, F., Nucci, C.A., Ianoz, M., and Mazzetti, C. (1996). Influence of a Lossy ground on lightning-induced voltages on overhead lines. *IEEE Transactions on Electromagnetic Compatibility* **38**: 250–263.

21 Poljak, D. and Kovac, N. (2009). Time domain modeling of a thin wire in a two-media configuration featuring a simplified reflection/transmission coefficient approach. *Engineering Analysis with Boundary Elements* **33**: 283–293.

22 Poljak, D., Sesnic, S., El Khamlichi Drissi, K. et al. (2016). Transient electromagnetic field coupling to buried thin wire configurations: antenna models versus transmission line approach in the time domain. *International Journal of Antennas and Propagation* 1–11.

23 Poljak, D., Doric, V., Antonijevic, S., et al. (2007). Electromagnetic field coupling to overhead wires: comparison of wire antenna and transmission line model in the frequency and time domain. Proceedings of the ICECOM 2007, Dubrovnik, Croatia (September 2007), 119–122.

24 Boegard, J.C., Tijhuis, A.G., and Klaasen, J.A. (1998). Electromagnetic excitation of a thin wire: a travelling wave approach. *IEEE Transactions on Antennas and Propagation* **48** (1): 1202–1211.

25 Poljak, D., Dorić, V., and Šesnić, S., (2009). Coupling from HF transmitter to power line communications system using antenna theory – analytical versus numerical approach. XV International Symposium on Theoretical Electrical Engineering ISTET'09, Lubeck, Germany.

26 Poljak, D., El Khamlichi Drissi, K., Kerroum, K., and Sesnic, S. (2011). Comparison of analytical and boundary element modeling of electromagnetic field coupling to overhead and buried wires. *Engineering Analysis with Boundary Elements* **35**: 555–563.

27 Takashima, T., Nakae, T., and Ishibasi, R. (1980). Calculation of complex fields in conducting media. *IEEE Transactions on Dielectrics and Electrical Insulation* **15**: 1–7.

28 Wallart, J., El Khamlichi Drissi, K., and Paladian, F. (1998). Study of propagation constant for a single buried wire in a lossy ground. Proceedings of the EMC '98 Symposium, vol. 2, Roma (September 1998), 557–562.

29 El Khamlichi Drissi, K., Kerroum, K., Paladian, F., and Wallart, J. (1997). Time domain study of Lossy multiconductor transmission line with nonlinear loads. EMC Zurich 97, Zurich (February 1997), 391–396.

30 Poljak, D., Tham, C.Y., McCowen, A., and Roje, V. (2000). Transient analysis of two coupled horizontal wires over a real ground. *IEE Proceedings - Microwaves, Antennas and Propagation* **147**: 87–94.
31 Fernandez Pantoja, M., Yarovoy, A.G., Rubio Bretones, A., and González García, S. (2009). Time domain analysis of thin wire antennas over Lossy ground using the reflection-coefficient approximation. *Radio Science* **44**.
32 Bourgeois, J.M. and Smith, G. (1996). A fully three-dimensional simulation of a ground-penetrating radar: FDTD theory compared with experiment. *IEEE Transactions on Geoscience and Remote Sensing* **34** (1): 36.
33 Gürel, I. and Oguz, U. (2000). Three-dimensional FDTD modeling of a ground-penetrating radar. *IEEE Transactions On Geoscience and Remote Sensing* **38** (4).
34 Poljak, D. and Dorić, V. (2015). Transmitted field in the Lossy ground from ground penetrating radar (GPR) dipole antenna. 17th International Conference on Computational Methods and Experimental Measurements (CMEM 2015), Opatija, Croatia (May 2015).
35 King, R.W.P. (1999). The electric field induced in the human body when exposed to electromagnetic fields at 1–30 MHz on shipboard. *IEEE Transactions on Biomedical Engineering* **46** (6): 747–751.
36 King, R.W.P. (1998). Fields and currents in the organs of the human body when exposed to power lines and VLF transmitters. *IEEE Transactions on Biomedical Engineering* **45** (4): 520–530.
37 Chang, D. and Wait, J. (1970). Theory of a vertical tubular antenna located above a conducting half-space. *IEEE Transactions on Antennas and Propagation* **18** (2): 182–188.
38 Poljak, D. and Doric, V. (2005). Time-domain modeling of electromagnetic field coupling to finite-length wires embedded in a dielectric half-space. *IEEE Transactions on Electromagnetic Compatibility* **47** (2): 247–253.
39 Poljak, D., Sesnic, S., Cavka, D., and El Khamlichi Drissi, K. (2015). On the use of the vertical straight wire model in electromagnetics and related boundary element solution. *Engineering Analysis with Boundary Elements* **50**: 19–28.
40 Harrington, R.F. (1967). Matrix methods for field problems. *Proceedings of the IEEE* **55** (2): 136–149.
41 Poljak, D., Doric, V., El Khamlichi Drissi, K., et al. (2012). Vertical dipole above a PEC ground representing a part of a PLC system – numerical versus analytical study. 2012 International Conference on Software, Telecommunications and Computer Networks, Split.
42 Li, X., El Khamlichi Drissi, K., and Paladian, F. (2004). Insulated vertical antennas above ground. *IEEE Transactions on Antennas and Propagation* **52** (1): 321–324.

43 Hallen, E. (1938). Theoretical investigations into the transmitting and receiving qualities of antennae. *Nova acta Regiae Societatis Scientiarum Upsaliensis* **11**: 3–44.

44 Pocklington, H.C. (1897). Electrical oscillations in wires. *Proceedings of the Cambridge Philosophical Society* **9**: 324–333.

45 Balanis, C.A. (2005). *Antenna Theory: Analysis and Design*. Hoboken, NJ: Wiley.

46 Gibson, W.C. (2008). *A Methods of Moments in Electromagnetics*. Chapman & Hall/CRC.

47 Junker, G.P., Kishk, A.A., and Glisson, A.W. (1995). A novel delta gap source model for center fed cylindrical dipoles. *IEEE Transactions on Antennas and Propagation* **43** (5): 537–540.

48 Werner, D.H. (1998). A method of moments approach for the efficient and accurate modeling of moderately thick cylindrical wire antennas. *IEEE Transactions on Antennas and Propagation* **46** (3): 373–382.

49 Braaten, B.D. and Nelson, R.M. (2008). An RGF delta-gap source for thin-wire EFIE. *IEEE Transactions on Electromagnetic Compatibility* **50** (1): 212–215.

50 Popovic, B.D. (1973). Thin monopole antenna: finite-size belt-generator representation of coaxial-line excitation. *Proceedings of the IEE* **120**: 544–550.

51 Cavka, D. and Poljak, D. (2015). Magnetic current loop as a source model for finite thin wire antennas. *International Journal of Numerical Modelling* **28**: 189–200.

52 Popovic, B.D., Dragovic, M.B., and Djordjević, A.R. (1982). *Analysis and Synthesis of Wire Antennas*. Wiley.

53 Fikioris, G. and Valagiannopoulos, C.A. (2005). Input admittances arising from explicit solutions to integral equations for infinite-length dipole antennas. *Progress in Electromagnetics Research, PIER* **55**: 285–306.

54 Otto, D.V. (1967). Cylindrical antenna theory. PhD dissertation. University of Auckland, New Zealand.

55 Otto, D.V. (1967). The admittance of cylindrical antennas driven from a coaxial line. *Radio Science (New Series)* **2** (9): 1031–1042.

56 Otto, D.V. (1968). Fourier transform method in cylindrical antenna theory. *Radio Science (New Series)* **3** (11): 1050–1057.

57 Kolundzija, B.M. and Djordjević, A.R. (2002). *Electromagnetic Modeling of Composite Metallic and Dielectric Structures*. Boston, London: Artech House.

58 Stutzman, W.L. and Thiele, G.A. (1981). *Antenna Theory and Design*. Hoboken, NJ: Wiley.

59 Fikioris, G., Lionas, G., and Lioutas, C.G. (2003). The use of the frill generator in thin-wire integral equations. *IEEE Transactions on Antennas and Propagation* **51** (8): 1847–1854.

60 Geranmayeh, A., Moini, R., Sadeghi, S.H.H., and Deihimi, A. (2006). A fast wavelet-based moment method for solving thin-wire EFIE. *IEEE Transactions on Magnetics* **42** (4): 575–578.

61 Chen, N.-W. (2007). A magnetic frill source model for time-domain integral-equation based solvers. *IEEE Transactions on Antennas and Propagation* **55** (11): 3093–3098.

62 Richmond, J.H. (1979). Admittance of infinitely long cylindrical wire with finite conductivity and magnetic-frill excitation. *IEEE Transactions on Antennas and Propagation* **ap-27** (2): 264–266.

63 Papakanellos, P.J., Fikioris, G., and Michalopoulou, A. (2010). On the oscillations appearing in numerical solutions of solvable and nonsolvable integral equations for thin-wire antennas. *IEEE Transactions on Antennas and Propagation* **58** (5): 1635–1644.

64 Jensen, M.A. and Rahmat-Samii, Y. (1994). Electromagnetic characteristics of superquadric wire loop antennas. *IEEE Transactions on Antennas and Propagation* **42** (2): 246–249.

65 Zhou, G. and Smith, G.S. An accurate theoretical model for the thin-wire circular half-loop antenna. *IEEE Transactions on Antennas and Propagation* **39** (8).

66 Tsai, L.L. (1972). A numerical solution for the near and far fields of an annular ring of magnetic current. *IEEE Transactions on Antennas and Propagation* **20** (5).

67 Miller, E.K. (1988). A selective survey of computational electromagnetics. *IEEE Transactions on Antennas and Propagation* **36** (9): 1281–1305.

68 Mack, R.B. (1963). A Study of Circular Arrays. Cruft Lab., Harvard University. *Tech. Rep. 382 and 383.*

69 King, R.W.P. (1971). *Tables of Antenna Characteristics.* New York: Plenum.

# 3

# Electromagnetic Field Coupling to Overhead Wires

Electromagnetic field coupling to overhead wire configurations of arbitrary shape is of great practical interest for many EMC applications [1–11], such as transient excitation of antennas, ground penetrating radar (GPR) power, or communications cables. The electromagnetic field coupling to finite length overhead wires can be analyzed via the transmission line (TL) model or the thin wire antenna theory (AT) in either frequency or time domain [1]. In particular, the transient response of a wire configuration of interest can be computed directly by solving the corresponding time domain equations, or by the indirect approach, i.e. by solving their frequency domain counterpart. Within the framework of the indirect approach the frequency spectrum has to be determined, and then the transient response is evaluated by using the inverse Fourier transform (IFT).

Many practical engineering problems pertaining to the electromagnetic field coupling to thin wires can be analyzed by using the TL models [1–6] including the study of the incident electromagnetic field exciting the line and the propagation of induced currents and voltages along the line.

The TL models give valid results provided the line length is appreciably larger than the separation between the wires, and also larger than the actual height above ground [8].

On the other hand, the TL approximation cannot provide a complete solution for the excitation of a given wire configuration by an incident field if the wavelength of the electromagnetic field exciting a wire structure is comparable to or less than the transverse electrical dimensions of the structure. Namely, the TL model fails to predict resonances and accounts for the presence of a lossy ground only approximately [1]. One of the serious problems with the TL approach occurs due to the fact that the current grows to infinity at resonant points as losses do not exist and there is no radiation resistance to limit its flow [8]. The full wave approach, based on the wire AT and related integral equations is more rigorous and should be used whenever the aboveground TLs of finite length are considered. However, a serious drawback of AT approach

*Computational Methods in Electromagnetic Compatibility: Antenna Theory Approach versus Transmission line Models*, First Edition. Dragan Poljak and Khalil El Khamlichi Drissi.

is the rather long computational time required for the calculations pertaining to long lines.

This chapter deals with the analysis of electromagnetic field coupling to overhead wires in either frequency or time domain by using both antenna model and TL approach, respectively.

Finally, a number of illustrative computational examples regarding electromagnetic coupling to overhead wires is given in this chapter.

The aboveground wires are subjected to electromagnetic fields arriving from a distant source and inducing current to flow along the wires. The key to understanding the behavior of induced fields is the knowledge of current distribution induced along the wires. These currents generate scattered fields propagating away from the equipment.

This chapter is organized as follows: Section 3.2 deals with frequency domain analysis followed by related numerical solution methods for overhead wires and with a set of illustrative examples related to the aboveground lines and PLC (power line communications) systems. Section 3.3 outlines time domain analysis and related method of solutions of governing equations. Some computational examples pertaining to multiconductor aboveground lines are given.

## 3.1 Frequency Domain Models and Methods

This section deals with the wire AT and TL approximation, respectively, for the analysis of electromagnetic field coupling to overhead lines of finite length in the frequency domain. The formulation arising from the wire AT is based on the set of coupled Pocklington's integro-differential equation for half-space problems. The effect of a two-media configuration is taken into account by means of the reflection coefficient (RC) approximation [12]. The resulting integro-differential expressions are numerically handled via the frequency domain Galerkin–Bubnov scheme of the indirect boundary element method (GB-IBEM) [8].

The TL model in the frequency domain is based on the corresponding telegrapher's equations, which are handled by using the chain matrix method [10].

### 3.1.1 Antenna Theory Approach: Set of Coupled Pocklington's Equations

Modeling of arbitrarily shaped wires located at different heights above a lossy ground is an important task in both antenna and electromagnetic compatibility (EMC) studies [1].

This section firstly deals with an assessment of the current induced along multiple wire configurations above a lossy ground. Once the currents along the wire array have been obtained, the components of the radiated field could be determined.

The set of Pocklington's equations for a configuration of overhead wires can be obtained as an extension of the Pocklington's integro-differential equation for a single wire of arbitrary shape. Pocklington's equation for a single wire above a lossy ground can be derived by enforcing the continuity conditions for the tangential components of the electric field along the perfectly conducting (PEC) wire surface. First, a single wire of arbitrary shape, insulated in free space, as shown in Figure 3.1, is considered.

For the PEC wire the total field composed from the excitation field $\vec{E}^{exc}$ and scattered field $\vec{E}^{sct}$ vanishes [1, 8]:

$$\vec{e}_x \cdot (\vec{E}^{exc} + \vec{E}^{sct}) = 0 \quad \text{on the wire surface.} \tag{3.1}$$

Starting from Maxwell's equations and Lorentz gauge the scattered electric field can be expressed in terms of the vector potential $\vec{A}$:

$$\vec{E}^{sct} = -j\omega\vec{A} + \frac{1}{j\omega\mu\varepsilon}\nabla(\nabla\vec{A}). \tag{3.2}$$

The vector potential is defined by the particular integral over a given path $C$ (considered conductive wire structure):

$$\vec{A}(s) = \frac{\mu}{4\pi}\int_C I(s')g_0(s,s',s^*)\vec{s}'\vec{s}'\,\mathrm{d}s', \tag{3.3}$$

where $I(s')$ is the induced current along the line, and $g_0(s, s')$ denotes the lossless medium Green function:

$$g_0(s,s') = \frac{\mathrm{e}^{-jkR}}{R}, \tag{3.4}$$

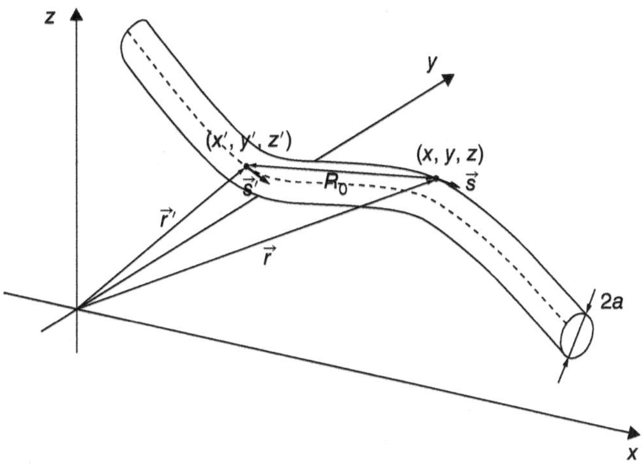


**Figure 3.1** Single wire of arbitrary shape in free space.

and $R$ is the distance from the source point to the observation point, respectively, while the propagation constant of the homogeneous medium is given by

$$k^2 = \omega^2 \mu_0 \varepsilon_0. \tag{3.5}$$

Inserting (3.3) into (3.2) yields the relation for the scattered electric field:

$$\vec{E}^{sct} = \frac{1}{j4\pi\omega\varepsilon_0} \int_{C'} I(s') \cdot \vec{s}' \cdot [k^2 + \nabla\nabla] g_0(s, s') ds'. \tag{3.6}$$

Combining (3.6) and (3.1) results in Pocklington's integral equation for the unknown current distribution along the wire of arbitrary shape insulated in free space:

$$E_{tan}^{exc}(s) = -\frac{1}{j4\pi\omega\varepsilon_0} \int_{C'} I(s') \cdot \vec{s} \cdot \vec{s}' \cdot [k^2 + \nabla\nabla] g_0(s, s') ds', \tag{3.7}$$

where $E_{tan}^{exc}$ denotes the tangential component of the electric field illuminating the wire.

The case of curved wire located above an imperfectly conducting ground can be analyzed by extending integro-differential equation (3.7) using the RC approach [12]. The geometry of an arbitrary wire and its image, respectively, is shown in Figure 3.2.

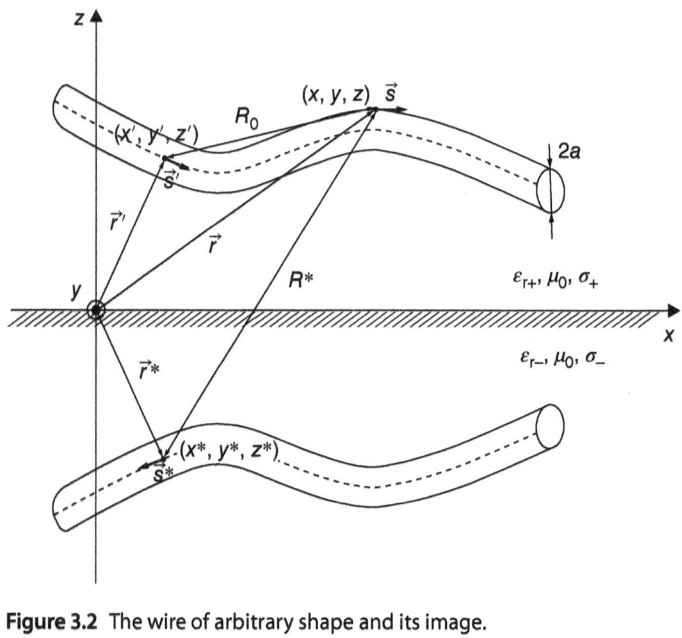


**Figure 3.2** The wire of arbitrary shape and its image.

The excitation function $E^{exc}$ is now composed from the incident and reflected field, respectively:

$$E^{exc} = E^{inc} + E^{ref}. \tag{3.8}$$

Performing certain mathematical manipulations, Pocklington's integro-differential equation for a curved wire above a lossy ground becomes [12]

$$\begin{aligned} E_s^{exc}(s) = \frac{j}{4\pi\omega\varepsilon_0} \int_0^L \Bigg\{ & \left[ k^2 \vec{e}_s \vec{e}_{s'} - \frac{\partial^2}{\partial s\, \partial s'} \right] g_0(s, s') \\ & + R_{TM} \left[ k^2 \vec{e}_s \vec{e}_{s^*} - \frac{\partial^2}{\partial s\, \partial s^*} \right] g_i(s, s^*) + \\ & + (R_{TE} - R_{TM}) \vec{e}_s \vec{e}_p \cdot \left[ k^2 \vec{e}_p \vec{e}_{s*} - \frac{\partial^2}{\partial p\, \partial s^*} \right] g_i(s, s^*) \Bigg\} I(s')\mathrm{d}s', \end{aligned} \tag{3.9}$$

where $\vec{e}_p$ is the unit vector normal to the incident plane, while $g_i(s, s^*)$ arises from the image theory and is given by

$$g_i(s, s^*) = \frac{\mathrm{e}^{-jkR^*}}{R^*}, \tag{3.10}$$

and $R^*$ is the distance from the image source point to the observation point, respectively.

An extension to the case of multiple curved wires is straightforward, i.e. it follows [12] that

$$\begin{aligned} E_{sm}^{exc}(s) = \frac{j}{4\pi\omega\varepsilon_0} \sum_{n=1}^{N_w} \int_0^{L_n} \Bigg\{ & \left[ k^2 \vec{e}_{s_m} \vec{e}_{s'_n} - \frac{\partial^2}{\partial s_m\, \partial s'_n} \right] g_{0n}(s_m, s'_n) + \\ & + R_{TM} \left[ k^2 \vec{e}_{s_m} \vec{e}_{s^*_n} - \frac{\partial^2}{\partial s_m\, \partial s^*_n} \right] g_{in}(s_m, s^*_n) \\ & + (R_{TE} - R_{TM}) \vec{e}_{s_m} \vec{e}_p \cdot \left[ k^2 \vec{e}_p \vec{e}_{s*} - \frac{\partial^2}{\partial p\, \partial s^*} \right] g_i(s_m, s^*_n) \Bigg\} \\ & I(s'_n)\mathrm{d}s', \end{aligned} \tag{3.11}$$

where $N_w$ is the total number of wires and $I_n(s'_n)$ is the unknown current distribution induced on the $n$th wire. Furthermore, $g_{0mn}(x, x')$ and $g_{imn}(s, s')$ are the Green functions of the form

$$g_{0mn}(s_m, s'_n) = \frac{\mathrm{e}^{-jkR_{1mn}}}{R_{1mn}}, \quad g_{imn}(s_m, s'_n) = \frac{\mathrm{e}^{-jkR_{2mn}}}{R_{2mn}}, \tag{3.12}$$

where $R_{1mn}$ and $R_{2mn}$ are distances from the source point and from the corresponding image, respectively to the observation point of interest.

The influence of a lossy half-space is taken into account via the Fresnel plane wave RC for TM and TE polarization, respectively [12]:

$$R'_{\mathrm{TM}} = \frac{\underline{n}\cos\theta' - \sqrt{\underline{n} - \sin^2\theta'}}{\underline{n}\cos\theta' + \sqrt{\underline{n} - \sin^2\theta'}}, \tag{3.13}$$

$$R_{\mathrm{TE}} = \frac{\cos\theta' - \sqrt{\underline{n} - \sin^2\theta'}}{\cos\theta' + \sqrt{\underline{n} - \sin^2\theta'}}, \tag{3.14}$$

where $\theta'$ is the angle of incidence and $\underline{n}$ is given by

$$\underline{n} = \frac{\varepsilon_{\mathrm{eff}}}{\varepsilon_0}, \quad \varepsilon_{\mathrm{eff}} = \varepsilon_{\mathrm{r}}\varepsilon_0 - j\frac{\sigma}{\omega}, \tag{3.15}$$

and $\varepsilon_{\mathrm{eff}}$ is the complex permittivity of the ground.

For the special case of single horizontal straight wire above a lossy half-space (Figure 3.3), integro-differential equation (3.9) simplifies into

$$E_x^{\mathrm{exc}} = j\omega\frac{\mu}{4\pi}\int_0^L I(x')g(x,x')\mathrm{d}x' - \frac{1}{j4\pi\omega\varepsilon}\frac{\partial}{\partial x}\int_0^L \frac{\partial I(x')}{\partial x'}g(x,x')\mathrm{d}x', \tag{3.16}$$

where $I(x')$ is the induced current along the horizontal wire, and $g(x,x')$ denotes the Green function given by

$$g(x,x') = g_0(x,x') - R_{\mathrm{TM}}g_i(x,x'). \tag{3.17}$$

Furthermore, if an array of multiple horizontal wires is considered (Figure 3.4), the system of equations (3.11) becomes

$$E_{xm}^{\mathrm{exc}} = -\frac{1}{j4\pi\omega\varepsilon_0}\sum_{n=1}^{N_w}\int_0^{L_n}\left[\frac{\partial^2}{\partial x^2} + k_1^2\right]g_{mn}(x,x')I_n(x')\mathrm{d}x',$$
$$m = 1, 2, \ldots, M, \tag{3.18}$$

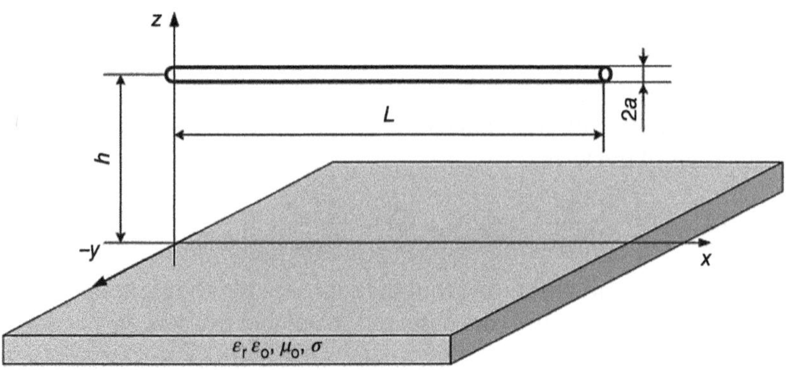


**Figure 3.3** Horizontal wire above a lossy ground.

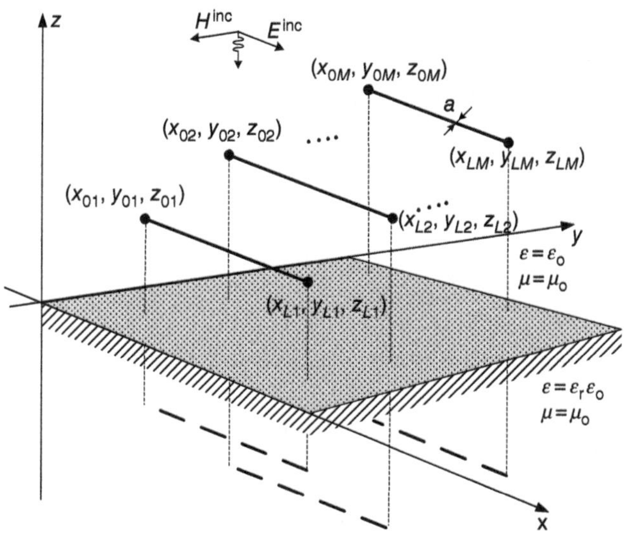


**Figure 3.4** Horizontal wires above a lossy half-space at different heights.

where $I_n(x')$ is the unknown current distribution induced along $n$th wire, $E_{xm}^{\text{exc}}$ is the known excitation field tangential to the $j$th wire surface, and $g_{mn}$ is the corresponding Green function:

$$g_{mn}(x, x') = g_{0mn}(x, x') - R'_{\text{TM}} g_{imn}(x, x'), \quad m = 1, 2, \ldots, M. \tag{3.19}$$

It is worth noting that a trade-off between the rigorous Sommerfeld integral approach and approximate RC approach is presented in [12]. Although RC approximation causes certain error (up to 10%) it takes a significantly less computational effort then a rigorous Sommerfeld approach [8].

The total electric field irradiated by configuration of multiple wires of arbitrary shape is given by [13, 14]

$$\vec{E} = \sum_{n=1}^{N_w} [\vec{E}_{0n} + R_{\text{TM}} \vec{E}_{in} + (R_{\text{TE}} - R_{\text{TM}})(\vec{E}_{in} \cdot \vec{e}_p)\vec{e}_p], \tag{3.20}$$

where

$$\vec{E}_{0n} = \frac{1}{j4\pi\omega\varepsilon_0} \begin{bmatrix} k_1^2 \int_0^{L_n} \vec{e}_{s'_n} I(s'_n) g_{0n} \left(\vec{r}, \vec{r}'\right) \mathrm{d}s'_n \\ + \int_0^{L} \dfrac{\partial I(s'_n)}{\partial s'_n} \nabla g_{0n} \left(\vec{r}, \vec{r}'\right) \mathrm{d}s'_n \end{bmatrix}, \tag{3.21}$$

$$\vec{E}_{in} = \frac{1}{j4\pi\omega\varepsilon_0} \begin{bmatrix} k_1^2 \int_0^{L_n} \vec{e}_{s_n*} I(s'_n) g_{in} \left(\vec{r}, \vec{r}^*\right) \mathrm{d}w' \\ - \int_0^{L_n} \dfrac{\partial I(s^*)}{\partial s_n^*} \nabla g_{in} \left(\vec{r}, \vec{r}^*\right) \mathrm{d}s' \end{bmatrix}. \tag{3.22}$$

Note that index 0 and $i$ are related to the source and image wire, respectively.

For the special case of single horizontal straight wire above a lossy half-space (Figure 3.3) it follows [15] that

$$E_x = \frac{1}{j4\pi\omega\varepsilon_0}\left[-\int_0^L \frac{\partial I(x')}{\partial x'}\frac{\partial g(x,x')}{\partial x'}\mathrm{d}x' + k^2\int_0^L g(x,x')I(x')\mathrm{d}x'\right], \tag{3.23}$$

$$E_y = \frac{1}{j4\pi\omega\varepsilon_0}\int_0^L \frac{\partial I(x')}{\partial x'}\frac{\partial g(x',y)}{\partial y}\mathrm{d}x', \tag{3.24}$$

$$E_z = \frac{1}{j4\pi\omega\varepsilon_0}\int_0^L \frac{\partial I(x')}{\partial x'}\frac{\partial g(x',z)}{\partial z}\mathrm{d}x'. \tag{3.25}$$

For the case of multiple horizontal wires, the expressions for electric field are given by [15]

$$E_x = \frac{1}{j4\pi\omega\varepsilon_0}\sum_{n=1}^{N_w}\begin{bmatrix} -\int_0^{L_n} \frac{\partial I_n(x')}{\partial x'}\frac{\partial g_{nm}(x,x')}{\partial x'}\mathrm{d}x' \\ +k^2\int_0^{L_n} g_{nm}(x,x')I_n(x')\mathrm{d}x' \end{bmatrix}, \tag{3.26}$$

$$E_y = \frac{1}{j4\pi\omega\varepsilon_0}\sum_{n=1}^{N_w}\int_0^{L_n} \frac{\partial I_n(x')}{\partial x'}\frac{\partial g_{nm}(x,x')}{\partial y}\mathrm{d}x', \tag{3.27}$$

$$E_z = \frac{1}{j4\pi\omega\varepsilon_0}\sum_{n=1}^{N_w}\int_0^{L_n} \frac{\partial I_n(x')}{\partial x'}\frac{\partial g_{nm}(x,x')}{\partial z}\mathrm{d}x'. \tag{3.28}$$

The radiated magnetic field of the curved wire system can be written as follows [13, 14]:

$$\vec{H} = \sum_{n=1}^{N_w}[\vec{H}_{Sn} + R_{\mathrm{TE}}\vec{H}_{In} + (R_{\mathrm{TM}} - R_{\mathrm{TE}})(\vec{H}_{In}\cdot\vec{e}_p)\vec{e}_p], \tag{3.29}$$

where

$$\vec{H}_{Sn} = -\frac{1}{4\pi}\int_0^{L_n} I(s_n')\vec{e}_{s'}\times\nabla g_{0n}(\vec{r},\vec{r}')\mathrm{d}s', \tag{3.30}$$

$$\vec{H}_{In} = -\frac{1}{4\pi}\int_0^{L_n} I(s_n')\vec{e}_{s*}\times\nabla g_{in}(\vec{r},\vec{r}^*)\mathrm{d}s'. \tag{3.31}$$

The reduction to the case of a single straight wire or straight wire array is straightforward, as in the case of the electric field given by Equations (3.23)–(3.28).

### 3.1.2 Numerical Solution

The set of Pocklington's integro-differential equations (3.11) has been solved by using the GB-IBEM. An outline of the method is given here, for the sake of completeness while the method has been presented in detail elsewhere, e.g. in [8].

Performing the Galerkin–Bubnov scheme of (GB-IBEM) in the frequency domain the set of coupled integro-differential equations (3.11) is transformed into the following matrix equation [13]:

$$\sum_{n=1}^{M}\sum_{i=1}^{N_n}[Z]_{ji}^{e}\{I\}_i^e = \{V\}_j^e, \tag{3.32}$$

where the mutual impedance matrix is given by [13]

$$\begin{aligned}
[Z]_{ij}^{e} = &-\int_{-1}^{1}\int_{-1}^{1}\{D\}_j\{D'\}_i^T g_{0nm}(s_n, s'_m)\frac{ds'_m}{d\xi'}d\xi'\frac{ds_n}{d\xi}d\xi \\
&+ k_1^2\vec{e}_{s_n}\vec{e}_{s_m}\int_{-1}^{1}\int_{-1}^{1}\{f\}_j\{f'\}_i^T g_{0nm}(s_n, s'_m)\frac{ds'_m}{d\xi'}d\xi'\frac{ds_n}{d\xi}d\xi- \\
&- R_{\mathrm{TM}}\int_{-1}^{1}\int_{-1}^{1}\{D\}_j\{D'\}_i^T g_{inm}(s_n, s^*_m)\frac{ds'_m}{d\xi'}d\xi'\frac{ds_n}{d\xi}d\xi \\
&+ R_{\mathrm{TM}}k_1^2\vec{e}_{s_n}\vec{e}_{s^*_m}\int_{-1}^{1}\int_{-1}^{1}\{f\}_j\{f'\}_i^T g_{inm}(s_n, s^*_m)\frac{ds'_m}{d\xi'}d\xi'\frac{ds_n}{d\xi}d\xi \\
&+ \frac{j}{4\pi\omega\varepsilon_0}\int_{-1}^{1}Z'_T\{f\}_j\{f'\}_j^T\frac{ds_n}{d\xi}d\xi,
\end{aligned} \tag{3.33}$$

while the voltage vector is given by [13]

$$\{V\}_j^n = -j4\pi\omega\varepsilon_0\int_{-1}^{1}E_{s_n}^{\mathrm{exc}}(s_n)f_{jn}(s_n)\frac{ds_n}{d\xi}d\xi_n. \tag{3.34}$$

Once the current distribution is obtained, the radiated field can be obtained applying the similar BEM formalism [13].

Thus, the total field is given by

$$\vec{E} = \sum_{k=1}^{N}\left[\vec{E}_{Sk}^e + R_{\mathrm{TM}}\vec{E}_{Ik}^e + (R_{\mathrm{TE}} - R_{\mathrm{TM}})(\vec{E}_{Ik}^e\cdot\vec{e}_p)\vec{e}_p\right], \tag{3.35}$$

where the field components due to a wire segment radiation are given by

$$\vec{E}_{Sk}^e = \frac{1}{j4\pi\omega\varepsilon_0}\sum_{i=1}^{n}\begin{bmatrix} k^2\int_{-1}^{1}\vec{e}_{ks'}I_{ik}^e f_i(\xi)g_{0k}\left(\vec{r},\vec{r}'\right)\frac{ds'_k}{d\xi}d\xi \\ +\int_{-1}^{1}I_{ik}^e\frac{\partial f_i(\xi)}{\partial\xi}\nabla g_{0k}(\vec{r},\vec{r}')\frac{ds'_k}{d\xi}d\xi\end{bmatrix}, \tag{3.36}$$

$$\vec{E}_I^e = \frac{1}{j4\pi\omega\varepsilon_0}\sum_{i=1}^{n}\begin{bmatrix} k^2\int_{-1}^{1}\vec{e}_{ks*}I_{ik}^e f_i(\xi)g_{ik}\left(\vec{r},\vec{r}^*\right)\frac{ds'_k}{d\xi}d\xi \\ -\int_{-1}^{1}I_{ik}^e\frac{\partial f_i(\xi)}{\partial\xi'}\nabla g_{ik}(\vec{r},\vec{r}^*)\frac{ds'_k}{d\xi}d\xi\end{bmatrix}. \tag{3.37}$$

The total magnetic field is given by [13]

$$\vec{H} = \sum_{k=1}^{N}[\vec{H}_{Sk}^e + R_{\mathrm{TE}}\vec{H}_{Ik}^e + (R_{\mathrm{TM}} - R_{\mathrm{TE}})(\vec{H}_{Ik}^e\cdot\vec{e}_p)\vec{e}_p], \tag{3.38}$$

while the magnetic field components are given by [13]

$$\vec{H}^{e}_{Sk} = -\frac{1}{4\pi}\sum_{i=1}^{n}\int_{-1}^{1} I_{ik} f_i(\xi)\vec{e}_{s'_k} \times \nabla g_{0k}(\vec{r},\vec{r}')\frac{\mathrm{d}s'_k}{\mathrm{d}\xi}\mathrm{d}\xi, \tag{3.39}$$

$$\vec{H}^{e}_{Ik} = -\frac{1}{4\pi}\sum_{i=1}^{n}\int_{-1}^{1} I^{e}_{ik} f_i(\xi)\vec{e}_{ks*} \times \nabla g_{ik}(\vec{r},\vec{r}^{*})\frac{\mathrm{d}s'_k}{\mathrm{d}\xi}\mathrm{d}\xi. \tag{3.40}$$

The reduction to the case of a single straight wire or straight wire array is straightforward, and can be found elsewhere, e.g. in [8].

### 3.1.3 Transmission Line Approximation: Telegrapher's Equations in the Frequency Domain

Voltages and currents along the multiconductor transmission line (MTL) shown in Figure 3.4 induced by an external field excitation can be obtained using the field-to-TL matrix equations in the frequency domain [10]:

$$\frac{\mathrm{d}}{\mathrm{d}x}[\hat{V}(x)] + [\hat{Z}]\cdot[\hat{I}(x)] = -j\omega\mu_0\int_0^h [\hat{H}^{\mathrm{exc}}_y(x,z)]\mathrm{d}z, \tag{3.41}$$

$$\frac{\mathrm{d}}{\mathrm{d}x}[\hat{I}(x)] + [\hat{Y}]\cdot[\hat{V}(x)] = -j\omega\mu_0\int_0^h [\hat{E}^{\mathrm{exc}}_z(x,z)]\mathrm{d}z, \tag{3.42}$$

where the longitudinal impedance matrix is given by

$$[\hat{Z}] = j\omega\,[L] + [\hat{Z}_{\mathrm{w}}] + [\hat{Z}_{\mathrm{g}}] \tag{3.43}$$

and the transversal admittance matrix can be written as

$$[\hat{Y}] = j\,\omega\,[C] + [G], \tag{3.44}$$

where $[L]$ is the per-unit-length longitudinal inductance matrix for a perfect soil, and $[C]$ and $[G]$ are the per-unit-length transverse capacitance and conductance matrix of the multiconductor line, respectively. Furthermore, $[\hat{Z}_{\mathrm{w}}]$ is the per-unit-length internal impedance matrix of the conductors and $[\hat{Z}_{\mathrm{g}}]$ is the per-unit-length ground impedance matrix. Finally, $[\hat{H}^{\mathrm{exc}}_y(x,z)]$ and $[\hat{E}^{\mathrm{exc}}_z(x,z)]$ are sources vectors expressed in terms of the incident magnetic and electric field, respectively [1, 8].

### 3.1.4 Computational Examples

The first computational example is related to the analysis of an overhead wire (Figure 3.3) of length $L = 20$ m, radius $a = 0.005$ m located at height $h = 1$ m above PEC ground and illuminated by the plane wave. The amplitude of the electric field is $E_0 = 1\ \mathrm{V\,m^{-1}}$ and it is parallel to $x$ axis. Figure 3.5 shows the frequency response at the center of the line. The results computed via GB-IBEM and TL are compared to the results obtained via NEC using RC and Sommerfeld

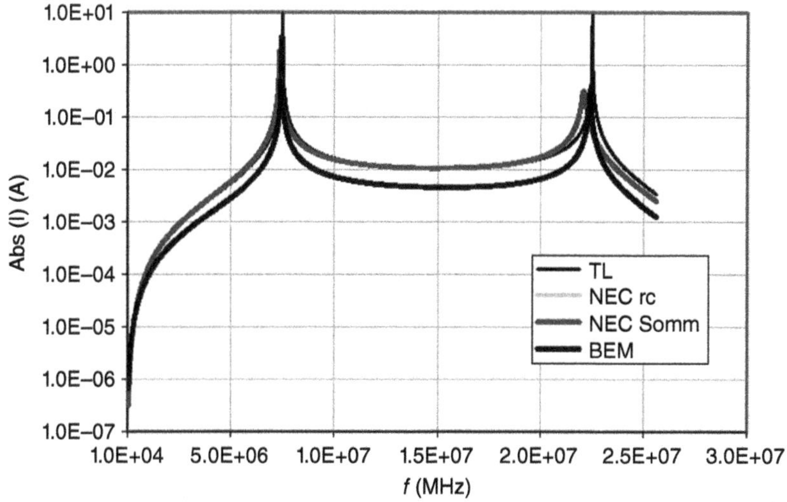


**Figure 3.5** Current induced at the center of the line above a PEC ground versus frequency.

integral approach, respectively, to account for the presence of a lossy half-space. The agreement between the results obtained via the different approaches is found to be satisfactory.

Figure 3.6 shows the frequency response for the same line located above an imperfectly conducting half-space for various values of ground conductivity $\sigma = 1\ \mathrm{mS\ m^{-1}}$. The results calculated via different approaches agree satisfactorily again.

The next computational example is related to a simple PLC system. PLC technology aims to provide users with the necessary communication means by using the already existing and widely distributed power line network and electrical installations in houses and buildings. However, one of the principal drawbacks of this technology is related to electromagnetic interference (EMI) problems, as overhead power lines at the PLC frequency range (1–30 MHz) act as transmitting or receiving antennas, respectively [13].

Figure 3.7 shows the geometry of a simple PLC system consisting of two conductors placed in parallel above each other at the distance *d*. The conductors are suspended between two poles of equal height, thus heaving the shape of the catenary.

The geometry of a catenary is fully defined by such parameters as the distance between the points of suspension, *L*, the sag of the conductor, *s*, and the height of the suspension point, *h*, as shown in Figure 3.7. The imperfectly conducting ground is characterized with electrical permeability $\varepsilon_r$ and conductivity $\sigma$.

The conductors are modeled as thin wire antennas excited by the voltage generator $V_g$ at one end, and terminated by the load impedance $Z_L$ at the other end.

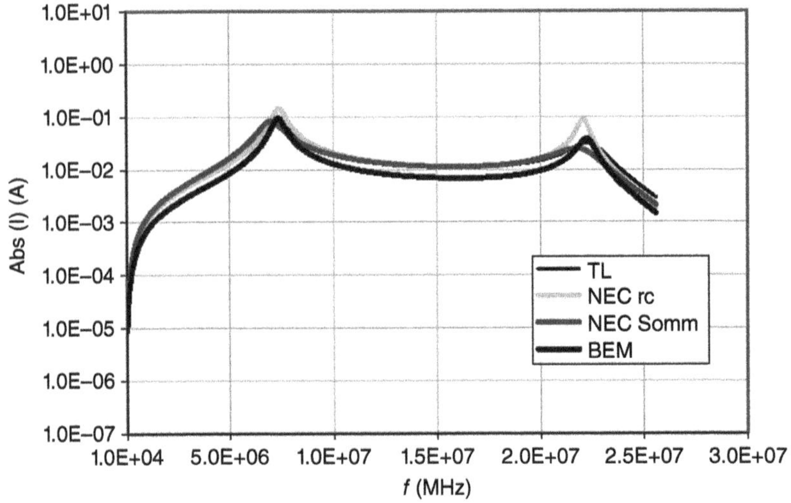


**Figure 3.6** Current induced at the center of the line above a lossy ground versus frequency ($\sigma = 0.001\ \text{S m}^{-1}$, $\varepsilon_r = 10$).

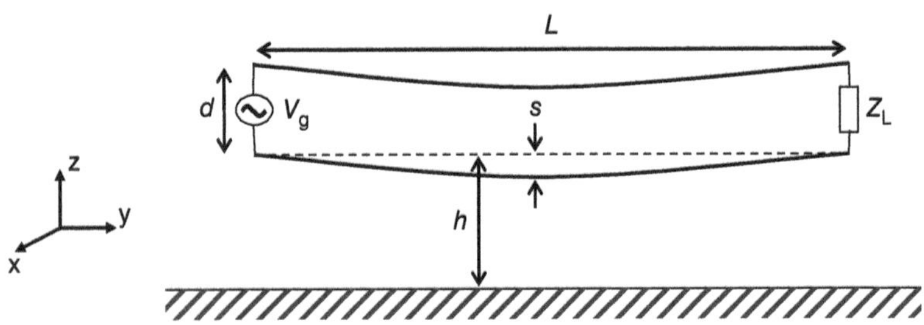


**Figure 3.7** Simple PLC circuit.

The influence of the load impedance is taken into account by modifying the continuity condition for the tangential components of the electric field at the wire surface:

$$E_s^{\text{inc}} + E_s^{\text{sct}} = Z_L' I(s), \tag{3.45}$$

where $Z_L'$ is the corresponding per-length impedance of the conductor.

The modified Pocklington's equation for the wire containing the load impedance is now given by

$$E^{\text{inc}}(s) = -\frac{1}{j4\pi\omega\varepsilon_0} \int_0^L \left\{ \begin{aligned} &\left[k_1^2 \vec{e}_s \vec{e}_{s'} - \frac{\partial^2}{\partial s\, \partial s'}\right] g_0(s, s') \\ &+ R_{\text{TM}} \left[k_1^2 \vec{e}_s \vec{e}_{s*} - \frac{\partial^2}{\partial s\, \partial s^*}\right] g_i(s, s^*) \end{aligned} \right\} I(s')\text{d}s' + Z_L' I(s). \tag{3.46}$$

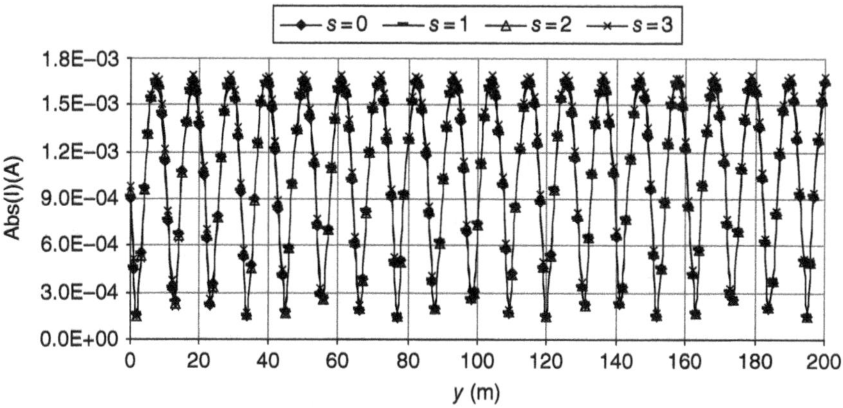


**Figure 3.8** The current distribution along a simple PLC system.

Integral equation (3.46) is numerically solved using the GB-IBEM.

The actual example is related to the simple PLC circuit shown in Figure 3.7. The distance between poles is $L = 200$ m, with the radii of wires $a = 6.35$ mm. The wires are suspended on the poles at heights $h_1 = 10$ m and $h_2 = 11$ m. The maximum sag of the conductor is assumed to be $s = 2$ m. Ground parameters are $\varepsilon_r = 13$ and $\sigma = 0.005$ S m$^{-1}$. The power of the applied voltage generator is 2.5 μW (minimum power required for the PLC system operation) and operating frequency is chosen to be 14 MHz. The value of the terminating load $Z_L$ is 500 Ω. Figure 3.8 shows the current distribution along the simple PLC system for different values of sag.

Radiated electric and magnetic fields at the distance of 30 m from the wires and 10 m above ground are shown in Figures 3.9 and 3.10, respectively.

Analysis of the radiated field distributions shows that the conductor sag does not influence the far field region significantly while the near field distribution is mainly determined by the conductor geometry. Finally, the power of the applied voltage generator is changed to 1 mW (average power used at the actual PLC systems) and operating frequency is varied between 1 and 30 MHz. The values of the terminating load $Z_L$ are chosen to be 50, 500, 5000 Ω, thus simulating different conditions within the power grid. The maximum values of the radiated electric field at the distance of 30 m for different arrangements are shown in Table 3.1.

According to the available international standards [16, 17], radiated electric fields should not exceed the level of 30 μV m$^{-1}$ at the distance of 30 m. Obviously, the radiated field levels are at best case more than 10 times higher than the proposed limit. The spatial distributions of the radiated electric field have been calculated for the number of frequencies in the frequency range from 1 to 30 MHz. Maximum levels of the calculated electric field values are shown to

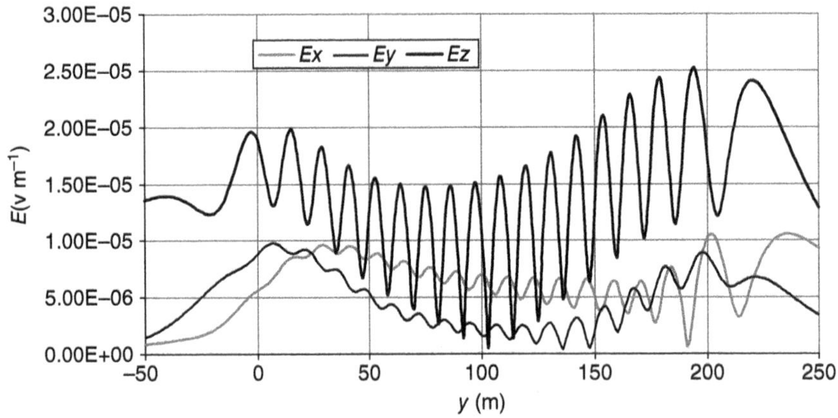


**Figure 3.9** Radiated electric field.

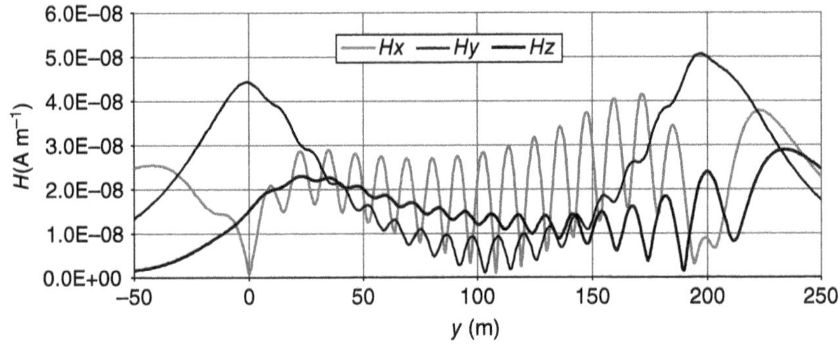


**Figure 3.10** Radiated magnetic field.

**Table 3.1** Maximum values of the radiated electric field at the 30 m distance.

| Frequency (MHz) | $Z_L$ (Ω) | $\|E\|_{max}$ (mV m$^{-1}$) |
|---|---|---|
| 7 | 50 | 0.459 |
| | 500 | 0.341 |
| | 5000 | 0.380 |
| 14 | 50 | 0.477 |
| | 500 | 0.458 |
| | 5000 | 0.541 |
| 28 | 50 | 2.394 |
| | 500 | 0.853 |
| | 5000 | 2.043 |

exceed the limits defined by the standard for the disturbances caused by information technology equipment.

## 3.2 Time Domain Models and Methods

This section deals with direct time domain analysis of transient electromagnetic field coupling to straight overhead wires using the wire AT and the TL method, respectively. The time domain AT formulation is based on a set of space–time Hallen integral equations. The TL approximation is based on the corresponding time domain telegrapher's equations. The space–time integral equations arising from the wire AT are handled by the time domain scheme of GB-IBEM. The time domain telegrapher's equations are solved using the finite difference time domain (FDTD) method. Time domain numerical results obtained with both approaches are compared to the results computed via NEC 2 code combined with IFT procedure. Some illustrative comparisons of results obtained by means of AT and TL approach are presented in this section.

It is worth mentioning that, for the sake of simplicity, only straight wires are analyzed in this chapter.

### 3.2.1 The Antenna Theory Model

Generally, a direct time domain analysis of thin wire in the presence of a lossy half-space can be carried out via the appropriate space–time integral equations of either Pocklington or Hallen type [1, 8]. When applied to the solution of the Hallen integral equation the GB-IBEM [8] results in relatively complex procedures compared to various procedures for the solution of Pocklington's equations, but, at the same time, it is proved to be highly efficient, accurate, and unconditionally stable [8, 18, 19]. On the other hand, the implementation of GB-IBEM to the solution of the Pocklington type equation is relatively simple, but suffers from serious numerical instabilities. The origin of these instabilities is the discretization of space–time differential operator [8]. The GB-IBEM solution of the Pocklington's equation in free space for certain values of time domain integration parameters has been discussed elsewhere, e.g. in [19], while the Hallen integral equation solution by means of GB-IBEM has been obtained for thin wire structures in the presence of a dielectric half-space, e.g. in [11]. In both cases, the influence of imperfect ground has been taken into account via the corresponding RC. The numerical solution was mostly limited to scenarios in which the finite conductivity of the ground could be ignored. This approximation involves cases where the wires are sufficiently far from the two-media interface, or where the ground conductivity is appreciably low or very high, i.e. where the approximation of pure dielectric medium or perfect ground is applied. Through these approximations the time dependent part of the RC function vanishes, and the resulting matrix equation simplifies significantly.

However, for the cases where these approximations are not valid, modifications to the original methods are required in order to include the ground conductivity [8]. Namely, the related RC is space–time dependent, and the resulting convolution integrals have to be included in the matrix system and numerically computed. This leads to a significant increase in the overall computational cost of the method, and consequently requires several modifications.

This section deals with the transient analysis of multiple horizontal wires above a lossy ground using the Hallen integral equation approach.

The set of space–time Hallen integral equations can be derived as an extension of the single wire case. First, a single wire insulated in free space is considered.

A thin wire antenna or scatterer of length $L$ and radius $a$, oriented along the $x$-axis, is considered. The wire is assumed to be PEC and excited by a plane wave electric field. For the sake of simplicity, the analysis is restricted to the case of a normally incident electric field.

The tangential component of the total field vanishes on the PEC wire surface, i.e.

$$E_x^{\mathrm{inc}}(x,t) + E_x^{\mathrm{sct}}(x,t) = 0, \tag{3.47}$$

where $E_x^{\mathrm{inc}}$ is the incident field and $E_x^{\mathrm{sct}}$ is the scattered field on the metallic wire surface. Starting from Maxwell's equations and obeying the Lorentz gauge one obtains a time domain version of Equation (3.2):

$$\left(\frac{\partial^2 \vec{A}}{\partial t^2} - \frac{1}{\mu\varepsilon}\nabla(\nabla\vec{A})\right)\Bigg|_{\tan} = \frac{\partial \vec{E}^{\mathrm{inc}}}{\partial t}\Bigg|_{\tan}, \tag{3.48}$$

where $\vec{A}$ is the space–time dependent vector potential.

According to the thin wire approximation, only the axial component of the vector potential exists, i.e. Equation (3.48) becomes

$$\frac{\partial^2 A_x}{\partial x^2} - \frac{1}{c^2}\frac{\partial^2 A_x}{\partial t^2} = -\frac{1}{c^2}\frac{\partial E_x^{\mathrm{inc}}}{\partial t}, \tag{3.49}$$

where $c$ denotes the velocity of light.

The corresponding solution of (3.49) can be expressed in terms of a sum of the general solution of the homogeneous equation and the particular solution of the inhomogeneous equation:

$$A_x(x,t) = A_x^h(x,t) + A_x^p(x,t). \tag{3.50}$$

The solution of the homogeneous wave equation is given as a superposition of incident and reflected waves [8]:

$$A_x^h(x,t) = F_1\left(t - \frac{x}{c}\right) + F_2\left(t + \frac{x}{c}\right), \tag{3.51}$$

while the particular solution is given by the integral [8]

$$A_x^p(x,t) = \frac{1}{2Z_0}\int_0^L E_x^{\text{inc}}\left(x', t - \frac{|x - x'|}{c}\right)\text{d}x', \tag{3.52}$$

where $L$ denotes the total antenna length.

On the other hand, the magnetic vector potential on the PEC wire surface is given by the particular integral

$$A_x(x,t) = \frac{\mu}{4\pi}\int_S \frac{I(x', t - R/c)}{R}\text{d}x'. \tag{3.53}$$

Combining Equations (3.50)–(3.53) yields the space–time Hallen equation:

$$\int_0^L \frac{I(x', t - R/c)}{4\pi R}\text{d}x' = F_0\left(t - \frac{x}{c}\right) + F_L\left(t - \frac{L-x}{c}\right) + \frac{1}{2Z_0}\int_0^L E_x^{\text{inc}}\left(x', t - \frac{|x - x'|}{c}\right)dx', \tag{3.54}$$

where $I(x')$ is the equivalent axial current to be determined, $E_x^{\text{inc}}$ is the known tangential incident field, $R = [(x - x')^2 + a^2]^{1/2}$ is the distance from the source point (the equivalent current in the antenna axis) to the observation point, and $Z_0$ is the wave impedance of a free space.

The unknown time dependent functions $F_0(t)$ and $F_L(t)$ account for the multiple reflections of the current at the free ends of the wire.

A direct time formulation for a straight thin wire above a dissipative half-space can be obtained as the extension of the free space Hallen equation (3.54).

The free space Hallen equation (3.54) is first transferred into the Laplace frequency domain:

$$\int_0^L \frac{I(x,s)\text{e}^{\frac{sR}{c}}}{4\pi R}\text{d}x' = F_0(s)\text{e}^{-\frac{sx}{c}} + F_L(s)\text{e}^{-s\frac{L-x}{c}} + \frac{1}{2Z_0}\int_0^L E_x^{\text{exc}}(x', s)\text{e}^{-s\frac{|x-x'|}{c}}\text{d}x', \tag{3.55}$$

where $s = j\omega$ is the Laplace variable.

According to the theory of images the free space integral equation (3.55) is extended by an additional term multiplying the Green function of the image source by space–frequency dependent RC $R_{\text{TM}}(\theta', s)$ for TM polarization. The integral equation in the frequency domain is given by

$$\int_0^L \frac{I(x,s)\text{e}^{\frac{sR}{c}}}{4\pi R}\text{d}x' - \int_0^L R_{\text{TM}}(\theta, s)\frac{I(x,s)\text{e}^{\frac{sR^*}{c}}}{4\pi R^*}\text{d}x' = F_0(s)\text{e}^{-\frac{sx}{c}} + F_L(s)\text{e}^{-s\frac{L-x}{c}} + \frac{1}{2Z_0}\int_0^L E_x^{\text{exc}}(x', s)\text{e}^{-s\frac{|x-x'|}{c}}\text{d}x', \tag{3.56}$$

where $R^* = \sqrt{(x-x')^2 + 4h^2}$ and $R_{TM}(\theta', s)$ are determined by the expression [1]

$$R_{TM}(\theta', s) = \frac{\varepsilon_r\left(1+\frac{\sigma}{\varepsilon s}\right)\cos\theta' - \sqrt{\varepsilon_r\left(1+\frac{\sigma}{\varepsilon s}\right) - \sin^2\theta'}}{\varepsilon_r\left(1+\frac{\sigma}{\varepsilon s}\right)\cos\theta' + \sqrt{\varepsilon_r\left(1+\frac{\sigma}{\varepsilon s}\right) - \sin^2\theta'}}, \tag{3.57}$$

where $\sigma$ and $\varepsilon$ are the lossy medium conductivity and permittivity respectively, and $\theta' = \mathrm{arctg}(|x-x'|/2h)$.

The RC approach is a satisfactory approximation in half-space calculations, as long as the field is calculated far away from the source and the imperfect ground, respectively to ensure $\theta' < \pi/2$ [8].

Performing the convolution, the time domain counterpart of Equation (3.57) is obtained in the form

$$\begin{aligned}\int_0^L \frac{I\left(x', t-\frac{R}{c}\right)}{4\pi R}\mathrm{d}x' - \int_{-\infty}^{t}\int_0^L r(\theta,\tau)\frac{I\left(x', t-\frac{R^*}{c}-\tau\right)}{4\pi R^*}\mathrm{d}x'\,\mathrm{d}\tau \\ = \frac{1}{2Z_0}\int_0^L E_x^{\mathrm{exc}}\left(x', t-\frac{|x-x'|}{c}\right)\mathrm{d}x' + F_0\left(t-\frac{x}{c}\right) \\ + F_L\left(t-\frac{L-x}{c}\right),\end{aligned} \tag{3.58}$$

where $r(\theta, t)$ is the space–time RC, which, for convenience, can be written in the form [18]

$$r(\theta,\tau) = r'(\theta,\tau) + r''(\theta,\tau), \tag{3.59}$$

where

$$r'(\theta, t) = K\delta(t), \tag{3.60}$$

$$r''(\theta, t) = \frac{4\beta}{1-\beta^2}\frac{e^{-\alpha t}}{t}\sum_{n=1}^{\infty}(-1)^{n+1} nK^n I_n(\alpha t), \tag{3.61}$$

$$\tau = \frac{\sigma}{\varepsilon_0\varepsilon_r}, \quad \beta = \frac{\sqrt{\varepsilon_r - \sin^2\theta}}{\varepsilon_r\cos\theta}, \quad \gamma = \frac{\tau}{1-\frac{\sin^2\theta}{\varepsilon_r}},$$

$$\theta = \mathrm{arctg}\frac{|x-x'|}{2h}, \quad K = \frac{1-\beta}{1+\beta}, \quad \alpha = \frac{\tau}{2}. \tag{3.62}$$

Note that $I_n$ is the modified Bessel function of the first order and $n$th degree.

If the case of normal incidence, (for the sake of simplicity) is considered, the excitation term is given by

$$E_x^{\mathrm{exc}}(t) = E_x^{\mathrm{inc}}(t) - E_x^{\mathrm{ref}}(t^*), \tag{3.63}$$

where $t^* = t - R^*/c$.

The transient ground reflected field is obtained as the convolution of the incident field and the space–time RC for the angle of incidence $\theta = 0$ (in accordance with the parallel incidence of the electric field), as is proposed in [20]:

$$E_x^{\text{ref}}(t) = \int_{-\infty}^{t} E_x^{\text{inc}}(t-\tau) r(\theta = 0,\ \tau)\mathrm{d}\tau \tag{3.64}$$

and the integral equation (3.58) becomes

$$\begin{aligned}
&\int_0^L \frac{I(x', t - R/c)}{4\pi R}\mathrm{d}x' - \int_{-\infty}^{t}\int_0^L r(\theta, \tau)\frac{I(x', t - R^*/c - \tau)}{4\pi R^*}\mathrm{d}x'\,\mathrm{d}\tau \\
&= F_0\left(t - \frac{x}{c}\right) + F_L\left(t - \frac{L-x}{c}\right) \\
&+ \frac{1}{2Z_0}\int_0^L E_x^{\text{inc}}\left(x', t - \frac{|x - x'|}{c}\right)\mathrm{d}x' \\
&- \frac{1}{2Z_0}\int_{-\infty}^{t}\int_0^L E_x^{\text{inc}}\left(x', t - \frac{|x - x'|}{c} - \tau\right) r(\theta = 0, \tau)\mathrm{d}x'\,\mathrm{d}\tau.
\end{aligned} \tag{3.65}$$

The unknown time functions $F_0(t)$, $F_\mathrm{L}(t)$, $F_0(t - L/c)$ and $F_\mathrm{L}(t - L/c)$ can be obtained in the same manner, as in the case of free space, in terms of auxiliary functions $K_0(t)$ and $K_\mathrm{L}(t)$ [8]:

$$F_0(t) = \sum_{n=0}^{\infty} K_0\left(t - \frac{2nL}{c}\right) - \sum_{n=0}^{\infty} K_L\left(t - \frac{(2n+1)L}{c}\right), \tag{3.66}$$

$$F_L(t) = \sum_{n=0}^{\infty} K_\mathrm{L}\left(t - \frac{2nL}{c}\right) - \sum_{n=0}^{\infty} K_0\left(t - \frac{(2n+1)L}{c}\right), \tag{3.67}$$

where

$$\begin{aligned}
K_L(t) = &\int_0^L \frac{I(x', t - R_0/c)}{4\pi R_0}\mathrm{d}x' \\
&- \int_{-\infty}^{t}\int_0^L r(\theta', \tau)\frac{I(x', t - R_0^*/c - \tau)}{4\pi R_0^*}\mathrm{d}x'\,\mathrm{d}\tau \\
&- \frac{1}{2Z_0}\int_0^L E_x^{\text{inc}}\left(x', t - \frac{x'}{c}\right)\mathrm{d}x',
\end{aligned} \tag{3.68}$$

$$\begin{aligned}
K_L(t) = &\int_0^L \frac{I(x', t - R_L/c)}{4\pi R_L}\mathrm{d}x' \\
&- \int_{-\infty}^{t}\int_0^L r(\theta', \tau)\frac{I(x', t - R_L^*/c - \tau)}{4\pi R_L^*}\mathrm{d}x'\,\mathrm{d}\tau \\
&- \frac{1}{2Z_0}\int_0^L E_x^{\text{exc}}\left(x', t - \frac{L - x'}{c}\right)\mathrm{d}x',
\end{aligned} \tag{3.69}$$

while $R_0$ and $R_L$ are the distances from the wire ends to the source point, and $R_0^*, R_L^*$ are the distances from the image wire ends to the image source point.

If the case of a perfect (ideal) dielectric half-space is considered the Hallen equation (3.58) simplifies into

$$\int_0^L \frac{I(x',t-R/c)}{4\pi R}\mathrm{d}x' - \int_0^L r(\theta)\frac{I(x',t-R^*/c)}{4\pi R^*}\mathrm{d}x'$$
$$= F_0\left(t-\frac{x}{c}\right) + F_L\left(t-\frac{L-x}{c}\right)$$
$$+\frac{1}{2Z_0}\int_0^L E_x^{\mathrm{exc}}\left(x',t-\frac{|x-x'|}{c}\right)\mathrm{d}x'. \tag{3.70}$$

Space–time integral equation (3.58) or (3.70), respectively, can be solved assuming the zero current at the free ends of the wire and with the initial conditions requiring the wire not to be excited before the certain instant $t=t_0$ [8].

The transient behavior of $M$ straight horizontal thin wires located at different heights above an infinite ground plane is determined by a set of the coupled space–time integral equations of the Hallen type [11]:

$$\sum_{s=1}^{M}\int_{x_{0s}}^{x_{Ls}} \frac{I_s\left(x',t-\frac{R_{vs}}{c}\right)}{4\pi R_{vs}}\mathrm{d}x' - \sum_{s=1}^{M}\int_{-\infty}^{t}\int_{x_{0s}}^{x_{Ls}} r_{vs}(\theta,\tau)\frac{I_s\left(x',t-\frac{R^*_{vs}}{c}-\tau\right)}{4\pi R^*_{vs}}\mathrm{d}x'\,\mathrm{d}\tau$$
$$= F_{0v}\left(t-\frac{x-x_{0v}}{c}\right) + F_{Lv}\left(t-\frac{x_{Lv}-x}{c}\right)$$
$$+\frac{1}{2Z_0}\int_{x_{0v}}^{x_{Lv}} E_{xv}^{\mathrm{exc}}\left(x',t-\frac{|x-x'|}{c}\right)\mathrm{d}x', \tag{3.71}$$

where $v, s=1,2,\ldots,M$ denote the index of the observed and source wire, respectively. Furthermore, $L_s$ and $L_v$ are the lengths of the $s$th and $v$th wire, and $x'$, $x$ are the $x$-coordinates of the source and observation points on the respective wires. The distances between observation point $(x, y, z)$ on the wire $v$ and source point $(x', y', z')$ on the wire $s$ are given by

$$R_{vs} = \begin{cases} \sqrt{(x-x')^2+(y-y')^2+(z-z')^2} & : v\neq s, \\ \sqrt{(x-x')^2+a^2} & : v=s, \end{cases}$$
$$R^*_{vs} = \sqrt{(x-x')^2+(y-y')^2+(z+z')^2}, \tag{3.72}$$

where the asterisk is related to source points located on the image wire.

Unknown time signals $F_{0v}(t)$ and $F_{Lv}(t)$ account for the multiple reflections of transient currents at the wire open ends and can be written in the form

$$F_{0v}(t) = \sum_{n=0}^{\infty} K_{0v}\left(t-\frac{2nL_v}{c}\right) - \sum_{n=0}^{\infty} K_{Lv}\left(t-\frac{(2n+1)L_v}{c}\right), \tag{3.73}$$

$$F_{Lv}(t) = \sum_{n=0}^{\infty} K_{Lv}\left(t-\frac{2nL_v}{c}\right) - \sum_{n=0}^{\infty} K_{0v}\left(t-\frac{(2n+1)L_v}{c}\right), \tag{3.74}$$

while the auxiliary functions $K$ are defined as follows:

$$K_{0v}(t) = \sum_{s=1}^{M} \int_{x_{0s}}^{x_{Ls}} \frac{I_s\left(x', t - \dfrac{R_{vs}^{(0)}}{c}\right)}{4\pi R_{vs}^{(0)}} \mathrm{d}x' - \sum_{s=1}^{M} \int_{-\infty}^{t} \int_{x_{0s}}^{x_{Ls}} r_{vs}(\theta, \tau) \frac{I_s\left(x', t - \dfrac{R_{vs}^{*(0)}}{c} - \tau\right)}{4\pi R_{vs}^{*(0)}} \mathrm{d}x' \, \mathrm{d}\tau - \frac{1}{2Z_0} \int_{x_{0v}}^{x_{Lv}} E_{xv}^{\mathrm{exc}}\left(x', t - \frac{|x - x'|}{c}\right) \mathrm{d}x', \tag{3.75}$$

$$K_{Lv}(t) = \sum_{s=1}^{M} \int_{x_{0s}}^{x_{Ls}} \frac{I_s\left(x', t - \dfrac{R_{vs}^{(L)}}{c}\right)}{4\pi R_{vs}^{(L)}} \mathrm{d}x' - \sum_{s=1}^{M} \int_{-\infty}^{t} \int_{x_{0s}}^{x_{Ls}} r_{vs}(\theta, \tau) \frac{I_s\left(x', t - \dfrac{R_{vs}^{*(L)}}{c} - \tau\right)}{4\pi R_{vs}^{*(L)}} \mathrm{d}x' \, \mathrm{d}\tau - \frac{1}{2Z_0} \int_{x_{0v}}^{x_{Lv}} E_{xv}^{\mathrm{exc}}\left(x', t - \frac{|x - x'|}{c}\right) \mathrm{d}x', \tag{3.76}$$

where $R_{vs}^{(0)}$ and $R_{vs}^{(L)}$ are distances from the considered source point on each wire $s$ to corresponding observation points at the ends of the wire $v$:

$$R_{vs}^{(0)} = R_{vs}|_{x=x_{0v}}, \quad R_{vs}^{(L)} = R_{vs}|_{x=x_{Lv}}, \tag{3.77}$$

while $R_{vs}^{*(0)}$ and $R_{vs}^{*(L)}$ are distances between the source point at the image of the wire $s$ and observation point located at the ends of the wire $v$:

$$R_{vs}^{*(0)} = R_{vs}^{*}|_{x=x_{0v}}, \quad R_{vs}^{*(L)} = R_{vs}^{*}|_{x=x_{Lv}}. \tag{3.78}$$

The space–time RC $r_{vs}(\theta, t)$ accounts for the influence of the interface, and is given by [11]

$$r_{vs}(\theta'_{vs}, t) = A\delta(t), \tag{3.79}$$

where

$$A = \frac{1-\beta}{1+\beta}, \quad \beta = \frac{\sqrt{\varepsilon_r - \sin^2\theta'}}{\varepsilon_r \cos\theta'},$$
$$\theta'_{vs} = \mathrm{arctg}\frac{\sqrt{(x'-x)^2 + (y'-y)^2}}{z'+z}. \tag{3.80}$$

The angle $\theta'_{vs}$ is the angle between the source point on the image of the wire $s$ ($x'$; $y'$; $-z'$) and the observation point ($x$, $y$, $z$) on wire $v$.

Substituting (3.79) into (3.71) yields

$$\sum_{s=1}^{M} \int_{x_{0s}}^{x_{Ls}} \frac{I_s\left(x', t - \frac{R_{vs}}{c}\right)}{4\pi R_{vs}} \mathrm{d}x' - \sum_{s=1}^{M} \int_{x_{0s}}^{x_{Ls}} r_{vs}(\theta) \frac{I_s\left(x', t - \frac{R^*_{vs}}{c}\right)}{4\pi R^*_{vs}} \mathrm{d}x'$$
$$= F_{0v}\left(t - \frac{x - x_{0v}}{c}\right) + F_{Lv}\left(t - \frac{x_{Lv} - x}{c}\right)$$
$$+ \frac{1}{2Z_0} \int_{x_{0v}}^{x_{Lv}} E^{\mathrm{exc}}_{xv}\left(x', t - \frac{|x - x'|}{c}\right) \mathrm{d}x', \tag{3.81}$$

where $E^{\mathrm{exc}}_{xv}$ is the space–time dependent tangential electric field on the $v$th wire.

For the case of normal incidence, the total excitation field $E^{\mathrm{exc}}_{xv}(x', t)$ is given as the sum of the incident field $E^{\mathrm{inc}}_{xv}(x', t)$ and the field reflected from the interface $E^{\mathrm{ref}}_{xv}(x', t)$ [11]:

$$E^{\mathrm{exc}}_{xv}(x', z, t) = E^{\mathrm{inc}}_{xv}(x', t - T) + E^{\mathrm{ref}}_{xv}(x', t - T). \tag{3.82}$$

The time shift $T$ represents the time required for the wave to travel from the highest wire to the height $z$ of the observed $v$th wire. Assigning the highest wire with index $U$, it can be written that

$$T = \frac{z_U - z}{c}, \quad z_U = \max(z_1, z_2, \dots, z, \dots, z_M). \tag{3.83}$$

The field reflected from the interface for the case of normal incidence, is given by

$$E^{\mathrm{ref}}_{xv}(x', t - T) = r(\theta = 0) \cdot E^{\mathrm{inc}}_{xv}\left(x', t - T - \frac{2z}{c}\right), \tag{3.84}$$

where $t - T - \frac{2z}{c}$ is the time needed for the wave to travel from the observed $v$th wire to the interface.

For the case of PEC ground plane, the space–time RC (3.59) simply becomes

$$r_{vs}(\theta'_{vs}, t) = 1. \tag{3.85}$$

Thus, the set of Equations (3.81) simplifies into

$$\sum_{s=1}^{M} \int_{x_{0s}}^{x_{Ls}} \frac{I_s\left(x', t - \frac{R_{vs}}{c}\right)}{4\pi R_{vs}} \mathrm{d}x' - \sum_{s=1}^{M} \int_{x_{0s}}^{x_{Ls}} \frac{I_s\left(x', t - \frac{R^*_{vs}}{c}\right)}{4\pi R^*_{vs}} \mathrm{d}x'$$
$$= F_{0v}\left(t - \frac{x - x_{0v}}{c}\right) + F_{Lv}\left(t - \frac{x_{Lv} - x}{c}\right)$$
$$+ \frac{1}{2Z_0} \int_{x_{0v}}^{x_{Lv}} E^{\mathrm{exc}}_{xv}\left(x', t - \frac{|x - x'|}{c}\right) \mathrm{d}x' \tag{3.86}$$

and the field reflected from PEC ground is simply given by

$$E_{xv}^{\mathrm{ref}}(x',t-T)=E_{xv}^{\mathrm{inc}}\left(x',t-T-\frac{2z}{c}\right). \tag{3.87}$$

Given the dielectric constant of the medium and the known excitation $E_{xv}^{\mathrm{inc}}(x',t)$, a system of $M$ coupled Hallen integral equations can be solved using the time domain version of GB-IBEM and by taking into account appropriate boundary and initial conditions. Boundary conditions are related to zero currents at the end of each wire, while initial conditions assume all the currents to be zero for $t\leq 0$.

### 3.2.2 The Numerical Solution

First, the numerical procedure for single wire Hallen equation is outlined. Applying the weighted residual approach in the spatial domain and GB-IBEM procedure [8], the following local matrix system is obtained:

$$\begin{aligned}
&[A]\{I\}_i\big|_{t-\frac{R}{c}}-[A^*]\{I\}_i\big|_{t-\frac{R^*}{c}}-\{\hat{A}\}\big|_{t-\frac{R^*}{c}}=[B]\{E\}\big|_{t-\frac{|x-x'|}{c}}\\
&+[C]\left\{\sum_{n=0}^{\infty}I^n\right\}_i\Bigg|_{t-\frac{R_0}{c}-\frac{2nL}{c}-\frac{x}{c}}-[C^*]\left\{\sum_{n=0}^{\infty}I^n\right\}_i\Bigg|_{t-\frac{R_0^*}{c}-\frac{2nL}{c}-\frac{x}{c}}\\
&-\left\{\sum_{n=0}^{\infty}\hat{C}^n\right\}\Bigg|_{t-\frac{R_0^*}{c}-\frac{2nL}{c}-\frac{x}{c}}-[B]\left\{\sum_{n=0}^{\infty}E^n\right\}\Bigg|_{t-\frac{x'}{c}-\frac{2nL}{c}-\frac{x}{c}}\\
&-[D]\left\{\sum_{n=0}^{\infty}I^n\right\}_i\Bigg|_{t-\frac{R_L}{c}-\frac{(2n+1)L}{c}-\frac{x}{c}}+[D^*]\left\{\sum_{n=0}^{\infty}I^n\right\}_i\Bigg|_{t-\frac{R_L^*}{c}-\frac{(2n+1)L}{c}-\frac{x}{c}}\\
&+[B]\left\{\sum_{n=0}^{\infty}E^n\right\}\Bigg|_{t-\frac{L-x'}{c}-\frac{(2n+1)L}{c}-\frac{x}{c}}+\left\{\sum_{n=0}^{\infty}\hat{D}^n\right\}\Bigg|_{t-\frac{R_L^*}{c}-\frac{(2n+1)L}{c}-\frac{x}{c}}\\
&+[D]\left\{\sum_{n=0}^{\infty}I^n\right\}_i\Bigg|_{t-\frac{R_L}{c}-\frac{2nL}{c}-\frac{L-x}{c}}-[D^*]\left\{\sum_{n=0}^{\infty}I^n\right\}_i\Bigg|_{t-\frac{R_L^*}{c}-\frac{2nL}{c}-\frac{L-x}{c}}\\
&+\left\{\sum_{n=0}^{\infty}\hat{D}^n\right\}\Bigg|_{t-\frac{R_L^*}{c}-\frac{2nL}{c}-\frac{L-x}{c}}-[B]\left\{\sum_{n=0}^{\infty}E^n\right\}\Bigg|_{t-\frac{L-x'}{c}-\frac{2nL}{c}-\frac{L-x}{c}}.\\
&-[C]\left\{\sum_{n=0}^{\infty}I^n\right\}_i\Bigg|_{t-\frac{R_0}{c}-\frac{(2n+1)L}{c}-\frac{L-x}{c}}+[C^*]\left\{\sum_{n=0}^{\infty}I^n\right\}_i\Bigg|_{t-\frac{R_0^*}{c}-\frac{(2n+1)L}{c}-\frac{L-x}{c}}\\
&+[B]\left\{\sum_{n=0}^{\infty}E^n\right\}\Bigg|_{t-\frac{x'}{c}-\frac{(2n+1)L}{c}-\frac{L-x}{c}}-\left\{\sum_{n=0}^{\infty}\hat{C}^n\right\}\Bigg|_{t-\frac{R_0^*}{c}-\frac{(2n+1)L}{c}-\frac{L-x}{c}}.
\end{aligned} \tag{3.88}$$

The space-dependent local matrices representing the interaction between $i$th source and $j$th observation element are defined as follows:

$$
\begin{aligned}
[A] &= \int_{\Delta l_j}\int_{\Delta l_i} \{f\}_j\{f\}_i^T \frac{1}{4\pi R} \mathrm{d}x'\,\mathrm{d}x;\\
[B] &= \frac{1}{2Z_0}\int_{\Delta l_j}\int_{\Delta l_i} \{f\}_j\{f\}_i^T \mathrm{d}x'\,\mathrm{d}x\\
[C] &= \int_{\Delta l_j}\int_{\Delta l_i} \{f\}_j\{f\}_i^T \frac{1}{4\pi R_0} \mathrm{d}x'\,\mathrm{d}x;\\
[D] &= \int_{\Delta l_j}\int_{\Delta l_i} \{f\}_j\{f\}_i^T \frac{1}{4\pi R_L} \mathrm{d}x'\,\mathrm{d}x\\
[A^*] &= \int_{\Delta l_j}\int_{\Delta l_i} \{f\}_j\{f\}_i^T \frac{r(\theta)}{4\pi R^*} \mathrm{d}x'\,\mathrm{d}x;\\
[C^*] &= \int_{\Delta l_j}\int_{\Delta l_i} \{f\}_j\{f\}_i^T \frac{r(\theta)}{4\pi R_0^*} \mathrm{d}x'\,\mathrm{d}x\\
[D^*] &= \int_{\Delta l_j}\int_{\Delta l_i} \{f\}_j\{f\}_i^T \frac{r(\theta)}{4\pi R_L^*} \mathrm{d}x'\,\mathrm{d}x;
\end{aligned}
\tag{3.89}
$$

where $\{f\}$ stands for the shape functions, while additional time dependent vectors are given by

$$
\begin{aligned}
\{\hat{A}\} &= \int_0^{t-\frac{R^*}{c}} \int_{\Delta l_j}\int_{\Delta l_i} \{f\}_j\{f\}_i^T H_1\,\mathrm{d}x'\,\mathrm{d}x\{I(\tau)\}_i \mathrm{d}\tau,\\
\{\hat{C}^n\} &= \int_0^{t-\frac{R_0^*}{c}-\frac{2nL}{c}-\frac{x}{c}} \int_{\Delta l_j}\int_{\Delta l_i} \{f\}_j\{f\}_i^T H_2\,\mathrm{d}x'\,\mathrm{d}x\{I(\tau)\}_i \mathrm{d}\tau,\\
\{\hat{D}^n\} &= \int_0^{t-\frac{R_L^*}{c}-\frac{(2n+1)L}{c}-\frac{x}{c}} \int_{\Delta l_j}\int_{\Delta l_i} \{f\}_j\{f\}_i^T H_3\,\mathrm{d}x'\,\mathrm{d}x\{I(\tau)\}_i \mathrm{d}\tau,
\end{aligned}
\tag{3.90}
$$

where

$$
\begin{aligned}
H_1 &= \frac{r''\left(\theta, t-\frac{R^*}{c}-\tau\right)}{4\pi R^*},\\
H_2 &= \frac{r''\left(\theta, t-\frac{R_0^*}{c}-\frac{2nL}{c}-\frac{x}{c}-\tau\right)}{4\pi R_0^*},\\
H_3 &= \frac{r''\left(\theta, t-\frac{R_L^*}{c}-\frac{(2n+1)L}{c}-\frac{x}{c}-\tau\right)}{4\pi R_0^*}.
\end{aligned}
\tag{3.91}
$$

Assembling the local matrices and vectors into the global ones the following global matrix system is formed:

$$[A]\{I\}\Big|_{t-\frac{R}{c}} = \{g\}\Bigg|_{\substack{\text{previous time}\\ \text{instants}}} + \{\hat{g}\}\Bigg|_{\substack{\text{previous time}\\ \text{instants}}}, \tag{3.92}$$

where

$$\begin{aligned}\{g\} &= [A^*]\{I\}\big|_{t-\frac{R^*}{c}} + [B]\{E\}\big|_{t-\frac{|x-x'|}{c}}\\
&+ [C]\left\{\sum_{n=0}^{\infty} I^n\right\}\Bigg|_{t-\frac{R_0}{c}-\frac{2nL}{c}-\frac{x}{c}} - [C^*]\left\{\sum_{n=0}^{\infty} I^n\right\}\Bigg|_{t-\frac{R_0^*}{c}-\frac{2nL}{c}-\frac{x}{c}}\\
&- [B]\left\{\sum_{n=0}^{\infty} E^n\right\}\Bigg|_{t-\frac{x'}{c}-\frac{2nL}{c}-\frac{x}{c}} - [D]\left\{\sum_{n=0}^{\infty} I^n\right\}\Bigg|_{t-\frac{R_L}{c}-\frac{(2n+1)L}{c}-\frac{x}{c}}\\
&+ [D^*]\left\{\sum_{n=0}^{\infty} I^n\right\}\Bigg|_{t-\frac{R_L^*}{c}-\frac{(2n+1)L}{c}-\frac{x}{c}} + [B]\left\{\sum_{n=0}^{\infty} E^n\right\}\Bigg|_{t-\frac{L-x'}{c}-\frac{(2n+1)L}{c}-\frac{x}{c}}\\
&+ [D]\left\{\sum_{n=0}^{\infty} I^n\right\}\Bigg|_{t-\frac{R_L}{c}-\frac{2nL}{c}-\frac{L-x}{c}} - [D^*]\left\{\sum_{n=0}^{\infty} I^n\right\}\Bigg|_{t-\frac{R_L^*}{c}-\frac{2nL}{c}-\frac{L-x}{c}}\\
&- [B]\left\{\sum_{n=0}^{\infty} E^n\right\}\Bigg|_{t-\frac{L-x'}{c}-\frac{2nL}{c}-\frac{L-x}{c}} - [C]\left\{\sum_{n=0}^{\infty} I^n\right\}\Bigg|_{t-\frac{R_0}{c}-\frac{(2n+1)L}{c}-\frac{L-x}{c}}\\
&+ [C^*]\left\{\sum_{n=0}^{\infty} I^n\right\}\Bigg|_{t-\frac{R_0^*}{c}-\frac{(2n+1)L}{c}-\frac{L-x}{c}} + [B]\left\{\sum_{n=0}^{\infty} E^n\right\}\Bigg|_{t-\frac{x'}{c}-\frac{(2n+1)L}{c}-\frac{L-x}{c}}\end{aligned} \tag{3.93}$$

and

$$\begin{aligned}\{\hat{g}\} &= \{\hat{A}\}\big|_{t-\frac{R^*}{c}} - \left\{\sum_{n=0}^{\infty} \hat{C}^n\right\}\Bigg|_{t-\frac{R_0^*}{c}-\frac{2nL}{c}-\frac{x}{c}} + \left\{\sum_{n=0}^{\infty} \hat{D}^n\right\}\Bigg|_{t-\frac{R_L^*}{c}-\frac{(2n+1)L}{c}-\frac{x}{c}}\\
&+ \left\{\sum_{n=0}^{\infty} \hat{D}^n\right\}\Bigg|_{t-\frac{R_L^*}{c}-\frac{2nL}{c}-\frac{L-x}{c}} - \left\{\sum_{n=0}^{\infty} \hat{C}^n\right\}\Bigg|_{t-\frac{R_0^*}{c}-\frac{(2n+1)L}{c}-\frac{L-x}{c}}.\end{aligned} \tag{3.94}$$

Applying the weighted residual approach in the time domain and using the Dirac impulses as weight functions the time sampling is provided, and the following recurrent formula is obtained:

$$I_j|_{t_k} = \frac{\sum_{i=1}^{N^s} a_{ji} I_j|_{t_k - \frac{R}{c}} - g_j|_{\substack{\text{previous time}\\ \text{instants}}} - \hat{g}_j|_{\substack{\text{previous time}\\ \text{instants}}}}{a_{jj}}, \tag{3.95}$$

where $I_j|_{t_k}$ is the current for the $j$th space node at $k$th time instant and $N$ is the total number of space segments, while the overbar indicates the absence of diagonal members.

It is worth noting that the numerical calculation of convolution integrals is a rather tedious task leading to a tremendously large computational time of the overall method. The main advantage of the method, on the other hand, is its unconditional stability.

Time domain GB-BEM procedure for the set of Hallen equations is undertaken in a similar manner as in the case of a single wire.

The solution of (3.81) and (3.86), respectively, is also carried out using the GB-IBEM technique.

Applying the boundary element discretization to (3.81) and (3.86), respectively, leads to a local system of linear equations for the $v$th observed wire:

$$\sum_{s=1}^{M} \left[ \begin{array}{l} \int_{\Delta l_i} \int_{\Delta l_j} \frac{1}{4\pi R_{vs}} \{f\}_j \{f\}_i^T \mathrm{d}x' \, \mathrm{d}x \{I_s\} \Big|_{t-\frac{R_{vs}}{c}} \\ - \int_{\Delta l_i} \int_{\Delta l_j} \frac{r_{vs}(\theta)}{4\pi R_{vs}^*} \{f\}_j \{f\}_i^T \mathrm{d}x' \, \mathrm{d}x \{I_s\} \Big|_{t-\frac{R_{vs}^*}{c}} \end{array} \right]$$

$$= \frac{1}{2Z_0} \int_{\Delta l_i} \int_{\Delta l_j} E_{xv}^{\text{exc}} \left( x', t - \frac{|x - x'|}{c} \right) \{f\}_j \mathrm{d}x' \, \mathrm{d}x$$

$$+ \int_{\Delta l_j} F_0 \left( t - \frac{x - x_{0v}}{c} \right) \{f\}_j \mathrm{d}x + \int_{\Delta l_j} F_L \left( t - \frac{x_{Lv} - x}{c} \right) \{f\}_j \mathrm{d}x, \tag{3.96}$$

where $i, j = 1, 2, \ldots, N$ denotes the index of the elements located on the $s$th source wire and the $v$th observed wire, respectively with $N$ as the total number of space segments, while $M$ is the actual number of wires.

Finally, substituting (3.73)–(3.76) into (3.96), the following local matrix system is obtained:

$$\sum_{s=1}^{M} [A_{vs}]\{I_s\} \Big|_{t-\frac{R_{vs}}{c}} - \sum_{s=1}^{M} [A_{vs}^*]\{I_s\} \Big|_{t-\frac{R_{vs}^*}{c}} = [B_v]\{E_v\} \Big|_{t-\frac{|x-x'|}{c}}$$

$$+ \sum_{s=1}^{M} [C_{vs}] \left\{ \sum_{n=0}^{\infty} I_s^n \right\} \Big|_{t-\frac{x-x_{0v}}{c} - \frac{2n}{c} L_v - \frac{R_{vs}^{(0)}}{c}}$$

$$- \sum_{s=1}^{M} [C_{vs}^*] \left\{ \sum_{n=0}^{\infty} I_s^n \right\} \Big|_{t-\frac{x-x_{0v}}{c} - \frac{2n}{c} L_v - \frac{R_{vs}^{*(0)}}{c}}$$

$$
\begin{aligned}
&- [D_v] \left\{ \sum_{n=0}^{\infty} E_v^n \right\} \Bigg|_{t - \frac{x - x_{0v}}{c} - \frac{2n}{c} L_v - \frac{|x' - x_{0v}|}{c}} \\
&- \sum_{s=1}^{M} [E_{vs}] \left\{ \sum_{n=0}^{\infty} I_s^n \right\} \Bigg|_{t - \frac{x - x_{0v}}{c} - \frac{2n+1}{c} L_v - \frac{R_{vs}^{(L)}}{c}} \\
&+ \sum_{s=1}^{M} [E_{vs}^*] \left\{ \sum_{n=0}^{\infty} I_s^n \right\} \Bigg|_{t - \frac{x - x_{0v}}{c} - \frac{2n+1}{c} L_v - \frac{R_{vs}^{*(L)}}{c}} \\
&+ [D_v] \left\{ \sum_{n=0}^{\infty} E_v^n \right\} \Bigg|_{t - \frac{x - x_{0v}}{c} - \frac{2n+1}{c} L_v - \frac{|x_{Lv} - x'|}{c}} \\
&+ \sum_{s=1}^{M} [E_{vs}] \left\{ \sum_{n=0}^{\infty} I_s^n \right\} \Bigg|_{t - \frac{x_{Lv} - x}{c} - \frac{2n}{c} L_v - \frac{R_{vs}^{(L)}}{c}} \\
&- \sum_{s=1}^{M} [E_{vs}^*] \left\{ \sum_{n=0}^{\infty} I_s^n \right\} \Bigg|_{t - \frac{x_{Lv} - x}{c} - \frac{2n}{c} L_v - \frac{R_{vs}^{*(L)}}{c}} \\
&- [D_v] \left\{ \sum_{n=0}^{\infty} E_v^n \right\} \Bigg|_{t - \frac{x_{Lv} - x}{c} - \frac{2n}{c} L_v - \frac{|x_{Lv} - x'|}{c}} \\
&- \sum_{s=1}^{M} [C_{vs}] \left\{ \sum_{n=0}^{\infty} I_s^n \right\} \Bigg|_{t - \frac{x_{Lv} - x}{c} - \frac{2n+1}{c} L_v - \frac{R_{vs}^{(0)}}{c}} \\
&+ \sum_{s=1}^{M} [C_{os}^*] \left\{ \sum_{n=0}^{\infty} I_s^n \right\} \Bigg|_{t - \frac{x_{Lo} - x}{c} - \frac{2n+1}{c} L_o - \frac{R_{os}^{*(0)}}{c}} \\
&+ [D_o] \left\{ \sum_{n=0}^{\infty} E_o^n \right\} \Bigg|_{t - \frac{x_{Lo} - x}{c} - \frac{2n+1}{c} L_o - \frac{|x' - x_{0o}|}{c}},
\end{aligned}
\tag{3.97}
$$

where $\{E\}$ denotes the excitation vector, and the space-dependent matrices are of the form

$$
\begin{aligned}
[A_{vs}] &= \int_{\Delta l_j} \int_{\Delta l_i} \frac{1}{4\pi R_{vs}} \{f\}_j \{f\}_i^T \mathrm{d}x' \,\mathrm{d}x, \\
[A_{vs}^*] &= \int_{\Delta l_j} \int_{\Delta l_i} \frac{r_{vs}(\theta)}{4\pi R_{vs}^*} \{f\}_j \{f\}_i^T \mathrm{d}x' \,\mathrm{d}x, \\
[B_v] &= \frac{1}{2Z_0} \int_{\Delta l_j} \int_{\Delta l_i} \{f\}_j \{f\}_i^T \mathrm{d}x' \,\mathrm{d}x, \\
[C_{vs}] &= \int_{\Delta l_j} \int_{\Delta l_i} \frac{1}{4\pi R_{vs}^{(0)}} \{f\}_j \{f\}_i^T \mathrm{d}x' \,\mathrm{d}x,
\end{aligned}
$$

$$[C_{vs}^*] = \int_{\Delta l_j} \int_{\Delta l_i} \frac{r_{vs}(\theta)}{4\pi R_{vs}^{*(0)}} \{f\}_j \{f\}_i^T \mathrm{d}x' \, \mathrm{d}x,$$

$$[D_v] = \frac{1}{2Z_0} \int_{\Delta l_j} \int_{\Delta l_i} \{f\}_j \{f\}_i^T \mathrm{d}x' \, \mathrm{d}x,$$

$$[E_{vs}] = \int_{\Delta l_j} \int_{\Delta l_i} \frac{1}{4\pi R_{vs}^{(L)}} \{f\}_j \{f\}_i^T \mathrm{d}x' \, \mathrm{d}x,$$

$$[E_{vs}^*] = \int_{\Delta l_j} \int_{\Delta l_i} \frac{r_{vs}(\theta)}{4\pi R_{vs}^{*(L)}} \{f\}_j \{f\}_i^T \mathrm{d}x' \, \mathrm{d}x. \tag{3.98}$$

Relations containing summations from $n = 0$ to infinity pertain to the reflections of transient current from the wire ends. Note that as the observed time interval is always finite, only a finite number of reflections occur within a given observation interval. A shorter observed interval requires a smaller number of summands, and vice versa.

According to GB-IBEM, a global matrix system is assembled from the local matrix systems for all wires $v = 1, 2, \ldots, M$. Finally, the resulting global matrix system can be written as follows:

$$[A]\{I\}\Big|_{t-\frac{R_{vs}}{c}} - [A^*]\{I\}\Big|_{t-\frac{R_{vs}^*}{c}} = \{g\}. \tag{3.99}$$

The time domain solution on the $i$th boundary element is given by

$$I_i(t) = \sum_{k=1}^{N_t} I_i^k T^k(t'), \tag{3.100}$$

where $I_i^k$ are unknown coefficients, $T_k$ are the linear time domain shape functions, and $N_t$ is the total number of time samples.

Applying the weighted residual approach to (3.99) leads to the expression

$$\int_{t_k}^{t_k+\Delta t} \left( [A]\{I\}\Big|_{t-\frac{R_{vs}}{c}} - [A^*]\{I\}\Big|_{t-\frac{R_{vs}^*}{c}} - \{g\}\theta_k \right) \mathrm{d}t = 0; \;\; k = 1, 2, \ldots, N_t, \tag{3.101}$$

where $\theta_k$ denotes the set of time domain weights.

Using the set of Dirac impulses for the test functions, time sampling is ensured and (3.101) becomes

$$[A]\{I\}\Big|_{t_k-\frac{R_{vs}}{c}} - [A^*]\{I\}\Big|_{t_k-\frac{R_{vs}^*}{c}} = \{g\}\Big|_{\substack{\text{all previous} \\ \text{discrete instants}}}. \tag{3.102}$$

If the space–time discretization is performed by satisfying the Courant condition, $\Delta x \geq c\Delta t$, the transient current for a $j$th space node and $k$th time node can be obtained from a recurrence formula. Separating the terms relating to the current induced at the instant $t_k$ in (3.102) yields

$$A_{jj}I_j|_{t_k} + [\overline{A}]\{I\}\Big|_{t_k-\frac{R_{vs}}{c}} - [A^*]\{I\}\Big|_{t_k-\frac{R^*_{vs}}{c}} = \{g\}|_{\substack{\text{all previous}\\ \text{discrete instants}}}, \tag{3.103}$$

where overbar indicates the absence of diagonal terms.

The first term in (3.103) pertains to the current at the $j$th space node and $k$th time node, i.e. the present instant. Other terms are related to all previous instants. Finally, the recurrence formula for the transient current at $j$th space node and $k$th time node is obtained in the form

$$I_j|_{t_k} = \frac{-\sum_{i=1}^{N}\left(\overline{A_{ji}}I_i|_{t_k-\frac{R_{vs}}{c}} + A^*_{ji}I_i|_{t_k-\frac{R^*_{vs}}{c}}\right) + g_j|_{\text{all previous discrete instants}}}{A_{jj}}, \tag{3.104}$$

where $N$ is the total number of space elements, $k = 1, 2, \ldots$, and $N_t$ is the index of the $k$th time instant.

### 3.2.3 The Transmission Line Model

The time domain field-to-TL coupling equations can be written in the matrix form [11]:

$$\frac{\partial}{\partial x}[V(x,t)] + [R][I(x,t)] + [L].\frac{\partial}{\partial t}[I(x,t)] = [E_F(x,t)] - [z'(t)] * [I(x,t)], \tag{3.105}$$

$$\frac{\partial}{\partial x}[I(x,t)] + [G].[V(x,t)] + [C].\frac{\partial}{\partial t}[V(x,t)] = [H_F(x,t)], \tag{3.106}$$

where '*' stands for the convolution product, $[z'(t)]$ is the transient IFT of the ground, conductors matrix $[\hat{Z}_w(s) + \hat{Z}_g(s)]$, and $s = j\omega$ is the Laplace variable.

$[E_F(x,t)]$and $[H_F(x,t)]$ are the excitation terms, given by [11]

$$[E_F(x,t)] = -\frac{\partial}{\partial x}[V_T(x,t)] + [E_L(x,t)], \tag{3.107}$$

$$[H_F(x,t)] = -[G][V_T(x,t)] - [C]\frac{\partial}{\partial t}[V_T(x,t)], \tag{3.108}$$

where $[V_T(x,t)]$ is the transverse voltage derived from the transverse incident field excitation [11], and $[E_L(x,t)]$ represents the longitudinal electrical field excitation.

The classical per-unit-length inductances of $[L]$ matrix have been used:

$$L_{ii} = \frac{\mu_0}{2\pi}\ln\left(\frac{2h_i}{a_i}\right); \quad L_{ij} = \frac{\mu_0}{2\pi}\ln\left(\frac{D_{ij}}{d_{ij}}\right), \tag{3.109}$$

where $h_i$, $a_i$, are respectively the height and the radius of the $i$th conductor.$D_{ij}$ is the distance between the $i$th conductor and the image of $j$th conductor and $d_{ij}$ is the distance between the $i$th conductor and the $j$th conductor.

The capacitance matrix $[C]$ is derived from the inductance matrix $[L]$ by $[C]=\varepsilon_0\mu_0[L]^{-1}$.

In case of PEC ground, the transient ground matrix $[z'(t)]$ is equal to $[Z_W(t)]$ while if a dielectric half-space is of interest, the ground matrix $[\hat{Z}_g(s)]$ is given by Carson integral [11].

It is worth emphasizing that for both cases, the conductivity matrix is neglected.

### 3.2.4 The Solution of Transmission Line Equations via FDTD

Discretizing each conductor of the MTL into $N_x$ sections, each of length $\Delta x$, and discretizing the entire time interval into increments of duration $\Delta t$, the FDTD method is applied to (3.105) and (3.106).

The solution for lossy MTL by FDTD is given by

$$\begin{aligned}[Z^+][I_k^{n+1/2}] &= -[Z^-][I_k^{n-1/2}] + [V_{k-1/2}^n] + [V_{\mathrm{T},k-1/2}^n] \\ &\quad - [V_{k+1/2}^n] - [V_{\mathrm{T},k+1/2}^n] + ([E_{\mathrm{L},k}^n] - [S_k^n])\ \Delta x \\ &\quad \text{for } 0 \le k \le N_x - 1,\ \ n \ge 1, \end{aligned} \tag{3.110}$$

where

$$[Z^\pm] = \frac{[R]}{2} \pm \frac{[L] + [Z_0(1)]\Delta t}{\Delta t}, \tag{3.111}$$

$$\begin{aligned}[Y^+]([V_{k+1/2}^{n+1}] + [V_{\mathrm{T},k+1/2}^{n+1}]) &= -[Y^-]([V_{k+1/2}^n] + [V_{\mathrm{T},k+1/2}^n]) \\ &\quad - [I_{k+1}^{n+1/2}] + [I_k^{n+1/2}] \\ &\quad \text{for } 0 \le k \le N_x - 1,\ \ n \ge 1, \end{aligned} \tag{3.112}$$

$$[Y^\pm] = \frac{[G]}{2} \pm \frac{[C]}{\Delta t}. \tag{3.113}$$

The expressions for $[V]$, $[I]$, $[V_\mathrm{T}]$, $[E_\mathrm{L}]$, $[S]$ are given for time $n \ge 1$ and space $k \ge 0$:

$$[V_{k+1/2}^n] = [V((k+1/2)\Delta x, n\Delta t)], \quad 0 \le k \le N_x - 1. \tag{3.114}$$

Voltage and adjacent current are interlaced in time and space respectively by $\Delta t/2$ and $\Delta x/2$; then

$$\begin{aligned}[I_k^{n-1/2}] &= [I(k\Delta x, (n-1/2)\Delta t)], \\ [V_{\mathrm{T},k+1/2}^n] &= [V_\mathrm{T}((k+1/2)\Delta x, n\Delta t)], \\ [E_{\mathrm{L},k}^n] &= [E_\mathrm{L}(k\Delta x, n\Delta t)], \end{aligned}$$

$$[S_k^n] = [S(k\Delta x, n\Delta t)],$$
$$0 \le k \le N_x. \tag{3.115}$$

The convolution product $[S_k^n]$ can be written as follows:

$$[S_k^n] = [Z_0(2)][I_k^{n-1/2}] + \sum_{l=1}^{n-1}[Z_0(l+1) - Z_0(l)][I_k^{n-l+1/2}] - [Z_0(n)][I_k^{1/2}], \tag{3.116}$$

where

$$[Z_0(l)] = [Z_{0\mathrm{w}}(l)] + [Z_{0\mathrm{g}}(l)] = \int_{l-1}^{l}[Z_\mathrm{w} + Z_\mathrm{g}](u\Delta t)\mathrm{d}u, \quad 1 \le l \le n, \tag{3.117}$$

$t = n\Delta t$, $N_x$ is the number of space steps, and $n$ is the number of time steps.

The corresponding components of the $M \times 1$ vectors $[E_{\mathrm{T},k}^n]$, $[E_{\mathrm{L},k}^n]$ are as follows:

$$[V_{\mathrm{T},k}^n]_i = y_i E_y^{\mathrm{inc}}(x_i = k\Delta x, y_i, z_i, n\Delta t) + z_i E_z^{\mathrm{inc}}(x_i = k\Delta x, y_i, z_i, n\Delta t), \tag{3.118}$$

$$[E_{\mathrm{L},k}^n]_i = E_x^{\mathrm{inc}}(x_i = k\Delta x, y_i, z_i, n\Delta t) - E_z^{\mathrm{inc}}(x_i = k\Delta x, 0, 0, n\Delta t). \tag{3.119}$$

Equations (3.117) and (3.118) are valid for $1 \le i \le M$, where $M$ is the number of conductors and $y_i$, $z_i$ are the positions of $i$th conductor. $E_x^{\mathrm{inc}}, E_y^{\mathrm{inc}}, E_z^{\mathrm{inc}}$ are the components of the incident electromagnetic field evaluated in the absence of conductors. $[R]$, $[L]$, $[G]$, and $[C]$ are respectively the per-unit-length resistance, inductance, conductance, and capacitance matrices of dimension $M \times M$.

$[Z_\mathrm{g}(t)]$ is the transient ground resistance and is equal to the inverse Fourier of $\left[\frac{Z_\mathrm{g}(s)}{s}\right]$:

$$Z_{\mathrm{g}_{ij}}(t) = F^{-1}\left(\frac{Z_{\mathrm{g}_{ij}}(s)}{s}\right), \tag{3.120}$$

where the ground impedance in frequency domain is given by Carson formula:

$$\frac{Z_{\mathrm{g}_{ij}}(s)}{s} = \frac{\mu_0}{\pi}\int_0^\infty \frac{e^{-(h_i+h_j)\lambda}}{\sqrt{\lambda^2 + \gamma_\mathrm{g}^2} + \lambda}.\cos(\mathrm{d}_{ij}\lambda)\,\mathrm{d}\lambda. \tag{3.121}$$

$[Z_\mathrm{w}(t)]$ is a diagonal matrix and corresponds to the transient conductor resistance, where each element is given by

$$Z_{\mathrm{w}i}(t) = \frac{1}{\pi\sigma_{\mathrm{w}_i}a_i^2}\sum_{m=1}^{\infty} e^{-x_m^2\frac{t}{\tau_{\mathrm{w}_i}}} \quad \text{where } \tau_{\mathrm{w}_i} = \mu_0\sigma_{\mathrm{w}_i}a_i^2, \tag{3.122}$$

and $h_i$, $h_j$, $d_{ij}$ are the corresponding heights of the two conductors $i$, $j$ and the distance between the two conductors in the horizontal plane.

Terms $x_m$ stand for the zeros of $J_1$, Bessel function of first kind.

Finally, $\gamma_g$ is the propagation constant defined as

$$\gamma_g^2 = s\mu_0(\sigma_g + s\varepsilon_0\varepsilon_{rg}), \tag{3.123}$$

where $\sigma_g$ and $\varepsilon_{rg}$ are respectively the ground conductivity and permittivity, respectively.

For the cases considered in this chapter, all conductors are in open circuit at both ends, so the current vanishes at the near and far ends.

By using the boundary formulation [11], the voltages at both ends are simply expressed as follows:

$$[V_0^n] = [V_{1/2}^n] \quad \text{and} \quad [V_{N_x}^n] = [V_{N_x-1/2}^n], \tag{3.124}$$

$$[I_0^n] = [0] \quad \text{and} \quad [I_{N_x}^n] = [0]. \tag{3.125}$$

### 3.2.5 Numerical Results

Figure 3.11 shows the transient response at the center of the straight wire $L = 20$ m, $a = 0.005$ m, located at height $h = 1$ m above a dielectric half-space ($\varepsilon_r = 10$) excited by the electromagnetic pulse (EMP):

$$E_x^{\text{inc}} = E_0(\mathrm{e}^{-at} - \mathrm{e}^{-bt}) \tag{3.126}$$

with $E_0 = 1.1\ \mathrm{V\,m^{-1}}$, $a = 7.92 * 10^4\ \mathrm{s^{-1}}$, $b = 4 * 10^4\ \mathrm{s^{-1}}$.

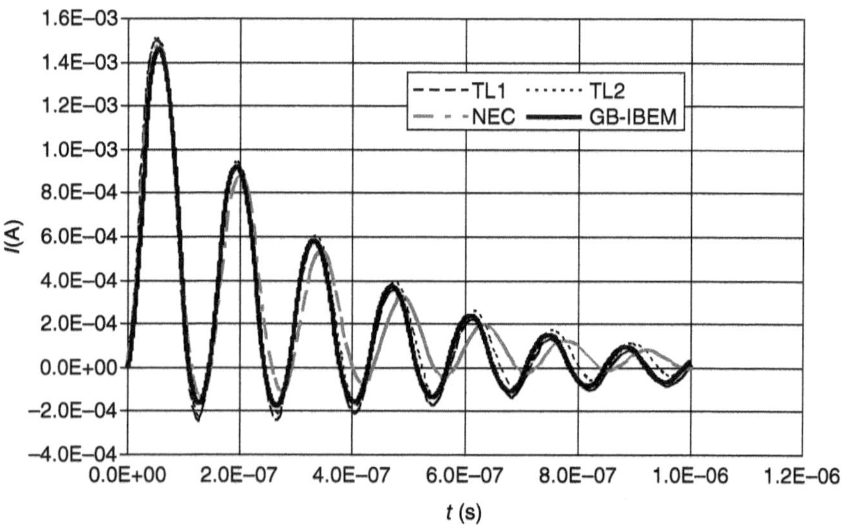


**Figure 3.11** The transient current induced at the center of the line above dielectric half-space ($\varepsilon_r = 10$).

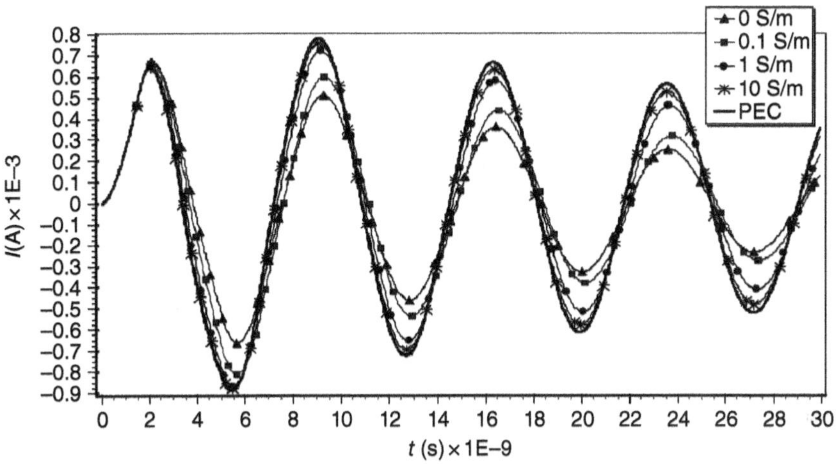


**Figure 3.12** Transient current at the wire center, $L = 1$ m, $a = 2$ mm, $h = 0.25$, $\varepsilon_r = 10$.

The next example is related to transient scattering from a straight thin wire of length is $L = 1$ m, radius $a = 2$ mm, located at height $h = 0.25$ m above ground with permittivity $\varepsilon_r = 10$, while the conductivity is varied. The wire is illuminated by the tangential EMP plane wave with $E_0 = 1$ V m$^{-1}$, $a = 4 \times 10^7$ s$^{-1}$, $b = 6 \times 10^8$ s$^{-1}$.

Figure 3.12 shows the transient current induced at the wire center for different ground conductivities.

The influence of the ground conductivity on the transient response is particularly visible from around 0.1 to 1 S m$^{-1}$.

The time domain results obtained via different approaches are found to agree satisfactorily.

The next set of examples is related to a two wire array above a PEC ground (Figure 3.13, Geometry No. 1) and dielectric half-space ($\varepsilon_r = 10$) (Figure 3.14, Geometry No. 2), respectively.

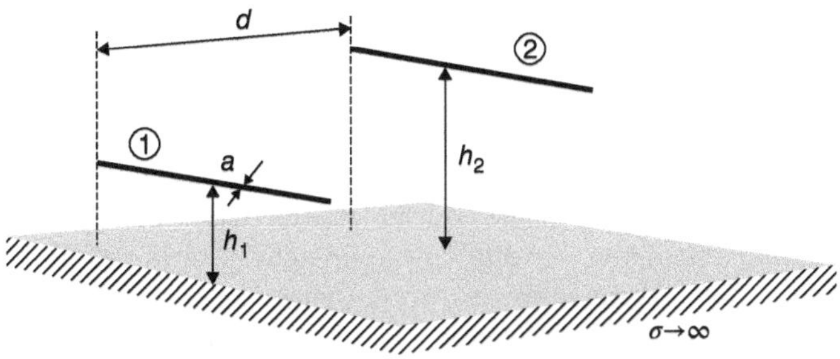


**Figure 3.13** Geometry No. 1: Two-wire array above a PEC ground ($a = 2$ cm, $L = 10$ m, $d = 1$ m, $h_1 = 1$ m, and $h_2 = 2$ m).

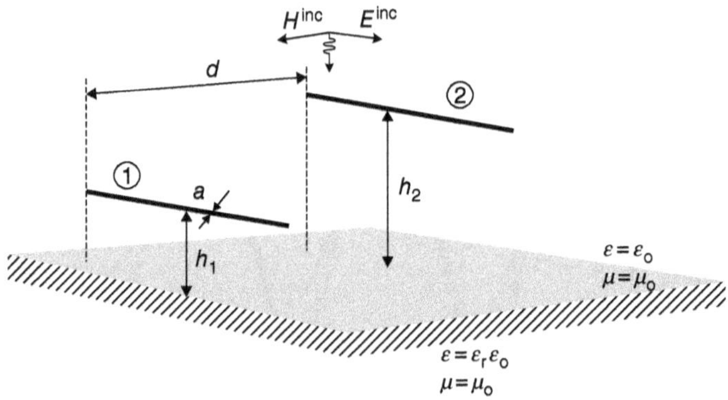


**Figure 3.14** Geometry No. 2: Two-wire array above a dielectric half-space ($\varepsilon_r = 10$, $a = 2$ cm, $L = 10$ m, $d = 1$ m, $h_1 = 1$.

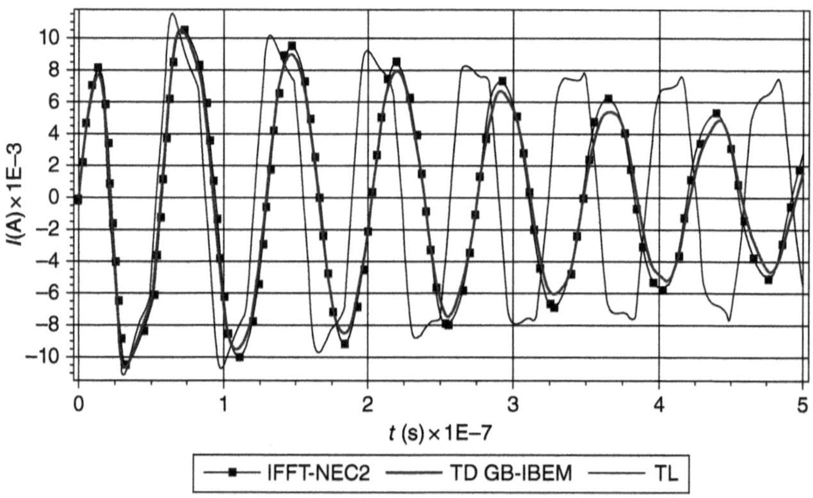


**Figure 3.15** Transient current induced at the center of wire 2 (Geometry No. 1) – comparison between IFFT-NEC2, TD GB-IBEM, and TL results.

Figures 3.15 and 3.16 show the transient current induced at the center of wire 2 for the case of Geometry No. 1 and 2, respectively, obtained via TD GB-IBEM, TL, and NEC 2 combined with inverse fast Fourier transform (IFFT).

Generally, the results calculated via different approaches are in relatively acceptable agreement. Nevertheless, some discrepancies can be noticed, in particular for the case of PEC ground. In this analysis, the applied TL model accounts not only for classical propagation effect but also for skin effects and for a correction resistance representing the radiation effect.

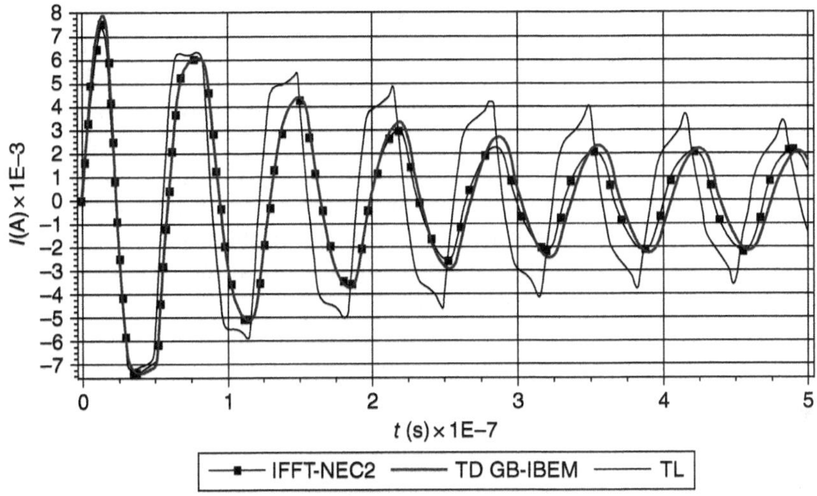


**Figure 3.16** Transient current induced at the center of wire 2 (Geometry No. 2) – comparison between IFFT-NEC2, TD GB-IBEM, and TL results.

In order to include the radiation effect in the TL model, a small DC resistance ($1\ \Omega\ \text{m}^{-1}$) has been used to represent the attenuation effect.

It is known that the TL model accounts for coupling between transverse cells only while AT takes into account mutual effects. This phenomenon is assumed to be the source of the differences in propagation velocities that are observed.

## 3.3 Applications to Antenna Systems

This section deals with some applications of thin wire modeling to various antenna systems. The specific applications are helix antennas, log-periodic dipole array (LPDA) systems, and ground penetrating dipole (GPR) antennas. The studies for helix antennas and LPDAs are carried out in the frequency domain.

### 3.3.1 Helix Antennas

Helical antennas have a number of applications in communication systems. Antennas used with portable transceivers at very high frequencies are usually in the form of helix, mounted on a radio case. Of great interest are various configurations of helical antennas either as individual elements or as parts of antenna arrays. In particular, helical antennas operate in various modes and the two principal modes are the normal (broadside) and axial (endfire) modes [21, 22].

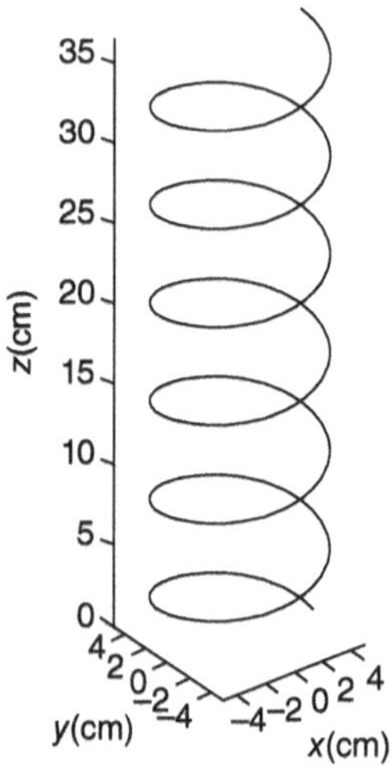


Figure 3.17 Geometry of cylindrical helix.

The normal mode radiation occurs when the helical antenna diameter is much smaller than the wavelength, while the axial mode that occurs when the helix circumference equals one wavelength ensures maximum radiation along the helical antenna axis. The helix axial mode is often of most importance and it is used in a wide range of applications [23, 24]. Antennas used with portable transceivers at very high frequencies are usually in the form of a helix, mounted on a radio case.

The boundary element modeling of certain curved wire configurations based on the solution of Pocklington's integro-differential equation for curved wires has been carried out [24].

The set of computational examples to follow deals with different geometry of helical antennas. First, the cylindrical helix, consisting of six turns, radius 5 cm, pitch angle 11°, and wire radius 1 mm, depicted in Figure 3.17, is of interest. The amplitude of current distribution at frequency $f = 30$ and 750 MHz, respectively is shown in Figures 3.18 and 3.19.

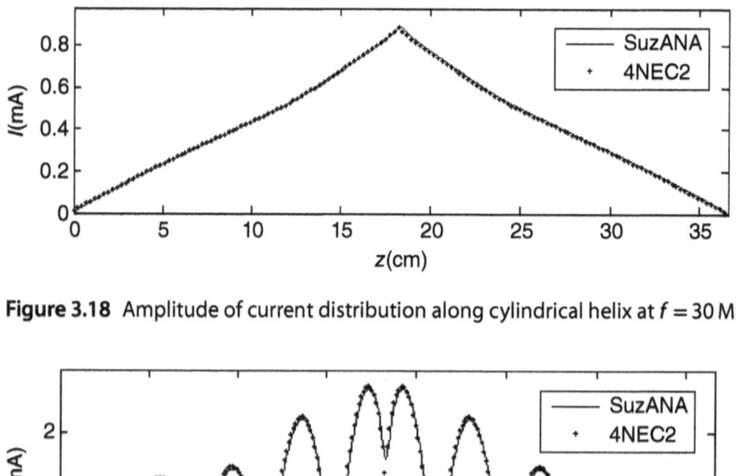


Figure 3.18 Amplitude of current distribution along cylindrical helix at $f = 30$ MHz.

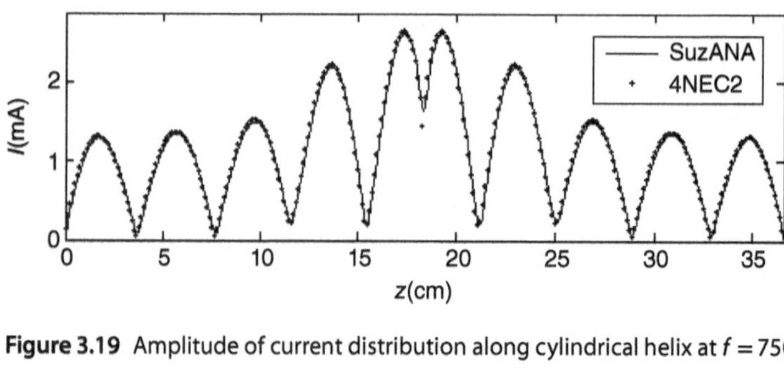


Figure 3.19 Amplitude of current distribution along cylindrical helix at $f = 750$ MHz.

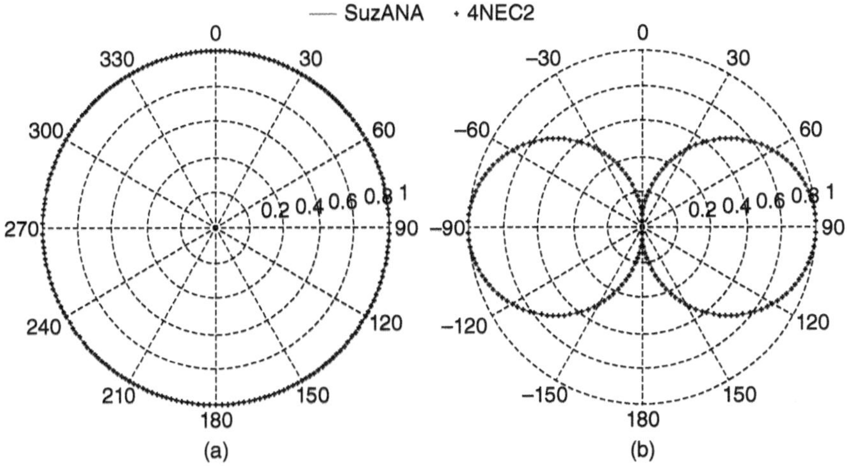


**Figure 3.20** Radiation pattern of cylindrical helix at $f = 30$ MHz. (a) Horizontal plane. (b) Vertical plane.

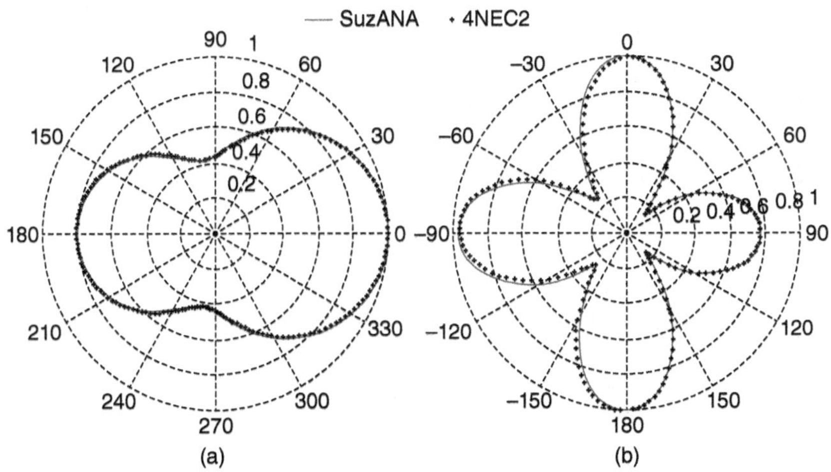


**Figure 3.21** Radiation pattern of cylindrical helix at $f = 750$ MHz. (a) Horizontal plane. (b) Vertical plane.

The related radiation pattern of cylindrical helix in horizontal and vertical plane, respectively, at frequency $f = 30$ and 750 MHz is shown in Figures 3.20 and 3.21, respectively.

The numerical results for the current distribution obtained via GB-IBEM (computed via SuzANA code [25]) are in satisfactory agreement with the results obtained by means of NEC code.

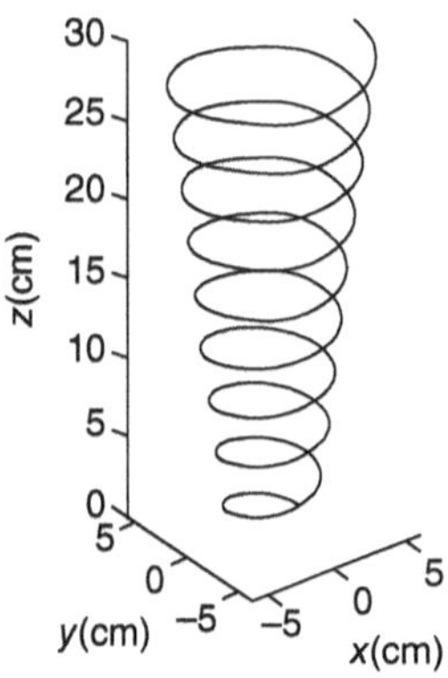


**Figure 3.22** Geometry of conical helix.

Furthermore, the conical helix consisting of nine turns, starting radius and pitch angle 5 cm and 12°, respectively, and wire radius 1 mm, shown in Figure 3.22, is of interest. The amplitude of current distribution at frequency $f = 1$ GHz is shown in Figure 3.23. The related radiation pattern is shown in Figure 3.24.

The results obtained via different approaches are in good agreement. The next example deals with the spherical helix with radius 7.5 cm and pitch angle 5°, and wire radius 0.2 mm, and is shown in Figure 3.25.

Figure 3.26 shows the amplitude of current distribution at frequency $f = 1$ GHz, while the related radiation pattern is shown in Figure 3.27. The results obtained via different approaches are in satisfactory agreement.

The last example deals with a multiple helix configuration presented in Figure 3.28. The helix at the center of the coordinate system is the active antenna. The distance between the wire axes is 0.5 m. The amplitudes of current distribution at frequency $f = 750$ MHz are shown in Figures 3.29–3.31.

The numerical results obtained by GB-IBEM are in good agreement with the results obtained via 4NEC2 code.

### 3.3.2 Log-Periodic Dipole Arrays

The LPDA consists of parallel straight wires having successively increasing lengths outward from the feeding point at the apex. Smaller elements in front of and larger elements behind each dipole generate a directional primary pattern as the Yagi–Uda array. However, in the case of Yagi–Uda antenna only one element is directly driven and the other wires operate in a parasitic mode, while all the LPDA elements are connected to feeder, i.e. excited by attaching dipole arms to the two-wire TL [26]. The TL feeder is crossed between each

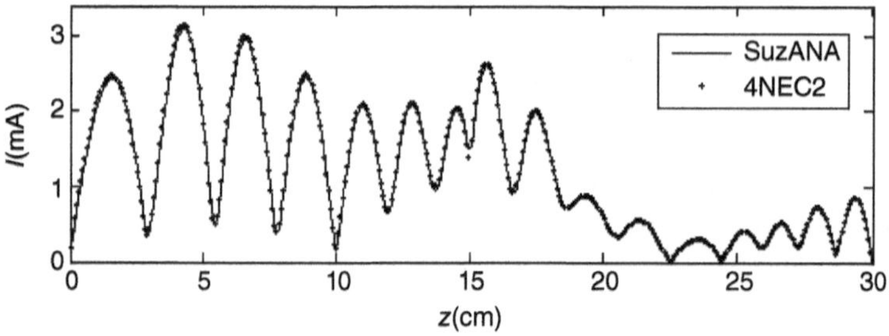


**Figure 3.23** Amplitude of current distribution along conical helix at $f = 1$ GHz.

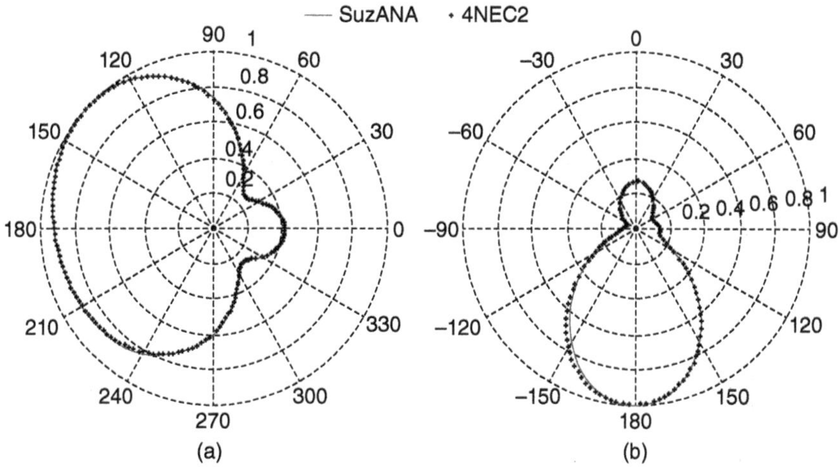


**Figure 3.24** Radiation pattern of conical helix at $f = 1$ GHz. (a) Horizontal plane. (b) Vertical plane.

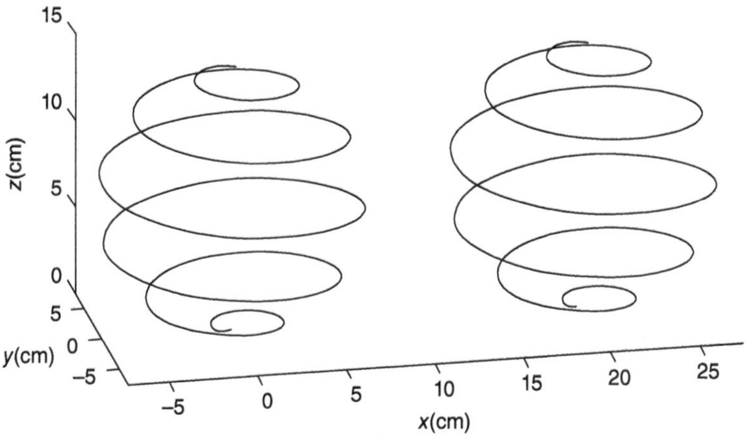


**Figure 3.25** Geometry of spherical helix.

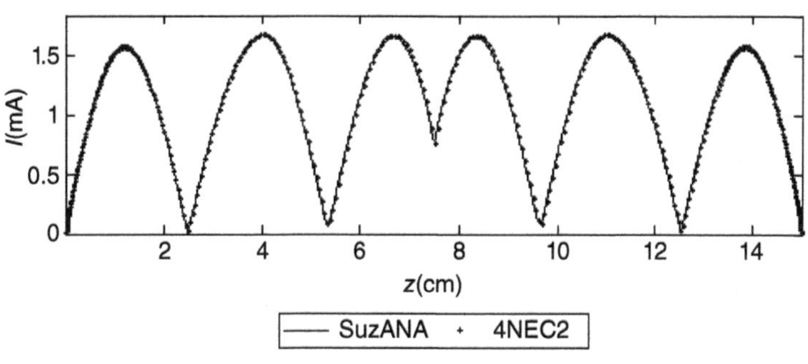


**Figure 3.26** Amplitude of current distribution along the spherical helix at $f = 500$ MHz.

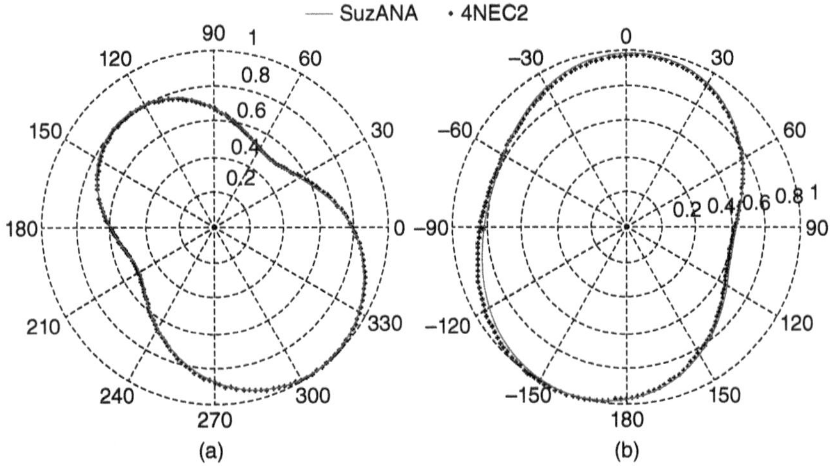


**Figure 3.27** Radiation pattern of spherical helix at $f = 500$ MHz. (a) Horizontal plane. (b) Vertical plane.

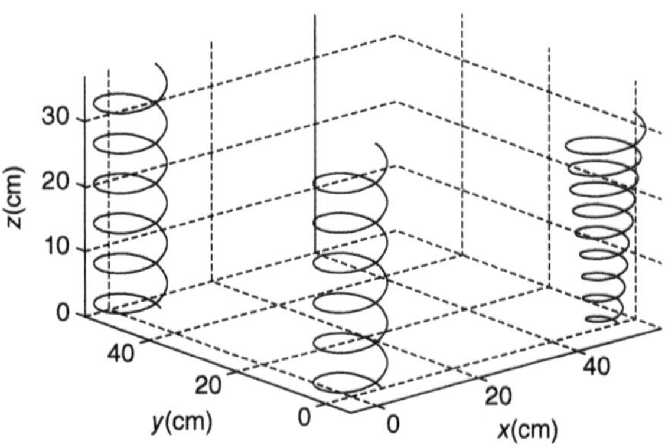


**Figure 3.28** System of multiple helical antennas.

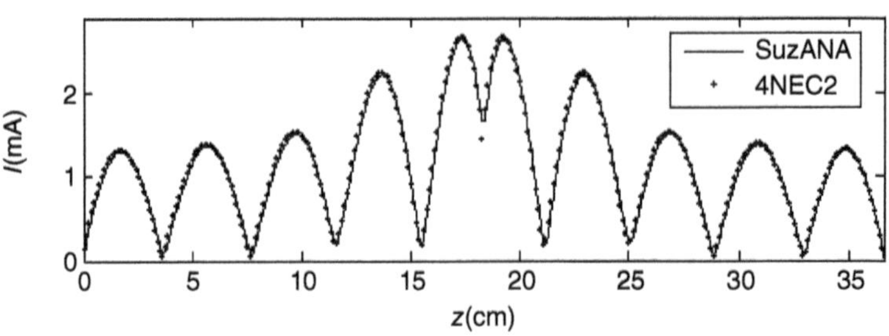


**Figure 3.29** Amplitude of current distribution along active cylindrical helix at $f = 750$ MHz.

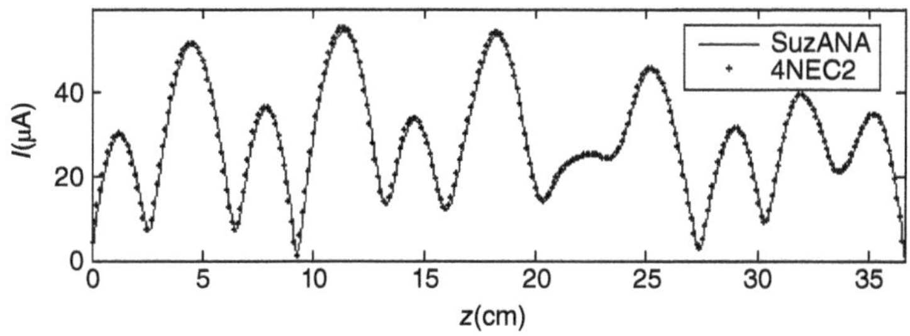


**Figure 3.30** Amplitude of current distribution along passive cylindrical helix at $f = 750$ MHz.

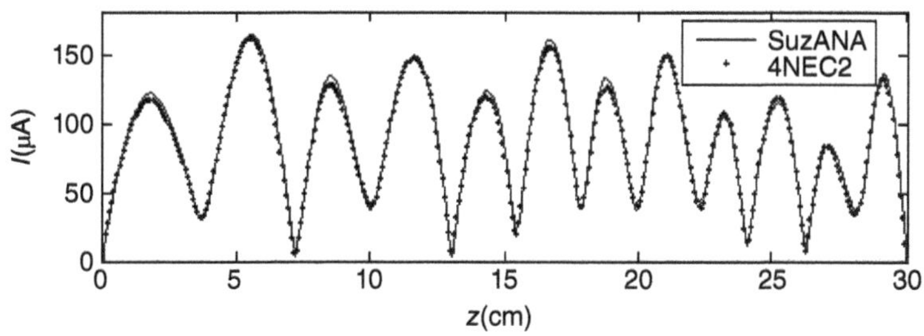


**Figure 3.31** Amplitude of current distribution along conical helix at $f = 750$ MHz.

dipole, thus reversing the fire direction. The LPDA antennas are easy to optimize, while the crossing of the feeder between each dipole element results in mutual cancellation of backlobe components from the individual elements yielding a very low level of backlobe radiation (around 25 dB below main lobe gain at HF and 35 dB at VHF and UHF). To achieve an ideal log-periodic configuration, an infinite array is required, while the practical broadband radiator is truncated at both ends, thus limiting the operating frequency to a certain bandwidth. The cutoff frequencies of such truncated structures are then determined by the electrical lengths of the largest and shortest element of the structure, respectively [22, 26].

Logarithmic antenna arrays are often used for electronic beam steering with an important application in air traffic, particularly in landing. Although the majority of landings can be performed via visual cues, aircraft frequently land in weather conditions requiring electronic assistance to the pilot, or autopilot, respectively. Thus, LPDA is an essential part of typical localizer antenna arrays (*Narrow Aperture* with 8 or 14 elements, respectively, and *Wide Aperture* 14 elements), which play an important role within the Instrumental Landing System (ILS). The purpose of the localizer is to shape a radiation pattern thus providing lateral guidance to the aircraft beginning its descent, intercepting the projected runway center line, and then making a final approach. Glide slope provides vertical guidance, while marker beacons alert pilots of their progress along the glide path.

Boundary element modeling of LPDA arrays, based on the set of coupled Pocklington's integro-differential equations for an arbitrary configuration of thin wires above a lossy ground, is reported in [27].

Computational examples are related to a single LPDA and *Wide Aperture* consisting of 14 elements per LPDA array, respectively. All numerical results have been obtained by using the advanced version of TWiNS (Thin Wire Numerical Solver) code [4].

First, a single LPDA, shown in Figure 3.32, is considered. The parameters of the LPDA are as follows: $\tau = 0.983$, $L_1 = 1.27$ m, $d_1 = 0.4756$ m (first distance between two dipoles), $N = 7$ (number of dipoles on LPDA), $a = 4$ mm (dipole radius), $N_{seg} = 11$ (number of segments per dipole) $h = 1.82$ m (height above ground), $\sigma = 5\,\text{mS m}^{-1}$ (ground conductivity), $\varepsilon_r = 13$ (relative permittivity of the ground), $f = 110$ MHz.

The absolute values of current distribution along the dipoles are shown in Figure 3.33.

The related radiation pattern of the structure is shown in Figures 3.34 and 3.35, respectively.

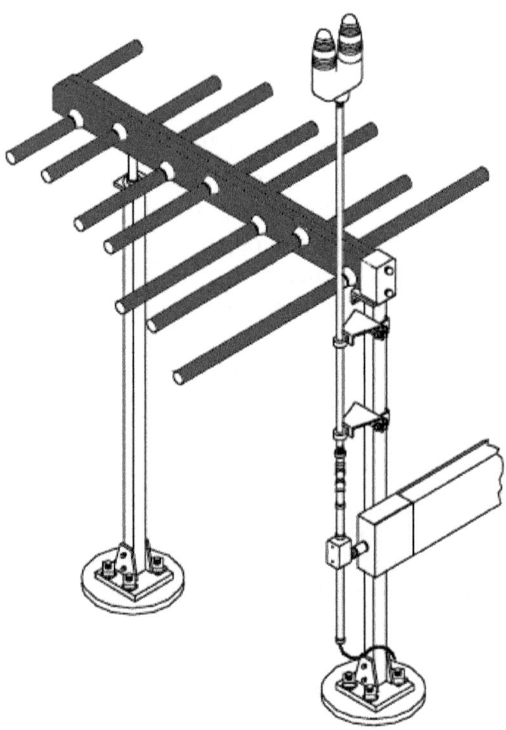

**Figure 3.32** The localizer antenna element.

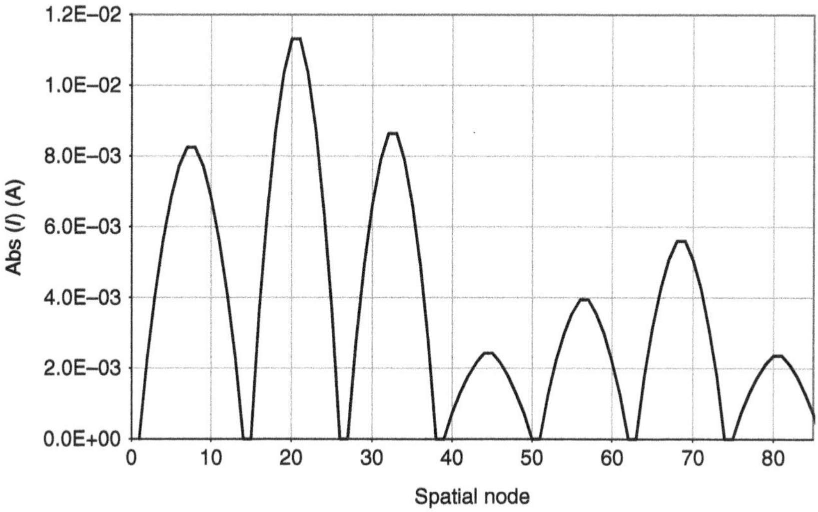


**Figure 3.33** The absolute values of currents induced along the single LPDA.

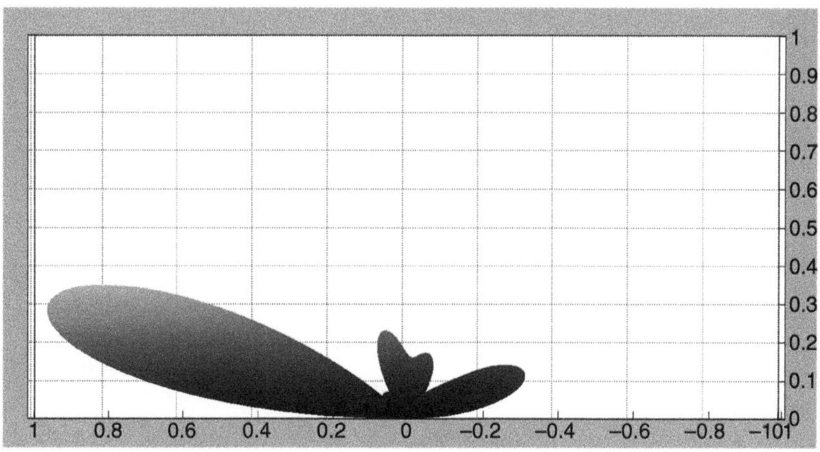


**Figure 3.34** The single LPDA radiation pattern (lateral view).

Figure 3.36 shows a realistic geometry of a localizer antenna system consisting of 14 LPDA elements. The related radiation pattern is presented in Figures 3.37 and 3.38, respectively.

A realistic antenna system used at the airport consists of 14 LPDA elements symmetrically placed over the $y$-axis. The coordinates of each element along with the corresponding magnitudes of the voltage sources excitations are given

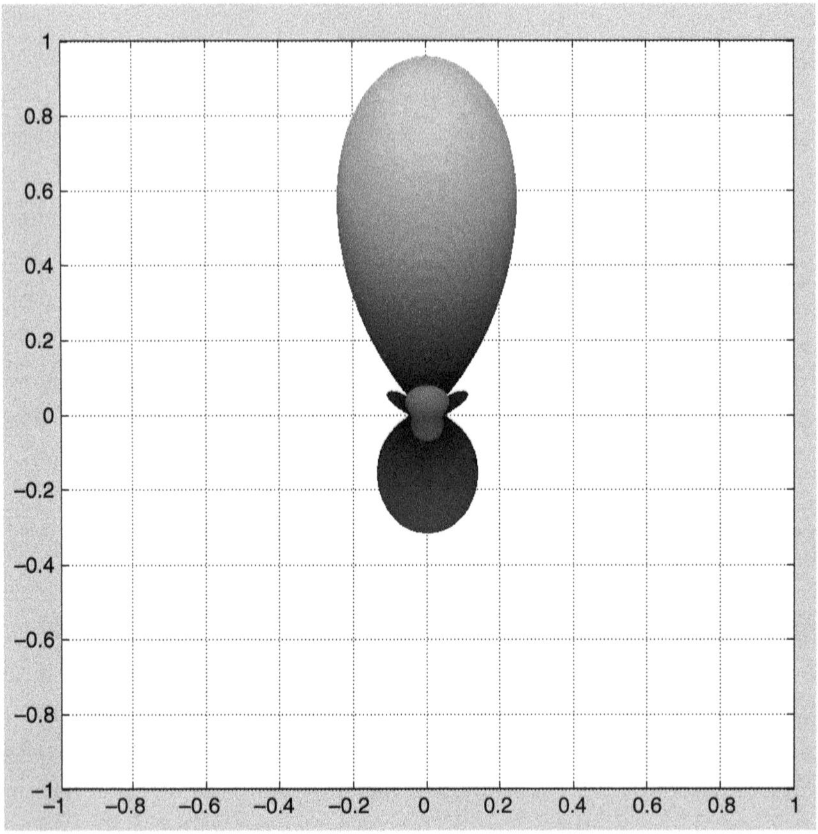


**Figure 3.35** The single LPDA radiation pattern (top view).

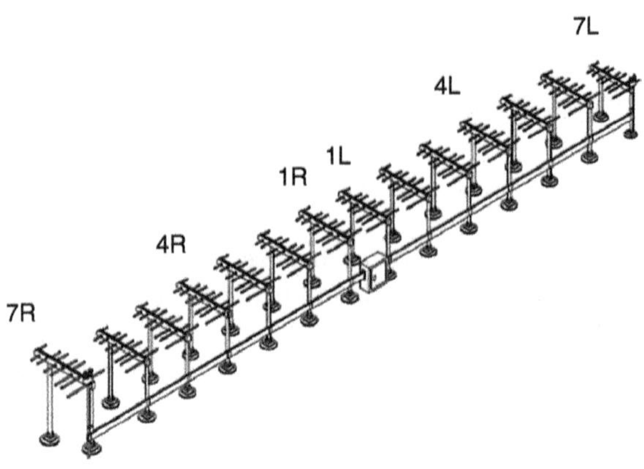


**Figure 3.36** Geometry of 14-element LPDA array.

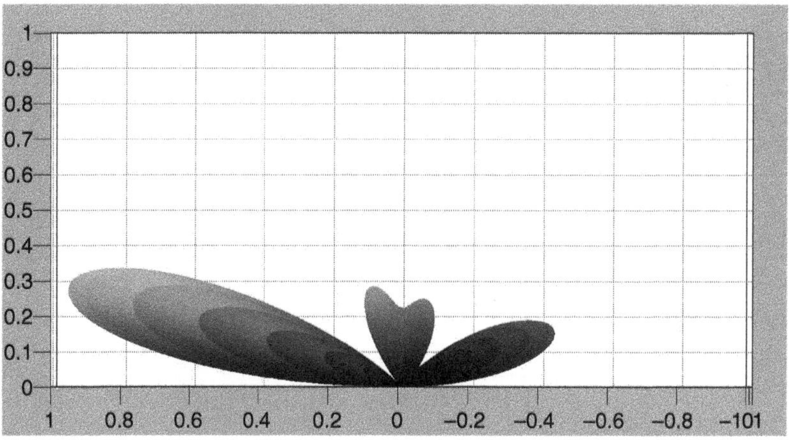

**Figure 3.37** The LPDA array radiation pattern (lateral view).

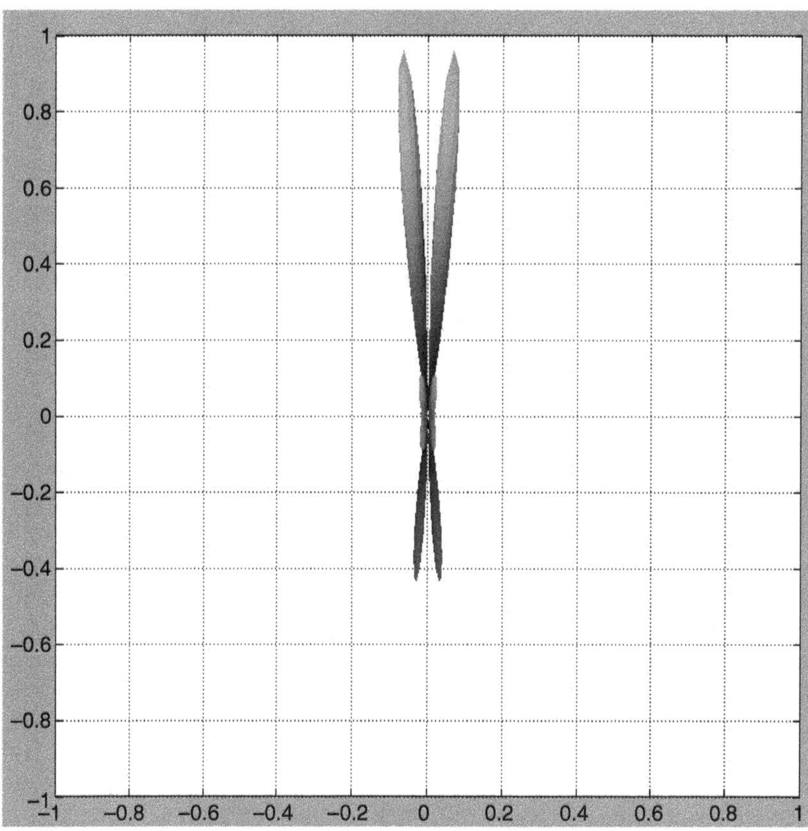

**Figure 3.38** The LPDA array radiation pattern (top view).

**Table 3.2** LPDA parameters.

| Antenna element | Y-coordinate (m) | $V_s$ (V) |
|---|---|---|
| 7L | −15.113 | −0.307 |
| 6L | −12.5476 | −0.605 |
| 5L | −10.16 | −0.8465 |
| 4L | −7.47 | −1 |
| 3L | −5.3594 | −0.95 |
| 2L | −2.9464 | −0.597 |
| 1L | −0.8128 | −0.122 |

in Table 3.2, where $V_s$ is the voltage source and $1L$ stands for the first antenna on the left side (closest to the center of the array).

Furthermore, $1R$ is then the first antenna on the right side of the array with $y = 0.8128$, $V_s = 0.122$ V, etc. Each pair of the antennas, e.g. $4L$ and $4R$, is fed by the voltage source having the same amplitude but opposite phase. This kind of antenna spatial arrangement, together with the voltage distribution along the array, results in the specific shape of radiation pattern enabling the airplane to line up with the airport runway.

The major limitation designing the localizer antenna systems is related to their high sensitivity to environment conditions. Namely, aircraft and vehicles on the airport surface, in particular on taxiways near the antennas, can reflect signals thus causing significant errors along the approach path. Advanced numerical modeling applications can be used for simulation of different solutions to predict such scenarios.

### 3.3.3 GPR Dipole Antennas

GPR is the geophysical system that utilizes the electromagnetic technique to probe the subsurface [28]. From its first successful application in measuring the ice thickness of polar caps in 1960s to date, the development of both hardware and software has enabled the use of GPR in numerous applications: archeology, earth sciences, civil engineering, underground engineering, utility detection etc. [28]. The GPR antenna is moved across the surface, at the same time transmitting and receiving the electromagnetic waves. The wave propagates through the lower medium at the velocity determined by the permittivity of the medium. Once reaching the target, the wave is scattered and detected by the antenna. The type and size of the antenna strongly depends on the specific application, but the most commonly used antennas are dipole and bowtie [28–30]. GPR operates over a finite range of frequencies. High frequency antennas have smaller size and ensure better resolution but at the cost of penetration depth. Physically larger, low frequency antennas provide a deeper penetration into the subsurface but are less easily packaged. Knowledge of the energy transmitted into the subsurface ensures better antenna design and gives deeper understanding of the target reflected wave [31].

Several numerical techniques in either frequency or time domain have been developed thus gaining a deeper insight into the transmitted field behavior [8, 31]. Time domain techniques are convenient particularly due to their ability to cover the large frequency spectrum within a single simulation. Some of the reported models can be found in [32], providing a comparison of the FDTD, finite integration technique (FIT), and time domain integral equation (TDIE) models. Frequency domain modeling has been reported in [31]. The main advantage of this approach is less computational effort, and it is considered to be a plausible choice if single frequency scenario is considered.

Comparison of the results for transmitted field, generated by the horizontal GPR dipole antenna above a dielectric half-space, obtained by solving the corresponding space–frequency Pocklington's integro-differential equation and space–time Hallen integral equations and related field formulas is reported in [33]. The corresponding transient response from the corresponding frequency response is obtained by using the IFFT.

The geometry of interest is a straight horizontal dipole antenna of length $L = 1$ m and radius $a = 6.74$ mm, radiating above a dielectric half-space ($\varepsilon_r = 10$) at height $h = 0.5$ m (Figure 3.39). The wire is considered as a perfect conductor and the voltage source is applied to the gap in the center.

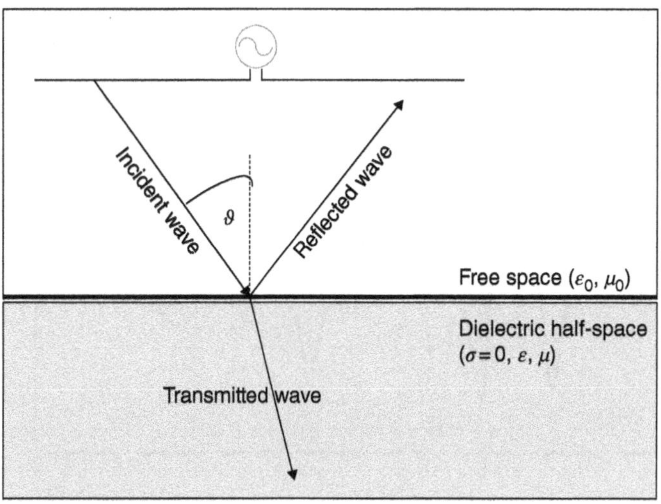


**Figure 3.39** GPR dipole antenna above a dielectric half-space.

The antenna is excited by the Gaussian pulse voltage source:

$$V(t) = V_0 \cdot \exp\left(-\left(\frac{t - t_0}{t_w}\right)^2\right), \tag{3.127}$$

where $V_0 = 1$ V is the amplitude, $t_0 = 1.43$ ns is the time delay of a pulse, and $t_w = 2/3$ ns is the half-width of a pulse in the time domain.

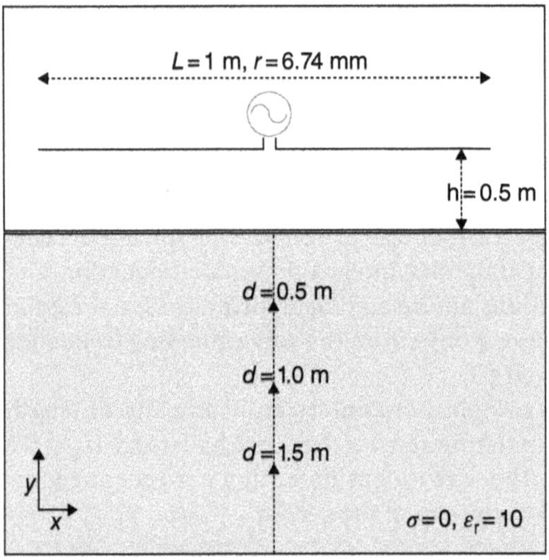


Figure 3.40 GPR dipole antenna above a dielectric half-space.

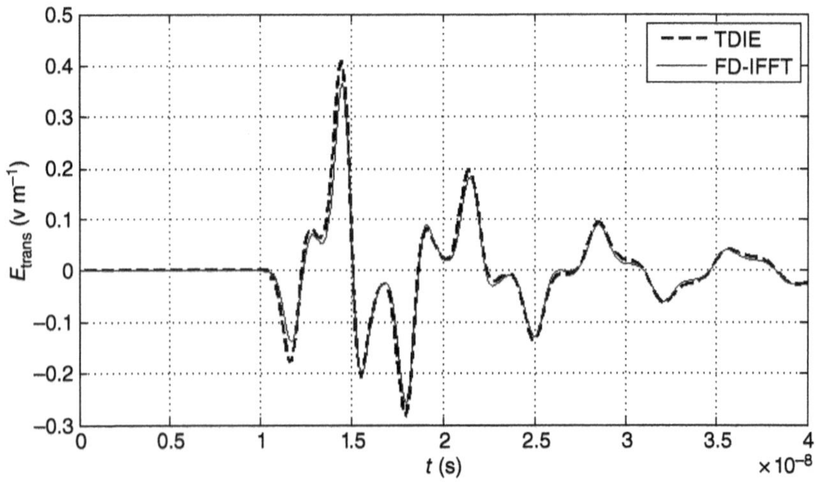


Figure 3.41 The horizontal ($E_z$) component of the transmitted electric field for penetration depth $d = 0.5$ m.

The frequency domain counterpart of the time domain Gaussian pulse is given by

$$V(f) = \sqrt{\pi}\ t_w\ \mathrm{e}^{-(\pi f t_w)^2} \mathrm{e}^{-f 2\pi f t_0}. \tag{3.128}$$

The field formulas in the frequency and time domain, respectively, are documented elsewhere, e.g. in [31, 34]. Comparison is carried out for the penetration depths of $d = 0.5$, 1.0 and 1.5 m, as depicted in Figure 3.40.

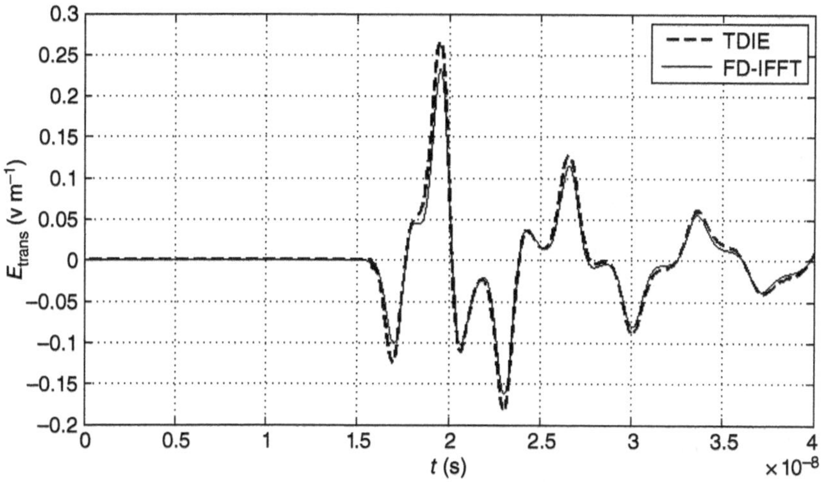


**Figure 3.42** The horizontal ($E_z$) component of transmitted electric field for penetration depth $d = 1$ m.

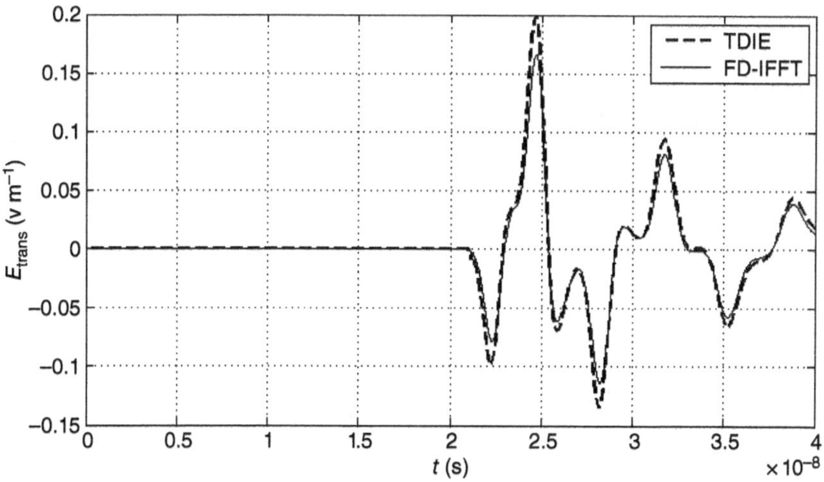


**Figure 3.43** The horizontal ($E_z$) component of transmitted electric field for penetration depth $d = 1.5$ m.

The results for the transmitted electrical field obtained via both frequency and time domain approaches are shown in Figures 3.41–3.43.

Note that the time domain method predicts somewhat higher values of the transmitted field. Nevertheless, the results show satisfactory agreement.

## References


1 Tesche, F., Ianoz, M., and Carlsson, F. (1997). *EMC Analysis Methods and Computational Models*. New York: Wiley.

2 Tkatchenko, S., Rachidi, F., and Ianoz, M. (1995). Electromagnetic field coupling to a line of a finite length: theory and fast iterative solutions in frequency and time domains. *IEEE Transactions on Electromagnetic Compatibility* **37** (4): 509–518.

3 Tkatchenko, S., Rachidi, F., and Ianoz, M. (2001). High frequency electromagnetic field coupling to long terminated lines. *IEEE Transactions on Electromagnetic Compatibility* **43** (2): 117–129.

4 Ianoz, M. (1999). Electromagnetic field coupling to lines, cables and networks, a review of problems and Solutions. *Proceedings of the International Conference on Electromagnetics in Advanced Applications, ICEAA'95*, Turin, Italy (12–15 September 1999), 75–80.

5 Degauque, P. and Zeddam, A. (1998). Remarks on the transmission approach to determining the current induced on above-ground cables. *IEEE Transactions on Electromagnetic Compatibility* **30** (1): 77–80.

6 Agrawal, A.K., Price, H.J., and Gurbaxani, S.H. (1980). Transient response of a multiconductor transmission line excited by a nonuniform electromagnetic field. *IEEE Transactions on Electromagnetic Compatibility* **EMC-22** (2): 119–129.

7 Poljak, D. and Roje, V. (1998). Time domain modeling of electromagnetic field coupling to transmission lines. *Proceedings of the 1988 IEEE EMC Symposium*, Denver, USA (August 1998), 1010–1013.

8 Poljak, D. (2007). *Advanced Modelling in Computational Electromagnetic Compatibility*. New York: Wiley.

9 Poljak, D., Rachidi, F., and Tkachenko, S. (2007). Generalized form of telegrapher's equations for the electromagnetic field coupling to finite length lines above a Lossy ground. *IEEE Transactions on Electromagnetic Compatibility* **49** (3): 689–697.

10 Poljak, D., Doric, V., Sesnic, S., et al. (2007). Electromagnetic field coupling to overhead wires: comparison of frequency domain wire antenna and transmission line model. *ICECom 2007, Conference Proceedings, 19th International Conference on Applied Electromagnetics and Communications*, 119–122.

11 Poljak, D., Antonijevic, S., El Khamlici Drissi, K., and Kerroum, K. (2010). Transient response of straight thin wires located at different heights above a ground plane using antenna theory and transmission line approach. *IEEE Transactions on Electromagnetic Compatibility* **52** (1): 108–1116.

12 Miller, E.K., Poggio, A.J., Burke, G.J., and Selden, E.S. (1972). Analysis of wire antennas in the presence of a conducting half-space, Part II: the horizontal antenna in free space. *Canadian Journal of Physics* **50**: 2614–2627.

13 Poljak, D., Doric, V., Milisic, M., and Birkic, M. (2010). Modeling of array of log-periodic dipole antennas array for air traffic applications. *ICECom 2010,* Dubrovnik, Croatia (September 2010).

14 Doric, V. and Poljak, D. (2010). EMC analysis of the PLC system based on the antenna theory. *ICECom 2010,* Dubrovnik, Croatia (September 2010).

15 Poljak, D., Doric, V., and Antonijevic, S. (2007). *Computer aided Design of Wire Structures: Frequency and Time Domain Analysis.* Southampton-Boston: WIT Press.

16 FCC Part 15 Subpart C – Intentional Radiators § 15.209 – Radiated Emission Limits, General Requirements.

17 CISPR 22/IEC *Information technology Equipment – Radio Disturbance Characteristics,* 3ee, 11–97. Limits and Methods of Measurements.

18 Antonijevic, S. and Poljak, D. (2010). On time domain numerical modeling of a thin wire above a Lossy ground. *SoftCOM 2010,* Split, Brac Island, Croatia (September 2010).

19 Poljak, D., Antonijevic, S., and Doric, V. (2011). *An Efficient Boundary Element Modeling of the Time Domain Integral Equations for Thin Wires Radiating in a Presence of a Lossy Media* BEM 2011. Southampton, UK: WIT Press.

20 Barnes, P. and Tesche, F. (1991). On the direct calculation of a transient plane wave reflected from a finitely conducting half-space. *IEEE Trans, EMC* **33** (2): 449–457.

21 Stutzman, W.L. (1981). *Antenna Theory and Design.* Wiley.

22 Balanis, C.A. (2005). *Antenna Theory: Analysis Design,* 3ee. Wiley.

23 Hoole, P.R.P. ed. (2001). Smart antennas and signal processing for communications. In: *Biomedical and Radar Systems.* Southampton-Boston: WIT Press.

24 Poljak, D., Doric, V., and Lerinc, F. (2015). *Boundary Element Modeling of Curved Wire Configurations,* BEM/MRM 39. New Forest, UK: WIT Press.

25 Poljak, D., Doric, V., and Antonijevic, S. (2009). *Computer Modeling of Wire Structures (in Croatian).* Zagreb: Kigen.

26 Burberry, R.A. (1992). *VHF and UHF Antennas, IEE Electromagnetics Series 35.* London, UK: Peter Peregrinus Ltd.

27 Poljak, D., Doric, V., Milisic, M., and Birkic, M. (2010). Modeling of array of log-periodic dipole antennas for air traffic applications. *ICECom 2010,* Dubrovnik (October 2010).

28 Takahashi, K., Igel, J., Preetz, H., and Kuroda, S. (2012). Basics and application of ground-penetrating radar as a tool for monitoring irrigation process, problems. In: *Perspectives and Challenges of Agricultural Water Management* (ed. M. Kumar). InTech. ISBN: 978-953-51-0117-8.

29 Pajewski, L., Benedetto, A., Derobert, X. et al. (2013, 2013). *"Applications of Ground Penetrating Radar in Civil Engineering", COST Action TUI208*. IEEE.

30 Warren, C., Chiwaridzo, N., and Giannopoulos, A. (2014). Radiation characteristics of a high-frequency antenna in different dielectric environments. 15th International Conference on Ground Penetrating Radar – GPR 2014, Brussels, Belgium, 796–801.

31 Poljak, D. and Dorić, V. (2015). Transmitted field in the Lossy ground from ground penetrating radar (GPR) dipole antenna. In: *Computational Methods and Experimental Measurements XVII*, 3, WIT Transactions on Modelling and Simulation, vol. **59**, 3–11. WIT Press.

32 Warren, C., Pajewski, L., Poljak, D., et al. (2016). A comparison of finite-difference, finite-integration, and integral-equation methods in the time domain for modelling ground penetrating radar antennas. Hong Kong, 16th International Conference of Ground Penetrating Radar.

33 Susnjara, A., Poljak, D., Sesnic, S., and Doric, V. (2016). Time domain and frequency domain integral equation method for the analysis of ground penetrating radar (GPR) antenna. *SoftCOM 2016*, Split (September 2016).

34 Poljak, D., Antonijevic, S., Šesnic, S. et al. (2016). On deterministic-stochastic time domain study of dipole antenna for GPR applications. *Engineering Analysis with Boundary Elements* **73**: 14–20.

# 4

# Electromagnetic Field Coupling to Buried Wires

Modeling of electromagnetic field coupling to thin wire configurations buried in a lossy medium is important in many electromagnetic compatibility (EMC) applications, e.g. communications and power cables, geophysical investigations, and grounding systems. The problem can be studied in either frequency or time domain by using the transmission line (TL) model, or thin wire antenna theory (AT) (full wave model) [1, 2] with the latter regarded as a more rigorous one. The TL approach is a valid approximation for long straight conductors with electrically small cross sections but it is not acceptable for finite length wires, particularly for wires of arbitrary shape and high frequency excitations. Consequently, the AT has to be used.

However, a principal drawback of the wire antenna models of buried conductors is the relatively high computational cost. On the other hand, using the enhanced TL model enables one to overcome some limitations of the model restrictions. Thus, a rigorous relationship between frequency domain TL equations and integral relationships arising from the wire AT for the single wire below ground is available [3].

The comparison between the frequency domain wire antenna model and TL model for a single buried conductor is given in [4, 5], and for multiple buried wires in [6]. The formulation used in [6], arising from the wire AT, is based on the set of the Pocklington's integro-differential equations for half-space problems. The TL model used in [4–6] deals with telegrapher's equations. The set of Pocklington's equations is numerically handled by means of the frequency domain Galerkin–Bubnov scheme of the indirect boundary element method (GB-IBEM) [2]. The telegrapher's equations, arising from TL model, are solved using the chain matrix method [4–6].

## 4.1 Frequency Domain Modeling

This section deals with comparison of the wire AT approach and the TL approximation both used in the analysis of electromagnetic coupling to buried

*Computational Methods in Electromagnetic Compatibility: Antenna Theory Approach versus Transmission line Models*, First Edition. Dragan Poljak and Khalil El Khamlichi Drissi.

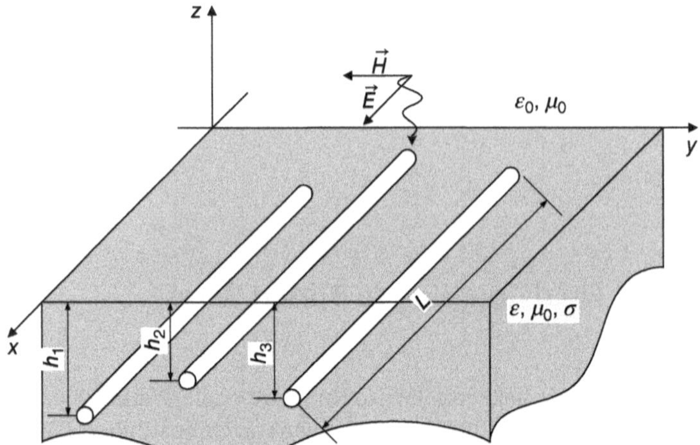


**Figure 4.1** The geometry of horizontal buried lines.

conductors in the frequency domain. The AT approach is based on the set of Pocklington's integro-differential equations for arbitrary wires. The effect of a lower lossy half-space is taken into account by means of approximate reflection coefficient (RC) approximation [7]. The resulting integro-differential equations are numerically solved via a frequency domain version of the GB-IBEM. The TL model in the frequency domain is based on the corresponding telegrapher's equations, which are treated using the chain matrix method.

The geometry of interest related to multiple horizontal conductors buried in a lossy ground is depicted in Figure 4.1.

The frequency response of buried conductors configurations using the AT and TL approach, respectively has been reported in [8].

### 4.1.1 Antenna Theory Approach: Set of Coupled Pocklington's Equations for Arbitrary Wire Configurations

The set of coupled Pocklington's equations for multiple buried wires of arbitrary shape can be readily derived as an extension of the Pocklington's integro-differential equation for a single buried wire enforcing the continuity condition for the tangential field component along the thin wires surface [8].

The wire placed in an unbounded medium is first considered, and then the formulation is extended to a corresponding half-space problem.

Assuming a perfectly conducting wire the total field composed from the excitation field $\vec{E}^{exc}$ and scattered field $\vec{E}^{sct}$ vanishes:

$$\vec{s} \cdot (\vec{E}^{exc} + \vec{E}^{sct}) = 0 \text{ on the wire surface,} \tag{4.1}$$

where $\vec{s}$ is the unit vector tangent at the observation point.

Starting from Maxwell's equations and Lorentz gauge the scattered field is expressed in terms of the magnetic vector potential $\vec{A}$:

$$\vec{E}^{\rm sct} = -j\omega\vec{A} + \frac{1}{j\omega\mu\varepsilon_{\rm eff}}\nabla(\nabla\vec{A}), \tag{4.2}$$

where the complex permittivity of the lossy ground $\varepsilon_{\rm eff}$ is given by

$$\varepsilon_{\rm eff} = \varepsilon_{\rm r}\varepsilon_0 - j\frac{\sigma}{\omega}, \tag{4.3}$$

where $\varepsilon_{\rm r}$ and $\sigma$ are the relative permittivity and conductivity of the ground, respectively, and $\omega$ is the operating frequency.

The magnetic vector potential is defined by the particular integral:

$$\vec{A}(s) = \frac{\mu}{4\pi}\int_C I(s')g_0(s,s')\vec{s}'\,{\rm d}s', \tag{4.4}$$

where $I(s')$ is the induced current along the line, $\vec{s}'$ is the unit vector tangent at the source point, and $g_0(s, s')$ is the Green's function of the form

$$g_0(s,s') = \frac{{\rm e}^{-jk_1R_0}}{R_0}, \tag{4.5}$$

where

$$k_1 = \omega\sqrt{\mu_0\varepsilon_{\rm eff}}, \tag{4.6}$$

with $k_1$ being the complex propagation constant of the lossy ground, while $R_0$ is the distance from the source to the observation point, respectively:

$$R_0 = \sqrt{(x-x')^2 + (y-y')^2 + (z-z')^2 + a^2}, \tag{4.7}$$

where $a$ denotes the wire radius.

Combining Equations (4.1)–(4.6) leads to the Pocklington's integro-differential equation for the unknown current distribution along the arbitrarily shaped wire insulated in an unbounded lossy medium [8]:

$$E^{\rm exc}(s) = -\frac{1}{j4\pi\omega\varepsilon_{\rm eff}}\int_{C'} I(s')\cdot\vec{s}\cdot\vec{s}'\cdot[k_1^2 + \nabla\nabla]\,g_0(s,s'){\rm d}s'. \tag{4.8}$$

Integral equation for an infinite lossy medium (4.8) can be extended to a case of a thin wire located near the interface between two media by modifying the kernel to account for the electric field reflecting from the interface.

The excitation field component can then be written as the sum of the incident field $\vec{E}^{\rm inc}$ and the field reflected from the interface $\vec{E}^{\rm ref}$, i.e.

$$\vec{E}^{\rm exc} = \vec{E}^{\rm inc} + \vec{E}^{\rm ref} \tag{4.9}$$

while the reflected field $\vec{E}^{\text{ref}}$ is

$$\vec{E}^{\text{ref}}(s) = \frac{1}{j4\pi\omega\varepsilon_{\text{eff}}} \left[ \frac{k_0^2 - k_1^2}{k_0^2 + k_1^2} \int_{C'} I(s') \cdot \vec{s}^* \cdot [k_1^2 + \nabla\nabla] g_i(s, s^*) \mathrm{d}s' + \int_{C'} I(s') \cdot \vec{s}^* \cdot G_s(s, s') \mathrm{d}s' \right]. \tag{4.10}$$

The Green function $g_i(s, s^*)$ arising from the image theory is

$$g_i(s, s^*) = \frac{\mathrm{e}^{-jk_1R_1}}{R_1}, \tag{4.11}$$

with $R_1$ being the distance from the image point to the observation point, respectively, and $\vec{s}^*$ is the unit vector tangential at the source point of the image wire, while $k_0$ is propagation constant of air:

$$k_0^2 = \omega^2 \mu_0 \varepsilon_0. \tag{4.12}$$

The kernel $G_s(s, s')$

$$\begin{aligned} \vec{G}_s(s, s') = {} & (\vec{e}_x \cdot \vec{s}') \cdot (G_\rho^H \cdot \vec{e}_\rho + G_\phi^H \cdot \vec{e}_\phi + G_z^H \cdot \vec{e}_z) \\ & + (\vec{e}_z \cdot \vec{s}') \cdot (G_\rho^V \cdot \vec{e}_\rho + G_z^V \cdot \vec{e}_z) \end{aligned} \tag{4.13}$$

is a correction (attenuation) term containing the Sommerfeld integrals involving the following components for horizontal and vertical dipoles [7, 9]:

$$G_\rho^V = \frac{\partial^2}{\partial\rho\, \partial z} k_0^2\, V^R, \tag{4.14}$$

$$G_z^V = \left( \frac{\partial^2}{\partial z^2} + k_1^2 \right) k_0^2\, V^R, \tag{4.15}$$

$$G_\rho^H = \cos\varphi \left( \frac{\partial^2}{\partial\rho^2} k_1^2\, V^R + k_1^2\, U^R \right), \tag{4.16}$$

$$G_\phi^H = -\sin\phi \left( \frac{1}{\rho}\frac{\partial}{\partial\rho} k_1^2\, V^R + k_1^2\, U^R \right), \tag{4.17}$$

$$G_z^H = -j4\pi\omega\varepsilon_{\text{efec}} \cos\phi\, G_\rho^V. \tag{4.18}$$

The Sommerfeld integral terms are

$$U^R = \int_0^\infty D_1(\lambda)\, \mathrm{e}^{-\gamma_1|z+z'|} J_0(\lambda\rho)\, \lambda\, \mathrm{d}\lambda, \tag{4.19}$$

$$V^R = \int_0^\infty D_2(\lambda) \mathrm{e}^{-\gamma_1|z+z'|} J_0(\lambda\rho)\, \lambda\ \mathrm{d}\lambda, \tag{4.20}$$

where

$$D_1(\lambda) = \frac{2}{\gamma_0 + \gamma_1} - \frac{2k_1^2}{\gamma_1(k_1^2 + k_0^2)}, \tag{4.21}$$

$$D_2(\lambda) = \frac{2}{k_1^2\gamma_0 + k_0^2\gamma_1} - \frac{2}{\gamma_1(k_1^2 + k_0^2)}, \tag{4.22}$$

and

$$\gamma_0 = \sqrt{\lambda^2 - k_0^2}; \quad \gamma_1 = \sqrt{\lambda^2 - k_1^2}. \tag{4.23}$$

Combining (4.1), (4.8)–(4.10) yields the Pocklington's integro-differential equation for the unknown current distribution along the single wire of arbitrary shape buried in a lossy ground:

$$E_s^{\text{exc}}(s) = -\frac{1}{j4\pi\omega\varepsilon_{\text{eff}}}\left[\begin{array}{l} \int_{C'} I(s')\cdot\vec{s}\cdot\vec{s}'\cdot[k_1^2+\nabla\nabla]\,g_0(s,s')ds'+ \\ +\dfrac{k_0^2-k_1^2}{k_0^2+k_1^2}\displaystyle\int_{C'} I(s')\cdot\vec{s}\cdot\vec{s}^*\cdot[k_1^2+\nabla\nabla]\,g_i(s,s^*)ds'+ \\ +\displaystyle\int_{C'} I(s')\cdot\vec{s}\cdot\vec{s}^*\cdot G_s(s,s')ds'. \end{array}\right] \tag{4.24}$$

The corresponding set of coupled Pocklington's accounts for the mutual influence of antennas:

$$E_{sm}^{\text{exc}}(s_m) = -\frac{1}{j4\pi\omega\varepsilon_{\text{eff}}} \times\sum_{n=1}^{N_W}\left[\begin{array}{l} \int_{C_n'} I_n(s_n')\cdot\vec{s}_m\cdot\vec{s}_n'\cdot[k_1^2+\nabla\nabla]\,g_{0n}(s_m,s_n')ds_n'+ \\ +\dfrac{k_0^2-k_1^2}{k_0^2+k_1^2}\displaystyle\int_{C_n'} I_n(s_n')\cdot\vec{s}_m\cdot\vec{s}_n^*\cdot[k_1^2+\nabla\nabla]\,g_{inm}(s_m,s_n^*)ds_n'+ \\ +\int_{C_n'} I_n(s_n')\cdot\vec{s}_m\cdot\vec{s}_n^*\cdot G_s(s_m,s_n')ds_n' \end{array}\right]$$

$$m = 1, 2, \ldots, N_W. \tag{4.25}$$

An approximate simplified form of Green function containing RC, deduced from the rigorous approach involving the Sommerfeld integrals, for overhead wires is given in [6]. In the same manner, for multiple buried wires the integral equation set (4.25) becomes [8]

$$E_{sm}^{\text{exc}}(s_m) = -\frac{1}{j4\pi\omega\varepsilon_{\text{eff}}}\sum_{n=1}^{N_w}\left[\begin{array}{l}\int_0^{L_n}\left\{\left[k_1^2\vec{s}_m\vec{s}_n^* - \frac{\partial^2}{\partial s_m \partial s_n'}\right] g_{0mn}(s_m, s_n')+\right.\\ +R_{\text{TM}}\left[k_1^2\vec{s}_m\vec{s}_n^* - \frac{\partial^2}{\partial s_m \partial s_n{}^*}\right] g_{imn}(s_m, s_n^*)+\\ +(R_{\text{TE}} - R_{\text{TM}})\vec{s}_m \cdot \vec{p}_m\left[k_1^2\vec{p}_m \cdot \vec{s}_n^* - \frac{\partial^2}{\partial p_m \partial s_n^*}\right]\\ \left.\times g_{imn}(s_m, s_n^*)\right\} I_n(s_n{}')\text{d}s'\end{array}\right]$$

$$m = 1, 2, \dots, N_{\text{W}}, \tag{4.26}$$

where $R_{\text{TM}}$ and $R_{\text{TE}}$ are the RCs for the case of transverse magnetic (TM) and transverse electric (TE) polarization, respectively, given by [6]

$$R_{\text{TM}} = \frac{\frac{1}{\underline{n}}\cos\theta' - \sqrt{\frac{1}{\underline{n}} - \sin^2\theta'}}{\frac{1}{\underline{n}}\cos\theta' + \sqrt{\frac{1}{\underline{n}} - \sin^2\theta'}}, \tag{4.27}$$

$$R_{\text{TE}} = \frac{\cos\theta' - \sqrt{\frac{1}{\underline{n}} - \sin^2\theta'}}{\cos\theta' + \sqrt{\frac{1}{\underline{n}} - \sin^2\theta'}}, \tag{4.28}$$

where $I_n(s_n')$ is unknown current distribution along the $n$th wire, $E_m^{\text{exc}}(s)$ is the excitation function on the $m$th wire, and $g_{0,nm}(s_m, s_n')$ is the free space Green function, while $g_{i,nm}(s_m, s_n^*)$ arises from the image theory and $\underline{n} = \frac{\varepsilon_{\text{eff}}}{\varepsilon_0}$.

The principal advantage of the RC approach versus rigorous Sommerfeld approach is the formulation simplicity and significantly less computational cost. Generally, the RC approach provides results roughly within 10% of those obtained via the rigorous Sommerfeld integral approach [8].

### 4.1.2 Antenna Theory Approach: Numerical Solution

The set of integral Equations (4.26) is handled via an efficient GB-IBEM. The essence of the method has been presented in detail in [2]. Some special features related to isoparametric elements implementation are discussed elsewhere, e.g. in [8].

The unknown current $I_n^e(\zeta)$ along the $n$th wire segment is expressed in terms of linearly independent basis functions $f_{ni}$, with unknown complex coefficients $I_{ni}$:

$$I_n^e(s') = \sum_{i=1}^{n} I_{ni} f_{ni}(s') = \{f\}_n^T \{I\}_n, \tag{4.29}$$

and the use of isoparametric elements yields

$$I_n^e(\zeta) = \sum_{i=1}^{n} I_{ni} f_{ni}(\zeta) = \{f\}_n^T \{I\}_n, \tag{4.30}$$

where $n$ is the number of local nodes per element.

A linear approximation over a wire segment is used in this work and shape functions given by

$$f_1 = \frac{1-\zeta}{2} \quad f_2 = \frac{1+\zeta}{2}, \tag{4.31}$$

as this choice was shown to be optimal for numerical treatment of various thin wire configuration [2].

Applying the weighted residual approach and implementing the Galekin–Bubnov procedure the set of Pocklington's equations is transformed into a system of algebraic equations.

Performing some mathematical manipulations, the following matrix equation is obtained:

$$\begin{aligned}
&\sum_{n=1}^{Nw}\sum_{i=1}^{Nn}\left\{\begin{aligned}
&-\int_{\Delta l_m}\int_{\Delta l_n}\frac{\mathrm{d}f_{jm}(s_m)}{ds_m}\frac{\mathrm{d}f_{in}(s_n')}{ds_n'}g_{0nm}(s_m,s_n')\mathrm{d}s_n'\,\mathrm{d}s_m\\
&+k_1^2\cdot\hat{s}_m\cdot\hat{s}_n'\int_{\Delta l_m}\int_{\Delta l_n}f_{jm}(s_m)f_{in}(s_n')g_{0nm}(s_m,s_n')\mathrm{d}s_n'\,\mathrm{d}s_m\\
&+\frac{k_0^2-k_1^2}{k_0^2+k_1^2}\left[\begin{aligned}
&-\int_{\Delta l_m}\int_{\Delta l_n}\frac{\mathrm{d}f_{jm}(s_m)}{\mathrm{d}s_m}\frac{\mathrm{d}f_{in}(s_n')}{\mathrm{d}s_n^*}g_{inm}(s_m,s_n^*)\mathrm{d}s_n'\,\mathrm{d}s_m\\
&+k_1^2\cdot\hat{s}_m\cdot\hat{s}_n^*\int_{\Delta l_m}\int_{\Delta l_n}f_{jm}(s_m)f_{in}(s_n')\\
&\times g_{inm}(s_m,s_n^*)\mathrm{d}s_n'\,\mathrm{d}s_m
\end{aligned}\right]+\\
&+\hat{s}_m\cdot\hat{s}_n'\int_{\Delta l_m}\int_{\Delta l_n}f_{jm}(s_m)f_{in}(s_n')G_{snm}(s_m,s_n')\mathrm{d}s_n'\,\mathrm{d}s_m
\end{aligned}\right\}\{I\}_{ni}\\
&= -j4\pi\omega\varepsilon_{\text{eff}}\int_{\Delta l_m}E_{sm}^{\text{tr}}(s_m)f_{jm}(s_m)\mathrm{d}s_m 0;\\
&\quad m=1,2,\ldots,N_W;\quad j=1,2,\ldots,N_m,
\end{aligned} \tag{4.32}$$

where $N_m$ is number of elements on the $m$th antenna and $N_n$ is the number of elements on the $n$th antenna.

Equation (4.32) can also be, for convenience, written in the matrix form:

$$\sum_{n=1}^{N_w}\sum_{i=1}^{N_n}[Z]_{ji}^{e}\{I_n\}_i = \{V_m\}_j, \quad m = 1, 2, \ldots, N_w; \quad j = 1, 2, \ldots, N_m, \quad (4.33)$$

where $[Z]_{ji}$ is the mutual impedance matrix for the $j$th observation segment on the $m$th wire and $i$th source segment on the $n$th antenna.

The use of isoparametric elements yields the following mutual impedance matrix:

$$\begin{aligned}
[Z]_{ji}^{e} = &-\int_{-1}^{1}\int_{-1}^{1}\{D\}_j\{D'\}_i^T g_{0nm}(s_m, s_n')\frac{ds_n}{d\zeta'}d\zeta'\frac{ds_m}{d\zeta}d\zeta \\
&+k_1^2\cdot\hat{s_m}\cdot\hat{s_n'}\int_{-1}^{1}\int_{-1}^{1}\{f\}_j\{f'\}_i^T g_{0nm}(s_m, s_n')\frac{ds_n}{d\zeta'}d\zeta'\frac{ds_m}{d\zeta}d\zeta+ \\
&+\frac{k_0^2-k_1^2}{k_0^2+k_1^2}\left[\begin{array}{l}-\int_{-1}^{1}\int_{-1}^{1}\{D\}_j\{D'\}_i^T g_{inm}(s_m, s_n^*)\frac{ds_n}{d\zeta'}d\zeta'\frac{ds_m}{d\zeta}d\zeta+ \\ +k_1^2\cdot\hat{s_m}\cdot\hat{s_n^*}\int_{-1}^{1}\int_{-1}^{1}\{f\}_j\{f'\}_i^T g_{inm}(s_m, s_n^*)\frac{ds_n}{d\zeta'}d\zeta'\frac{ds_m}{d\zeta}d\zeta\end{array}\right]+ \\
&+\hat{s_m}\cdot\hat{s_n'}\int_{-1}^{1}\int_{-1}^{1}\{f\}_j\{f'\}_i^T G_{snm}(s_m, s_n')\frac{ds_n}{d\zeta'}d\zeta'\frac{ds_m}{d\zeta}d\zeta. \quad (4.34)
\end{aligned}$$

Note that matrices $\{f\}$ and $\{f'\}$ contain the shape functions while $\{D\}$ and $\{D'\}$ contain their derivatives.

The voltage vector is given by

$$\begin{aligned}
\{V\}_j^m = -j4\pi\omega\varepsilon_{\text{eff}}\int_{-1}^{1}E_{s_m}^{\text{inc}}(s_m)f_{jm}(s_m)\frac{ds_m}{d\xi}d\xi_m, \\
m = 1, 2, \ldots, N_W; \quad j = 1, 2, \ldots, N_m, \quad (4.35)
\end{aligned}$$

and can be evaluated in the close form [8].

### 4.1.3 Transmission Line Approximation:

Voltages and currents along the multiple buried conductors shown in Figure 4.1 induced by an external field excitation can be determined from the field-to-TL matrix equations in the frequency domain [5]:

$$\frac{d}{dx}[\hat{V}(x)] + [\hat{Z}].[\hat{I}(x)] = [\hat{V}_F(x)], \quad (4.36)$$

$$\frac{d}{dx}[\hat{I}(x)] + [\hat{Y}].[\hat{V}(x)] = [\hat{I}_F(x)]. \quad (4.37)$$

The procedures for the assessment of longitudinal impedance matrix $[\hat{Z}]$ and the transversal admittance matrix $[\hat{Y}]$ are discussed in detail in [1]. The solution of the frequency domain TL equations is based on the chain matrix discussed in [3]. The per-unit-length parameters $R$, $L$, $C$, and $G$ of buried conductors are evaluated using the modal equation available in [10] and are frequency dependent. Such an approach is more accurate than use of the well-known Polaczeck formulas [1]. The standard TL and modified TL (MTL) [5] approach are both used, respectively.

The TL theory can handle the problem of electromagnetic coupling directly in time but without considering the impact of frequency on the parameters per unit length (in the expression of $Z$ impedance). The direct time domain solution of TL equations requires the numerical calculation of a convolution integral, which is a rather tedious task as the correction term includes the finite conductivity of the soil within the impedance $Z$. Moreover, this term is not available in the closed form and the use of inverse Fourier transform algorithm becomes necessary.

### 4.1.4 Computational Examples

Numerical results presented in this section are related to various configurations of three buried conductors.

The electric field of the transmitted plane wave exciting multiple wires configuration at certain burial depth $z$ is [6]

$$E^{\mathrm{tr}} = E_0 \Gamma_{\mathrm{TM}} \mathrm{e}^{-jk_1|z|}, \tag{4.38}$$

where $E_0$ is the field amplitude and the point of reference is located at the interface of two media $z = 0$. Note that $E_0 = 1\ \mathrm{V\,m^{-1}}$ in all examples to follow. Also, all numerical results obtained via GB-IBEM (abbreviated as BEM in figures to follow), standard TL and MTL, respectively are compared to the results calculated via NEC (Numerical Electromagnetic Code) [11] with Sommerfeld approach and RC approximation, respectively. Figure 4.2 shows the first configuration of interest. The radius of all conductors is 10.25 mm, the distance between neighboring conductors is 106 mm, and the burial depth is 1 m.

The numerical results for the current induced at the center of the middle wire (configuration No. 1) are shown in Figure 4.3. The length of conductors is 50 m and the ground parameters are $\sigma = 0.001\ \mathrm{S\,m^{-1}}$ and $\varepsilon_{\mathrm{r}} = 10$.

Figure 4.4 shows the current induced at the center of the middle wire (configuration No. 1) for the conductor length of 50 m with higher ground conductivity ($\sigma = 0.01\ \mathrm{S\,m^{-1}}$), while the permittivity is the same ($\varepsilon_{\mathrm{r}} = 10$).

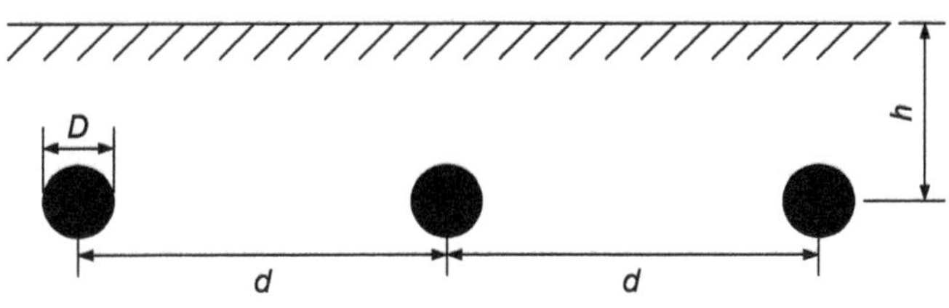


**Figure 4.2** Configuration No. 1: $D = 20.5$ mm, $d = 106$ mm, $h = 1$ m.

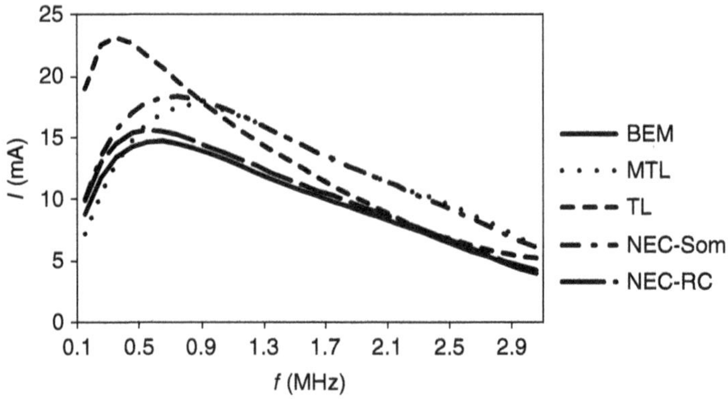


**Figure 4.3** The frequency response at the center of the middle wire (configuration No. 1, $L = 50$ m, $\sigma = 0.001$ S m$^{-1}$).

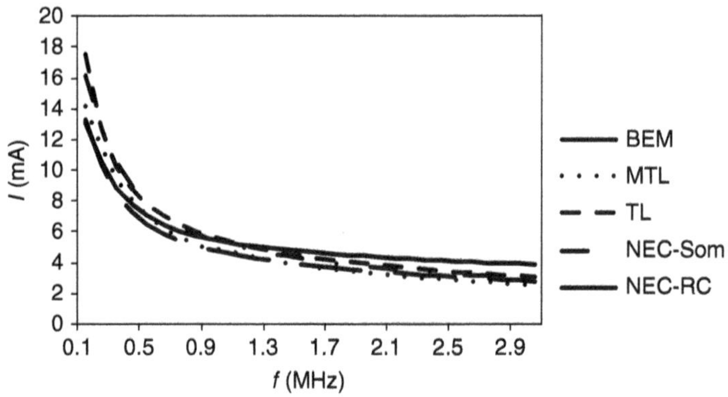


**Figure 4.4** The frequency response at the center of the middle wire (configuration No. 1, $L = 50$ m, $\sigma = 0.01$ S m$^{-1}$).

The second configuration of interest is shown in Figure 4.5. The radius of all conductors is 10.25 mm, the distances between neighboring conductors are $d_1 = 36$ mm, $d_2 = 18$ mm, while the burial depths are $h_1 = 1$ m, $h_2 = 0.97$ m.

The numerical results for the current induced at the center of the middle wire (configuration No. 2) are shown in Figure 4.6 for the conductor length of 50 m with the conductivity $\sigma = 0.001$ S m$^{-1}$ and permittivity $\varepsilon_r = 10$.

The numerical results for the current induced at the center of the middle wire (configuration No. 2) are shown in Figure 4.7 for the conductor length 50 m. The ground parameters are $\sigma = 0.01$ S m$^{-1}$ and $\varepsilon_r = 10$.

Although all waveforms are alike, the magnitudes vary significantly. Generally, the numerical results obtained via different approaches agree better for higher values of ground conductivity and longer wires.

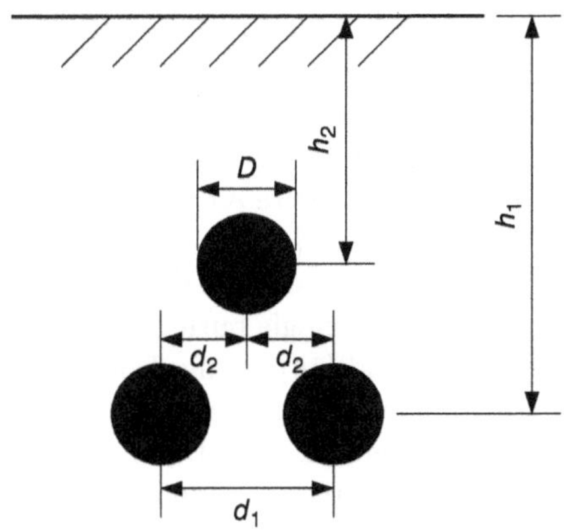


**Figure 4.5** Configuration No. 2: $D = 20.5$ mm, $d_1 = 36$ mm, $d_2 = 18$ mm, $h_1 = 1$ m, $h_2 = 0.97$ m.

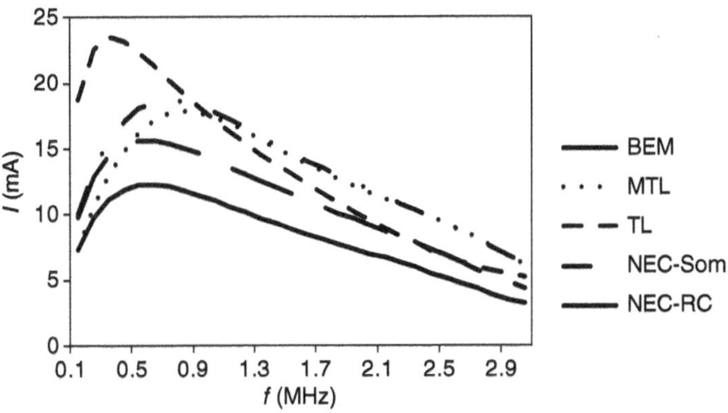


**Figure 4.6** The frequency response at the center of the middle wire (configuration No. 2, $L = 50$ m, $\sigma = 0.001$ S m$^{-1}$).

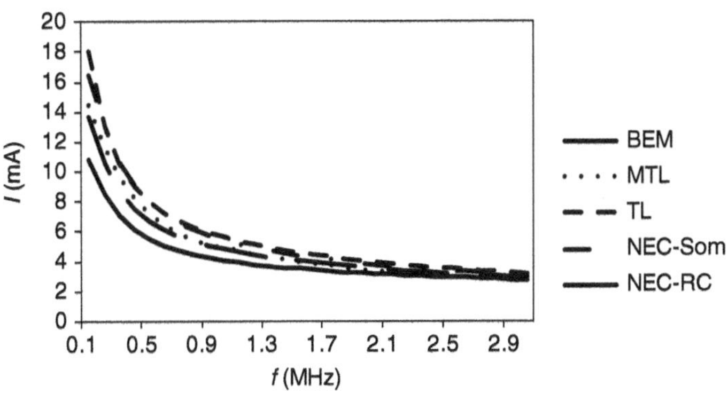


**Figure 4.7** The frequency response at the center of the middle wire (configuration No. 3, $L = 50$ m, $\sigma = 0.01$ S m$^{-1}$).

## 4.2 Time Domain Modeling

Although the analysis of the transient electromagnetic field coupling to buried wire configurations can be posed in either frequency or time domain by using the TL model or wire AT, direct time domain modeling offers some computational advantages [1, 2].

A comparison between the time domain wire antenna model and TL model pertaining to a single buried conductor is given in [12]. Time domain formulation [13] and [14] arising from the wire AT is based on the corresponding time domain Pocklington's integro-differential equations for thin wires buried in a lossy medium. The TL formulation is based on the telegrapher's equations. The Pocklington's equation is solved analytically [14], while the telegrapher's equations are treated using the modified transmission line method (MTLM) [12].

This section outlines the full wave (AT) analysis methods of transient electromagnetic coupling to a straight buried wire scatterer in the time domain reported in [10]. A comparison of the AT results with TL results is given as well.

The AT approach is based on the space–time variant of the Pocklington's integro-differential approach. The effect of the earth–air interface is taken into account via the simplified RC arising from the modified image theory (MIT) [10, 11].

### 4.2.1 Antenna Theory Approach

The geometry of interest, related to the horizontal thin wire buried in a lossy ground, is shown in Figure 4.8.

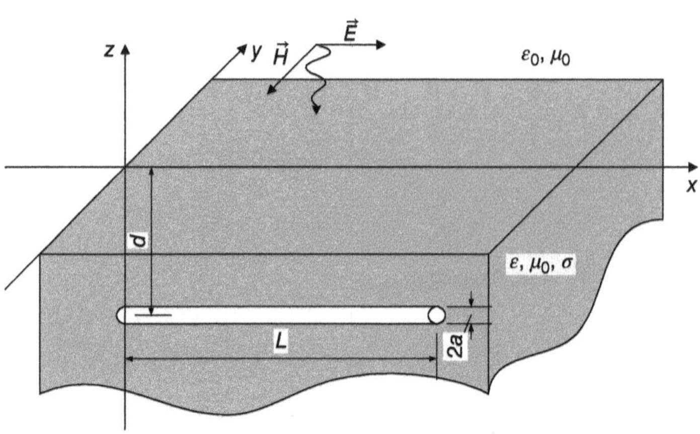


**Figure 4.8** A horizontal thin wire buried in a lossy medium.

Direct time domain formulation for the transient analysis of horizontal straight buried wire is based on the space–time Pocklington's integro-differential equation [5]

$$\left(\mu\varepsilon\frac{\partial}{\partial t}+\mu\sigma\right)E_x^{\mathrm{tr}}(t)=-\left(\frac{\partial^2}{\partial x^2}-\mu\sigma\frac{\partial}{\partial t}-\mu\varepsilon\frac{\partial^2}{\partial t^2}\right)\cdot\left[\begin{array}{l}\dfrac{\mu}{4\pi}\displaystyle\int_0^L I\left(x',t-\frac{R}{v}\right)\frac{e^{-\frac{1}{\tau_g}\frac{R}{v}}}{R}\mathrm{d}x'-\\ \dfrac{\mu}{4\pi}\displaystyle\int_0^t\int_0^L \Gamma_{\mathrm{ref}}^{\mathrm{MIT}}(\tau)I\left(x',t-\frac{R^*}{v}-\tau\right)\frac{e^{-\frac{1}{\tau_g}\frac{R^*}{v}}}{R^*}\mathrm{d}x'\,\mathrm{d}\tau\end{array}\right],\qquad(4.39)$$

where $I\left(x',t-\frac{R}{v}\right)$ is the space–time dependent current to be determined, $E_x^{\mathrm{tr}}$ is the tangential transmitted field, and $\Gamma_{\mathrm{ref}}^{\mathrm{MIT}}$ is the corresponding RC arising from the MIT [11, 13].

The distance from the source point in the wire axis to the observation point located on the wire surface is

$$R=\sqrt{(x-x')^2+a^2},\qquad(4.40)$$

while the distance from the source point on the image wire to the observation point on the original wire, according to the image theory, is

$$R^*=\sqrt{(x-x')^2+4d^2}.\qquad(4.41)$$

The time constant $\tau_g$ and propagation velocity in the lossy medium $v$ are

$$\tau_g=\frac{2\varepsilon}{\sigma},\qquad(4.42)$$

$$v=\frac{1}{\sqrt{\mu\varepsilon}}.\qquad(4.43)$$

The effects of the earth–air interface are taken into account via the RC arising from the MIT given by [11, 13]

$$\Gamma_{\mathrm{ref}}^{\mathrm{MIT}}(t)=-\left[\frac{\tau_1}{\tau_2}\delta(t)+\frac{1}{\tau_2}\left(1-\frac{\tau_1}{\tau_2}\right)e^{-\frac{t}{\tau_2}}\right],\qquad(4.44)$$

where the corresponding time constants are

$$\tau_1=\frac{\varepsilon_0(\varepsilon_r-1)}{\sigma},\qquad(4.45)$$

$$\tau_2=\frac{\varepsilon_0(\varepsilon_r+1)}{\sigma}.\qquad(4.46)$$

Note that the RC (4.44) represents rather a simple characterization of the earth–air interface, taking into account only medium properties. Accuracy of (4.44) is reported elsewhere, e.g. in [13] and [14].

The analytical procedure presented in [6] for the case of impulse excitation yields the following formula:

$$I(x,t) = \frac{4\pi}{\mu}\left\{\begin{array}{l} R(s_\Psi)\left[1-\dfrac{\cosh\left(\gamma_\Psi\left(\frac{L}{2}-x\right)\right)}{\cosh\left(\gamma_\Psi\frac{L}{2}\right)}\right]\mathrm{e}^{\left(t+\frac{a}{v}\right)s_\Psi} \\ -\dfrac{\pi}{\mu\varepsilon L^2}\displaystyle\sum_{n=1}^{\infty}\dfrac{2n-1}{\pm\sqrt{b^2-4c_n}s_{1,2n}\Psi(s_{1,2n})} \\ \times\sin\dfrac{(2n-1)\pi x}{L}\mathrm{e}^{\left(t+\frac{d}{v}\right)s_{1,2n}} \end{array}\right\}, \tag{4.47}$$

where

$$\begin{aligned} R(s_\Psi) &= \frac{1}{2\ln\dfrac{L}{2d}\dfrac{s_\Psi}{s_\Psi\tau_2+1}\left(\tau_1-\tau_2\dfrac{s_\Psi\tau_1+1}{s_\Psi\tau_2+1}\right)}, \\ s_\Psi &= -\frac{\ln\dfrac{L}{a}+\ln\dfrac{L}{2d}}{\tau_1\ln\dfrac{L}{a}+\tau_2\ln\dfrac{L}{2d}}, \end{aligned} \tag{4.48}$$

$$\begin{aligned} \gamma_\Psi &= \sqrt{\mu\varepsilon(s_\Psi^2+bs_\Psi)}, \\ s_{1,2n} &= \frac{1}{2}(-b\pm\sqrt{b^2-4c_n}), \\ b &= \frac{\sigma}{\varepsilon}, \\ c_n &= \frac{(2n-1)^2\pi^2}{\mu\varepsilon L^2},\quad n=1,2,3,\ldots \end{aligned} \tag{4.49}$$

Note that coefficients $R(s_\Psi)$ and $s_\Psi$ account for the properties of the medium, the dimensions of the wire, and the distance from the interface.

Expression (4.47) represents the impulse response of the wire scatterer. Consequently, the response to an arbitrary excitation requires convolution. In this chapter the normal incidence is considered, i.e. the plane wave in the form of the double exponential function is assumed:

$$E_x(t) = E_0(\mathrm{e}^{-\alpha t}-\mathrm{e}^{-\beta t}). \tag{4.50}$$

The transmitted electric field exciting the buried wire in the Laplace domain is given by

$$E_x^{\mathrm{tr}}(s) = \Gamma_{tr}(s)E_x(s)\mathrm{e}^{-\gamma d}, \tag{4.51}$$

where $\Gamma_{tr}(s)$ is the Fresnel transmission coefficient [10]

$$\Gamma_{tr}(s) = \frac{2\sqrt{s\varepsilon_0}}{\sqrt{s\varepsilon + \sigma} + \sqrt{s\varepsilon_0}}. \tag{4.52}$$

As the analytical convolution, i.e. the time domain counterpart of (4.51), would be too complex to perform, numerical convolution is carried out, as reported in [14].

### 4.2.2 Transmission Line Model

Frequency domain analysis of a horizontal buried wire in a lossy medium and excited via plane wave can be carried out by the transmissions line equations in the frequency domain [1]:

$$\frac{\partial U(x,\omega)}{\partial x} + ZI(x,\omega) = E_x^{tr}(x,\omega), \tag{4.53}$$

$$\frac{\partial I(x,\omega)}{\partial x} + YU(x,\omega) = 0, \tag{4.54}$$

where, $U(x,\omega), I(x,\omega)$ are the induced voltage and current along the conductor, respectively, $Z(\omega)$ is the per-unit-length impedance, and $Y(\omega)$ is the effective per-unit-length admittance of the conductor, respectively.

The set of telegrapher's equations (4.35) and (4.54) is solved using the MTLM. The solution procedure is outlined in [10].

Space–time voltage and current $u(x,t)$ and $i(x,t)$ are obtained as the inverse Fourier transform of space–frequency response $U(x,\omega)$ and $I(x,\omega)$.The per-unit line parameters of buried horizontal wires can be calculated by using the approach proposed in [1].

The MTLM solution of telegrapher's equations starts with the set of telegrapher's equations [1]

$$\frac{\partial v(x,t)}{\partial x} + z(t) * i(x,t) = E_x^{tr}(x,t), \tag{4.55}$$

$$\frac{\partial i(x,t)}{\partial x} + y(t) * v(x,t) + C\frac{\partial v(x,t)}{\partial t} = 0, \tag{4.56}$$

where * is the convolution product. In the frequency domain, these can be written as follows [1]:

$$\frac{\partial U(x,\omega)}{\partial x} + ZI(x,\omega) = E_x^{tr}(x,\omega), \tag{4.57}$$

$$\frac{\partial I(x,\omega)}{\partial x} + YU(x,\omega) = 0. \tag{4.58}$$

The solution of TL equations for a buried wire excited by an external field $E_x^{tr}(\omega)$ transmitted into the ground, given by (4.51), can be written in the form

$$I(x,\omega) = A(\omega)e^{-\Gamma x} + B(\omega)e^{\Gamma x} + \frac{E_x^{tr}(\omega)}{Z(\omega)}, \tag{4.59}$$

where $A(\omega)$ and $B(\omega)$ are determined from the zero-current conditions at the wire ends

$$I(0,\omega) = I(L,\omega) = 0 \tag{4.60}$$

and are given by

$$A(\omega) = \frac{E_x^{tr}(\omega)}{Z(\omega)} \frac{(1 - e^{\Gamma L})}{(e^{\Gamma L} - e^{-\Gamma L})}, \tag{4.61}$$

$$B(\omega) = -\frac{E_x^{tr}(\omega)}{Z(\omega)} \frac{(1 - e^{-\Gamma L})}{(e^{\Gamma L} - e^{-\Gamma L})}. \tag{4.62}$$

Per-unit-length parameters $Z(\omega)$ and $Y(\omega)$ are defined by the relation

$$Z[\Gamma(\omega)] \cdot Y[\Gamma(\omega)] = [\Gamma(\omega)]^2, \tag{4.63}$$

where [7]

$$Z(\Gamma) = \frac{j\omega\mu_0}{2\pi}[K_0(\gamma_1 a) - K_0(\gamma_1(2d - a)) + I_1(\Gamma)] \tag{4.64}$$

and

$$Y(\Gamma) = \frac{j2\pi\omega\varepsilon_{\text{eff}}}{K_0(\gamma_1 a) - K_0(\gamma_1(2d - a)) + k_1^2 I_2(\Gamma)}. \tag{4.65}$$

Furthermore, $I_1$ and $I_2$ are determined by integrals

$$I_1(\Gamma) = \int_{-\infty}^{+\infty} \frac{e^{-2u_1 d}}{u_1 + u_2} d\lambda, \tag{4.66}$$

$$I_2(\Gamma) = \int_{-\infty}^{+\infty} \frac{e^{-2u_1 d}}{k_2^2 u_1 + k_1^2 u_2} d\lambda, \tag{4.67}$$

where $K_0$ is zero order Bessel function of the second kind, while $u_1$ and $u_2$ are given by

$$u_1 = (\lambda^2 - \Gamma^2 - k_1^2)^{\frac{1}{2}} = (\lambda^2 + \gamma_1^2)^{\frac{1}{2}}, \tag{4.68}$$

$$u_2 = (\lambda^2 - \Gamma^2 - k_2^2)^{\frac{1}{2}} = (\lambda^2 + \gamma_2^2)^{\frac{1}{2}}, \tag{4.69}$$

where $k_1$ and $k_2$ are the propagation constants of a lossy ground and air, respectively, and are expressed as follows:

$$k_1^2 = \omega^2 \mu_0 \varepsilon_{\text{eff}}, \tag{4.70}$$

$$k_2^2 = \omega^2 \mu_0 \varepsilon_0, \tag{4.71}$$

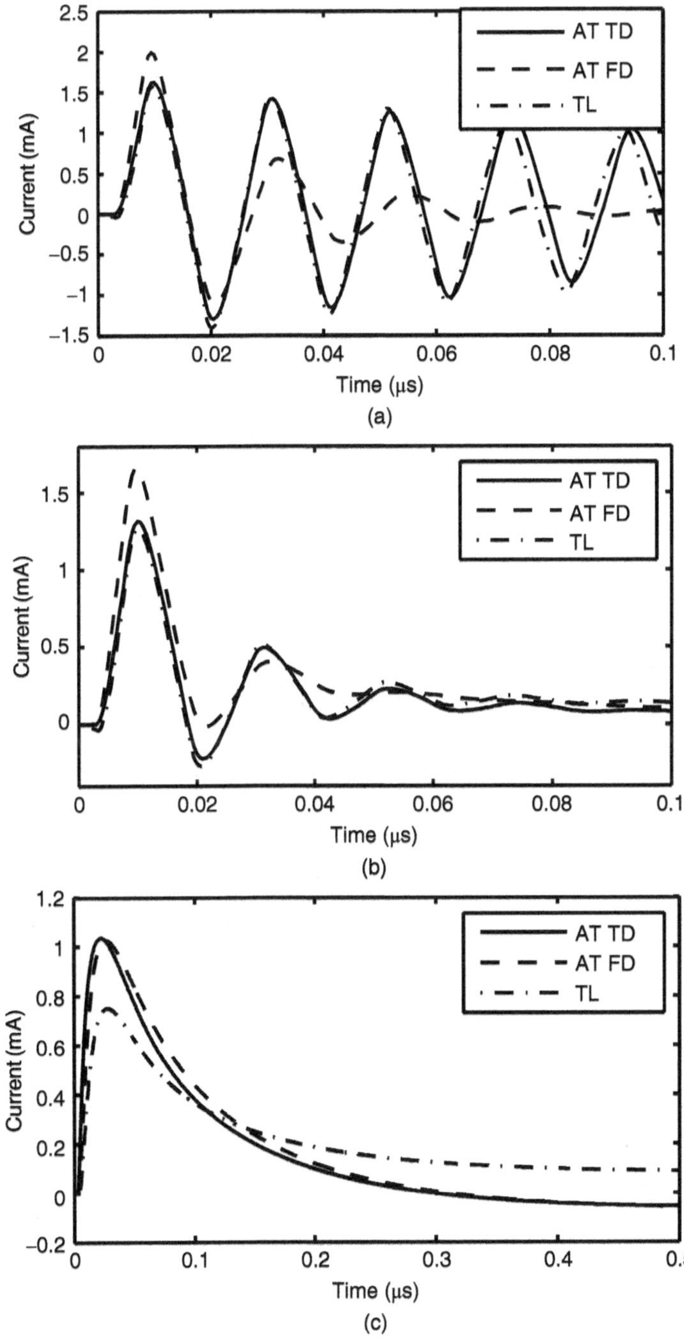


**Figure 4.9** Transient current at the center of the straight wire, $L = 1$ m, $d = 30$ cm. (a) $\sigma = 1$ mS m$^{-1}$. (b) $\sigma = 10$ mS m$^{-1}$. (c) $\sigma = 100$ mS m$^{-1}$.

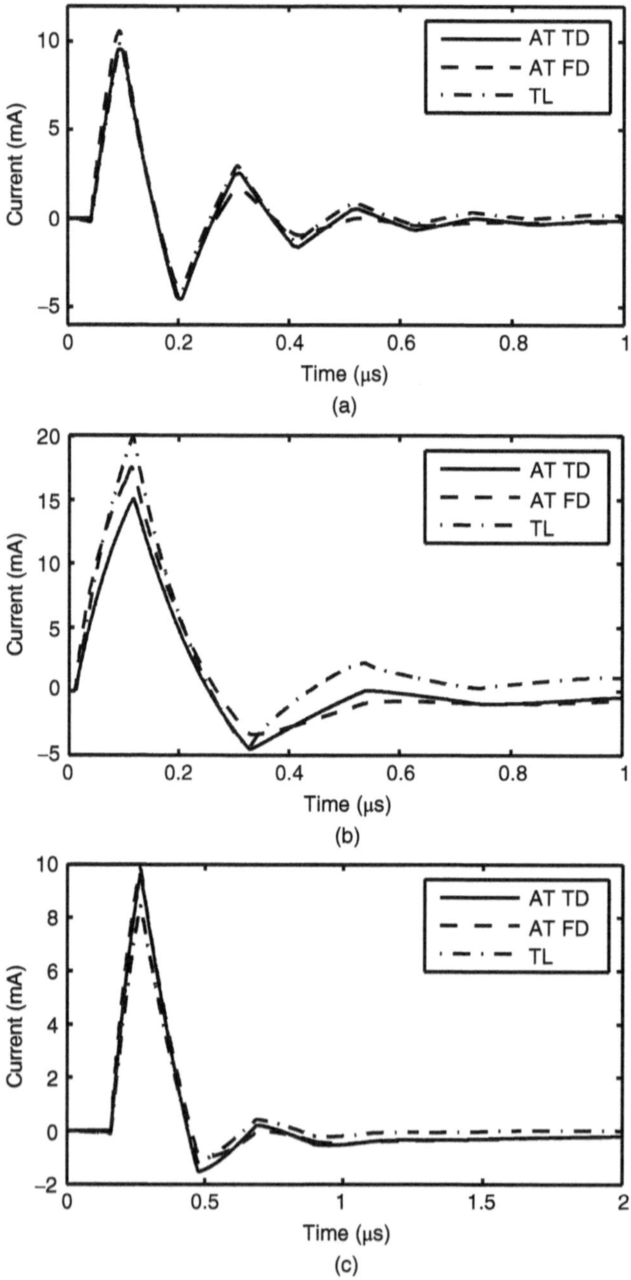


**Figure 4.10** Transient current at the center of the straight wire, $L = 10$ m, $d = 4$ m. (a) $\sigma = 1$ mS m$^{-1}$. (b) $\sigma = 10$ mS m$^{-1}$. (c) $\sigma = 100$ mS m$^{-1}$.

with

$$\varepsilon_{\text{eff}} = \varepsilon_0\varepsilon_r - j\frac{\sigma}{\omega}. \tag{4.72}$$

Space–time current $I(x, t)$ is obtained as the inverse Fourier transform of space–frequency response $I(x, \omega)$.

### 4.2.3 Computational Examples

Some illustrative computational examples for the transient response of a buried wire illuminated by the transient plane wave (normal incidence) transmitted into the ground are presented in Figures 4.9 and 4.10.

The results obtained using the AT approach in both frequency (FD) and time domain (TD), respectively, are compared to the results obtained by the MTLM approach. Figure 4.2 shows the transient current induced at the center of the short wire ($L = 1$ m, $d = 30$ cm, $a = 5$ mm, $\varepsilon_r = 10$) for different values of ground conductivity.

Some discrepancies can be observed between the results obtained via AT–TD and TL methods versus AT–FD method, which is expected since the two former methods have a similar approximation, which is not valid for shorter wires.

Figure 4.10 shows the transient current induced at the center of a longer wire ($L = 10$ m). Again, there is good agreement between the results obtained via different techniques.

The results calculated via different approaches agree satisfactorily in most cases. Some bigger differences exist only for the case of shorter wires and lower ground conductivity, which was expected due to the nature of the TL model. The mathematical details regarding the AT solution method could be found elsewhere, e.g. in [8].

## References


1 Tesche, F., Ianoz, M., and Carlsson, F. (1997). *EMC Analysis Methods and Computational Models*. New York: Wiley.

2 Poljak, D. (2007). *Advanced Modelling in Computational Electromagnetic Compatibility*. New York: Wiley.

3 Poljak, D., Doric, V., Rachidi, F. et al. (2009). Generalized form of telegrapher's equations for the electromagnetic field coupling to buried wires of finite length. *IEEE Transactions on Electromagnetic Compatibility* **51** (2): 331–337.

4 Poljak, D., Doric, V., Sesnic, S., et al. (2007). Electromagnetic field coupling to buried wires: comparison of frequency domain wire antenna and transmission line model, ICECom 2007. Conference Proceedings, 19th

International Conference on Applied Electromagnetics and Communications, 123–126.
5 Poljak, D., El Khamlichi Drissi, K., Kerroum, K., and Šesnić, S. (2011). Comparison of analytical and boundary element modeling of electromagnetic field coupling to overhead and buried wires. *Engineering Analysis with Boundary Elements* **35** (3): 555–563.
6 Poljak, D., Sesnic, S., El-Khamlichi Drissi, K., and Kerroum, K. (2011). Electromagnetic field coupling to multiple buried thin wires – antenna model versus transmission line approach. Proceedings of EMC Europe 2011 York, 10th International Symposium on Electromagnetic Compatibility, York, EMC Europe, York, 272–277.
7 Miller, E.K., Poggio, A.J., Burke, G.J., and Selden, E.S. (1972). Analysis of wire antennas in the presence of a conducting half-space, Part II: the horizontal antenna in free space. *Canadian Journal of Physics* **50**: 2614–2627.
8 Poljak, D., El Khamlichi Drissi, K., and Nekhoul, B. (2013). Electromagnetic field coupling to arbitrary wire configurations buried in a lossy ground. *International Journal of Computational Methods and Experimental Measurements* **1**: 142–163.
9 Burke, G.J. and Miller, E.K. (1984). Modeling antennas near to and penetrating a lossy interface. *IEEE Transactions on Antennas and Propagation* **AP-32** (10): 1040–1049.
10 Poljak, D., Šesnić, S., El-Khamlichi Drissi, K. et al. (2016). Transient electromagnetic field coupling to buried thin wire configurations: antenna model versus transmission line approach in the time domain. *International Journal of Antennas and Propagation* 1–11.
11 Takashima, T., Nakae, T., and Ishibashi, R. (1980). Calculation of complex fields in conducting media. *IEEE Transactions on Electrical Insulation* **15** (1): 1–7.
12 Poljak, D., Sesnic, S., El Khamlichi Drissi, K., Kerroum, K. (2013). Transient response of a buried wire. Conference Proceedings of the ICECom 2013.
13 Poljak, D. and Kovac, N. (2009). Time domain modeling of a thin wire in a two-media configuration featuring a simplified reflection/transmission coefficient approach. *Engineering Analysis with Boundary Elements* **33**: 283–293.
14 Šesnić, S., Poljak, D., and Tkachenko, S. (2011). Time domain analytical modeling of a straight thin wire buried in a lossy medium. *Progress in Electromagnetics Research* **121**: 485–504.

# 5

# Lightning Electromagnetics

The Chapter deals with certain vertical antenna models applied to lightning channel analysis and lightning rod. The formulation is based on the corresponding electric field integral equation and related numerical solution methods.

## 5.1 Antenna Model of Lightning Channel

There are many models for the lightning return stroke and they can be classified as follows [1, 2]:

- Gas dynamic models
- Distributed circuit models [3–6]
- Engineering models [7, 8]
- Electromagnetic models [9–15]

Contrary to most of the simple lightning channel models in which the current distribution is assumed the electromagnetic models are the most rigorous as they stem from the solution of Maxwell's equations. These models provide the current distribution along the lightning channel featuring the use of numerical techniques such as method of moments (MoM) or finite difference method (FDM) [16]. When using these models, the lightning channel is usually represented by an equivalent monopole antenna above a PEC ground. In most of the scenarios, the antenna is energized either by an ideal current source (ICS) [9, 14, 17] or by a delta gap voltage source [10–13, 15]. The formulation is derived in either frequency [9, 11, 13–15, 17] or time domain [10, 12, 17], respectively. Numerical Electromagnetic Code (NEC) is usually used for current distribution calculation [13, 15] in the frequency domain, while the corresponding transient response is obtained by means of the inverse fast Fourier transform (IFFT). The principal drawback related to the IFFT implementation is the mutual dependence of all the parameters of the FFT algorithm (duration of a signal in the time domain, sampling frequency, sampling time, Nyquist frequency, number of frequency samples, number of time samples, etc.).

*Computational Methods in Electromagnetic Compatibility: Antenna Theory Approach versus Transmission line Models*, First Edition. Dragan Poljak and Khalil El Khamlichi Drissi.

In this section, the lightning channel is represented by a finitely conducting monopole antenna above a PEC ground represented by an equivalent dipole antenna immersed in a dielectric medium to reduce the current propagation velocity. The dipole is energized by an ICS and by a voltage delta gap source, respectively. The current distribution is governed by the corresponding Pocklington's integro-differential equation for imperfectly conducting dipole antenna. An adaptive sampling technique, based on the work presented in [18, 19], is used in [20] to reduce the number of frequency samples required for a satisfactory convergence of the results in the time domain. The idea is to carry out the calculation at frequencies significantly contributing to the transient behavior of the current, while for frequencies having low impact on time domain behavior of the signal the interpolation schemes can be used. To achieve these goals the hybrid inverse Fourier transform (HIFT) is implemented, instead of standard IFFT. More details on the implementation of HIFT approach to evaluate the lightning current spectrum can be found in [20].

### 5.1.1 Integral Equation Formulation

The lightning channel is represented by the monopole antenna, i.e. equivalent center-fed imperfectly conducting dipole antenna of a length $2L$ and radius $a$, located in an unbounded medium with dielectric constant $\varepsilon_r > 1$ chosen to decrease the velocity of the current along the channel. The dipole is excited by a current or a voltage source, respectively, as illustrated in Figure 5.1.

The lightning channel current is governed by the integro-differential equation of the Pocklington type [20, 21].

In the case of an equivalent voltage source the excitation is expressed by an incident electric field $E_z^{\text{inc}}(z)$ and the unknown current distribution $I(z')$ along

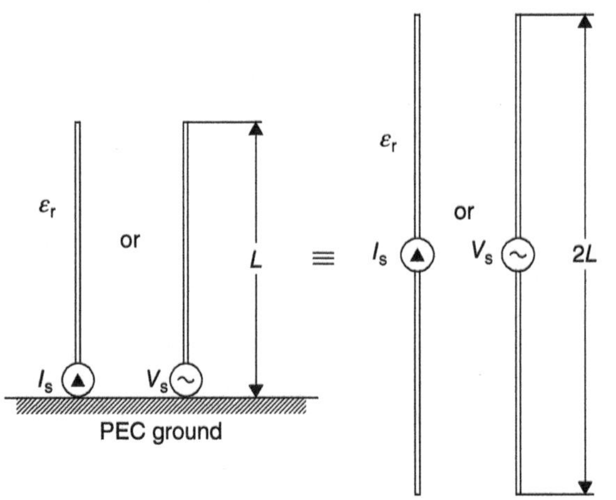


**Figure 5.1** Monopole and dipole representation of the lightning channel energized by current source or voltage source.

the channel is governed by the following Pocklington's integro-differential equation [20]

$$E_z^{\text{inc}}(z) = -\frac{1}{j4\pi\omega\varepsilon_0\varepsilon_r}\int_{-L}^{L} I(z')\left[k^2 + \frac{d^2}{dz^2}\right] g_0(z,z')dz' + Z_S I(z), \quad (5.1)$$

where $g_0(z,z')$ is the homogenous medium Green function, $k$ is the propagation constant, and $Z_S$ is the impedance term to account for the losses.

On the other hand, if the current source is used to excite the channel base, the incident field $E_z^{\text{inc}}(z)$ is set to zero. Thus, in the case of a current source at the dipole center, integral equation (5.1) simplifies to a homogenous one:

$$-\frac{1}{j4\pi\omega\varepsilon_0\varepsilon_r}\int_{-L}^{L} I(z')\left[k^2 + \frac{d^2}{dz^2}\right] g_0(z,z')dz' + Z_S I(z) = 0. \quad (5.2)$$

In this scenario, the excitation is incorporated into the formulation through the forced condition within the numerical solution procedure [20]. In addition to excitation types in terms of the voltage source and current source, the magnetic current loop (MCL) approach, recently reported in [22], is applied to modeling of lightning channel as promoted in [23]. MCL source model is based on the magnetic current density existing at the source area and generating an axial excitation electric field. The comparison of the lightning channel, represented by a straight monopole antenna vertically mounted on a perfectly conducting (PEC) ground, is depicted in Figure 5.2.

While in the case of an ICS excitation at the channel base (Figure 5.2a), the incident field is set to zero and the ICS is incorporated into the formulation as the forced condition within the numerical solution, in the case of voltage source excitation or MCL (Figure 5.2b), $E_z^{\text{inc}}(z)$ is expressed in terms of input voltage and feed-gap width [23].

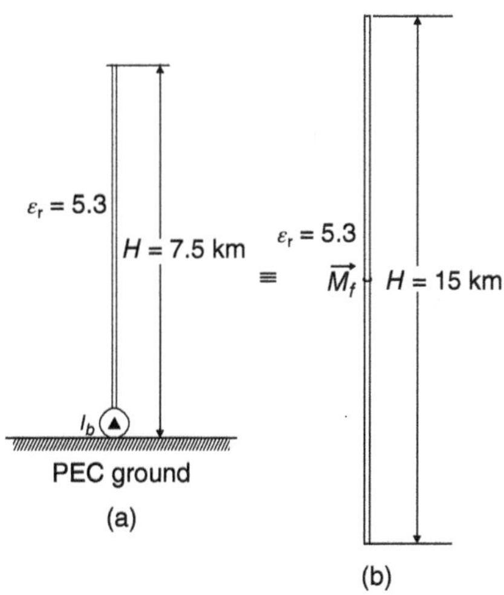


**Figure 5.2** Antenna model of the lightning channel. (a) ICS model. (b) MCL model.

### 5.1.2 Computational Examples

The first set of figures is related to the use of ICS and ideal voltage source (IVS) models. The lightning channel is represented by the equivalent monopole, which corresponds to center-fed dipole antenna of length $2L = 4$ km, pertaining to the actual lightning channel of 2 km length, with the radius $a = 0.05$ m and distributed resistance $R_\mathrm{D} = 0.07\,\Omega\,\mathrm{m}^{-1}$. The waveform of the channel-base current is determined by a peak value of about 11 kA and a peak current rise rate of about 105 kA $\mu$s$^{-1}$ [20]. However, the dielectric constant is set to be $\varepsilon_\mathrm{r} = 5.3$, thus reducing the current propagation velocity to about 0.43$c$ ($c$ – velocity of light in vacuum). The equivalent dipole is discretized into segments of about 3 m in length.

Figure 5.3 shows the waveforms of the lightning channel current at different heights from the channel base obtained via the ICS model [23]. Frequency domain calculations are carried out for 500 frequencies up to 4 MHz. A comparison with the results obtained in [10] is shown Figure 5.3 as well.

Figures 5.4 and 5.5 show frequency spectra of the lightning channel current at the height of 500 m for the unit current and voltage source, respectively.

It is visible from Figures 5.4 and 5.5 that both spectra are highly resonant. However, the model with the IVS shows higher resonant behavior. Namely, in the frequency range between 0 Hz and 4 MHz ICS spectrum has 183 extreme values (minimum and maximum) while IVS spectrum has 30% more (239). On the other hand, ICS spectrum begins to oscillate later than the IVS (first peek occurs at 2.9 kHz for the current, 15.6 kHz for voltage). Nevertheless, there are no significant differences in the convergence rate for the time domain results between models with voltage and current source.

The next set of figures is related to the use of the MCL model.

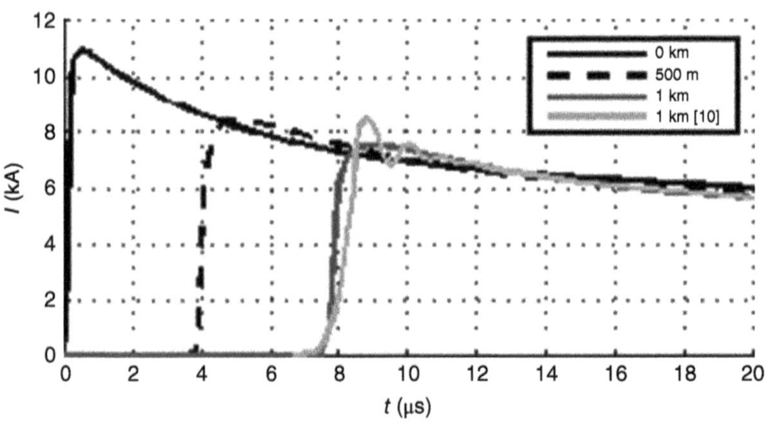


**Figure 5.3** Current waveforms at different heights along the channel.

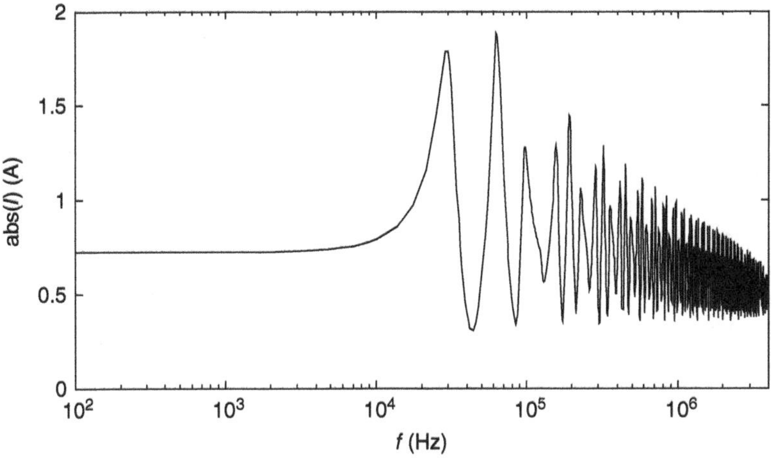


**Figure 5.4** Frequency spectrum of current at height $h = 500$ m (unit current source) – 2 km channel.

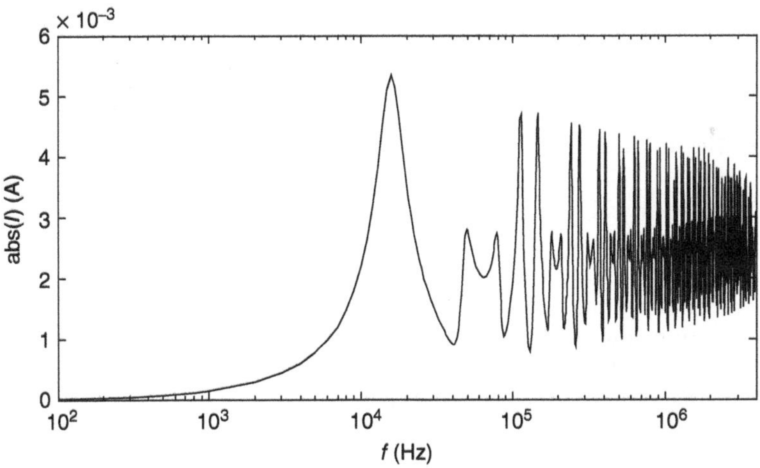


**Figure 5.5** Frequency spectrum of current at height $h = 500$ m (unit voltage source) – 2 km channel.

In this case the lightning channel is represented by a 7.5 km long monopole antenna (15-km long equivalent dipole antenna) with a wire radius $a = 0.05$ m, excited by a magnetic loop at the channel base. As in the previous example, the wire is immersed in a virtual homogeneous dielectric space with $\varepsilon_r = 5.3$, to adjust the signal propagation velocity to realistic values [24, 25] ($v = 0.43c$). It is worth noting that the virtual dielectric space is only considered to obtain the spatial–temporal distribution of the current along the channel. For the calculation of related fields, the virtual dielectric space is removed and replaced by free space. PEC and finitely conducting wire, respectively, are considered. The latter model assumes the channel losses $R = 0.07\,\Omega\ \mathrm{m}^{-1}$. The source voltage in the MCL model is adjusted to ensure the specified channel-base current. The input current waveform is assumed to have a peak value of 11 kA and a maximum steepness of 105 kA μs$^{-1}$ [9, 10]. The transient lightning channel current at different locations along the channel is shown in Figures 5.6 and 5.7 (0 m, 100 m, 500 m, 1 km, 3 km, and 5 km) for two wire models.

The channel is represented by a finitely conducting wire ($R = 0.07\,\Omega\ \mathrm{m}^{-1}$). The results at 1 km, 3 km, and 5 km are compared with those of [10]. Figure 5.6 shows a slight attenuation of the current along the channel while more rapid attenuation is visible in Figure 5.7, as the current moves upward along the channel. Some small ringing can be observed, due to numerical instabilities [9, 14]. The results obtained via the MCL model agree satisfactorily with the results published in [9, 10]. Figures 5.8 and 5.9 show the vertical electric field at 500 m and 5 km from the channel base for the PEC wire model.

The results are comparable with the results published in [9, 10]. The differences in late times are due to numerical instabilities arising from the current calculation procedure [10]. The segment length in [10] is 15 m, while in this chapter, 3 m long segments are used.

## 5.2 Vertical Antenna Model of a Lightning Rod

The model of a vertical wire penetrating the ground can be used to analyze various EMC problems such as ground stakes, typical metal rods used with many HF and VHF antennas, and for grounding systems, as well.

The behavior of a straight vertical wire in the presence of a lossy half-space, located above or below an interface, is to a certain degree affected by the presence of the other medium. Vertical wire above or below the interface, respectively, can be regarded as a special case of penetrating wire.

Significant changes in the current distribution, input impedance, and near field for an antenna approaching the interface are visible [21, 26–32]. Of particular importance for the analysis of lightning rods is the special case of a straight wire penetrating the interface. As long as the wire is located above or below the interface, respectively, it is not necessary to consider the transmitted fields within the integral equation formulation for the determination of the current distribution. On the other hand, when the wire penetrates the interface there are "two" wires on the opposite sides of the interface and the transmitted fields must be incorporated into the formulation.

The model presented in [21] is based on the rigorous formulation derived from Maxwell's equations, with the ground effects taken into account via the exact Sommerfeld integral approach. Two types of wire excitation are used: the

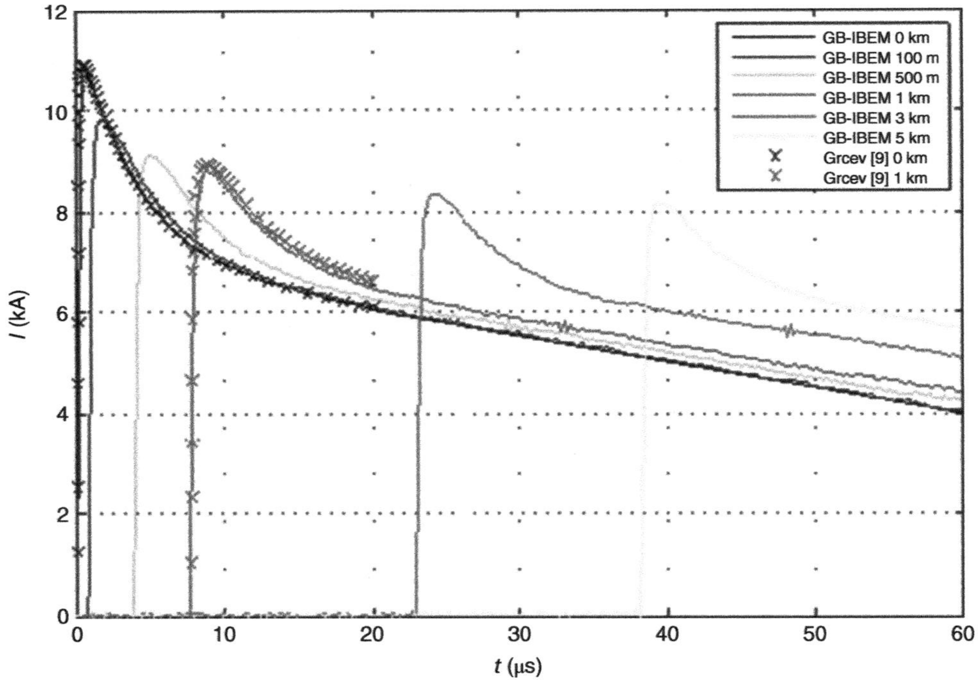


**Figure 5.6** Lightning return stroke channel current versus time for different heights (0, 100 m, 500 m, 1 km, 3 km, 5 km). The channel is considered as a PEC wire. The results at 1 km height are compared with those of [11]. *Source*: From [11]. Reproduced with permission of IEEE.

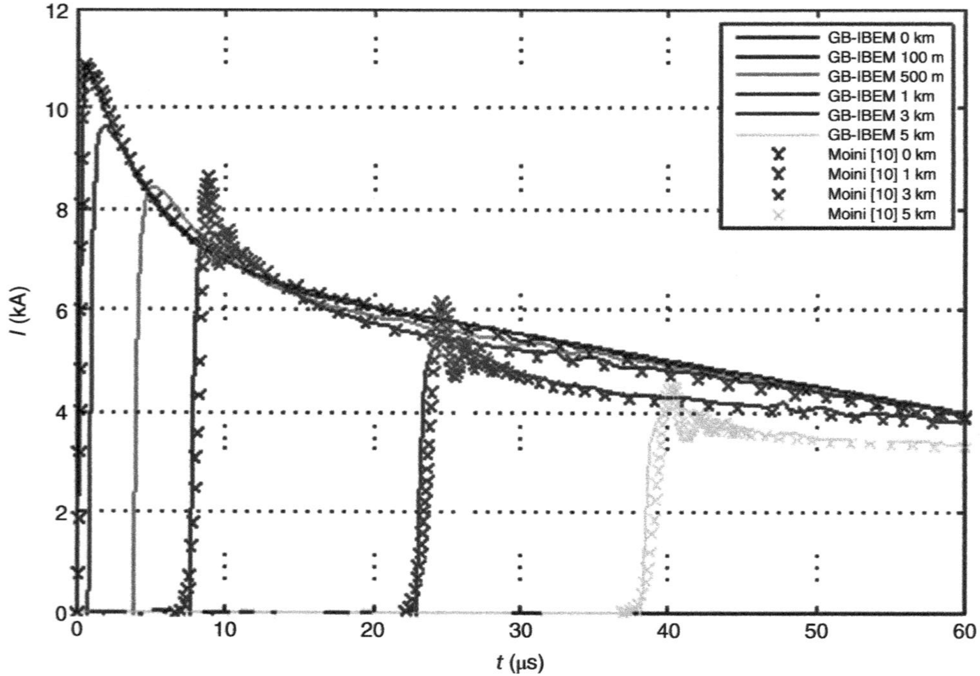


**Figure 5.7** Lightning channel current versus time for different heights (0, 100 m, 500 m, 1 km, 3 km, 5 km).

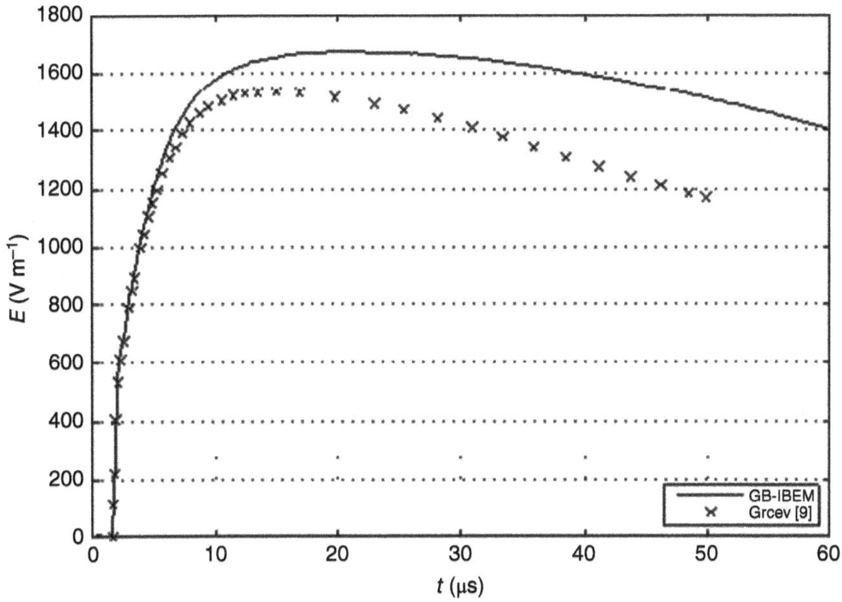


Figure 5.8 Vertical field at 500 m (PEC wire model).

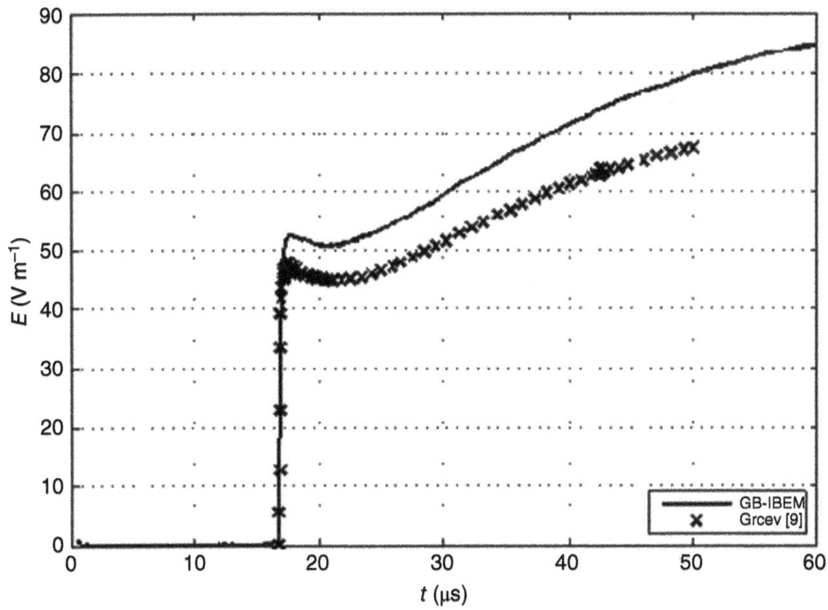


Figure 5.9 Vertical field at 5 km (PEC wire model).

voltage source placed in the upper part of the wire [21] and current generator mounted on the top of the wire [2].

### 5.2.1 Integral Equation Formulation

Figure 5.10 shows the configuration of interest, pertaining to thin vertical wires of radius $a$ penetrating the horizontal air–ground interface. The wire extends from height $L_a$, to depth $L_g$. The upper medium is free space while the lower half-space is a lossy ground with relative permittivity $\varepsilon_r$ and conductivity $\sigma$. There are two cases that need to be analyzed. The first case, shown in Figure 5.10a, deals with the wire driven by a unit voltage source $V_0 = 1\,\text{V}$ located at the distance $L_s < L_a$ above the interface. In the second case, the excitation is the current source at one end of the wire, as shown in Figure 5.10b. The current source is included into the integral equation scheme via the boundary condition

$$I(L_s) = I_g, \tag{5.3}$$

where $I_g$ is the current source.

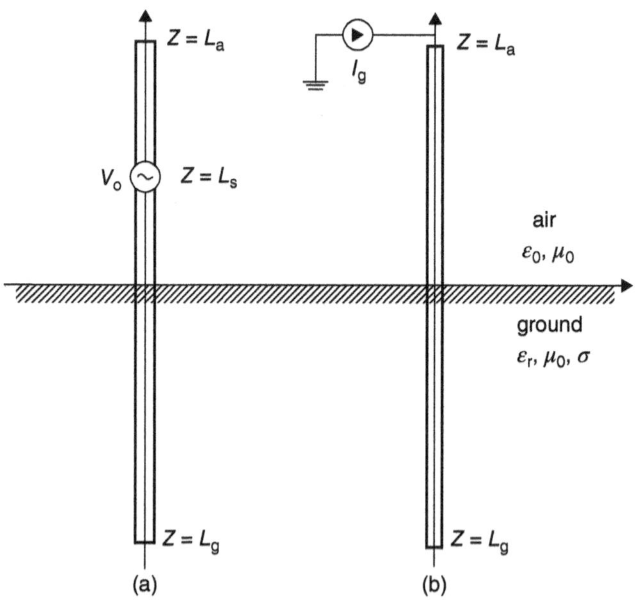


**Figure 5.10** Vertical antenna penetrating the ground excited by a voltage source (a) and by a current source (b).

The current flowing along the straight thin vertical wire penetrating the lossy ground is given by the corresponding set of the integro-differential equations [33, 34].

$$E_z^{\text{inc}} = -\frac{1}{j4\pi\omega\varepsilon_{\text{eff}}}\left[\begin{array}{l}\int_{-L_g}^{0}\left(\frac{\partial^2}{\partial z^2}+k_2^2\right)G^{22}(\rho,z,z')I(z')\mathrm{d}z'+\\ \int_{0}^{L_a}\left(\frac{\partial^2}{\partial z^2}+k_2^2\right)G^{12}(\rho,z,z')I(z')\mathrm{d}z'\end{array}\right];\quad z\le 0, \tag{5.4}$$

$$E_z^{\text{inc}} = -\frac{1}{j4\pi\omega\varepsilon_0}\left[\begin{array}{l}\int_{-L_g}^{0}\left(\frac{\partial^2}{\partial z^2}+k_1^2\right)G^{21}(\rho,z,z')I(z')\mathrm{d}z'+\\ \int_{0}^{L_a}\left(\frac{\partial^2}{\partial z^2}+k_1^2\right)G^{11}(\rho,z,z')I(z')\mathrm{d}z'\end{array}\right];\quad z> 0, \tag{5.5}$$

where

$$G^{11}(\rho,z,z') = g_0(z,z') - g_i(z,z') + g_{Sa}(\rho,z,z'),\quad \text{for points } z>0 \text{ and } z'>0 \tag{5.6}$$

while $g_{Sa}$ is given by

$$g_{Sa}(\rho,z,z') = 2\int_0^\infty J_0(\lambda\rho)\mathrm{e}^{-\mu(z+z')}\frac{\varepsilon_{\text{eff}}}{\varepsilon_{\text{eff}}\mu+\varepsilon_0\mu_E}\lambda\,\mathrm{d}\lambda \tag{5.7}$$

with

$$\mu=\sqrt{\lambda^2-k^2};\quad \mu_E=\sqrt{\lambda^2-k_2^2};\quad k_2^2=\underline{n}k^2, \tag{5.8}$$

where $\underline{n}$ is the relative complex permittivity of the air–ground interface, $\varepsilon_{\text{eff}}$ is the complex permittivity of the ground, and $k$ is wave propagation of free space.

Furthermore,

$$G^{22}(\rho,z,z') = g_0(z,z') - g_i(z,z') + g_{Su}(\rho,z,z'),\quad \text{for points } z<0 \text{ and } z'<0 \tag{5.9}$$

while $g_0$, $g_i$, $g_{Su}$ are given by

$$g_0(z,z')=\frac{\mathrm{e}^{-jk_2R}}{R},\quad g_i(z,z')=\frac{\mathrm{e}^{-jk_2R_i}}{R_i}, \tag{5.10}$$

$$g_{Su}(\rho,z,z') = 2\int_0^\infty J_0(\lambda\rho)\mathrm{e}^{-\mu_E(z+z')}\frac{\varepsilon_0}{\varepsilon_0\mu_E+\varepsilon_{\text{eff}}\mu}\lambda\,\mathrm{d}\lambda. \tag{5.11}$$

The Green functions related to transmitted field are

$$G^{12}(\rho, z, z') = 2\int_0^{\infty} J_0(\lambda\rho)\mathrm{e}^{-\mu_E|z|}\mathrm{e}^{-\mu|z'|}\frac{\varepsilon_{\mathrm{eff}}}{\varepsilon_{\mathrm{eff}}\mu + \varepsilon_0\mu_E}\lambda\,\mathrm{d}\lambda,$$
$$\text{for points } z < 0 \text{ and } z' > 0, \tag{5.12}$$

$$G^{21}(\rho, z, z') = 2\int_0^{\infty} J_0(\lambda\rho)\mathrm{e}^{-\mu_E|z|}\mathrm{e}^{-\mu|z'|}\frac{\varepsilon_0}{\varepsilon_{\mathrm{eff}}\mu + \varepsilon_0\mu_E}\lambda\,\mathrm{d}\lambda,$$
$$\text{for points } z > 0 \text{ and } z' < 0. \tag{5.13}$$

Sommerfeld integrals (5.7), (5.11)–(5.13) are evaluated numerically using Simpson adaptive quadrature in complex plane [21]. Furthermore, certain continuity conditions have to be satisfied at the air–ground interface, i.e.

$$I(z = 0^+) = I(z = 0^-), \tag{5.14}$$

$$\frac{\partial I(z = 0^+)}{\partial z}\varepsilon_{\mathrm{eff}} = \frac{\partial I(z = 0^-)}{\partial z}\varepsilon_0, \tag{5.15}$$

where (+) and (−) denote above and below the interface, respectively.

### 5.2.2 Computational Examples

The computational example deals with a penetrating vertical wire driven by the unit voltage source placed on the wire part in the air. The part of the wire in the

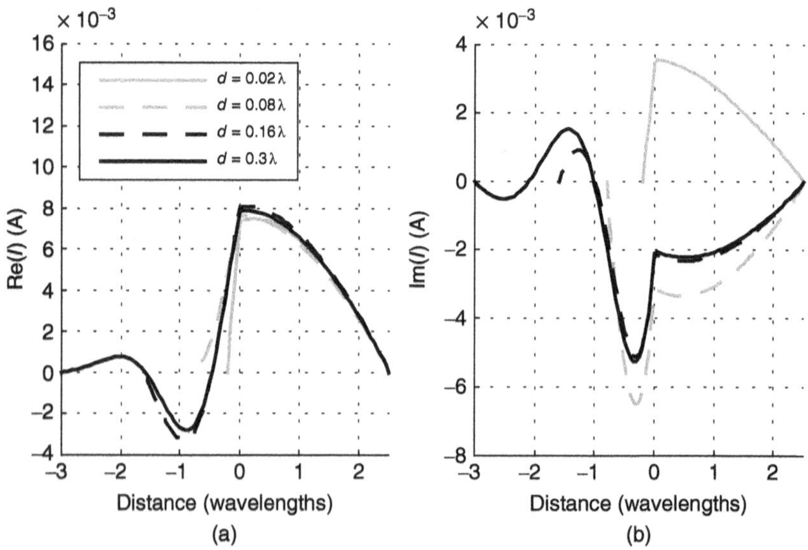


**Figure 5.11** Current distribution induced along the vertical wire penetrating a ground for various lengths of ground stake and voltage source at bottom end of the air stake. (a) Real part. (b) Imaginary part.

air is set to be $0.25\lambda_0$, while the buried part is varied as follows: $0.02\lambda_0$, $0.08\lambda_0$, $0.16\lambda_0$ and $0.3\lambda_0$. The wire radius is $0.00025\lambda_0$, while the relative complex permittivity is $\varepsilon/\varepsilon_0 = 16 - j16$.

Figure 5.11 shows the real part and the imaginary part of the actual current distribution. The results are in satisfactory agreement with the results computed via subsectional collocation variant of the MoM [33].

## 5.3 Antenna Model of a Wind Turbine Exposed to Lightning Strike

A strong request for clean and renewable energy sources has caused the growth of wind turbines (WTs). Lightning strikes are particularly dangerous for WTs due to their special shape and isolated locations mainly in high altitude areas. Consequently, development and installation of an integral lightning protection system for WT is of continuous interest [35–42].

Therefore, studies on WT's transient behavior due to a direct lightning strike, which involves the assessment of the transient current distribution along WT configuration, and analysis and design of an efficient low impedance grounding system, as a major prerequisite for an effective protection of WT from lightning strikes, are of interest.

The lightning discharge effects on WTs have attracted the attention of many prominent researchers in the last few decades (e.g. [43]). There are several models for the assessment of the current distribution along the structure and lightning channel. The representations are often based on the extension of certain return stroke models initially developed for the case of return stroke initiated at the ground level. The presence of a tall structure has been included in two basic classes of return stroke models: engineering models and antenna theory (AT) models [44]. Within the framework of engineering models the presence of a tall object has been considered using a uniform, lossless transmission line representation [8, 45]. The AT models [11, 46–50] have been mostly applied in the analysis of lightning strikes to CN tower in Toronto, e.g. [11], or similar towers, e.g. [50]. The formulation can be posed in either the time [11, 50] or the frequency domain [46–49], respectively. NEC is commonly used for the calculation of current distribution assuming the ground to be PEC [11].

This section reviews the analysis of a direct lightning strike to the WT using the AT model, i.e. WT is represented by a simple PEC wire configuration, consisting of a tower and three blades, while the lightning channel is represented by a lossy vertical wire attached to WT. The lightning channel current is energized by an ideal voltage generator at the tip of the WT blade. The current distribution along the WT and lightning channel is obtained by solving the set of Pocklington's integro-differential equations for arbitrarily shaped wires in the frequency domain. The coupled Pocklington's equations are solved by means of the Galerkin–Bubnov variant of indirect boundary element method (GB-IBEM) [51].

The antenna model presented in this section is based on the set of homogeneous integro-differential equations of Pocklington type, with air–ground interface effects taken into account via the exact Sommerfeld integral formulation. The transient response is obtained by means of the IFFT. More details can be found in [51].

Once the currents along the multiple wire configuration are evaluated, other parameters of interest could be calculated. The wire of arbitrary shape radiating above or below ground is treated by means of the image theory, as depicted by Figure 5.12, and Sommerfeld integral approach [51].

The set of Pocklington's equations for a configuration of multiple wires in the presence of a two media configuration is obtained as an extension of the Pocklington's integro-differential equation for a single wire of arbitrary shape, which can be derived by enforcing the continuity conditions for the tangential components of the electric field along the PEC wire surface [51].

First, a single wire of an arbitrary shape, insulated in an unbounded medium, as shown in Figure 5.13, is considered.

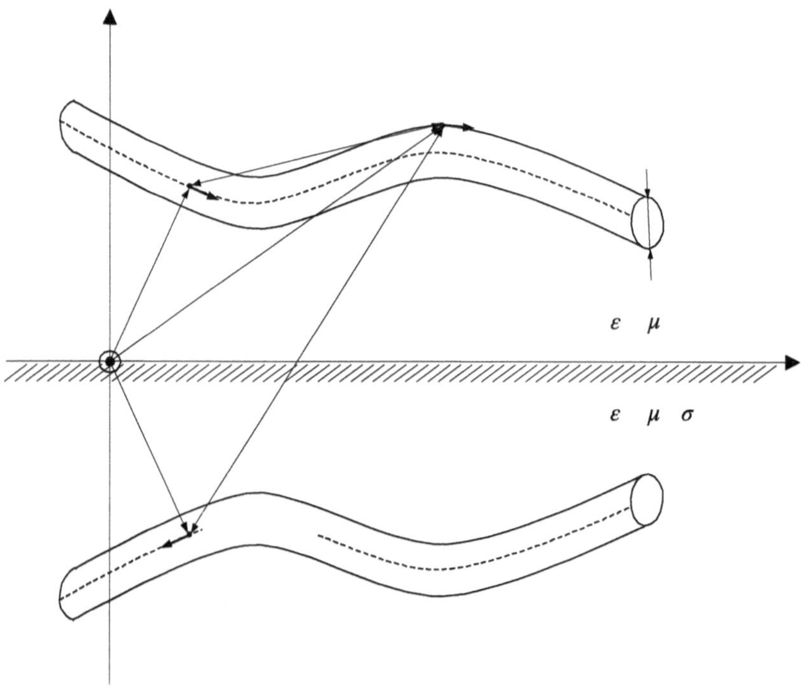


**Figure 5.12** A wire of arbitrary shape and its image.

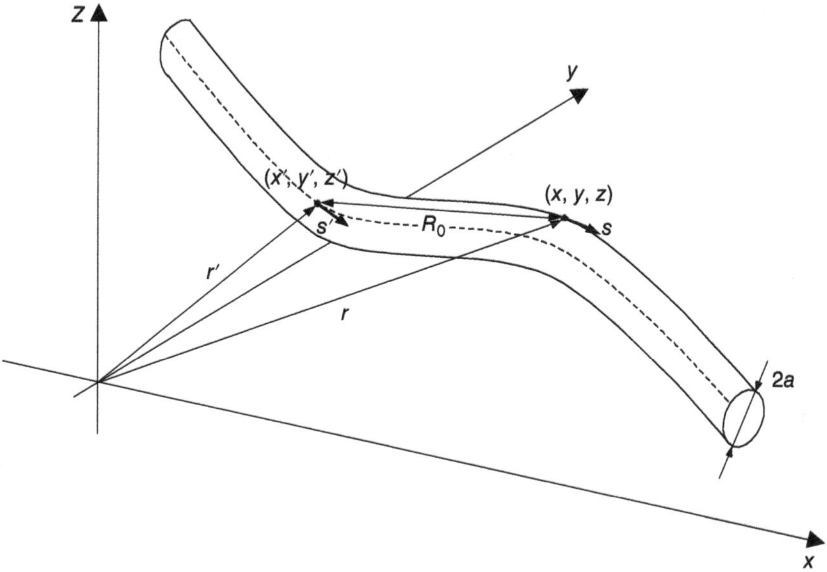


**Figure 5.13** Single wire of an arbitrary shape in a homogeneous medium.

For the PEC wire the total field composed from the excitation field $\vec{E}^{\text{exc}}$ and scattered field $\vec{E}^{\text{sct}}$ vanishes [51, 52]:

$$\vec{e}_x \cdot (\vec{E}^{\text{exc}} + \vec{E}^{\text{sct}}) = 0 \quad \text{on the wire surface.} \tag{5.16}$$

Combining Maxwell's equations and Lorentz gauge the scattered electric field can be expressed in terms of the vector potential $\vec{A}$:

$$\vec{E}^{\text{sct}} = -j\omega\vec{A} + \frac{1}{j\omega\mu\varepsilon}\nabla(\nabla\vec{A}). \tag{5.17}$$

The vector potential is defined by the particular integral along a wire configuration:

$$\vec{A}(s) = \frac{\mu}{4\pi}\int_C I(s')g_0(s,s',s^*)\vec{s}'\,\mathrm{d}s', \tag{5.18}$$

where $I(s')$ is the current distribution along the wire and $g_0(s, s')$ is the homogeneous medium Green function:

$$g_0(s,s') = \frac{e^{-jkR}}{R}, \tag{5.19}$$

where $R$ is the distance from the source point to the observation point, respectively, while $k$ is the propagation constant of the homogeneous medium (lossless or lossy medium, respectively).

Inserting (5.18) into (5.17) yields the integral relation for the scattered electric field:

$$\vec{E}^{\text{sct}} = \frac{1}{j4\pi\omega\varepsilon_0}\int_{C'} I(s')\cdot\vec{s}'\cdot[k^2+\nabla\nabla]g_0(s,s')ds'. \tag{5.20}$$

Combining (5.20) and (5.16) results in Pocklington's integral equation for the unknown current distribution along the wire of an arbitrary shape insulated in free space:

$$E_{\text{tan}}^{\text{exc}}(s) = -\frac{1}{j4\pi\omega\varepsilon_0}\int_{C'} I(s')\cdot\vec{s}\cdot\vec{s}'\cdot[k^2+\nabla\nabla]g_0(s,s')ds', \tag{5.21}$$

where $E_{\text{tan}}^{\text{exc}}$ denotes the tangential component of the electric field illuminating the wire.

### 5.3.1 Integral Equation Formulation for Multiple Overhead Wires

For the case of a wire configuration above a lossy half-space the excitation function $E^{\text{exc}}$ is composed from the incident and reflected fields, respectively:

$$E^{\text{exc}} = E^{\text{inc}} + E^{\text{ref}}. \tag{5.22}$$

The corresponding set of integral equations is derived by extending the expression (5.21) to a multiple wire configuration and is given by

$$E_{sm}^{\text{exc}}(s) = -\frac{1}{j4\pi\omega\varepsilon_0}\sum_{n=1}^{N_W}\left[\begin{array}{c}\int_{C_n'} I_n(s')\cdot\vec{s}\cdot\vec{s}'\cdot[k_0^2+\nabla\nabla]g_{0n}(s_m,s_n')ds'+\\ \frac{k_g^2-k_0^2}{k_g^2+k_0^2}\int_{C_n'} I_n(s_n')\cdot\vec{s}\cdot\vec{s}^*\cdot[k_0^2+\nabla\nabla]g_{\text{in}}(s_m,s_n^*)ds'+\\ \int_{C_n'} I_n(s')\cdot\vec{s}\cdot\vec{G}_s(s_m,s_n')ds'\end{array}\right]$$
$$+Z_S\cdot I_m(s)$$
$$m = 1,2,\ldots,N_W; \tag{5.23}$$

where $N_w$ is the total number of wires and $I_n(s_n')$ is the unknown current distribution induced on the $n$th wire.

Furthermore, $g_{0mn}(x,x')$ and $g_{imn}(s,s')$ are the Green functions of the form:

$$g_{0mn}(s_m,s_n') = \frac{e^{-jk\,R_{1mn}}}{R_{1mn}},\quad g_{imn}(s_m,s_n') = \frac{e^{-jk\,R_{2mn}}}{R_{2mn}}, \tag{5.24}$$

where $R_{1mn}$ and $R_{2mn}$ are distances from the source point and from the corresponding image, respectively to the observation point of interest.

Furthermore, $k_0$ and $k_g$ are propagation constants of air and lossy ground:

$$k_0^2 = \omega^2\mu_0\varepsilon_0, \tag{5.25}$$

$$k_g^2 = \omega^2\mu_0\varepsilon_{\text{efec}} = \omega^2\mu_0\left(\varepsilon_0\varepsilon_{\text{rg}} - j\frac{\sigma_g}{\omega}\right), \tag{5.26}$$

where $\varepsilon_{rg}$ and $\sigma_g$ are the relative permittivity and conductivity of the ground, respectively, and $\omega$ is the operating frequency.

The term $\vec{G}_s(s,s')$ contains the Sommerfeld integrals and is constructed from the vector components for horizontal and vertical dipoles [34]:

$$\vec{G}_s(s,s') = (\hat{x}\cdot\hat{s}')\cdot(G_\rho^H\cdot\vec{\rho} + G_\phi^H\cdot\vec{\phi} + G_z^H\cdot\vec{z}) + (\hat{z}\cdot\hat{s}')\cdot(G_\rho^V\cdot\vec{\rho} + G_z^V\cdot\vec{z}), \tag{5.27}$$

where

$$G_\rho^V = \frac{\partial^2}{\partial\rho\,\partial z}k_g^2\,V^R, \tag{5.28}$$

$$G_z^V = \left(\frac{\partial^2}{\partial z^2} + k_0^2\right)k_g^2\,V^R, \tag{5.29}$$

$$G_\rho^H = \cos\phi\left(\frac{\partial^2}{\partial\rho^2}k_0^2\,V^R + k_0^2\,U^R\right), \tag{5.30}$$

$$G_\phi^H = -\sin\phi\left(\frac{1}{\rho}\frac{\partial}{\partial\rho}k_0^2\,V^R + k_0^2\,U^R\right), \tag{5.31}$$

$$G_z^H = -j4\pi\omega\varepsilon_0\cos\phi\,G_\rho^V, \tag{5.32}$$

$$U^R = \int_0^\infty D_1(\lambda)\,\mathrm{e}^{-\gamma_0|z+z'|}\,J_0(\lambda\rho)\,\lambda\,\mathrm{d}\lambda, \tag{5.33}$$

$$V^R = \int_0^\infty D_2(\lambda)\,\mathrm{e}^{-\gamma_0|z+z'|}\,J_0(\lambda\rho)\,\lambda\,\mathrm{d}\lambda, \tag{5.34}$$

$$D_1(\lambda) = \frac{2}{\gamma_0+\gamma_g} - \frac{2k_0^2}{\gamma_0(k_0^2+k_g^2)}, \tag{5.35}$$

$$D_2(\lambda) = \frac{2}{k_g^2\gamma_0 + k_0^2\gamma_g} - \frac{2}{\gamma_0(k_0^2+k_g^2)}, \tag{5.36}$$

$$\gamma_0 = \sqrt{\lambda^2 - k_0^2};\quad \gamma_g = \sqrt{\lambda^2 - k_g^2}, \tag{5.37}$$

$$\phi = \mathrm{arctg}\frac{\rho}{z+z'}. \tag{5.38}$$

### 5.3.2 Numerical Solution of Integral Equation Set for Overhead Wires

The set of Pocklington's integro-differential equations (5.23) is numerically handled via the GB-IBEM [51, 52].

As a first step, at a wire segment the current is expressed in terms of a linear combination of shape functions:

$$I_n^e(s') = \sum_{i=1}^{n} I_{ni} f_{ni}(s') = \{f\}_n^T\{I\}_n \tag{5.39}$$

and the implementation of the isoparametric parameters yields

$$I_n^e(\zeta) = \sum_{i=1}^{n} I_{ni} f_{ni}(\zeta) = \{f\}_n^T \{I\}_n, \tag{5.40}$$

where $n$ is the number of local nodes per element.

Furthermore, applying the weighted residual approach and utilizing the Galerkin–Bubnov procedure the set of Pocklington's integro-differential equations is transformed into a system of algebraic equations, the matrix form of which is given by

$$\sum_{n=1}^{N_w} \sum_{i=1}^{N_n} [Z]_{ji}^e \{I\}_i^e = \{V\}_j^e, \quad m = 1, 2, \ldots, N_w; \quad j = 1, 2, \ldots, N_m, \tag{5.41}$$

where $[Z]_{ji}$ is the mutual impedance matrix for the $j$th observation segment on the $m$th wire and $i$th source segment on the $n$th antenna:

$$\begin{aligned}
[Z]_{ji}^e = &- \int_{-1}^{1} \int_{-1}^{1} \{D\}_j \{D'\}_i^T g_{0nm}(s_n, s'_m) \frac{ds'_m}{d\xi'} d\xi' \frac{ds_n}{d\xi} d\xi + \\
&+ k_0^2 \vec{s}_n \cdot \vec{s}'_m \int_{-1}^{1} \int_{-1}^{1} \{f\}_j \{f'\}_i^T g_{0nm}(s_n, s'_m) \frac{ds'_m}{d\xi'} d\xi' \frac{ds_n}{d\xi} d\xi - \\
&- \frac{k_g^2 - k_0^2}{k_g^2 + k_0^2} \int_{-1}^{1} \int_{-1}^{1} \{D\}_j \{D'\}_i^T g_{inm}(s_n, s_m^*) \frac{ds'_m}{d\xi'} d\xi' \frac{ds_n}{d\xi} d\xi + \\
&+ \frac{k_g^2 - k_0^2}{k_g^2 + k_0^2} k_0^2 \vec{s}_n \cdot \vec{s}'_m \int_{-1}^{1} \int_{-1}^{1} \{f\}_j \{f'\}_i^T g_{inm}(s_n, s_m^*) \frac{ds'_m}{d\xi'} d\xi' \frac{ds_n}{d\xi} d\xi + \\
&+ \vec{s}'_m \int_{-1}^{1} \int_{-1}^{1} \{f\}_j \{f'\}_i^T \vec{G}_{snm}(s_n, s_m^*) \frac{ds'_m}{d\xi'} d\xi' \frac{ds_n}{d\xi} d\xi + \\
&+ \frac{j}{4\pi\omega\varepsilon_0} \int_{-1}^{1} Z'_T \{f\}_j \{f'\}_j^T \frac{ds_n}{d\xi} d\xi.
\end{aligned} \tag{5.42}$$

Note that matrices $\{f\}$ and $\{f'\}$ contain the shape functions while $\{D\}$ and $\{D'\}$ contain their derivatives.

The voltage vector is given by

$$\begin{aligned}
&\{V\}_j^m = -j4\pi\omega\varepsilon_{\text{eff}} \int_{-1}^{1} E_{s_m}^{\text{inc}}(s_m) f_{jm}(s_m) \frac{ds_m}{d\xi} d\xi_m, \\
&m = 1, 2, \ldots, N_W; \quad j = 1, 2, \ldots, N_m
\end{aligned} \tag{5.43}$$

and can be evaluated in the close form [52].

### 5.3.3 Computational Example: Transient Response of a WT Lightning Strike

A WT struck by lightning is depicted in Figure 5.14a, while the corresponding wire antenna representation is shown in Figure 5.14b. WT configuration is modeled by a simple configuration of four PEC wires representing the tower

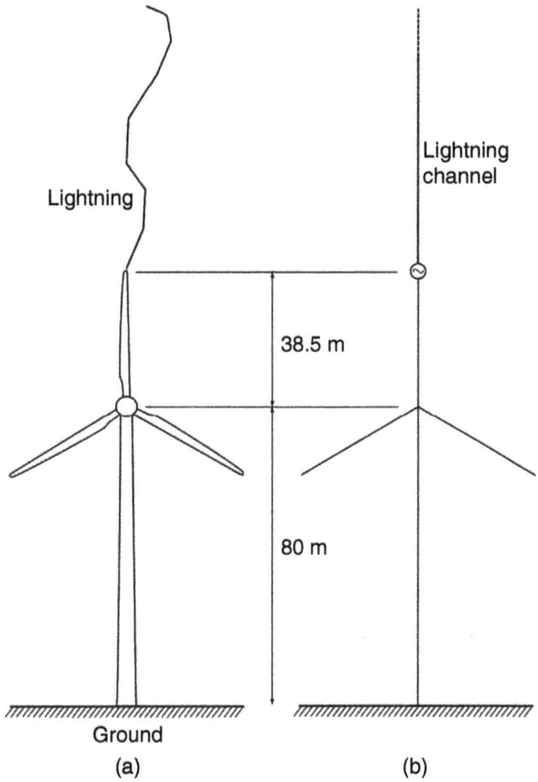


**Figure 5.14** Wind turbine struck by lightning and the related wire antenna model.

and three blades. The lightning channel is modeled as a lossy vertical wire antenna, neglecting the corona effect [51]. The resistance per unit channel length is assumed to be 0.07 Ω m$^{-1}$, as in [50], while the wire radius is $a = 10$ cm. For the sake of simplicity the current wave along the channel is assumed to propagate at the velocity of light although the realistic current propagation velocity is by the factor of 2 or 3 less than in the present model used in [51]. The lightning return stroke current is injected by using the equivalent voltage source on the tip of the WT blade.

The channel-base current shown in Figure 5.15 is approximated by the Heidler's function. The current and current derivative peaks are assumed to be 4.7 kA and 25 kA μs$^{-1}$, respectively [50]. The simulations are carried out for the different ground conductivities: PEC ground, 0.01 S m$^{-1}$, 0.001 S m$^{-1}$, 0.1 m S m$^{-1}$, and $\varepsilon_r = 10$. The grounding wires are neglected, as it was shown that they do not significantly affect the current distribution [49, 53].

The transient current is analyzed at the characteristic points along the WT; strike blade tip, middle of the strike blade, middle of the side blade, and WT base, as shown in Figure 5.16.

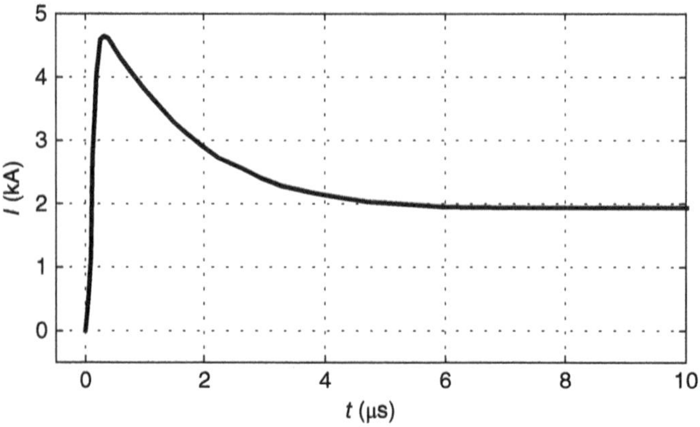


Figure 5.15 The source current waveform.

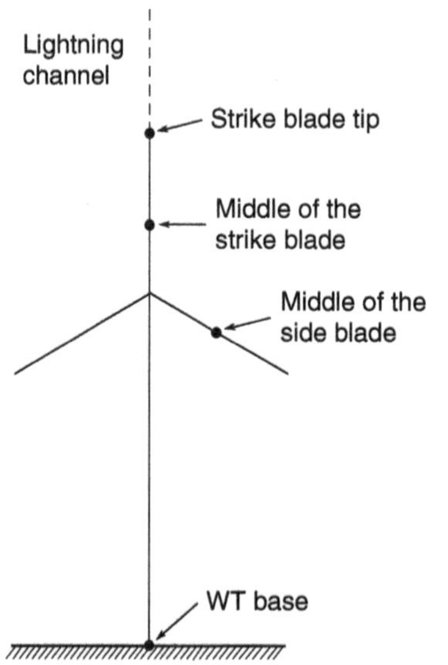


Figure 5.16 Characteristic points along the WT.

Figure 5.17 shows the transient current induced at different points along the WT (strike blade tip, middle of the strike blade, middle of the side blade, WT base) for the case of PEC ground.

All major reflections are clearly visible, e.g. the maximum value of the transient current at the side blades is five times less than at the strike blade.

It can also be seen that the transient at the side blade is of significantly shorter duration than in other WT points.

Figures 5.18–5.21 show transient current waveforms induced at different points along the WT: strike blade tip, middle of the strike blade, middle of the

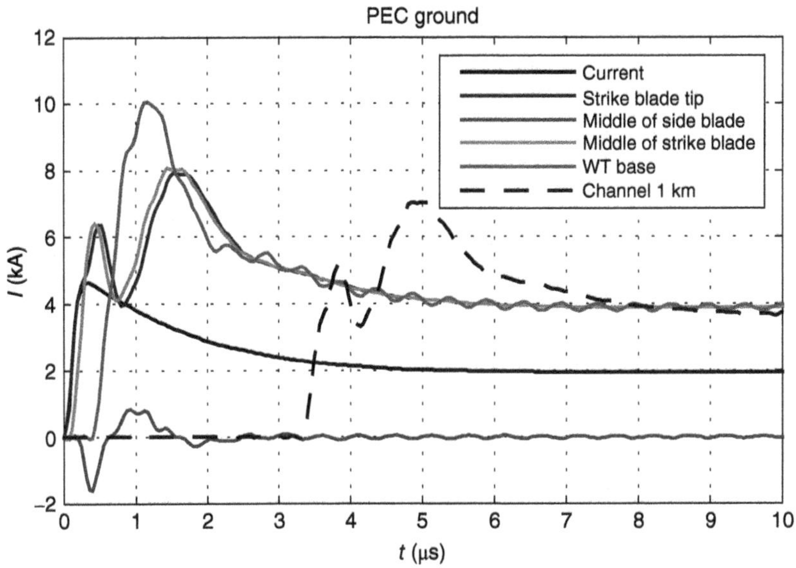


**Figure 5.17** Transient induced at different points along WT for the case of PEC ground.

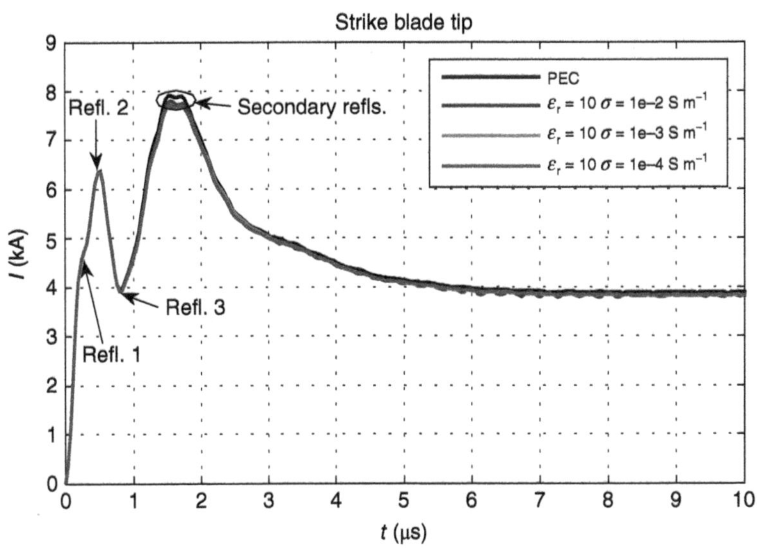


**Figure 5.18** Transient current induced at the strike blade tip.

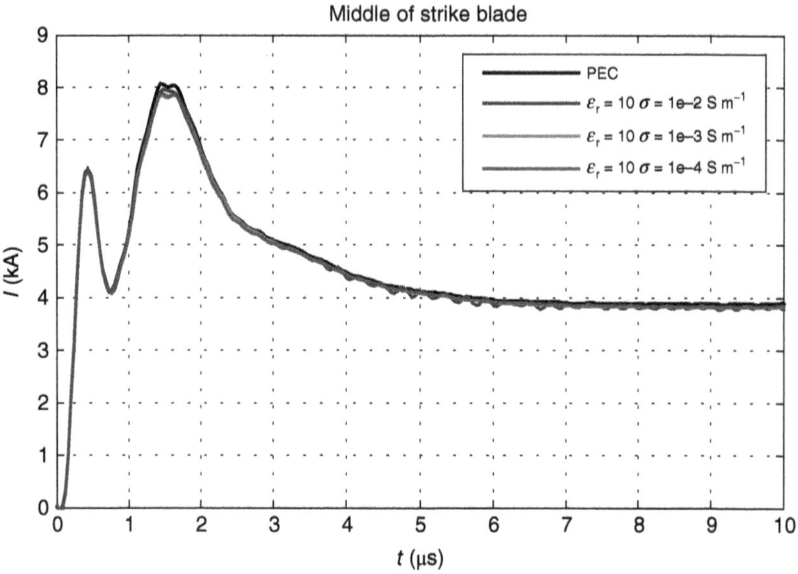


**Figure 5.19** Transient current induced at the middle of the strike blade.

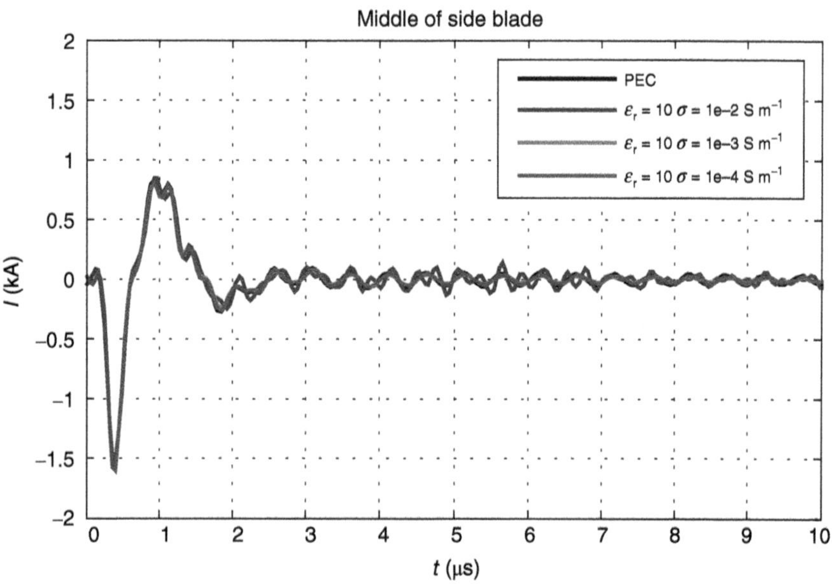


**Figure 5.20** Transient current induced at the middle of the side blade.

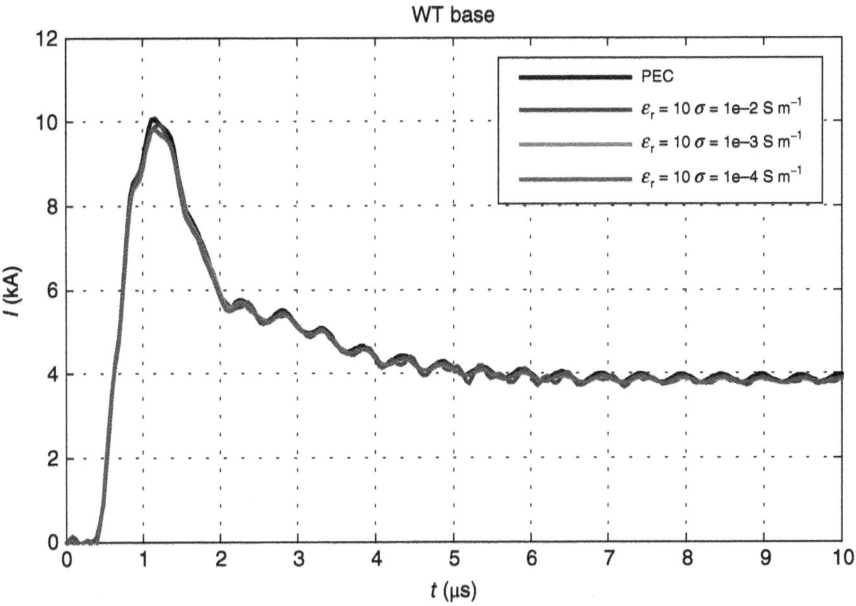


**Figure 5.21** Transient current induced at WT base.

side blade, and WT base, obtained for different values of ground conductivity. In all cases, the permittivity is assumed to be $\varepsilon_r = 10$.

Analyzing small differences between current waveforms for different ground conductivities, which is consistent with the results published in [48] for the CN tower, it can be considered that the influence of a finite ground conductivity to the transient current induced along the WT is rather negligible.

Figure 5.18 shows all major reflection. Thus, the curve assigned as Refl.1 represents the first reflection of the transient current from the wire junction occurring after 0.253 μs. Furthermore, Refl. 2 represents the reflection from the ends of the side blades. This reflection occurs after 0.507 μs. The reflection from the ground appears after 0.78 μs and is assigned as Refl. 3. Finally, due to secondary reflections, the transient current reaches a maximum around 1.6 μs due to secondary reflections, which is followed by the small oscillations.

## References

1 Olsen, R.G. and Willis, M.C. (1996). A comparison of exact and Quasi-static methods for evaluating grounding systems at high frequencies. *IEEE Transactions on Power Delivery* **11** (2): 1071–1081.

2 Grcev, L. and Toseva, V.A. (2000). Computer model for electromagnetic transients in thin wire structures above, below and penetrating the earth. Proceedings of the EMF 2000, Gent, Belgium.

3 Uman, M.A. and McLain, D.K. (1969). Magnetic field of lightning return stroke. *Journal of Geophysical Research* **74**: 6899–6910.

4 Nucci, C.A., Diendorfer, G., Uman, M.A. et al. (1990). Lightning return stroke current models with specified channel base current: a review and comparison. *Journal of Geophysical Research* **95**: 20395–20408.

5 Thottappillil, R., Rakov, V.A., and Uman, M.A. (1997). Distribution of charge along the lightning channel: relation to remote electric and magnetic fields and to return-stroke models. *Journal of Geophysical Research* **102**: 6987–7006.

6 Rakov, V.A. and Dulzon, A.A. (1991). A modified transmission line model for lightning return stroke field calculation. *Proceedings of the International Zurich Symposium on EMC* **9**: 229–235.

7 Diendorfer, G. and Uman, M.A. (1990). An improved return-stroke model with specified channel-base current. *Journal of Geophysical Research* **95**: 13621–13644.

8 Rachidi, F., Rakov, V.A., Nucci, C.A., and Bermudez, J.L. (2002). Effect of vertically extended strike object on the distribution of current along the lightning channel. *Journal of Geophysical Research* **107** (D23): 4699. doi: 10.1029/2002JD002119.

9 Grčev, L., Rachidi, F., and Rakov, V.A. (2003). Comparison of electromagnetic models of lightning return strokes using current and voltage sources. International Conference on Atmospheric Electricity, ICAE'03 Versailles, France (June 2003).

10 Moini, R., Kordi, B., Rafi, G.Z., and Rakov, V.A. (2000). A new lightning return stroke model based on antenna theory model. *Journal of Geophysical Research* **105** (D24): 29693–29702.

11 Baba, Y. and Ishii, M. (2001). Numerical electromagnetic field analysis of lightning current in tall structures. *IEEE Transactions on Power Delivery* **16** (2): 324–328.

12 Kordi, B., Moini, R., and Rakov, V.A. (2003). Comparison of lightning return stroke electric fields predicted by the transmission line and antenna theory models. Proceedings of the 15th International Zurich Symposium on Electromagnetic Compatibility, Zurich, Switzerland (February 2003), 551–556.

13 Baba, Y. and Ishii, M. (2003). Characteristics of electromagnetic return-stroke models. *IEEE Transactions on Electromagnetic Compatibility* **45** (1): 129–135.

14 Shoory, A., Moini, R., Sadeghi, S.H.H., and Rakov, V.A. (2005). Analysis of lightning-radiated electromagnetic fields in the vicinity of lossy ground. *IEEE Transactions on Electromagnetic Compatibility* **47** (1): 131–145.

**15** Pokharel, R.K., Ishii, M., and Baba, Y. (2003). Numerical electromagnetic analysis of lightning-induced voltage over ground of finite conductivity. *IEEE Transactions on EMC* **45** (4): 651–656.

**16** Tesche, F.M., Ianoz, M., and Karlsson, T. (1997). *EMC Analysis Methods and Computational Models*. New York: Wiley.

**17** Baba, Y. and Rakov, V.A. (2005). On the mechanism of attenuation of current waves propagating along a vertical perfectly conducting wire above ground: application to lightning. *IEEE Transactions on Electromagnetic Compatibility* **47**: 521–532.

**18** Antonijevic, S., Doric, V., and Poljak, D. (2010). Optimized transient analysis using GB-BEM. XI-th International Workshop on Optimization and Inverse Problems in Electromagnetism, Sofia, Bulgaria (4–18 September 2010).

**19** Cavka, D., Poljak, D., Doric, V., and Goic, R. (2012). Transient analysis of wind turbines. *Renewable Energy* **43**: 284–291.

**20** Čavka, D., Poljak, D., Dorić, V., and Antonijević, S. (2012). Some computational aspects of using current and voltage sources in electromagnetic models of lightning return strokes. ICLP 2012, Vienna, Austria.

**21** Poljak, D., Sesnic, S., Cavka, D., and El Khamlichi Drissi, K. (2015). On the use of the vertical straight wire model in electromagnetics and related boundary element solution. *Engineering Analysis with Boundary Elements* **50**: 19–28.

**22** Cavka, D. and Poljak, D. (2015). Magnetic current loop as a source model for finite thin wire antennas. *International Journal for Numerical Modeling* **28**: 189–200.

**23** Poljak, D., Cavka, D., and Rachidi, F. (2017). On the use of magnetic current loop source model in lightning electromagnetics. 32nd URSI GASS, Montreal (19–26 August 2017).

**24** Baba, Y. and Rakov, V.A. (2008). Application of electromagnetic models of the lightning return stroke. *IEEE Transactions on Power Delivery* **23**: 800–811.

**25** Karami, H., Rachidi, F., and Rubinstein, M. (2016). On practical implementation of electromagnetic models of lightning return-strokes. *Atmosphere* **7**: 135. doi: 10.3390/atmos7100135.

**26** Miller, E.K., Poggio, A.J., Burke, G.J., and Selden, E.S. (1972). Analysis of wire antennas in the presence of a conducting half-space. Part I. The vertical antenna in free space. *Canadian Journal of Physics* **50**: 332–341.

**27** Li, X., El Khamlichi Drissi, K., and Paladian, F. (2004). Insulated vertical antennas above ground. *IEEE Transactions on Antennas and Propagation* **52** (1): 321–324.

**28** Poljak, D., El Khamlichi Drissi, K., Kerroum, K., and Sesnic, S. (2011). Comparison of analytical and boundary element modeling of electromagnetic

field coupling to overhead and buried wires. *Engineering Analysis with Boundary Elements* **35**: 555–563.
**29** Poljak, D., Rachidi, F., and Tkachenko, S.V. (2007). Generalized form of telegrapher's equations for the electromagnetic field coupling to finite-length lines above a lossy ground. *IEEE Transactions on Electromagnetic Compatibility* **49** (3): 689–697.
**30** Poljak, D., Doric, V., Rachidi, F. et al. (2009). Generalized form of telegrapher's equations for the electromagnetic field coupling to buried wires of finite length. *IEEE Transactions on Electromagnetic Compatibility* **51** (2): 331–337.
**31** Poljak, D. and Doric, V. (2005). Time-domain modeling of electromagnetic field coupling to finite-length wires embedded in a dielectric half-space. *IEEE Transactions on Electromagnetic Compatibility* **47** (2): 247–253.
**32** Miller, E.K., Poggio, A.J., Burke, G.J., and Selden, E.S. (1972). Analysis of wire antennas in the presence of a conducting half-space. Part II. The horizontal antenna in free space. *Canadian Journal of Physics* **50**: 2614–2627.
**33** Burke, G.J., Johnson, W.A., and Miller, E.K. (1983). Modeling of simple antennas near to and penetrating an interface. *Proceedings of the IEEE* **71** (1): 174–175.
**34** Burke, G.J. and Miller, E.K. (1984). Modeling antennas near to and penetrating a Lossy interface. *IEEE Transactions on Antennas and Propagation* **AP-32** (10): 1040–1049.
**35** IEC International Standard IEC 61400-24 (2010). *Wind turbine generation system – 24: lightning protection*. Geneva: International Electro-Technical Commission.
**36** IEA (1997). Recommended practices for wind turbine testing and evaluation, 9. Lightning Protection for Wind Turbine Installations. Ed. 1997.
**37** IEE Professional Group S1 (1997). (New concepts in the generation, distribution and use of electrical energy): half-day colloquium on "Lightning protection of wind turbines", (1997, 11).
**38** Sorensen, T., Sorensen, J.T., and Nielsen, H. (1998). Lightning damages to power generating wind turbines. Proceedings of the 24th International Conference on Lightning Protection (ICLP98), 176–179.
**39** McNiff, B. (2002). Wind Turbine Lightning Protection Project 1999–2001. NREL Subcontractor Report. *SR-500-31115*.
**40** Rachidi, F., Rubinstein, M., Montanya, J. et al. (2008). A review of current issues in lightning protection of new-generation wind turbine blades. *IEEE Transactions on Industrial Electronics* **55** (6): 2489–2496.
**41** Yoh, Y., Toshiaki, F., and Toshiaki, U. (2007). How does ring earth electrode effect to wind turbine? 42nd International Universities Power Engineering Conference, UPEC 2007, 796–799.
**42** Glushakow, B. (2007). Effective lightning protection for wind turbine generators. *IEEE Transactions on Energy Conversion* **22** (1): 214–222.

43 Rakov, V.A. (2001). Transient response of a tall object to lightning. *IEEE Transactions on Electromagnetic Compatibility* **43**: 654–661.
44 Rakov, V.A. and Uman, M.A. (1998). Review and evaluation of lightning return stroke models including some aspects of their application. *IEEE Transactions on Electromagnetic Compatibility* **40**: 403–426.
45 Pavanello, D., Rachidi, F., Rakov, V.A. et al. (2007). Return stroke current profiles and electromagnetic fields associated with lightning strikes to tall towers: comparison of engineering models. *Journal of Electrostatics* **65**: 316–321.
46 Podgorski, S. and Landt, J.A. (1987). Three dimensional time domain modeling of lightning. *IEEE Transactions on Power Delivery* **2**: 931–938.
47 Petrache, E., Rachidi, F., Pavanello, D. et al (2005). Lightning strikes to elevated structures: influence of grounding conditions on currents and electromagnetic fields. Presented at IEEE International Symposium on Electromagnetic Compatibility, Chicago.
48 Petrache, E., Rachidi, F., Pavanello, D. et al. (2005). Influence of the finite ground conductivity on the transient response to lightning of a tower and its grounding. Presented at 28th General Assembly of International Union of Radio Science (URSI), New Delhi, India.
49 Podgorski, S. and Landt, J.A. (1985). Numerical analysis of the lightning-CN tower interaction. Presented at 6th Symposium and Technical Exhibition on Electromagnetic Compatibility, Zurich, Switzerland.
50 Kordi, B., Moini, R., Janischewskyj, W. et al. (2003). Application of the antenna theory model to a tall tower struck by lightning. *Journal of Geophysical Research* **108**.
51 Poljak, D. and Čavka, D. (2015). Electromagnetic compatibility issues of wind turbine analysis and design. *International Journal of Computational Methods and Experimental Measurements* **3** (3): 250–268.
52 Poljak, D., El Khamlichi Drissi, K., and Nekhoul, B. (2013). Electromagnetic field coupling to arbitrary wire configurations buried in a lossy ground: a review of antenna model and transmission line approach. *International Journal of Computational Methods and Experimental Measurements* **1** (2): 142–163.
53 Rachidi, F. (2005). Modeling lightning return strokes to tall structures: recent development. VIII International Symposium on Lightning Protection, Sao Paulo, Brazil.

# 6

# Transient Analysis of Grounding Systems

Transient analysis of grounding systems is a rather important issue in the design of lightning protection systems (LPS). Particularly important application is related to LPS for environmentally attractive wind turbines (WTs). In general, analysis of grounding systems can be carried out by means of the transmission line (TL) model [1–3] or the full-wave model, also referred to as the antenna theory (AT) model (AM) [4–6]. The latter is considered to be the rigorous one, while the principal advantage of the TL approach is simplicity [7]. Both TL and AT models can be formulated in either frequency domain (FD) or time domain (TD) [8].

This chapter deals with both FD–AT and TD–AT approaches, respectively. First, analysis of horizontal and vertical grounding electrodes, respectively, being an important component in many realistic grounding systems of complex shape, is carried out, which is followed by sections pertaining to more realistic grounding systems, such as rectangular grounding grids or complex grounding systems for WTs. The key parameter in the study of horizontal grounding electrode is the equivalent current distribution along the electrode. Once the current distribution along the electrode is determined, other parameters of interest, such as voltage distribution or transient impedance, can be calculated.

Within the AT approach the effect of an earth–air interface is first taken into account via the corresponding reflection coefficient (RC) thus avoiding the rigorous approach based on the Sommerfeld integrals. Furthermore, the rigorous integral equation formulation featuring the Sommerfeld integral approach is presented. Finally, the TL approximation is used. The space–frequency and space–time integro-differential expressions arising from the AT model are numerically treated by means of the Galerkin–Bubnov scheme of the indirect boundary element method (GB-IBEM) [8]. In addition to the numerical solution, an approximate analytical solution of FD and TD expressions is carried out as well.

*Computational Methods in Electromagnetic Compatibility: Antenna Theory Approach versus Transmission line Models*, First Edition. Dragan Poljak and Khalil El Khamlichi Drissi.

## 6.1 Frequency Domain Analysis of Horizontal Grounding Electrode

Figure 6.1 shows the configuration of interest: the horizontal grounding electrode of length $L$ and radius $a$, buried in a lossy medium at depth $d$ and energized by an equivalent current generator $I_g$.

The integral relationships for the induced current and voltage induced along the electrode can be derived by enforcing the continuity conditions for the tangential components of the electric field along the electrode surface.

First, an integral equation formulation/RC approximation is considered. Then the integral equation including rigorous Sommerfeld integral approach presented. Finally, the TL approximation is used.

### 6.1.1 Integral Equation Formulation/Reflection Coefficient Approach

The total tangential electric field at the buried conductor surface given by a sum of the excitation field $\vec{E}^{\text{exc}}$ and the scattered field $\vec{E}^{\text{sct}}$ is equal to the product of the current along the electrode $I(x)$ and surface internal impedance $Z_s(x)$ per unit length of the conductor [9]:

$$\vec{e}_x \cdot (\vec{E}^{\text{exc}} + \vec{E}^{\text{sct}}) = Z_s(x)I(x), \tag{6.1}$$

where the surface internal impedance $Z_s(x)$ is given by [7, 10]

$$Z_s(x) = \frac{Z_{cw}}{2\pi a}\frac{I_0(\gamma_{wa})}{I_1(\gamma_{wa})}. \tag{6.2}$$

While $I_0(\gamma_w)$ and $I_1(\gamma_w)$ denote modified Bessel functions of the zero and first order respectively, $Z_{cw}$ and $\gamma_w$ are given by [7, 10]

$$Z_{cw} = \sqrt{\frac{j\omega\mu_w}{\sigma_w + j\omega\varepsilon_w}}, \tag{6.3}$$

$$\gamma_w = \sqrt{j\omega\mu(\sigma_w + j\omega\varepsilon_w)}. \tag{6.4}$$

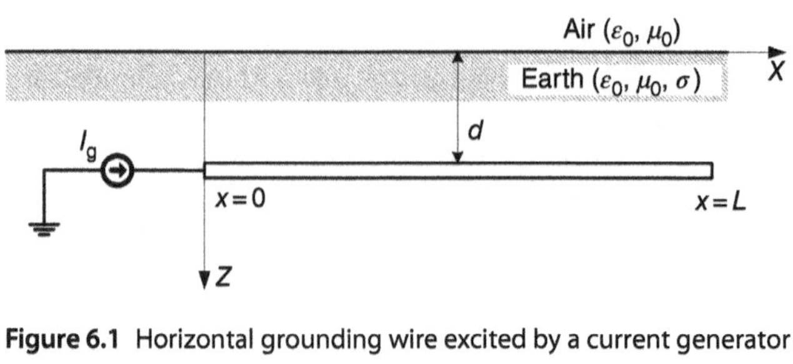


**Figure 6.1** Horizontal grounding wire excited by a current generator $I_g$.

The surface impedance $Z_s(x)$ can be neglected if good conductors are considered. The scattered electric field can be expressed in terms of the vector potential $\vec{A}$ and the scalar potential. According to the thin wire approximation [8, 11] only the axial component of the scattered field exists, i.e. it follows that

$$E_x^{\text{sct}} = -j\omega A_x - \frac{\partial \phi}{\partial x}, \tag{6.5}$$

where the vector and scalar potential are given by

$$A_x = \frac{\mu}{4\pi} \int_0^L I(x')g(x,x')\mathrm{d}x', \tag{6.6}$$

$$\phi(x) = \frac{1}{4\pi\varepsilon_{\text{eff}}} \int_0^L q(x')g(x,x')\mathrm{d}x'. \tag{6.7}$$

While $q(x)$ denotes the charge distribution along the electrode, $I(x')$ is the induced current along the electrode.

The complex permittivity of the lossy ground $\varepsilon_{\text{eff}}$ is

$$\varepsilon_{\text{eff}} = \varepsilon_r\varepsilon_0 - j\frac{\sigma}{\omega}, \tag{6.8}$$

where $\varepsilon_r$ and $\sigma$ denote the corresponding permittivity and conductivity, respectively.

The Green function $g(x, x')$ is given by

$$g(x,x') = g_0(x,x') - \Gamma_{\text{ref}} g_i(x,x'), \tag{6.9}$$

where $g_0(x, x')$ is the lossy medium Green function

$$g_0(x,x') = \frac{\mathrm{e}^{-\gamma R_1}}{R_1} \tag{6.10}$$

and $g_i(x, x')$, due to the image electrode in the air, is

$$g_i(x,x') = \frac{\mathrm{e}^{-\gamma R_2}}{R_2}. \tag{6.11}$$

The propagation constant of the lower medium is defined as

$$\gamma = \sqrt{j\omega\mu\sigma - \omega^2\mu\varepsilon} \tag{6.12}$$

and $R_1$ and $R_2$ are given by

$$R_1 = \sqrt{(x-x')^2 + a^2}, \quad R_2 = \sqrt{(x-x')^2 + 4d^2}. \tag{6.13}$$

The effect of a ground–air interface is taken into account in terms of the RC [6]:

$$\Gamma_{\text{ref}} = \frac{\frac{1}{\underline{n}}\cos\theta - \sqrt{\frac{1}{\underline{n}} - \sin^2\theta}}{\frac{1}{\underline{n}}\cos\theta + \sqrt{\frac{1}{\underline{n}} - \sin^2\theta}}; \quad \theta = \text{arctg}\frac{|x-x'|}{2d}; \quad \underline{n} = \frac{\varepsilon_{\text{eff}}}{\varepsilon_0}. \tag{6.14}$$

The main advantage of the RC approach versus rigorous Sommerfeld integral approach is the simplicity of the formulation and the significantly less computational cost required for the numerical solution of the related integral expression [8, 11].

Combining the continuity equation [8]

$$q = -\frac{1}{j\omega}\frac{\mathrm{d}I}{\mathrm{d}x} \tag{6.15}$$

with (6.7) yields

$$\phi(x) = -\frac{1}{j4\pi\omega\varepsilon_{\mathrm{eff}}}\int_0^L \frac{\partial I(x')}{\partial x'} g(x,x')\mathrm{d}x'. \tag{6.16}$$

Furthermore, inserting (6.6) and (6.16) into (6.5) gives an integral relationship for the scattered field

$$E_x^{\mathrm{sct}} = -j\omega\frac{\mu}{4\pi}\int_0^L I(x')g(x,x')\mathrm{d}x' + \frac{1}{j4\pi\omega\varepsilon_{\mathrm{eff}}}\frac{\partial}{\partial x}\int_0^L \frac{\partial I(x')}{\partial x'} g(x,x')\mathrm{d}x'. \tag{6.17}$$

For the case of grounding electrodes, the excitation is given in the form of a current source and the tangential field at the electrode surface does not exist, i.e. [4]

$$E_x^{\mathrm{exc}} = 0. \tag{6.18}$$

Combining Equations (6.1), (6.17) and (6.18) leads to the homogeneous Pocklington's integro-differential equation for the unknown current distribution along the wire:

$$j\omega\frac{\mu}{4\pi}\int_0^L I(x')g(x,x')\mathrm{d}x' - \frac{1}{j4\pi\omega\varepsilon_{\mathrm{eff}}}\frac{\partial}{\partial x}\int_0^L \frac{\partial I(x')}{\partial x'} g(x,x')\mathrm{d}x' + Z_s(x)I(x) = 0. \tag{6.19}$$

Once the current along the electrode is determined the scattered voltage can be obtained from the line integral of a scattered vertical field from the remote soil to the electrode surface [9]:

$$V^{\mathrm{sct}}(x) = -\int_{\infty}^{d} E_z^{\mathrm{sct}}(x,z)\mathrm{d}z. \tag{6.20}$$

The vertical field component is expressed by the scalar potential gradient

$$E_z^{\mathrm{sct}} = -\frac{\partial\phi}{\partial z}, \tag{6.21}$$

and the scattered voltage along the electrode can be written as

$$V^{\mathrm{sct}}(x) = \int_{-\infty}^{d} \frac{\partial\varphi}{\partial z}\mathrm{d}z = \frac{\mathrm{d}}{\mathrm{d}z}\int_{-\infty}^{d} \varphi(x,z)\mathrm{d}z. \tag{6.22}$$

Integrating the scattered field from the infinite soil to the electrode surface and assuming the scalar potential in the remote soil to be zero [9] it follows that

$$V^{\text{sct}}(x) = -\frac{1}{j4\pi\omega\varepsilon_{\text{eff}}}\int_0^L \frac{\partial I(x')}{\partial x'} g(x,x')\text{d}x'. \tag{6.23}$$

The grounding electrode is excited by an equivalent ideal current generator with one terminal connected to the grounding electrode and the other one grounded at infinity, as shown in Figure 6.1.

The ideal current generator is inserted into the integro-differential equation formulation through the forced boundary conditions [6]:

$$I(0) = I_g, \quad I(L) = 0, \tag{6.24}$$

where $I_g$ is the impressed unit current generator.

To calculate the transient impedance of the grounding electrode, the time transient voltage at the injection point has to be determined by applying the IFFT algorithm. Once the scattered voltage is obtained, the transient impedance can be expressed as [4]

$$z(t) = \frac{v(0,t)}{i(0,t)}, \tag{6.25}$$

where voltage $v(0, t)$ and $i(0, t)$ represent their values at the injection point of the electrode, i.e. at $x = 0$. The equivalent current source $i(0, t)$ is a lightning strike current, usually represented by a double exponential pulse [5]

$$i(0,t) = I_0(\text{e}^{-\alpha t} - \text{e}^{-\beta t}), \tag{6.26}$$

with its FD counterpart

$$I(0,\omega) = I_0\left(\frac{1}{\alpha + j\omega} - \frac{1}{\beta + j\omega}\right). \tag{6.27}$$

Having performed extensive numerical experiments, it has been found that optimal parameters for performing discrete sampling and subsequent IFFT include $2^{16}$ samples and a maximum frequency of 500 MHz. Namely, these parameters provide accurate results within a reasonable time frame.

### 6.1.2 Numerical Solution

The current $I^e(x)$ along the wire segment can expressed as follows:

$$I^e(x') = \{f\}^T\{I\}. \tag{6.28}$$

Using the weighted residual approach, performing certain mathematical manipulations and eventually assembling the contributions from all segments, the integro-differential equation (6.19) is transferred into the following matrix equation [8]:

$$\sum_{j=1}^{M} [Z]_{ji}\{I\}_i = 0, \quad \text{and} \quad j = 1, 2, \dots, M, \tag{6.29}$$

where $M$ is the total number of segments and $[Z]_{ji}$ is the mutual impedance matrix representing the interaction of the $i$th source with the $j$th observation segment, respectively:

$$[Z]_{ji} = -\frac{1}{4j\pi\omega\varepsilon_{\text{eff}}}\begin{pmatrix} \int_{\Delta l_j} \{D\}_j \int_{\Delta l_i} \{D'\}_i^T g(x,x')\text{d}x'\,\text{d}x + \\ k^2 \int_{\Delta l_j} \{f\}_j \int_{\Delta l_i} \{f\}_i^T g(x,x')\text{d}x'\,\text{d}x \end{pmatrix}$$
$$+ \int_{\Delta l_j} Z_{\text{L}}(x)\{f\}_j\{f\}_i^T \text{d}x. \tag{6.30}$$

Matrices $\{f\}$ and $\{f'\}$ contain the shape functions, while $\{D\}$ and $\{D'\}$ contain their derivatives, and $\Delta l_i$, $\Delta l_j$ are the widths of $i$th and $j$th boundary elements.

A linear approximation over a boundary element is often used:

$$f_i = \frac{x_{i+1} - x'}{\Delta x}, \quad f_{i+1} = \frac{x' - x_i}{\Delta x}, \tag{6.31}$$

as this choice was proved to be optimal one in modeling various wire structures [8].

The excitation function in the form of the current generator $I_g$ is taken into account through the forced boundary condition at the first node of the solution vector, i.e.

$$I_1 = I_g \quad \text{and} \quad I_g = 1\text{e}^{j0}. \tag{6.32}$$

Once the current distribution is obtained, the scattered voltage (6.23) can be readily evaluated using the boundary element formalism.

As the current distribution derivative on the segment is simply given by

$$\frac{\partial I(x')}{\partial x'} = \frac{I_{i+1} - I_i}{\Delta x}, \tag{6.33}$$

the scattered voltage can be computed from the following formula:

$$V^{\text{sct}}(x) = -\frac{1}{j4\pi\omega\varepsilon_{\text{eff}}} \sum_{i=1}^{M} \frac{I_{i+1} - I_i}{\Delta x} \int_{x_i}^{x_{i+1}} g(x,x')\text{d}x'. \tag{6.34}$$

The integral on the right-hand side of (6.34) is calculated via the standard Gaussian quadrature.

### 6.1.3 Integral Equation Formulation/Sommerfeld Integral Approach

The homogeneous Pocklington's integro-differential equation for the current distribution along the horizontal grounding electrode, utilizing Sommerfeld integrals to account for an interface effects, is derived by expressing the electric field in terms of the Hertz vector potential and by satisfying certain boundary conditions for the tangential field components at the electrode surface [4–6, 8].

The current distribution induced along the electrode is governed by the following integro-differential equation [6]:

$$E_x^{\mathrm{exc},H} = -\frac{1}{j4\pi\omega\varepsilon_{\mathrm{eff}}}\left\{\begin{array}{l}\int_{-L/2}^{L/2}\frac{\partial^2}{\partial x^2}\left[g_0^H(xx') - g_i^H(x,x') + k_1^2 V_{11}\right]I(x')\mathrm{d}x' \\ +k_1^2\int_{-L/2}^{L/2}[g_0^H(xx') - g_i^H(x,x') + U_{11}]I(x')\mathrm{d}x'\end{array}\right\}, \tag{6.35}$$

where $I(x')$ is the unknown current distribution along the wire, $E_x^{\mathrm{exc},H}$ is the excitation function, and $g_0^H(x,x',z)$ is the earth Green function of the form

$$g_0^H(x,x',z) = \frac{\mathrm{e}^{-jk_1R_{1h}}}{R_{1h}}, \tag{6.36}$$

while $g_i^H(x,x',z)$ arises from the image theory and is given by

$$g_i^H(x,x',z) = \frac{\mathrm{e}^{-jk_1R_{2h}}}{R_{2h}}, \tag{6.37}$$

where $R_{1h}$ and $R_{2h}$ are the distances from the horizontal wire in the lossy ground and its image in the air to the observation point in the lower lossy medium, respectively.

Furthermore, $k_1$ and $k_2$ are the propagation constants of a lossy ground and air respectively:

$$k_1^2 = \omega^2\mu_0\varepsilon_{\mathrm{eff}}, \tag{6.38}$$

$$k_2^2 = \omega^2\mu_0\varepsilon_0. \tag{6.39}$$

The effects of the imperfectly conducting half-space are taken into account by Sommerfeld integrals $U_{11}$ and $V_{11}$, [4]:

$$U_{11} = 2\int_0^\infty \frac{\mathrm{e}^{-\mu_1(d-z)}}{\mu_1+\mu_2}J_0(\lambda\rho)\lambda\,\mathrm{d}\lambda, \tag{6.40}$$

$$V_{11} = 2\int_0^\infty \frac{\mathrm{e}^{-\mu_1(d-z)}}{k_2^2\mu_1+k_1^2\mu_2}J_0(\lambda\rho)\lambda\,\mathrm{d}\lambda, \tag{6.41}$$

where $J_0(\lambda\rho)$ is the zero order Bessel function of the first kind, $d > 0, z < 0$ ,while $\mu_1, \mu_2,$ and $\rho$ are given by

$$\mu_1 = (\lambda^2 - k_1^2)^{1/2}, \quad \mu_2 = (\lambda^2 - k_2^2)^{1/2}, \quad \rho = |\,x - x'\,|\,. \tag{6.42}$$

To calculate the frequency spectrum, a repeated evaluation of the Sommerfeld integrals and matrix inversion on several frequencies is required. The numerical solution procedure is similar to the one presented in Section 6.1.2.

### 6.1.4 Analytical Solution

For convenience, the homogeneous integral equation (6.19) can be written as follows:

$$-\frac{1}{j4\pi\omega\varepsilon_{\text{eff}}}\left(\frac{\partial^2}{\partial x^2}-\gamma^2\right)\int_0^L I(x')g(x,x')\mathrm{d}x' = 0. \tag{6.43}$$

Note, that the assumption $Z_s = 0$ is adopted, for simplicity.

Now, the integral in (6.43) can be written as

$$\int_0^L I(x')g(x,x')\mathrm{d}x' = I(x)\int_0^L g(x,x')\mathrm{d}x' + \int_0^L [I(x')-I(x)]g(x,x')\mathrm{d}x'. \tag{6.44}$$

The second integral on the right-hand side of (6.44) can be neglected [12, 13], and (6.43) becomes

$$-\frac{1}{j4\pi\omega\varepsilon_{\text{eff}}}\left(\frac{\partial^2}{\partial x^2}-\gamma^2\right)I(x)\int_0^L g(x,x')\mathrm{d}x' = 0. \tag{6.45}$$

This approximation, however, results in the loss of information about radiated energy because of the current reflection from the electrode ends. However, it has been shown that the approximation is quite sufficient for the case of lossy medium [13].

The integral in (6.45) can be readily calculated as documented in [12, 13]:

$$\int_0^L g(x,x')\mathrm{d}x' = 2\left(\ln\frac{L}{a} - \Gamma_{\text{ref}}\ln\frac{L}{2d}\right) = \Psi. \tag{6.46}$$

Now, the homogeneous Pocklington's equation becomes

$$\left(\frac{\partial^2}{\partial x^2}-\gamma^2\right)I(x) = 0. \tag{6.47}$$

Equation (6.47) is solved by prescribing the boundary conditions (6.24) and is given by

$$I(x) = I_g\frac{\sinh[\gamma(L-x)]}{\sinh(\gamma L)}. \tag{6.48}$$

The expression for scattered voltage can be obtained by substituting (6.48) into (6.47), which yields

$$V^{\text{sct}}(x) = \frac{\gamma I_g}{j4\pi\omega\varepsilon_{\text{eff}}\sinh(\gamma L)}\int_0^L \cosh[\gamma(L-x)]g(x,x')\mathrm{d}x'. \tag{6.49}$$

The integral in (6.49) is computed by means of standard numerical integration.

Once the results for the scattered voltage in the FD are obtained, the TD scattered voltage is obtained by means of IFFT.

### 6.1.5 Modified Transmission Line Method (TLM) Approach

The TL equations for a horizontal grounding wire excited by the current generator can be derived from Maxwell's equations and expressed in terms of voltage and current induced along the electrode [14]:

$$\frac{\partial V(x,\omega)}{\partial x} + Z(\omega)\cdot I(x,\omega) = E_x^{\text{ext}}, \tag{6.50}$$

$$\frac{\partial I(x,\omega)}{\partial x} + Y(\omega)\cdot V(x,\omega) = 0. \tag{6.51}$$

The solution of the FD TL equations is based on the chain matrix [14]. The per-unit-length parameters Z($\omega$) and Y($\omega$) for a buried conductor can be computed using modal equation [14] and are frequency dependent:

$$Z(\gamma(\omega))\cdot Y(\gamma(\omega)) = (\gamma(\omega))^2, \tag{6.52}$$

where

$$Z(\gamma(\omega)) = \frac{j\omega\mu_0}{2\pi}[K_0(\gamma_1 a) - K_0(\gamma_1(2d-a)) + I_1(\gamma(\omega))] \tag{6.53}$$

and

$$Y(\gamma(\omega)) = \frac{j2\pi\omega\varepsilon_{\text{eff}}}{K_0(\gamma_1 a) - K_0(\gamma_1(2d-a)) + k_1^2 I_2(\gamma(\omega))}. \tag{6.54}$$

$\gamma_1$ and $\gamma$ are related as follows:

$$\gamma_1^2 = -(\gamma(\omega)^2 + k_1^2). \tag{6.55}$$

$I_1$ and $I_2$ are given by

$$I_1(\gamma) = \int_{-\infty}^{+\infty} \frac{\exp(-2u_1 d)}{u_1 + u_2} \mathrm{d}\lambda, \tag{6.56}$$

$$I_2(\gamma) = \int_{-\infty}^{+\infty} \frac{\exp(-2u_1 d)}{k_2^2 u_1 + k_1^2 u_2} \mathrm{d}\lambda. \tag{6.57}$$

where $K_0$ is zero order Bessel function of the second kind, while $u_1$ and $u_2$ are given by

$$u_1 = (\lambda^2 - \gamma^2 - k_1^2)^{\frac{1}{2}} = (\lambda^2 + \gamma_1^2)^{\frac{1}{2}}, \tag{6.58}$$

$$u_2 = (\lambda^2 - \gamma^2 - k_2^2)^{\frac{1}{2}} = (\lambda^2 + \gamma_2^2)^{\frac{1}{2}}. \tag{6.59}$$

$k_1$ and $k_2$ are the propagation constants respectively in lossy ground.

### 6.1.6 Computational Examples

Figures 6.2–6.4 show the frequency response at the center of the electrode with $L = 20$ m, $d = 1$ m, $a = 5$ mm, and $I_g = 1$ A. Three values of ground conductivity with constant ground permittivity $\varepsilon_r = 10$ are considered ($\sigma = 0.1$ S m$^{-1}$,

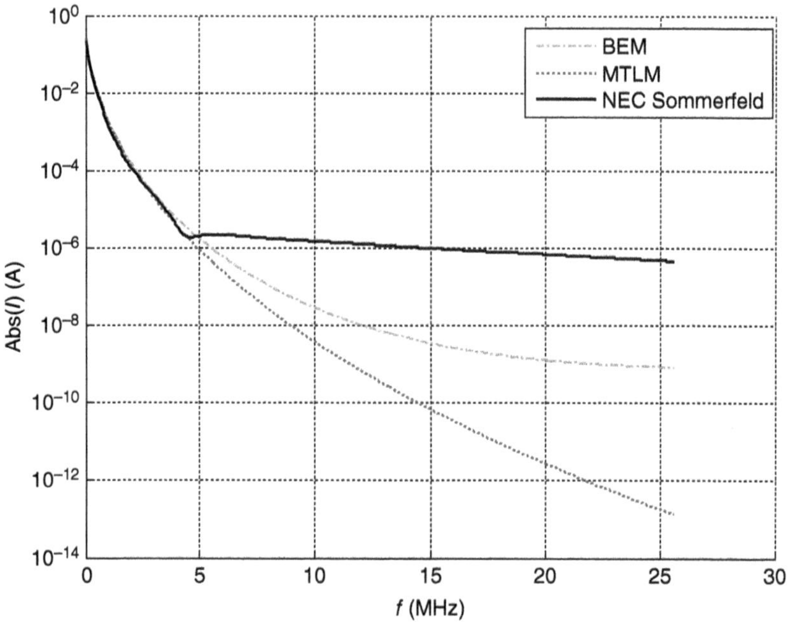


**Figure 6.2** Current induced at the center of the grounding wire versus frequency ($L = 20$ m, $d = 1$ m, $a = 5$ mm, $\sigma = 0.1$ S m$^{-1}$, $\varepsilon_r = 10$).

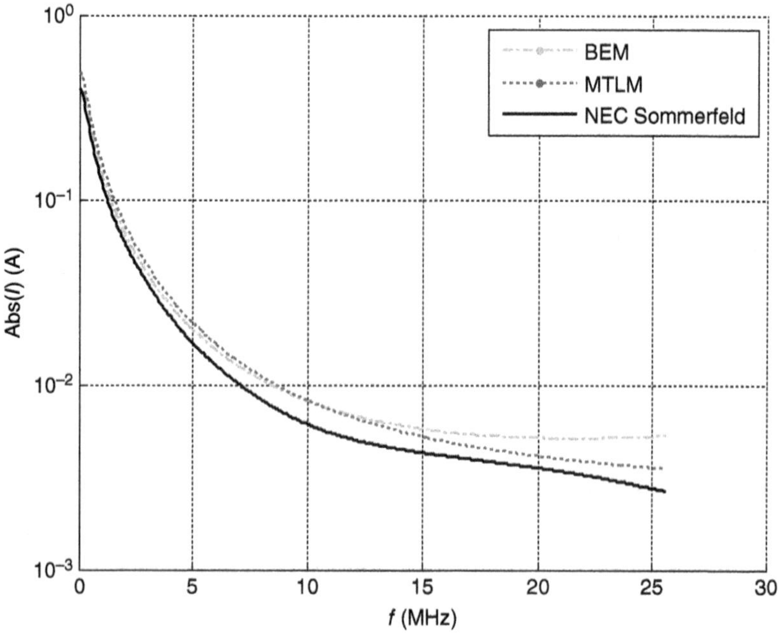


**Figure 6.3** Current induced at the center of the grounding wire versus frequency ($L = 20$ m, $d = 1$ m, $a = 5$ mm, $\sigma = 0.01$ S m$^{-1}$, $\varepsilon_r = 10$).

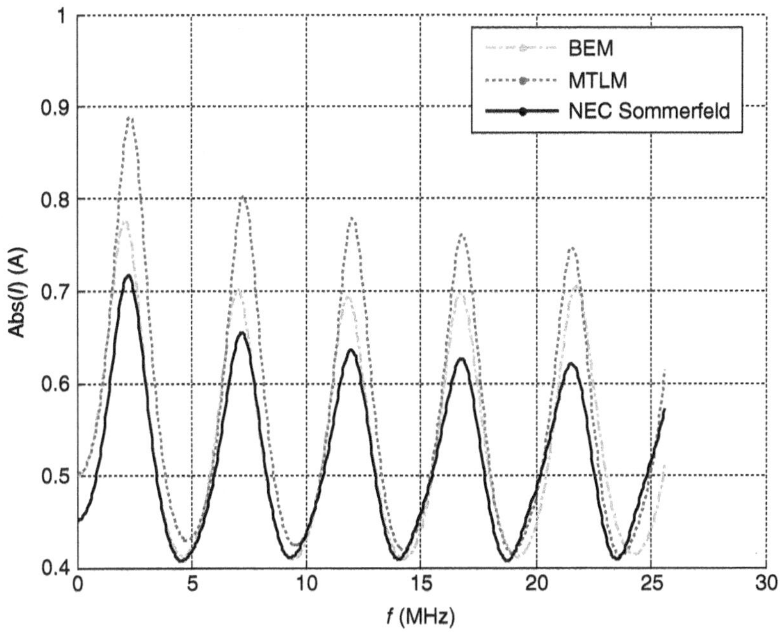


**Figure 6.4** Current induced at the center of the grounding wire versus frequency ($L = 20$ m, $d = 1$ m, $a = 5$ mm, $\sigma = 0.001$ S m$^{-1}$, $\varepsilon_r = 10$).

$\sigma = 0.01$ S m$^{-1}$, and $\sigma = 0.001$ S m$^{-1}$). The results computed via GB-IBEM are compared to the results obtained via NEC using Sommerfeld integral approach and the modified transmission line model (MTLM) [15]. The results obtained via different approaches agree satisfactorily for the given set of parameters.

The agreement between the results obtained via the different approaches is found to be satisfactory for lower values of ground conductivity ($\sigma = 0.01$ S m$^{-1}$, $\sigma = 0.001$ S m$^{-1}$) while certain discrepancies are observed at higher frequencies for the case of the high ground conductivity ($\sigma = 0.1$ S m$^{-1}$).

In this case, all approaches provide rather similar results up to the 5 MHz. At this point, the NEC curve rapidly changes shape while the BEM and MTLM curves agree satisfactory. Obviously, MTLM approximation is not valid for very high frequencies.

Namely, at about this frequency the 20 m long line acts as an antenna and the TL approach fails to account for existing radiation and resonance effects.

Finally, it is worth noting that the BEM solution for $\sigma = 0.1$ S m$^{-1}$ provides, contrary to NEC results, smooth decay of the current for higher frequencies.

Figure 6.5 shows the voltage spectrum at the injection point of the electrode with $L = 10$ m, $d = 1$ m, $a = 5$ mm, and $I_g = 1$ A. The ground conductivity is $\sigma = 0.01$ S m$^{-1}$, while the permittivity is $\varepsilon_r = 10$.

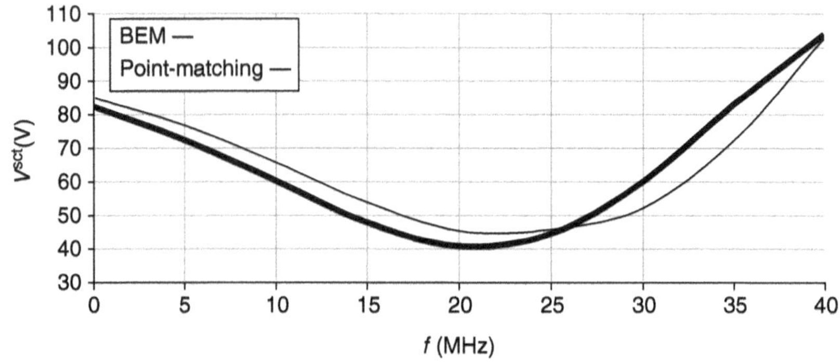


**Figure 6.5** Voltage spectrum at the grounding electrode driving point ($L = 10$m, $d = 1$ m, $a = 5$ mm, $\sigma = 0.01$ S m$^{-1}$, $\varepsilon_r = 10$).

The results obtained via GB-IBEM with linear approximation is in good agreement with the results calculated via the point-matching technique.

The next set of results deals with comparison of BEM results with the analytical results. First, the induced current along the electrode is analyzed, and then the results for the scattered voltage. Figure 6.6 shows the current distribution along the electrode characterized by the following parameters: $L = 10$ m, $d = 0.3$ m, $a = 5$ mm, and $I_g = 1$ A at the operating frequency $f = 1$ MHz. The ground conductivity is $\sigma = 0.01$ S m$^{-1}$ while the permittivity is $\varepsilon_r = 10$. The results computed via BEM and analytically are in satisfactory agreement.

Figure 6.7 shows the spatial current distribution along the same electrode, $\sigma = 0.001$ S m$^{-1}$. A satisfactory agreement between different approaches is achieved again. Figures 6.8 and 6.9 show the current distribution along the same electrode at $f = 10$ MHz for two values of conductivity $\sigma = 0.01$ S m$^{-1}$ $\sigma = 0.001$ S m$^{-1}$, respectively.

The waveforms obtained via different approaches are still alike. However, for the lower value of conductivity ($\sigma = 0.001$ S m$^{-1}$) differences in the current distribution are higher.

Figures 6.10–6.13 show the spatial distribution of scattered voltage induced along the same electrode for different values of earth conductivity at two different frequencies.

Again, bigger differences between analytical and numerical results exist at $f = 10$ MHz for lower values of ground conductivity. In Figures 6.10–6.12 bigger differences in voltage waveforms appear near the electrode lengths.

The next set of figures is related to the grounding electrode of radius $a = 5$ mm, buried at depth $d = 1$ m into a lossy ground with relative permittivity $\varepsilon_r = 10$, with the electrode conductivities of copper ($\sigma_w = 55$ MS m$^{-1}$) and aluminum ($\sigma_w = 37$ MS m$^{-1}$).

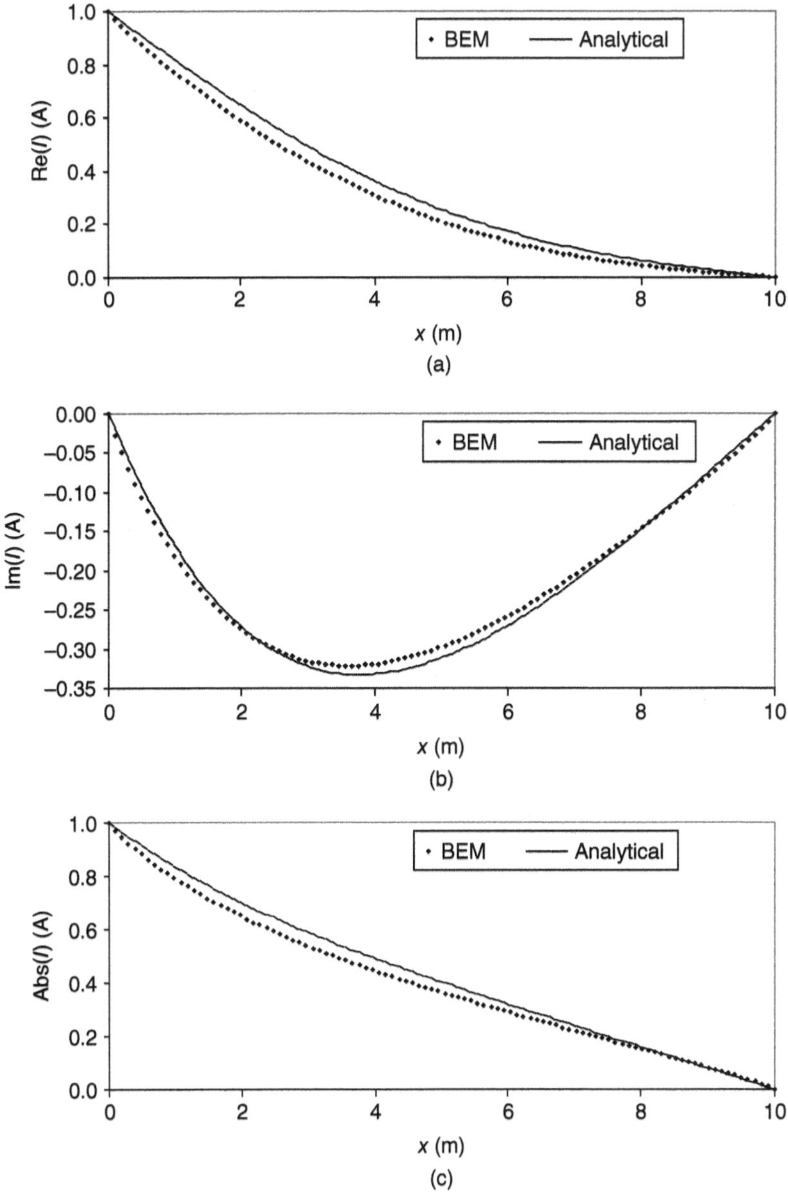


**Figure 6.6** Current distribution along the horizontal electrode ($f = 1$ MHz, $L = 10$ m, $a = 0.005$ m, $d = 0.3$ m, $\sigma = 0.01$ S m$^{-1}$, $\varepsilon_r = 10$). (a) Real part, (b) imaginary part, (c) absolute value.

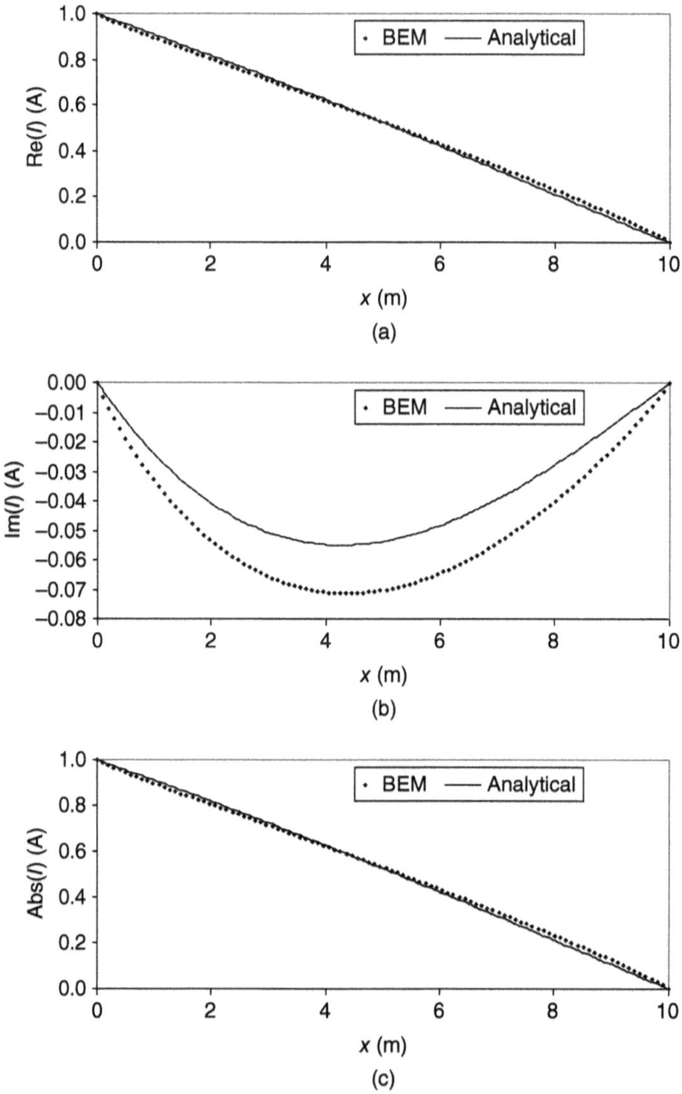


**Figure 6.7** Current distribution along the horizontal electrode ($f = 1$ MHz, $L = 10$ m, $a = 0.005$ m, $d = 0.3$ m, $\sigma = 0.001$ S m$^{-1}$, $\varepsilon_r = 10$). (a) Real part, (b) Imaginary part, (c) absolute value.

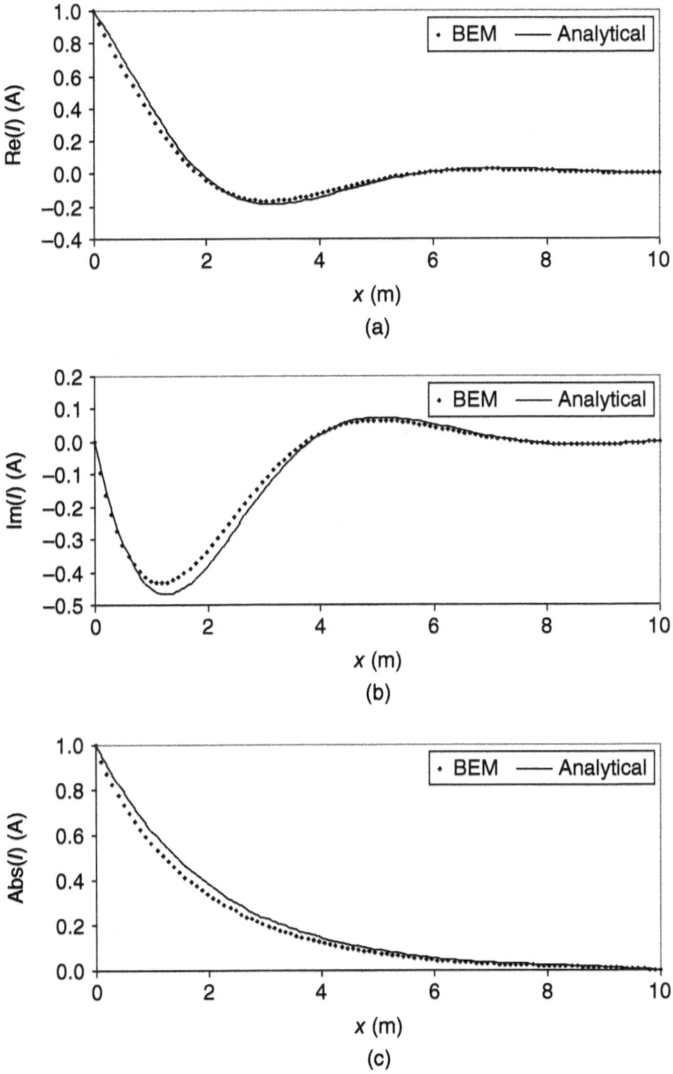


**Figure 6.8** Current distribution along the horizontal electrode ($f = 10$ MHz, $L = 10$ m, $a = 0.005$ m, $d = 0.3$ m, $\sigma = 0.01$ S m$^{-1}$, $\varepsilon_r = 10$). (a) Real part, (b) imaginary part, (c) absolute value.

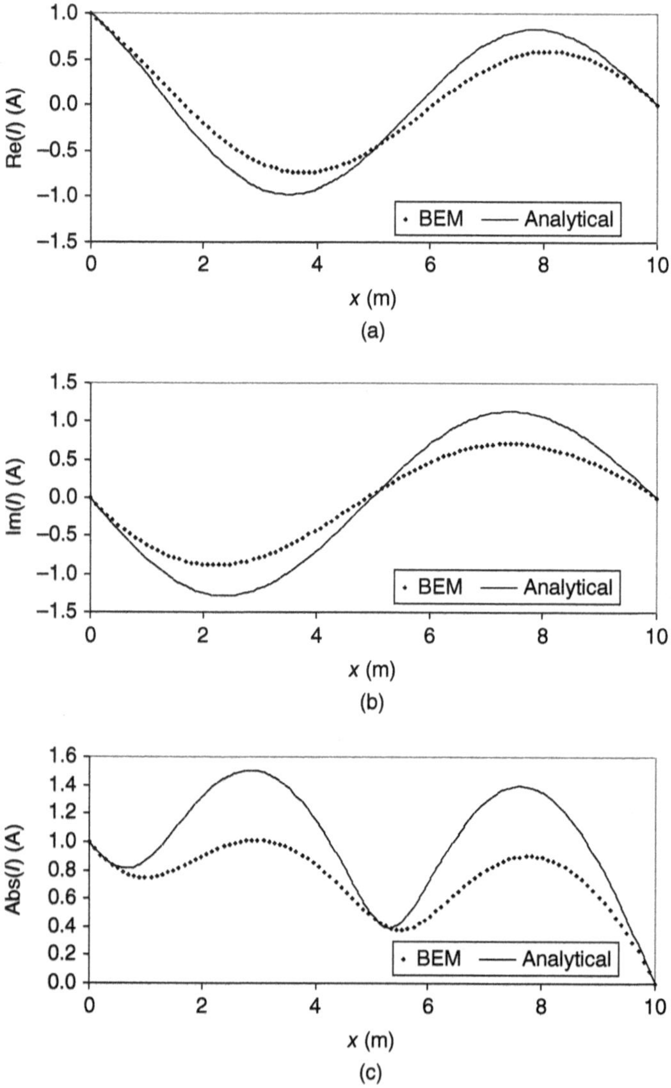


**Figure 6.9** Current distribution along the horizontal electrode ($f = 10$ MHz, $L = 10$ m, $a = 0.005$ m, $d = 0.3$ m, $\sigma = 0.001$ S m$^{-1}$, $\varepsilon_r = 10$). (a) Real part, (b) imaginary part, (c) absolute value.

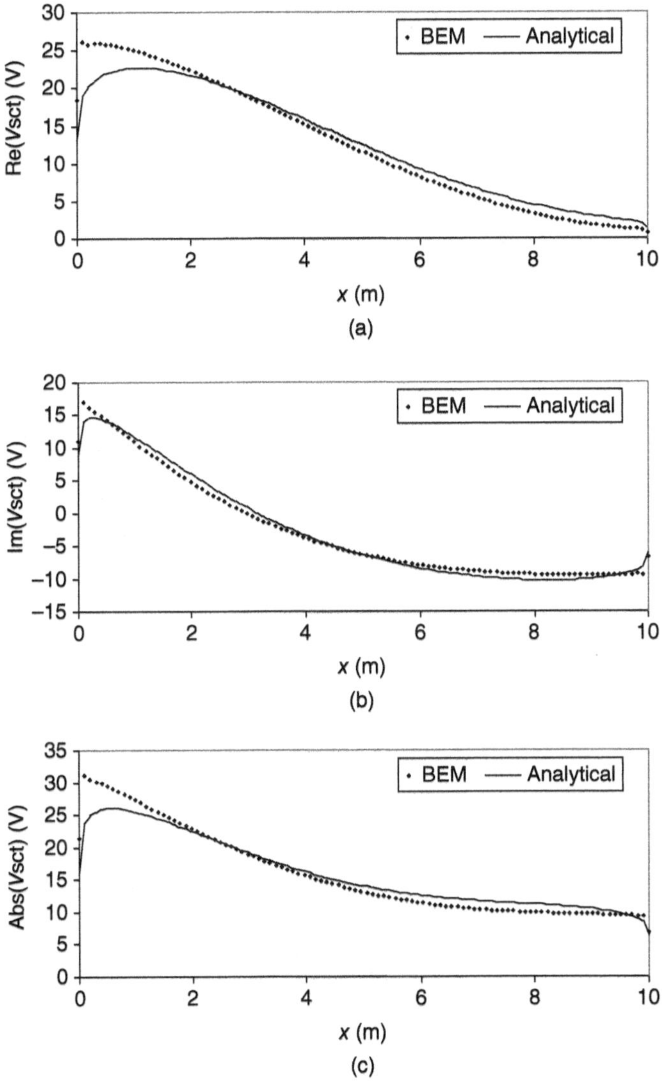


**Figure 6.10** Scattered voltage distribution along the horizontal electrode ($f = 1$ MHz, $L = 10$ m, $a = 0.005$ m, $d = 0.3$ m, $\sigma = 0.01$ S m$^{-1}$, $\varepsilon_r = 10$). (a) Real part, (b) imaginary part, (c) absolute value.

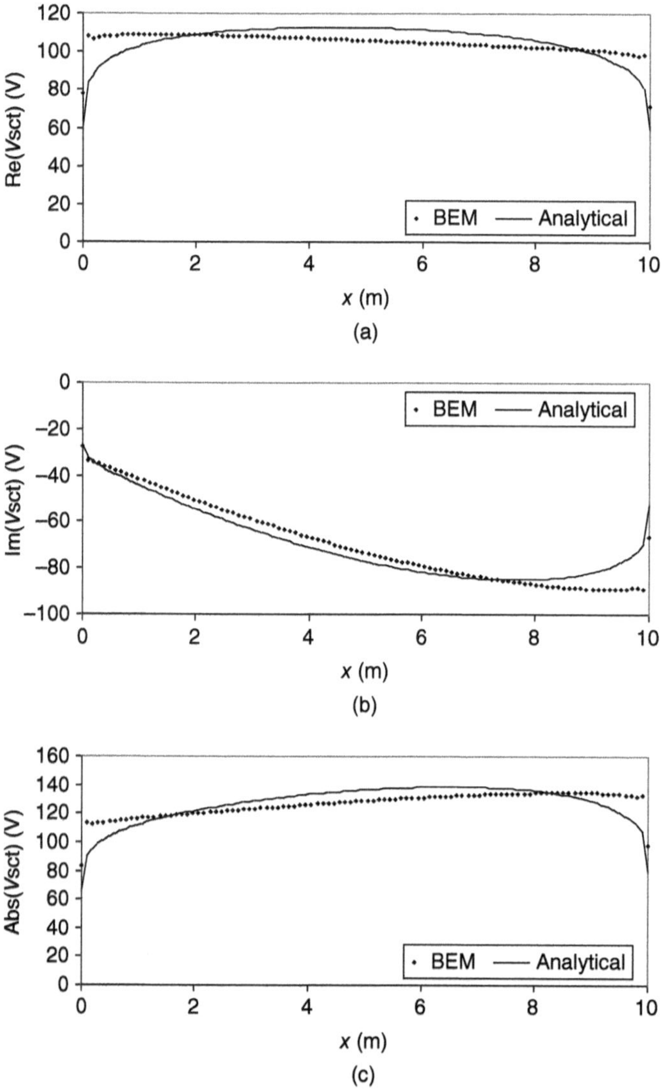


**Figure 6.11** Scattered voltage distribution along the horizontal electrode ($f = 1$ MHz, $L = 10$ m, $a = 0.005$ m, $d = 0.3$ m, $\sigma = 0.001$ S m$^{-1}$, $\varepsilon_r = 10$). (a) Real part, (b) imaginary part, (c) absolute value.

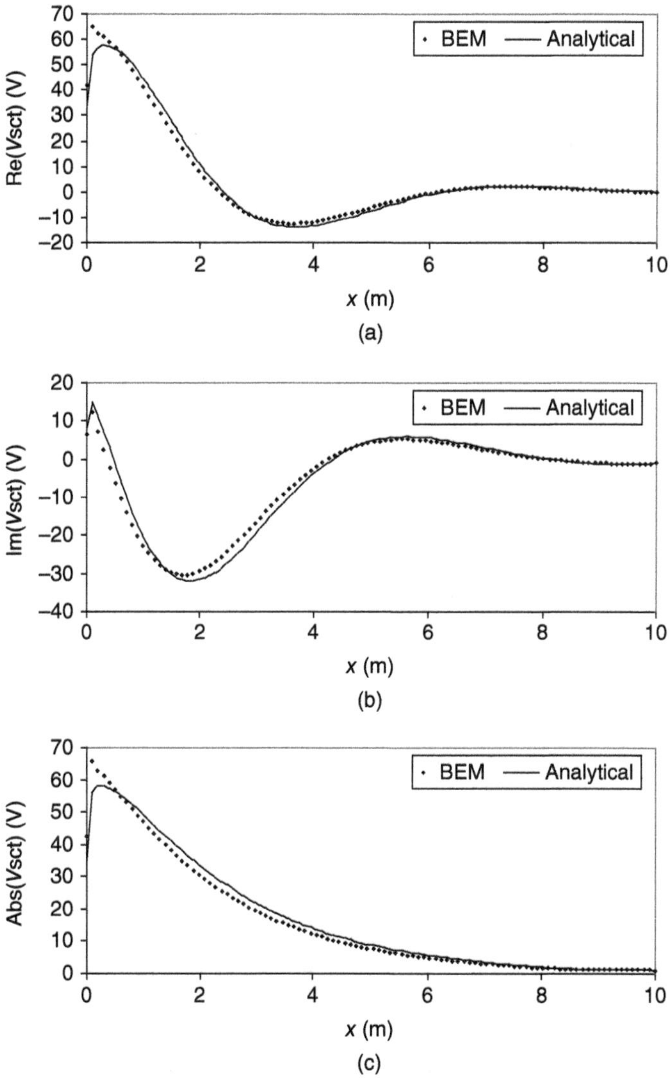


**Figure 6.12** Scattered voltage distribution along the horizontal electrode ($f = 10$ MHz, $L = 10$ m, $a = 0.005$ m, $d = 0.3$ m, $\sigma = 0.01$ S m$^{-1}$, $\varepsilon_r = 10$). (a) Real part, (b) imaginary part, (c) absolute value.

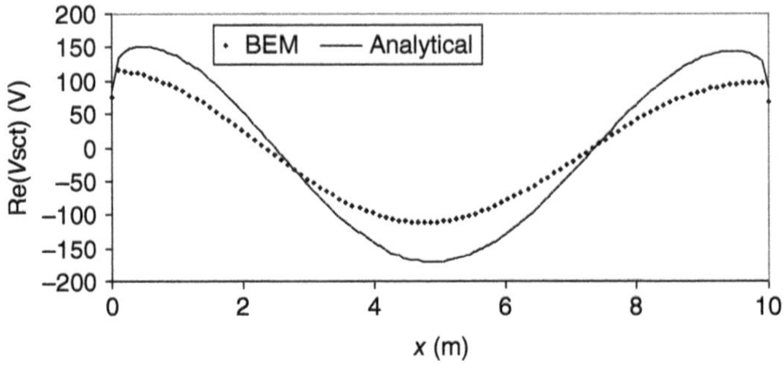


(a)

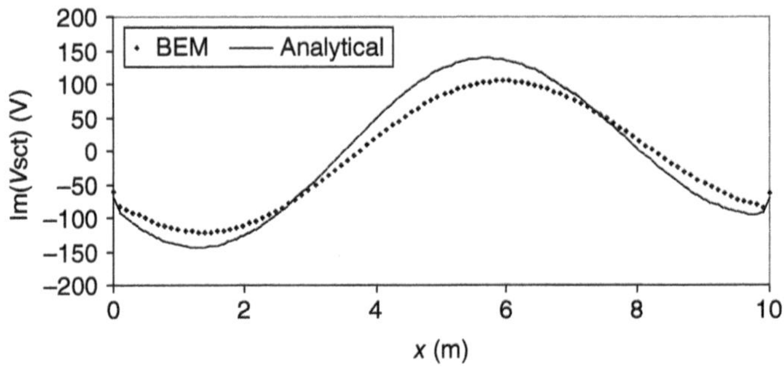


(b)

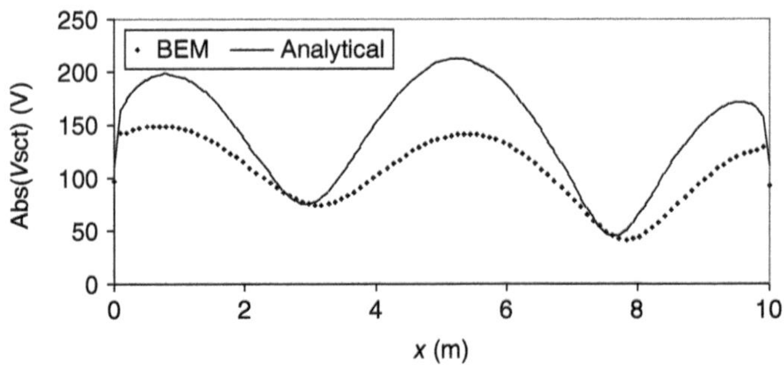


(c)

**Figure 6.13** Scattered voltage distribution along the horizontal electrode ($f = 10$ MHz, $L = 10$ m, $a = 0.005$ m, $d = 0.3$ m, $\sigma = 0.001$ S m$^{-1}$, $\varepsilon_r = 10$). (a) Real part, (b) imaginary part, (c) absolute value.

Figures 6.14–6.18 show the comparison of the real and imaginary parts of the induced current as well as absolute value of the current distribution along the electrode for different parameters of ground conductivity and operating frequency. The magnitude of the current source at a given frequency is $I_g(\omega) = 1$ A,

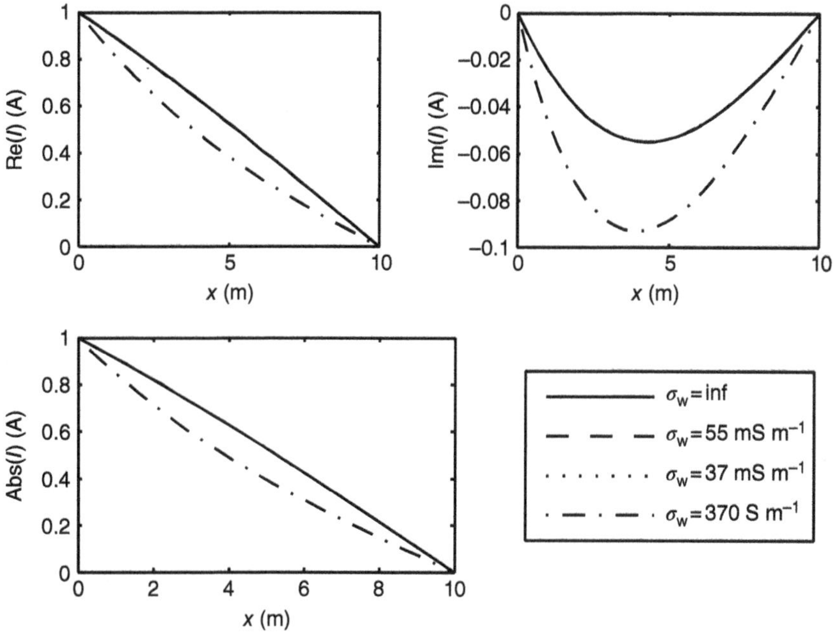


**Figure 6.14** Real part, imaginary part, and absolute value of the current induced along the electrode ($\sigma = 1$ mS m$^{-1}$, $f = 1$ MHz).

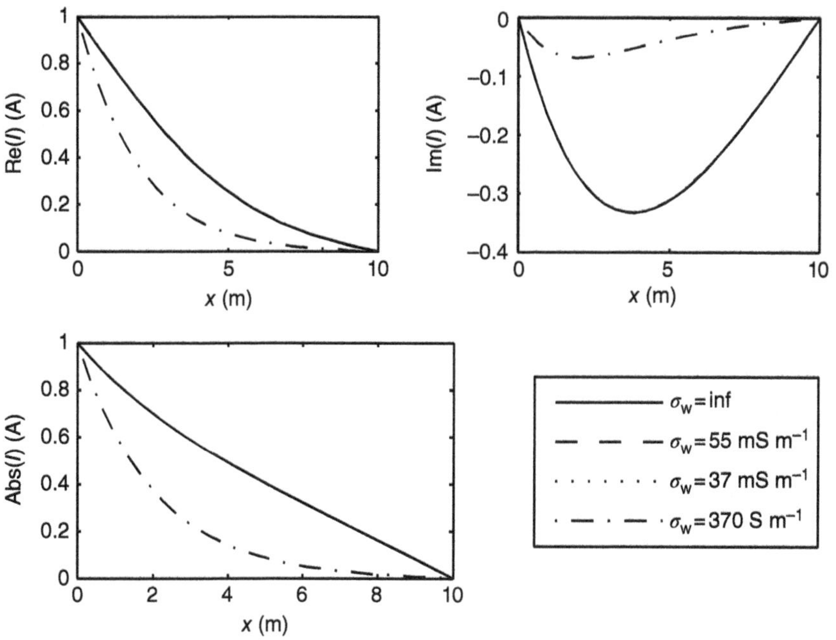


**Figure 6.15** Real part, imaginary part and absolute value of the current induced along the electrode ($\sigma = 10$ mS m$^{-1}$, $f = 1$ MHz).

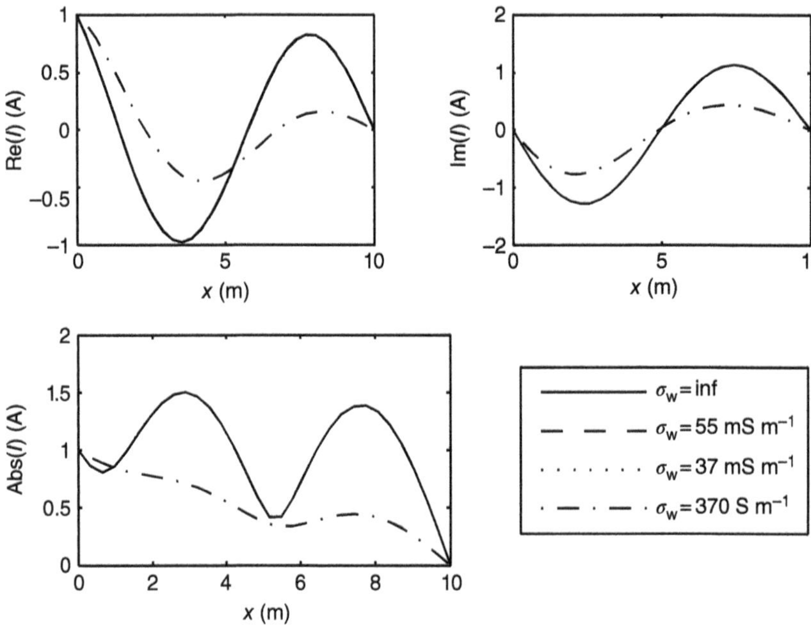


**Figure 6.16** Real part, imagin ary part, and absolute value of the current induced along the electrode ($\sigma = 1\ \text{mS}\ \text{m}^{-1}, f = 10\ \text{MHz}$).

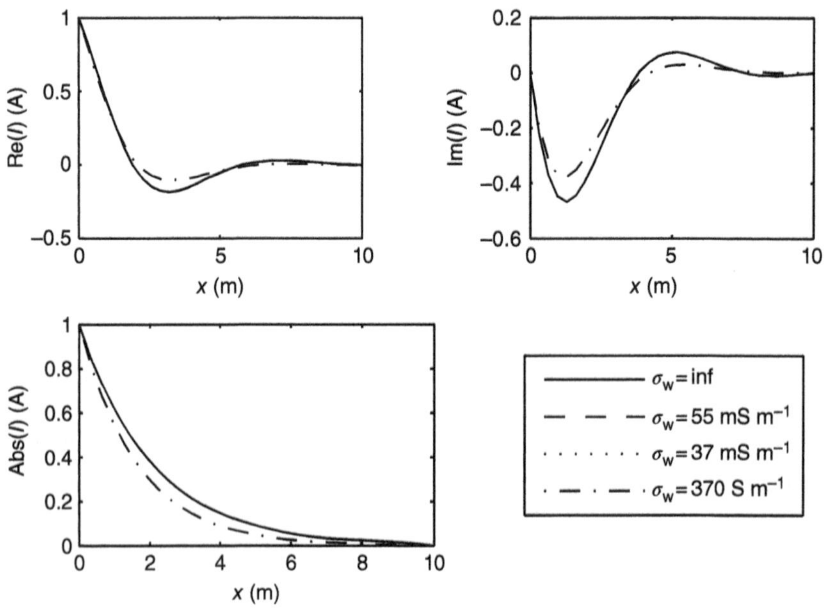


**Figure 6.17** Real part, imaginary part, and absolute value of the current induced along the electrode ($\sigma = 10\ \text{mS}\ \text{m}^{-1}, f = 10\ \text{MHz}$).

while the electrode length is $L = 10$ m. The values of current distribution for different electrode conductivities agree rather satisfactorily. Furthermore, current distribution for the electrode with extremely low conductivity $\sigma_w = 370$ S m$^{-1}$ is determined to stress out the discrepancy of the results. The value of current monotonously decreases from the current source to the electrode end at $f = 1$ MHz and $f = 10$ MHz; the oscillatory behavior of the current distribution is shown.

Figures 6.18–6.21 show spectra of the current absolute value for various conductivities of the grounding electrode. As can be seen from Figures 6.18 and 6.19, the resonant frequency for electrode length $L = 1$ m is around $f = 50$ MHz and is almost independent of ground conductivity.

On the other hand, Figures 6.20 and 6.21 show the frequency spectrum for the electrode length $L = 10$ m. For the ground conductivity $\sigma = 1$ mS m$^{-1}$, resonant frequencies of the grounding electrode are observed to be odd multiplicities of $f = 5$ MHz. It is visible from Figure 6.19 that there is no resonant frequency $\sigma = 10$ mS m$^{-1}$.

The numerical results to follow are related to the transient behavior of the current induced at the center of the grounding electrodes. The following set of results is calculated for the case of lightning pulse with parameters $\alpha = 0.07924 \times 10^6$ s$^{-1}$ and $\beta = 4.0011 \times 10^6$ s$^{-1}$, which represent the 1/10 μs pulse [15]. The calculations are carried out for the electrode length $L = 1$ m

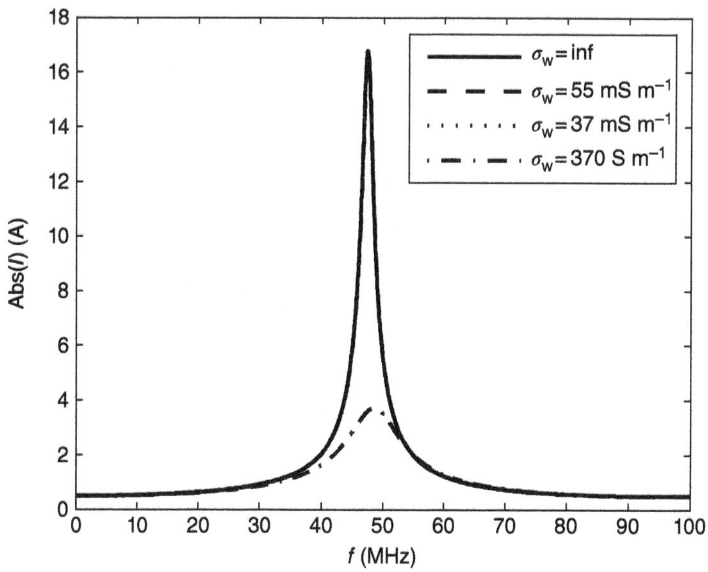


**Figure 6.18** Frequency spectrum of the current induced at the center of the grounding electrode ($L = 1$ m, $\sigma = 1$ mS m$^{-1}$).

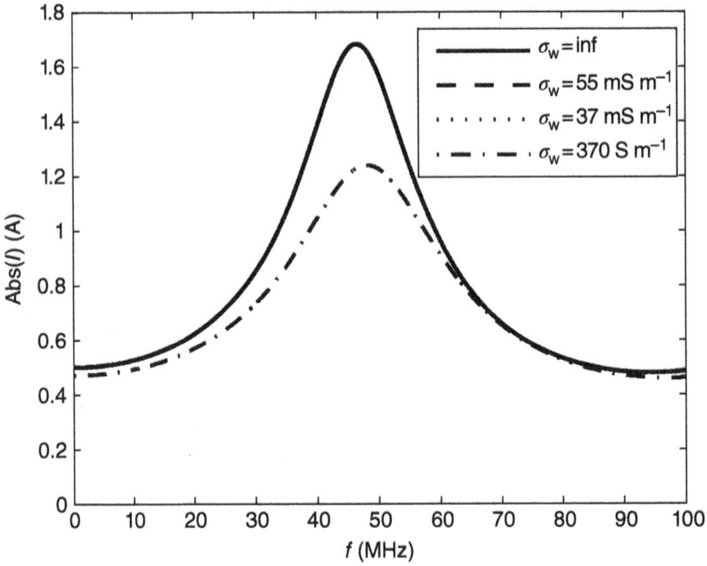


**Figure 6.19** Frequency spectrum of the current induced at the center of the grounding electrode ($L = 1$ m, $\sigma = 10$ mS m$^{-1}$).

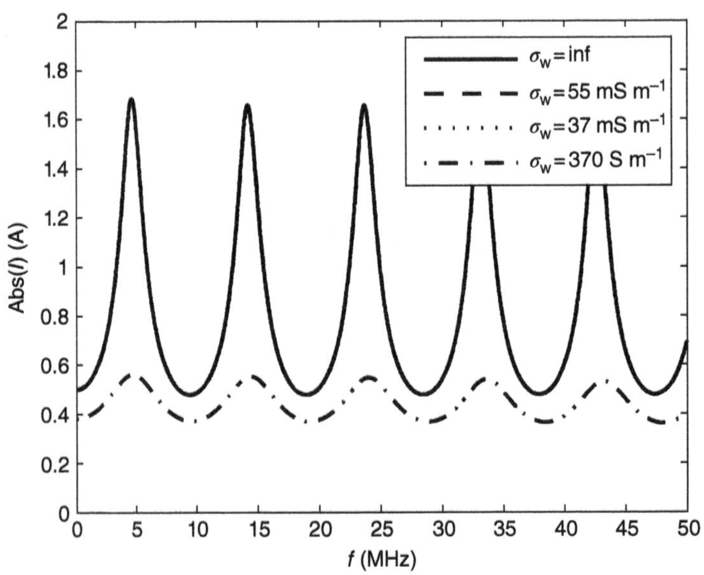


**Figure 6.20** Frequency spectrum of the current induced at the center of the grounding electrode ($L = 10$ m, $\sigma = 1$ mS m$^{-1}$).

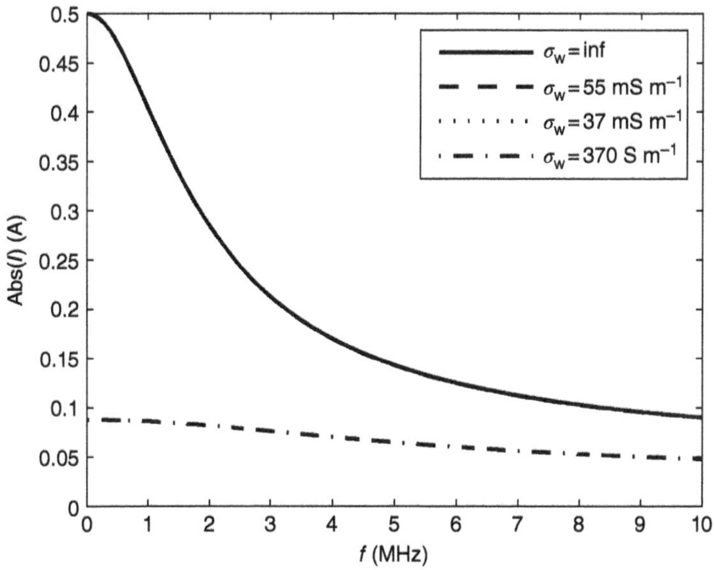


**Figure 6.21** Frequency spectrum of the current induced at the center of the grounding electrode ($L = 10$ m, $\sigma = 10$ mS m$^{-1}$).

and $L = 10$ m, respectively, and for the soil conductivity $\sigma = 1$ mS m$^{-1}$ and $\sigma = 10$ mS m$^{-1}$, respectively. Figures 6.22–6.24 show the transient currents induced at the center of the electrode for various conductivities.

Note that some discrepancies appear for low conducting materials. The following set of results, shown in Figures 6.25–6.27, are calculated for the current pulse with parameters $\alpha = 0.07924 \times 10^7$ s$^{-1}$, and $\beta = 4.0011 \times 10^7$ s$^{-1}$, which represent the 0.1/1 μs pulse. The length of the electrode is $L = 5$ m and $L = 10$ m, respectively, while the soil conductivity is $\sigma = 1$ mS m$^{-1}$ and $\sigma = 10$ mS m$^{-1}$, respectively. The results for the PEC electrode and realistic electrodes are in good agreement. The discrepancies appear only for extremely low conducting electrode, i.e. $\sigma_w = 370$ S m$^{-1}$.

Analyzing the results presented in Figures 6.14–6.27, it can be concluded that the concept of the PEC electrode is an acceptable approximation in an engineering sense.

In the examples to follow 1/10 μs pulse excitation is used to calculate the transient impedance of the grounding electrode. Figure 6.28 shows the transient impedance of the grounding electrode with length $L = 10$ m, radius $a = 5$ mm, buried horizontally at depth $d = 0.5$ m in the ground with $\varepsilon_r = 10$ and $\sigma = 10$ mS m$^{-1}$.

The results obtained via different techniques agree satisfactorily with maximum discrepancy of around 7%. The impedance in the early time period

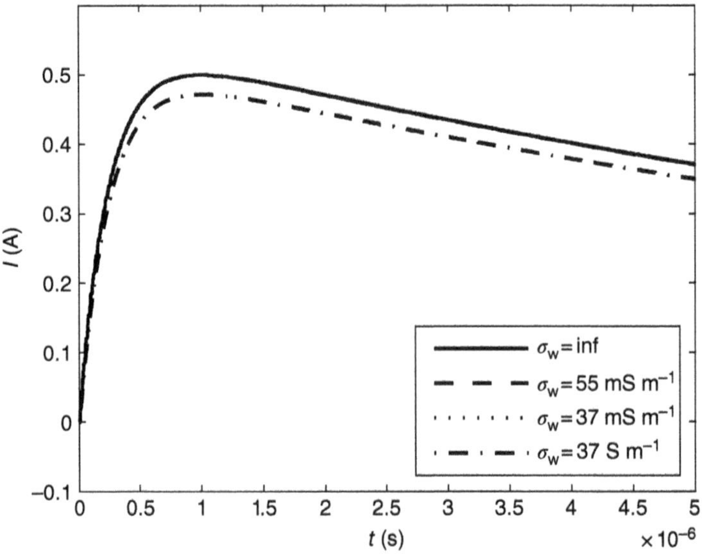


**Figure 6.22** Transient current at the center of the electrode, $L = 1$ m, $\sigma = 1$ mS m$^{-1}$, 1/10 μs pulse.

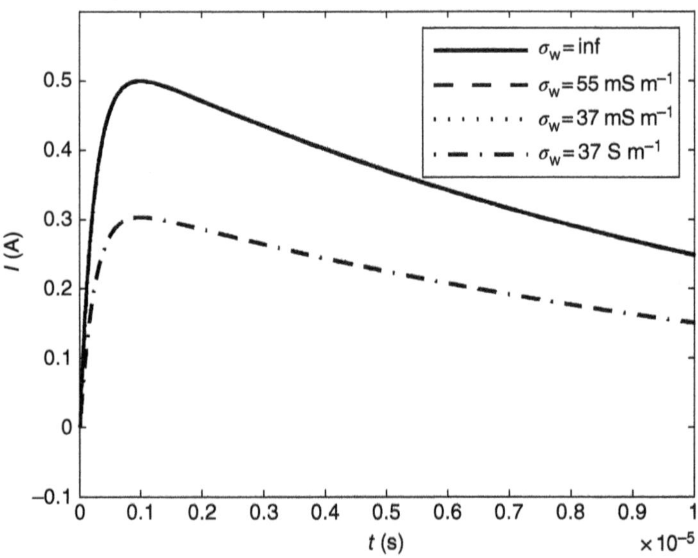


**Figure 6.23** Transient current at the center of the electrode, $L = 1$ m, $\sigma = 10$ mS m$^{-1}$, 1/10 μs pulse.

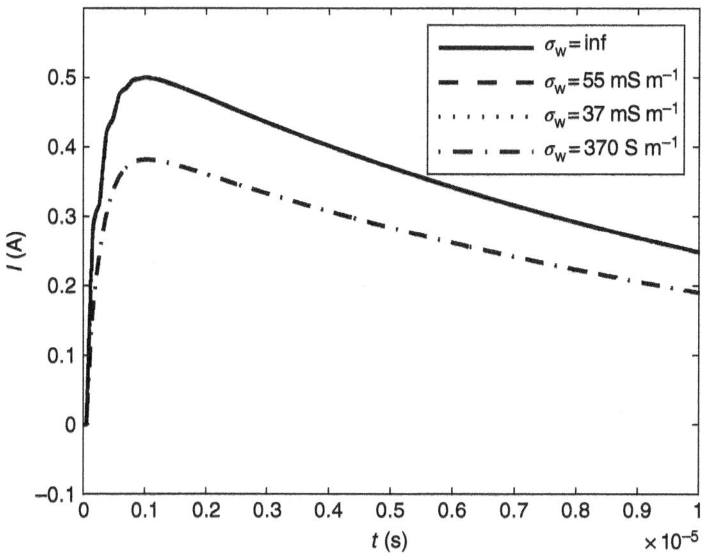


**Figure 6.24** Transient current at the center of the electrode, $L = 10$ m, $\sigma = 1$ mS m$^{-1}$, 1/10 μs pulse.

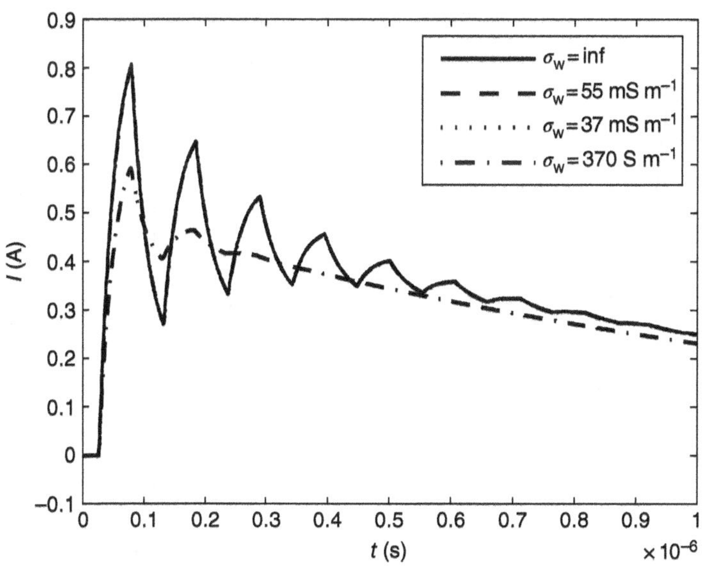


**Figure 6.25** Transient current at the center of the electrode, $L = 5$ m, $\sigma = 1$ mS m$^{-1}$, 0.1/1 μs pulse.

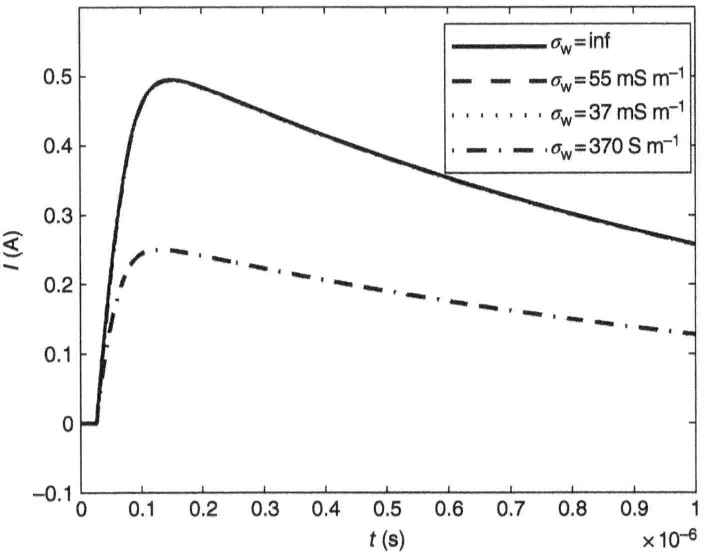


**Figure 6.26** Transient current at the center of the electrode, $L = 5$ m, $\sigma = 10$ mS m$^{-1}$, 0.1/1 μs pulse.

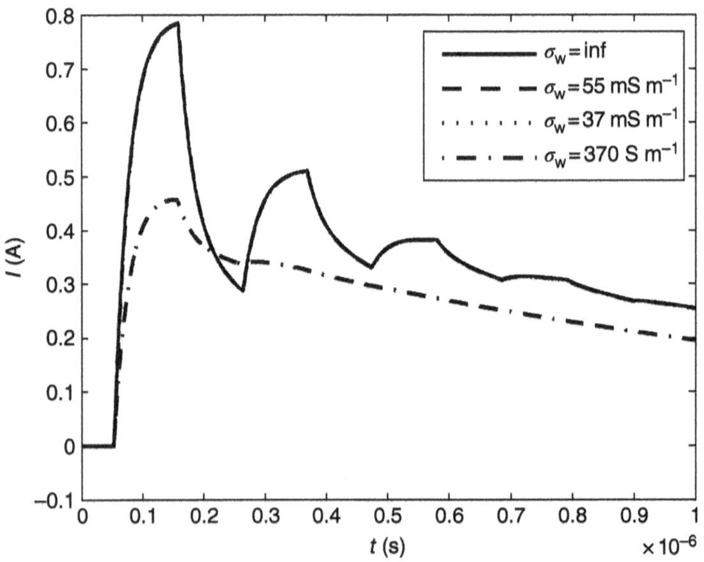


**Figure 6.27** Transient current at the center of the electrode, $L = 10$ m, $\sigma = 1$ mS m$^{-1}$, 0.1/1 μs pulse.

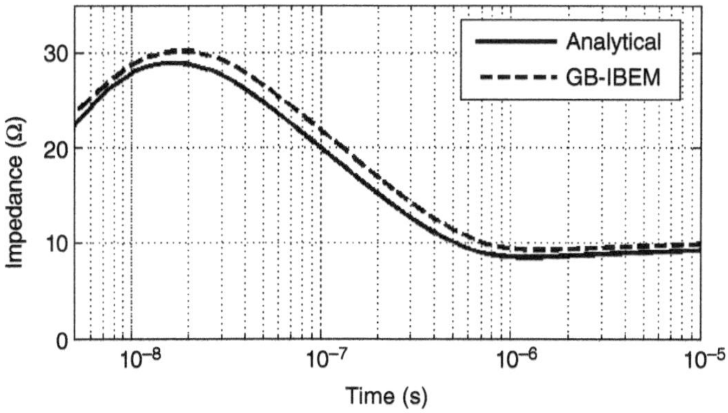


**Figure 6.28** Transient impedance of the grounding electrode (1/10 μs pulse, $L = 10$ m, $\sigma = 10$ mS m$^{-1}$).

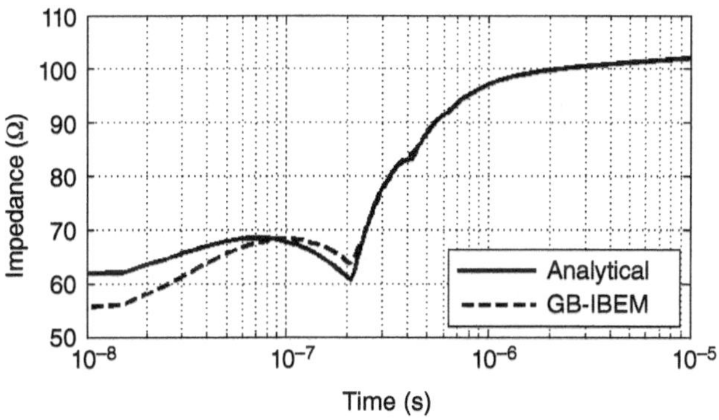


**Figure 6.29** Transient impedance of the grounding electrode (1/10 μs pulse, $L = 10$ m, $\sigma = 1$ mS m$^{-1}$).

(around 20 ns) is three times larger than the steady state impedance values, thus emphasizing the importance of the accurate assessment of transient impedance, since grounding systems are usually designed from the data set arising from steady state study.

Figure 6.29 shows the results for the same electrode, but for lower ground conductivity ($\sigma = 1$ mS m$^{-1}$). The discrepancy of around 10% can be observed in the early time behavior. The transient impedance is higher than in the previous example, increasing with time, which is expected for the lower ground conductivity.

The next three examples for the 1/10 μs pulse correspond to the configurations presented in [15], for a very low ground conductivity of $\sigma = 0.19$ mS m$^{-1}$. Figure 6.30 shows the transient impedance for electrode of length $L = 1$ m.

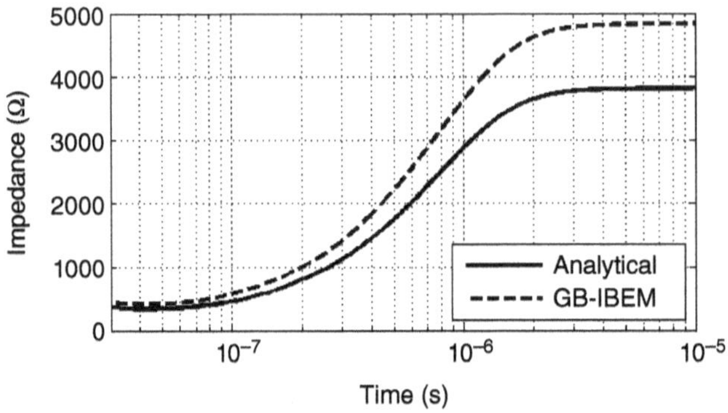


**Figure 6.30** Transient impedance of the grounding electrode (1/10 μs pulse, $L = 1$ m, $\sigma = 0.19$ mS m$^{-1}$).

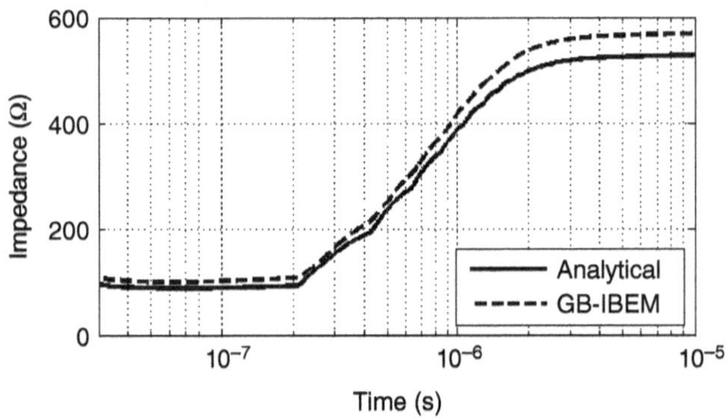


**Figure 6.31** Transient impedance of the grounding electrode (1/10 μs pulse, $L = 10$ m, $\sigma = 0.19$ mS m$^{-1}$).

Some discrepancy is observed between analytical and GB-IBEM solution in the steady state (20%).

Figure 6.31 considers the electrode of length $L = 10$ m. The agreement between the results is more satisfactory and the discrepancy in the steady state is smaller (7%) than for the case of a shorter electrode. Therefore, the approximation adopted for analytical solution is much more dependent on the length of the electrode, making such a solution more convenient for longer electrodes. Also, it can be observed that the steady state impedance is an order of magnitude lower than for the electrode of length $L = 1$ m, thus emphasizing the importance of the electrode length in the design of a grounding system.

A longer electrode ($L = 30$ m) case is shown in Figure 6.32. The discrepancy between the results is negligible, thus justifying the use of analytical solution when transient impedance of longer electrode is calculated. Furthermore,

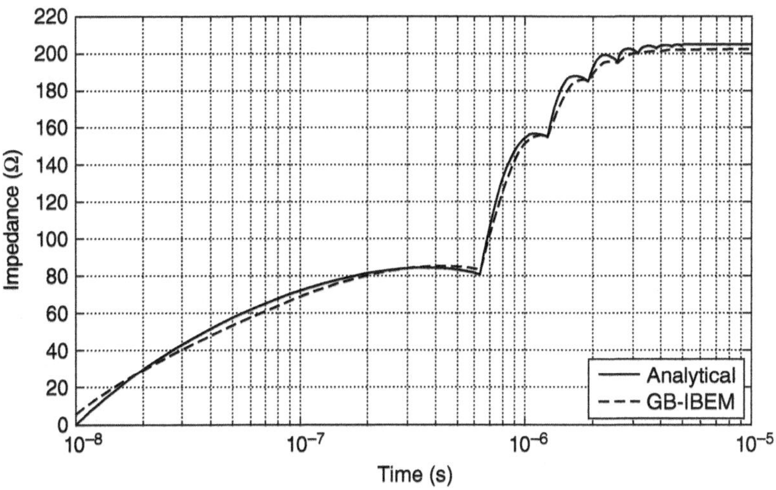


**Figure 6.32** Transient impedance of the grounding electrode (1/10 μs pulse, $L = 30$ m, $\sigma = 0.19$ mS m$^{-1}$).

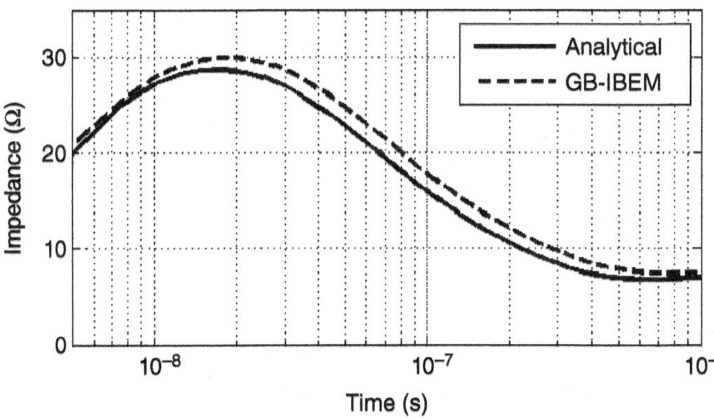


**Figure 6.33** Transient impedance of the grounding electrode (0.1/1 μs pulse, $L = 10$ m, $\sigma = 10$ mS m$^{-1}$).

appreciably lower impedance is observed when compared to the electrodes of smaller length.

Numerical results shown in Figures 6.33 and 6.34 are obtained for the current pulse with parameters $I_0 = 1.1043$ kA, $\alpha = 0.07924 \times 10^7$, $\beta = 4.0011 \times 10^7$, i.e. the 0.1/1 μs pulse. Figure 6.33 shows the transient impedance of the electrode of length $L = 10$ m, radius $a = 5$ mm, buried horizontally at depth $d = 0.5$ m in the ground with $\varepsilon_r = 10$ and $\sigma = 10$ mS m$^{-1}$. The results agree satisfactorily. The values of impedance are lower than in the case of 1/10 μs pulse.

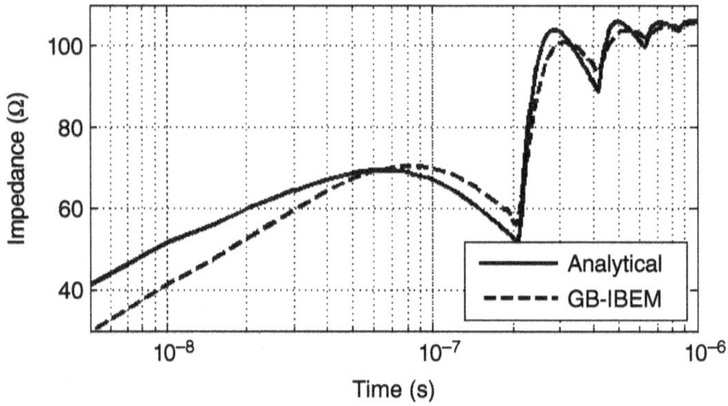


**Figure 6.34** Transient impedance of the grounding electrode (0.1/1 μs pulse, $L = 10$ m, $\sigma = 1$ mS m$^{-1}$).

Figure 6.34 shows the results for the same configuration, but with a lower ground conductivity ($\sigma = 1$ mS m$^{-1}$).

The agreement between the results seems plausible, with small discrepancy in the early time behavior. As expected, the value of impedance is higher due to the lower ground conductivity.

### 6.1.7 Application of Magnetic Current Loop (MCL) Source model to Horizontal Grounding Electrode

The use of MCL to model the excitation for the horizontal grounding electrodes has been recently investigated in [16]. The main advantage over the traditional ideal current source (ICS) model in grounding applications is that no additional calculation of the input voltage is required for the input impedance assessment, as the input voltage is already set by the source itself.

In this case, the current distribution along the electrode $I(x')$ is governed by the following Pocklington's integro-differential equation [4, 16]:

$$-\frac{1}{j4\pi\omega\varepsilon_{\text{eff}}}\int_0^L I(x')\cdot\left[k^2+\frac{\mathrm{d}^2}{\mathrm{d}x^2}\right]G(x,x')\mathrm{d}x' = \vec{E}^{\text{exc}}(x), \tag{6.60}$$

where $G(x, x')$ is the lossy medium Green function [16], and $k$ and $\varepsilon_{\text{eff}}$ are the complex propagation constant and complex permittivity of the lossy ground, respectively.

Although the ICS model has been widely used for over three decades, its use might suffer from difficulties in defining the input voltage being necessary to obtain the input impedance of the grounding system [4, 15]. The ICS model as well as MF and MCL source models (both based on the magnetic current density existing at the source area and generating axial excitation electric field) are illustrated in Figure 6.35. While in antenna applications, the source

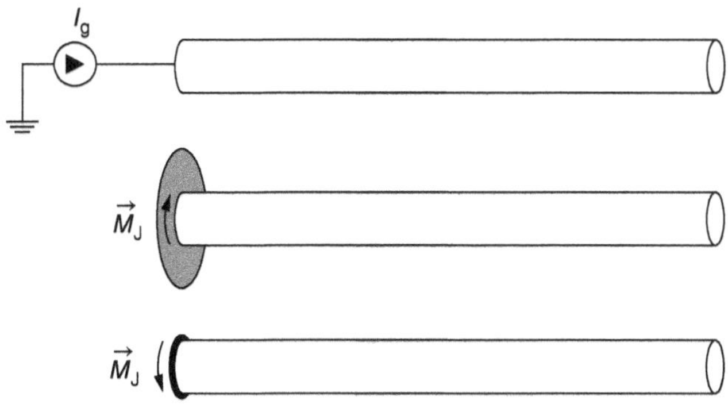


**Figure 6.35** ICS, MF, and MCL source at the open wire end.

is placed in the feed-gap area with input terminals put close together, in grounding applications input terminals are between two distant points (one at the electrode, the other in the remote soil), thus causing a problem with the implementation of classical delta gap voltage source particularly in the numerical sense. On the other hand, MF and MCL are planar perpendicularly to the wire axis and can be readily applied to the open wire thus making them rather convenient for grounding applications.

The relation for excitation field $\vec{E}^{\text{exc}}$ for the MF model is

$$E_x^{\text{exc}}(x) = \frac{V_g}{\ln\left(\frac{b}{a}\right)}\left[\frac{e^{-jk\sqrt{a^2+(x-x')^2}}}{\sqrt{a^2+(x-x')^2}} - \frac{e^{-jk\sqrt{b^2+(x-x')^2}}}{\sqrt{b^2+(x-x')^2}}\right], \tag{6.61}$$

while for the MCL model it follows that

$$E_x^{\text{exc}}(x) = V_g a^2\left[jk + \frac{1}{\sqrt{a^2+(x-x')^2}}\right]\frac{e^{-jk\sqrt{a^2+(x-x')^2}}}{a^2+(x-x')^2}, \tag{6.62}$$

where $V_g$ is the imposed input voltage and $a$ is the wire radius.

In the MF source model, $b$ denotes the outer radius of the magnetic current ring. As this radius is not clearly defined in the antenna analysis, $b$ is often determined from the characteristic impedance of the TL exciting the antenna [16]. Figures 6.36 and 6.37 show the comparison of the input impedance spectra determined by different models. The differences between the results increase with frequency with maximal difference below 2% and within 0.5% appearing in the range relevant for lightning (0 Hz–10 MHz).

The TD results are obtained for a channel-base current expressed by Heidler function with $I_0 = 1$ A, $k = 0.93$, $\tau_1 = 19$ μs, $\tau_2 = 485$ μs [16] (modified to the unit amplitude). Figure 6.38 shows the transient current at the electrode center calculated via different models. No appreciable discrepancies between the results are noticed.

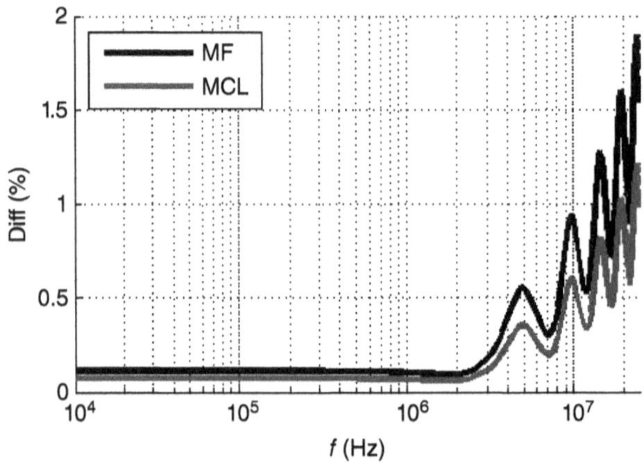


**Figure 6.36** Percentage difference from ICS ($L = 10$ m; $a = 0.005$ m; $\sigma = 0.001$ S m$^{-1}$).

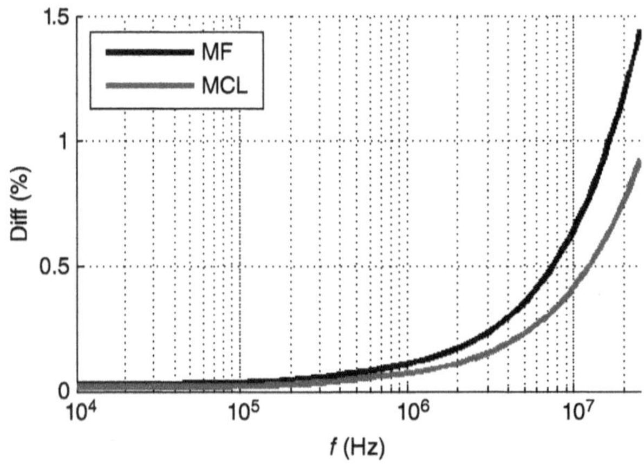


**Figure 6.37** Percentage difference from ICS ($L = 50$ m; $a = 0.005$ m; $\sigma = 0.001$ S m$^{-1}$).

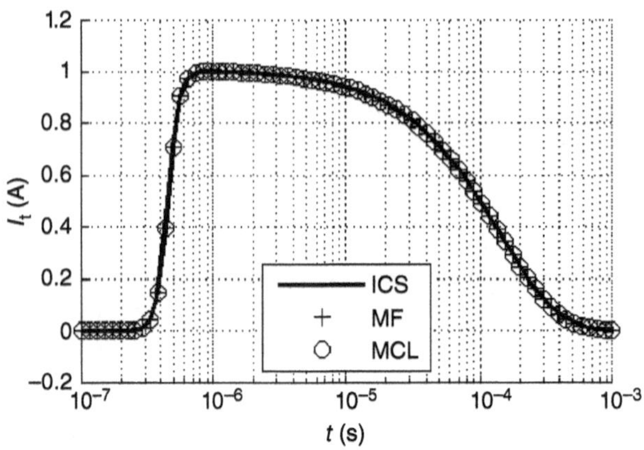


**Figure 6.38** Current at the center of grounding wire ($L = 50$ m; $a = 0.01$ m; $\sigma = 0.001$ S m$^{-1}$).

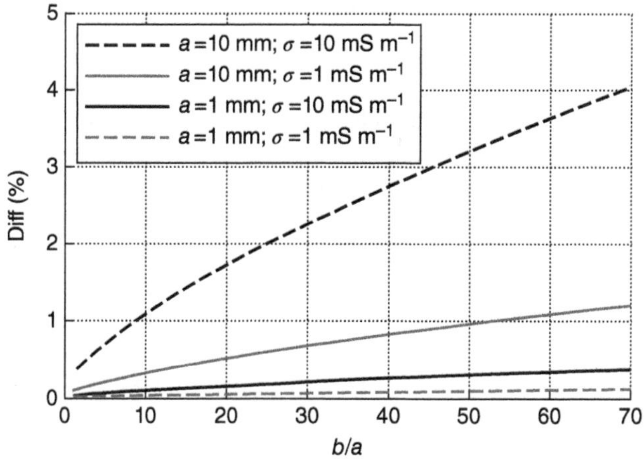


**Figure 6.39** Percentage difference (MF results with respect to ICS results; $L = 20$ m).

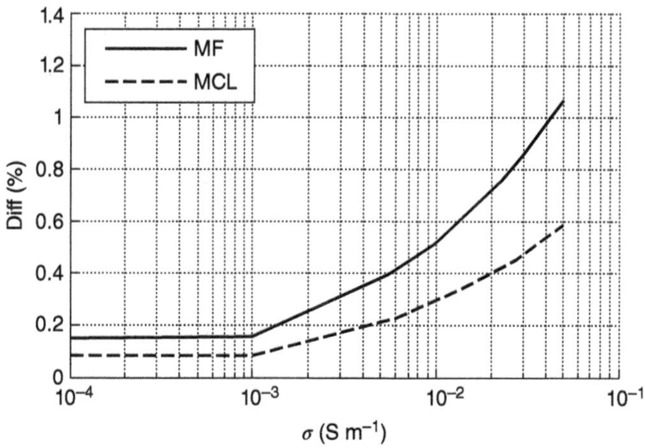


**Figure 6.40** Percentage difference (with respect to ICS results; $L = 20$ m, $b/a = 3$, $a = 0.001$ m).

Figures 6.39 and 6.40 deal with the maximal transient input voltage, which, for the case of the unit input current, corresponds to the impulse impedance. Figure 6.39 shows the percentage difference between the results obtained by means of the MF source with different values of $b/a$ and the results obtained using ICS for various values of the wire radius and ground conductivity. For a ratio $b/a$ less than 10, the errors remain within 1% regardless of the value of the ground conductivity. Figure 6.40 shows the results obtained via different models for various conductivities. Again, MCL and MF provide accurate results for typical values of ground conductivity.

## 6.2 Frequency Domain Analysis of Vertical Grounding Electrode

Transient analysis of the vertical grounding electrode, as an important component in many realistic grounding systems, such as WTs, is always of interest in electromagnetic compatibility (EMC) [17, 18]. FD numerical/analytical AT study of vertical electrodes as individual elements, using GB-IBEM and analytical solution, respectively, has been reported in [17], while a transient response of a straight vertical electrode has been presented in [18]. This section first deals with how the current distribution along the electrode is obtained by using both the GB-IBEM and the analytical solution. The transient response of the vertical electrode is obtained by means of inverse fast Fourier transform (IFFT). Some illustrative computational results are given in the paper.

### 6.2.1 Integral Equation Formulation/Reflection Coefficient Approach

Vertical grounding electrode of length $L$ and radius $a$, buried in a lossy medium at depth $d$ and excited by an equivalent current source is shown in Figure 6.41.

The governing equations for the current and scattered voltage induced along the vertical electrode can be readily derived from Maxwell's equations by enforcing the continuity conditions for the tangential components of the electric field along the wire [17, 18].

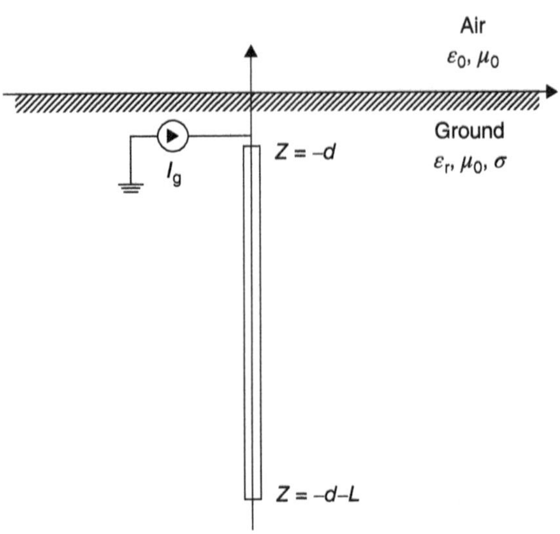


**Figure 6.41** Vertical grounding electrode excited by the current source.

The homogeneous integro-differential equation of the Pocklington type for the current distribution induced along the electrode is given by

$$j\omega\frac{\mu}{4\pi}\int_{-d-L}^{-d} I(z')g(z,z')\mathrm{d}z' - \frac{1}{j4\pi\omega\varepsilon_{\mathrm{eff}}}\frac{\partial}{\partial z}\int_{-d-L}^{-d}\frac{\partial I(z')}{\partial z'}g(z,z')\mathrm{d}z' + Z_s(z)I(z) = 0. \tag{6.63}$$

The Green function is of the form

$$g(z,z') = g_0(z,z') - \Gamma_{\mathrm{ref}} g_i(z,z'), \tag{6.64}$$

while $g_0(z,z')$ is the lossy medium Green function:

$$g_0(z,z') = \frac{\mathrm{e}^{-\gamma R_1}}{R_1} \tag{6.65}$$

and $g_i(z,z')$ is, according to the image theory, given by

$$g_i(z,z') = \frac{\mathrm{e}^{-\gamma R_2}}{R_2}, \tag{6.66}$$

and $R_1$, $R_2$ are the distances from the source and the image to the observation point, respectively. The alternative approach uses the approximate RC [8].

Within the numerical solution the Fresnel RC is used [17, 18]:

$$\Gamma_{\mathrm{ref}}^{\mathrm{MIT}} = \frac{\frac{1}{\underline{n}}\cos\theta - \sqrt{\frac{1}{\underline{n}} - \sin^2\theta}}{\frac{1}{\underline{n}}\cos\theta + \sqrt{\frac{1}{\underline{n}} - \sin^2\theta}}; \quad \theta = \mathrm{arctg}\frac{|x-x'|}{2d}; \quad \underline{n} = \frac{\varepsilon_{\mathrm{eff}}}{\varepsilon_0}. \tag{6.67}$$

Furthermore, a simplified RC arising from modified image theory is used within the analytical solution [17, 18]:

$$\Gamma_{\mathrm{ref}}^{\mathrm{MIT}} = -\frac{\varepsilon_{\mathrm{eff}} - \varepsilon_0}{\varepsilon_{\mathrm{eff}} + \varepsilon_0}. \tag{6.68}$$

The electrode is energized by means of an equivalent ICS with one terminal connected to the electrode and the other one grounded at infinity, as presented in Figure 6.41. This current source is incorporated into Pocklington's equation formulation via the following conditions at the electrode ends [19]:

$$I(-d) = I_g, I(-d-L) = 0, \tag{6.69}$$

where $I_g$ denotes the impressed unit current source.

Knowledge of the electrode current enables the assessment of the scattered voltage.

The scattered voltage along the vertical electrode is defined by a line integral of the horizontal component of the scattered field (normal to the electrode and tangential to the ground–air interface) from the remote soil to the electrode surface [9]:

$$V^{\mathrm{sct}}(z) = -\int_{\infty}^{a} E_x^{\mathrm{sct}}(x,z)\mathrm{d}x. \tag{6.70}$$

As the horizontal field component can be expressed in terms of the scalar potential gradient

$$E_x^{\text{sct}} = -\frac{\partial \phi}{\partial x}, \tag{6.71}$$

the scattered voltage along the electrode can be written as follows:

$$V^{\text{sct}}(z) = \int_{\infty}^{a} \frac{\partial \varphi(x,z)}{\partial x} \mathrm{d}x = \frac{\mathrm{d}}{\mathrm{d}z} \int_{\infty}^{a} \varphi(x,z)\mathrm{d}x, \tag{6.72}$$

where $\varphi(x, z)$ is defined by the following particular integral:

$$\varphi(z) = \frac{1}{4\pi\varepsilon_{\text{eff}}} \int_{-d-L}^{-d} q(z')g(z,z')\mathrm{d}z', \tag{6.73}$$

while the charge density and the current distribution are related through the continuity equation [17]:

$$q = -\frac{1}{j\omega}\frac{\mathrm{d}I}{\mathrm{d}z}. \tag{6.74}$$

Substituting continuity equation (6.74) into (6.73) yields

$$\varphi(z) = -\frac{1}{j4\pi\omega\varepsilon_{\text{eff}}} \int_{-d-L}^{-d} \frac{\partial I(z')}{\partial z'} g(z,z')\mathrm{d}z'. \tag{6.75}$$

Inserting (6.75) into (6.72) and assuming the zero-value of scalar potential in the remote soil [13] gives the scattered voltage along the electrode:

$$V^{\text{sct}}(z) = -\frac{1}{j4\pi\omega\varepsilon_{\text{eff}}} \int_{-d-L}^{-d} \frac{\partial I(z')}{\partial z'} g(z,z')\mathrm{d}z'. \tag{6.76}$$

It is worth noting that the tedious integration from infinity to the electrode surface is avoided by taking advantage of the very definition of voltage along the electrode (6.69).

Pocklington's equation (6.63) is solved numerically (GB-IBEM) and analytically, respectively.

### 6.2.2 Numerical Solution

The unknown current $I^e(z')$ along the straight wire segment can be written as

$$I^e(z') = \{f\}^T\{I\}. \tag{6.77}$$

Assembling the contributions from each segment, Pocklington's equation (6.63) is transferred into the matrix equation:

$$\sum_{j=1}^{M} [Z]_{ji}\{I\}_i = 0, \text{and } j = 1, 2, \ldots, M, \tag{6.78}$$

where $M$ is the total number of segments and $[Z]_{ji}$ given by

$$[Z]_{ji} = -\int_{\Delta l_j} \{D\}_j \int_{\Delta l_i} \{D'\}_i^T g(z,z')\mathrm{d}z' + \int_{\Delta l_j} \{D\}_j \int_{\Delta l_i} \{D'\}_i^T \Gamma_{\mathrm{ref}}^{\mathrm{Fr}} g_i(z,z')\mathrm{d}z' + \gamma^2 \int_{\Delta l_j} \{f\}_j \int_{\Delta l_i} \{f\}_i^T g(z,z')\mathrm{d}z'\,\mathrm{d}z \tag{6.79}$$

is the mutual impedance matrix representing the interaction of the $i$th source with the $j$th observation segment, respectively. The matrices $\{f\}$ and $\{f'\}$ contain the shape functions while $\{D\}$ and $\{D'\}$ contain their derivatives, and $\Delta l_i$, $\Delta l_j$ are the widths of $i$th and $j$th boundary elements.

A linear approximation over a wire segment is used in this work, as it was shown to be optimal in various EMC problems including thin wire configurations.

### 6.2.3 Analytical Solution

Pocklington's equation (6.63) can be solved analytically under certain conditions. It is convenient to write (6.63) in the form

$$-\frac{1}{j4\pi\omega\varepsilon_{\mathrm{eff}}}\left[\frac{\partial^2}{\partial z^2} - \gamma^2\right]\int_{-d-L}^{-d} \frac{\partial I(z')}{\partial z'} g(z,z')dz' + Z_{\mathrm{s}}(z)I(z) = 0. \tag{6.80}$$

The first step in the analytical solution of (6.80) is to write the integral on the left-hand side of (6.80) as follows:

$$\int_{-d-L}^{-d} I(z')g(z,z')\mathrm{d}z' = I(z)\int_{-d-L}^{-d} g(z,z')\mathrm{d}z' + \int_{-d-L}^{-d} [I(z') - I(z)]g(z,z')\mathrm{d}z'. \tag{6.81}$$

The integral on the left-hand side of (6.81) is now approximated by the first term on the right-hand side of (6.81), i.e. the second integral on the right-hand side of (6.81) is neglected.

Furthermore, the solution of characteristic integral from (6.81) is

$$\psi_{\mathrm{g}} = \int_{h-\frac{L}{2}}^{h+\frac{L}{2}} g(z,z')\mathrm{d}z' = \int_{h-\frac{L}{2}}^{h+\frac{L}{2}} \frac{\mathrm{e}^{-jkR}}{R}\mathrm{d}z' + \Gamma_{\mathrm{ref}}^{\mathrm{MIT}} \int_{h-\frac{L}{2}}^{h+\frac{L}{2}} \frac{\mathrm{e}^{-jkR_i}}{R_i}\mathrm{d}z' = 2\left(\ln\frac{L}{a} - \Gamma_{\mathrm{ref}}^{\mathrm{MIT}} \ln\frac{L}{2d}\right). \tag{6.82}$$

Consequently, Pocklington's equation (6.80) simplifies into the partial differential equation of the form:

$$\left[\frac{\partial^2}{\partial z^2} - \gamma_g^2\right] I(z) = 0. \tag{6.83}$$

The analytical solution of (6.83) is

$$I(z) = I_g \frac{sh[\gamma_{eq}(d + L - z)]}{sh(\gamma_{eq} L)}, \tag{6.84}$$

where

$$\gamma_{eq}^2 = \gamma^2 + j4\pi\omega \frac{\varepsilon_{eff}}{\psi_g} Z_s. \tag{6.85}$$

Furthermore, the scattered voltage along the electrode (6.76) is given by

$$\begin{aligned} V^{sct}(z) &= -\frac{1}{j4\pi\omega\varepsilon_{eff}} \int_{-d-L}^{-d} \frac{\partial I(z')}{\partial z'} g(z, z') dz', \\ &= \frac{\gamma_{eq} I_g}{j4\pi\omega\varepsilon_{eff} sh(\gamma_{eq} L)} \int_{-d-L}^{-d} ch\gamma_{eq}(d + L - z) g(z, z') dz', \end{aligned} \tag{6.86}$$

and can be computed by means of numerical integration procedures.

The space–time dependent current and voltage along the electrode are obtained by using numerical/analytical solution procedures. The mathematical description of the solution procedures can be found elsewhere, e.g. in [17]. The transient response of the vertical electrode is calculated by using the IFFT procedure. The discrete frequency response is obtained by sampling the analytical solution in the FD and is given by

$$I(x, \omega_i) = \sum_{i=1}^{N} I(x, \omega)\delta(\omega - \omega_i). \tag{6.87}$$

Transient current induced at the center of the electrode is evaluated via the following IFFT formula:

$$I(x, t_j) = \frac{1}{N} \sum_{i=1}^{N} I(x, \omega_i)\omega_N^{-(j-1)(i-1)}. \tag{6.88}$$

Having performed extensive numerical testing, $2^{16}$ samples with maximum frequency of 500 MHz were found to be optimal to ensure accurate results within a plausible time frame.

### 6.2.4 Examples

The first set of the results is related to the spatial distribution of induced current and scattered voltage computed numerically and analytically. The results are obtained for the following parameters of the electrode: $L = 10$ m, $d = 0.3$ m, $a = 5$ mm. The excitation is the ICS $I_g = 1$ A. Ground permittivity is $\varepsilon_r = 10$.

The results are related to the spatial distribution of the current and the scattered voltage induced along the electrode, computed numerically and analytically, respectively. The results are obtained for the following parameters of the electrode: $L = 10$ m, $d = 0.3$ m, $a = 5$ mm. The excitation is the current source $I_g = 1$ A, while the ground permittivity is $\varepsilon_r = 10$.

Figure 6.42 shows the real and imaginary part of the current and the scattered voltage distribution along the electrode at the operating frequency $f = 1$ MHz for the ground conductivity $\sigma = 0.01$ S m$^{-1}$.

The results computed analytically and via GB-IBEM agree rather satisfactorily. Figures 6.43–6.46 show the absolute value of the spatial distribution of the current and scattered voltage along the electrode.

Figure 6.43 shows the current distribution along the electrode at the operating frequency $f = 10$ MHz for the ground conductivity $\sigma = 0.01$ S m$^{-1}$. The results computed analytically and via GB-IBEM agree rather satisfactorily.

Figure 6.44 shows the current distribution along the same electrode at $f = 10$ MHz for the ground conductivity $\sigma = 0.001$ S m$^{-1}$. The waveforms obtained via different approaches are still alike. However, for the lower value of conductivity ($\sigma = 0.001$ S m$^{-1}$) differences in the results for the current distribution are slightly higher.

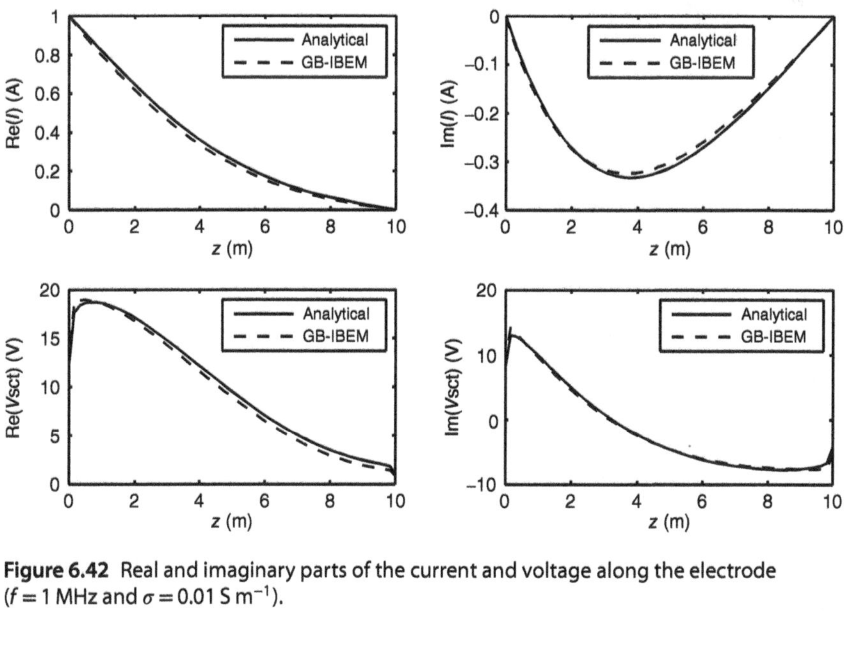


**Figure 6.42** Real and imaginary parts of the current and voltage along the electrode ($f = 1$ MHz and $\sigma = 0.01$ S m$^{-1}$).

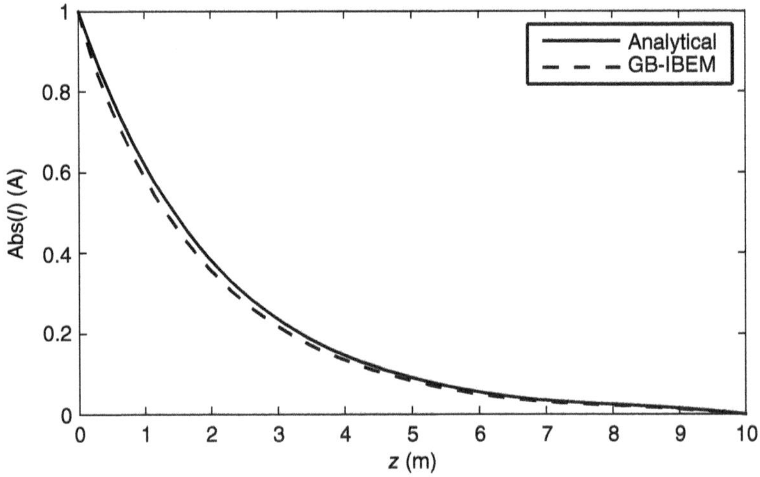


**Figure 6.43** Absolute value of current distribution along the electrode ($f = 10$ MHz and $\sigma = 0.01$ S m$^{-1}$).

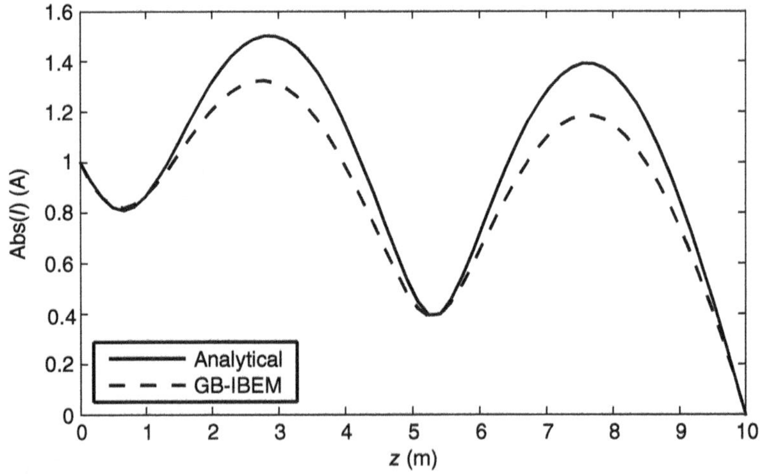


**Figure 6.44** Absolute value of current distribution along the electrode ($f = 10$ MHz and $\sigma = 0.001$ S m$^{-1}$).

Figures 6.45 and 6.46 show the distribution of the scattered voltage induced along the same electrode for different values of ground conductivity at $f = 10$ MHz.

A satisfactory agreement between analytical and numerical results is achieved. Again, certain differences are noticeable at $f = 10$ MHz for lower value of ground conductivity.

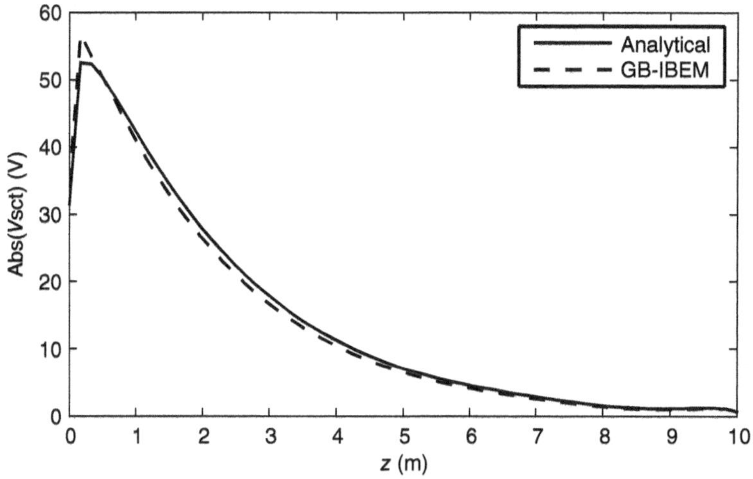


**Figure 6.45** Absolute value of the scattered voltage along the electrode ($f = 10$ MHz and $\sigma = 0.01$ S m$^{-1}$).

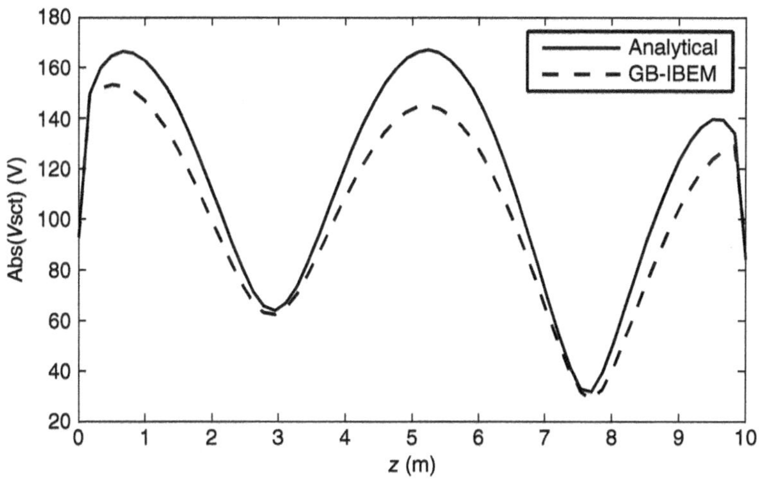


**Figure 6.46** Absolute value of the scattered voltage along the electrode ($f = 10$ MHz and $\sigma = 0.001$ S m$^{-1}$).

Finally, some illustrative computational examples for the transient current induced at the center of $L = 1$ m and $L = 10$ m long electrode obtained by means of analytical and numerical approach, respectively, are presented in Figures 6.47 and 6.48. The radius of the electrode is $a = 5$ mm, ground relative permittivity is $\varepsilon_r = 10$, burial depth is $d = 0.5$ m. The electrode is excited by the double exponential 0.1/1 μs pulse.

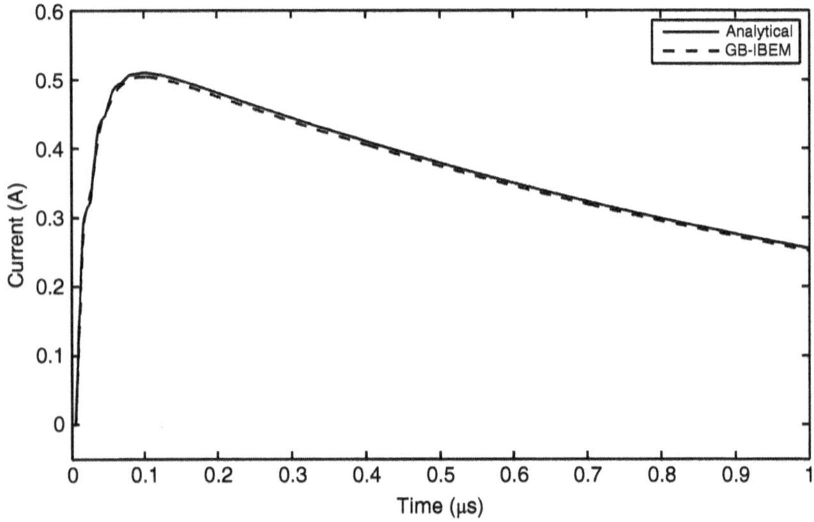


**Figure 6.47** Transient current at the center of the grounding electrode, $L = 1$ m, $\sigma = 10$ mS m$^{-1}$.

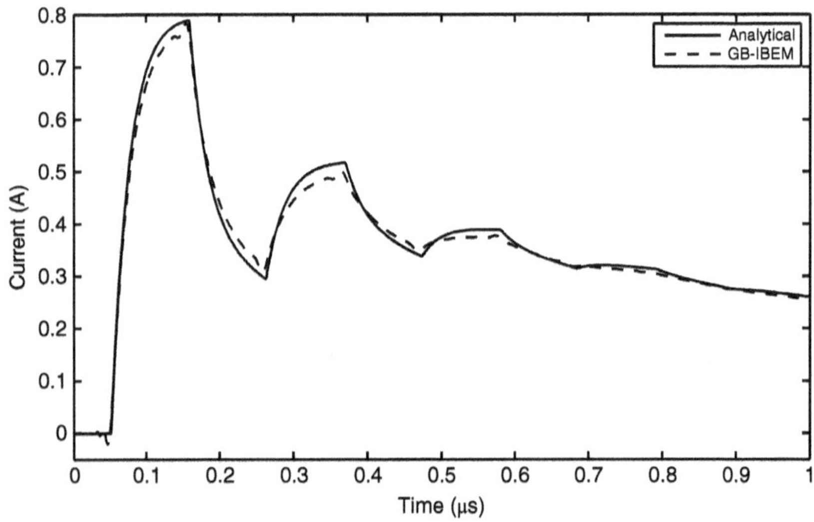


**Figure 6.48** Transient current at the center of the grounding electrode, $L = 10$ m, $\sigma = 1$ mS m$^{-1}$.

Note that in both cases the results obtained via different approaches agree quite satisfactorily.

## 6.3 Frequency Domain Analysis of Complex Grounding Systems

This section deals with an assessment of the transient behavior of different grounding grid configurations using both the AT and modified TL model, respectively.

The AT approach is based on the set of homogeneous FD Pocklington's integro-differential equations, with ground–air interface effects taken into account through the rigorous Sommerfeld integral formulation. The set of homogeneous Pocklington's equations is handled via the GB-IBEM [20] featuring linear isoparametric elements. The corresponding transient response is obtained by means of IFFT algorithm.

The MTL model is based on the corresponding telegrapher's equations. The partial differential equation for the transient voltage, arising from the TD–TL equations, is solved via the finite difference technique directly in TD.

The physical problem of interest shown in Figure 6.49 is related to the square grounding grid energized by a lightning channel current.

Several grounding grid configurations with dimensions varying from $10 \times 10\ \text{m}^2$ to $30 \times 30\ \text{m}^2$, with or without additional vertical electrodes, are considered in this section. All grids consist of wire conductors with radius $a = 5$ mm, and buried at $d = 1.5$ m depth.

Figure 6.50 shows various grid configurations. Two values of soil conductivity are considered: $\sigma_1 = 0.001\ \text{S m}^{-1}$ and $\sigma_2 = 0.01\ \text{S m}^{-1}$. In both cases, the relative permittivity is $\varepsilon_r = 9$. In all cases, the current injection point is located at the center of the grid.

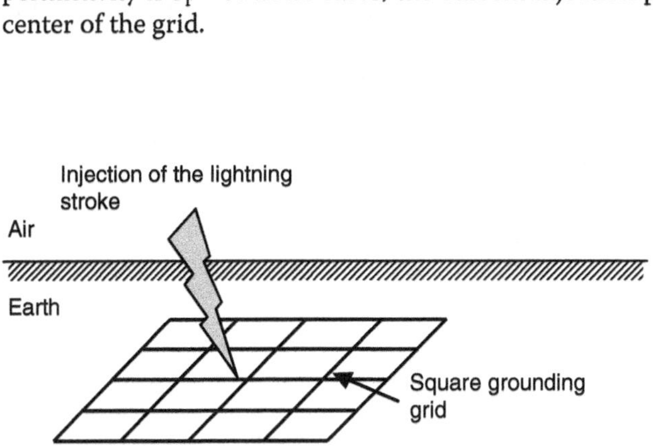


**Figure 6.49** Square grounding grid subjected to a lightning stroke.

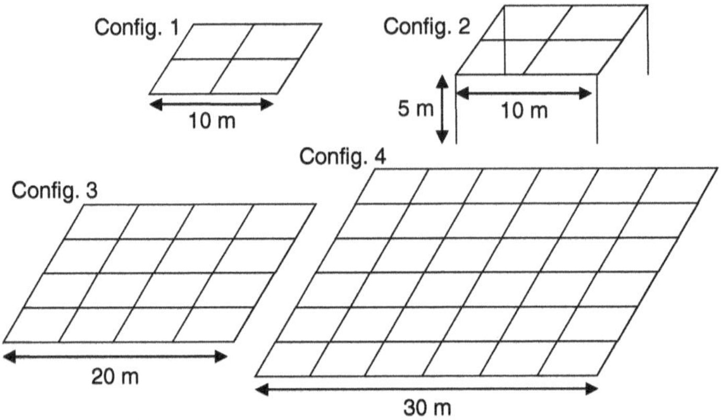


**Figure 6.50** Different grounding grid configurations.

### 6.3.1 Antenna Theory Approach: Set of Homogeneous Pocklington's Integro-Differential Equations for Grounding Systems

The currents flowing along the grounding grid are governed by the set of coupled Pocklington's integro-differential equations for wires of arbitrary shape [20]:

$$\sum_{n=1}^{N_W} \left[ \begin{array}{l} \int_{C'_n} I_n(s'_n) \cdot \vec{s}_m \cdot \vec{s}_n{}' \cdot [k_1^2 + \nabla\nabla] g_{0n}(s_m, s'_n) ds'_n \\ + \dfrac{k_0^2 - k_1^2}{k_0^2 + k_1^2} \displaystyle\int_{C'_n} I_n(s'_n) \cdot \vec{s}_m \cdot \vec{s}^{\,*}_n \cdot [k_1^2 + \nabla\nabla] g_{inm}(s_m, s_n^*) ds'_n \\ + \int_{C'_n} I_n(s'_n) \cdot \vec{s}_m \cdot \vec{s}_n{}^* \cdot G_s(s_m, s'_n) ds'_n \end{array} \right] = 0,$$

$$m = 1, 2, \ldots, N_W. \tag{6.89}$$

Note that the excitation is taken into account in the formulation through the forcing condition [20]:

$$I_1 = I_g, \tag{6.90}$$

where $I_g$ denotes current generator and $I_1$ current at the injection point.

Furthermore, at a junction consisting of two or more segments, the continuity properties of the electric field must be satisfied [20], which is governed by applying the Kirchhoff current law:

$$\sum_{k=1}^{n} I_k = 0, \tag{6.91}$$

and the continuity equation

$$\left[\frac{\partial I_1}{\partial s'_1}\right]\Bigg|_{\text{at junctuion}} = \left[\frac{\partial I_2}{\partial s'_2}\right]\Bigg|_{\text{at junctuion}} = \cdots = \left[\frac{\partial I_n}{\partial s_n{}'}\right]\Bigg|_{\text{at junctuion}}. \tag{6.92}$$

The condition (6.92) ensures that the discontinuities in charge per unit length are ruled out in passing from one conductor to another across the junction. On the other hand, at the conductor free ends, the total current vanishes.

The input impedance is given by the ratio

$$Z_{\text{in}} = \frac{V_{\text{g}}}{I_{\text{g}}}, \tag{6.93}$$

where $V_{\text{g}}$ and $I_{\text{g}}$ are the values of the voltage and the current at the driving point.Once the current distribution is calculated, a feeding point voltage is obtained by integrating the normal electric field component from infinity to the electrode surface:

$$V_{\text{g}} = -\int_{\infty}^{r} \vec{E}\,\mathrm{d}\vec{l}. \tag{6.94}$$

The direct calculation of (6.94) is very time consuming. However, by carefully choosing an integration path, the computational cost can be significantly reduced. Thus, in the case of horizontal arrangement of wires, the best path is vertical, i.e. over the $z$ axis.

The input impedance spectrum is multiplied with the current spectrum and the frequency response of the grounding system is obtained. Finally, the transient response is calculated by means of the inverse Fourier transform (IFT).

In addition, once the axial current induced along the wires is obtained, a scattered voltage between the point on the wire structure ($\vec{r}$) and remote soil can be calculated as a line integral over the scattered electric field:

$$V^{\text{sct}} = -\int_{\infty}^{\vec{r}} \vec{E}^{\text{sct}}\,\mathrm{d}\vec{l}. \tag{6.95}$$

Assuming the potential in the remote soil to be zero ($\varphi_{\infty} = 0$), and by using certain mathematical manipulations, the following integral formula for the voltage between the point on the wire ($\vec{r}$) and remote soil is obtained [21]:

$$V(\vec{r}) \cong \varphi(\vec{r}) = -\frac{1}{j4\pi\omega\varepsilon_{\text{eff}}} \sum_{i=1}^{N_{\text{w}}} \left\{ \begin{array}{l} \displaystyle\int_{C'} \frac{\partial I(s')}{\partial s'} \cdot g_0(\vec{r}, s')\mathrm{d}s' + \\ \displaystyle\frac{k_1^2 - k_2^2}{k_1^2 + k_2^2} \int_{C'} \frac{\partial I(s')}{\partial s^*} \cdot g_i(\vec{r}, s^*)\mathrm{d}s' \end{array} \right\}, \tag{6.96}$$

where $\vec{r}$ stands for the observation point, while (*) denotes the quantity along the image.

Expression (6.96) can be considered as an extension of the generalized telegrapher's equation for a straight wire [9] to the case of curved wires.

### 6.3.2 Antenna Theory Approach: Numerical Solution

The set of Pocklington's integro-differential equations (6.89) is numerically solved via the GB-IBEM. The boundary element solution technique used in this work is an extension of the method applied to single wire cases and presented elsewhere, e.g. in [20].

Undertaking the procedure already presented in Section II.B the set of Pocklington's equations (6.89) is transferred to the system of equations [20]:

$$\sum_{n=1}^{N_w}\sum_{i=1}^{N_n}[Z]_{ji}^{e}\{I_n\}_i = 0, \quad m = 1,2,\dots,N_w;\ j = 1,2,\dots,N_m, \tag{6.97}$$

where $[Z]_{ji}$ is the mutual impedance matrix for the $j$th observation segment on the $m$th antenna and $i$th source segment on the $n$th antenna. $N_w$ is the total number of wires, $N_m$ is the number of elements on the $m$th conductor and $N_n$ is the number of segments on the $n$th conductor.

Implementation of isoparametric elements yields the following expression for the mutual impedance matrix:

$$\begin{aligned}
[Z]_{ji}^{e} = &-\int_{-1}^{1}\int_{-1}^{1}\{D\}_j\{D'\}_i^T g_{0nm}(s_m,s'_n)\frac{\mathrm{d}s_n}{\mathrm{d}\zeta'}\mathrm{d}\zeta'\frac{\mathrm{d}s_m}{\mathrm{d}\zeta}\mathrm{d}\zeta \\
&+k_1^2\cdot s_m\cdot s'_n\int_{-1}^{1}\int_{-1}^{1}\{f\}_j\{f'\}_i^T g_{0nm}(s_m,s'_n)\frac{\mathrm{d}s_n}{\mathrm{d}\zeta'}\mathrm{d}\zeta'\frac{\mathrm{d}s_m}{\mathrm{d}\zeta}\mathrm{d}\zeta \\
&+\frac{k_0^2-k_1^2}{k_0^2+k_1^2}\left[\begin{array}{l}
-\int_{-1}^{1}\int_{-1}^{1}\{D\}_j\{D'\}_i^T g_{inm}(s_m,s_n^*)\frac{\mathrm{d}s_n}{\mathrm{d}\zeta'}\mathrm{d}\zeta'\frac{\mathrm{d}s_m}{\mathrm{d}\zeta}\mathrm{d}\zeta+ \\
+k_1^2\cdot s_m\cdot s_n^*\int_{-1}^{1}\int_{-1}^{1}\{f\}_j\{f'\}_i^T g_{inm}(s_m,s_n^*) \\
\frac{\mathrm{d}s_n}{\mathrm{d}\zeta'}\mathrm{d}\zeta'\frac{\mathrm{d}s_m}{\mathrm{d}\zeta}\mathrm{d}\zeta
\end{array}\right]+ \\
&+s_m\cdot s'_n\int_{-1}^{1}\int_{-1}^{1}\{f\}_j\{f'\}_i^T G_{snm}(s_m,s'_n)\frac{\mathrm{d}s_n}{\mathrm{d}\zeta'}\mathrm{d}\zeta'\frac{\mathrm{d}s_m}{\mathrm{d}\zeta}\mathrm{d}\zeta.
\end{aligned} \tag{6.98}$$

Matrices $\{f\}$ and $\{f'\}$ contain the shape functions while $\{D\}$ and $\{D'\}$ contain their derivatives.

The excitation function in the form of the current source $I_g$ is taken into account as a forced condition at the certain node $i$ of the grounding system [9]:

$$I_i = I_g. \tag{6.99}$$

The wire junctions are treated through Kirchhoff's current law in its integral and differential form, respectively, related to the continuity of induced currents and charges at the junction, (6.91) and (6.92).

Once the current distribution along the wire configuration is determined, it is possible to compute the corresponding scattered voltage defined with (6.96).

Thus, the use of GB-IBEM yields

$$V(\vec{r}) = -\frac{1}{j4\pi\omega\varepsilon_{\text{eff}}}\sum_{i=1}^{N_W}\sum_{n=1}^{N_g}\sum_{k=1}^{nl}\int_{-1}^{1} I_k^{ei}\frac{\partial f_k^e(\zeta')}{\partial\zeta'}\cdot\begin{bmatrix} g_0^i(\vec{r},s') \\ -\frac{k_1^2-k_2^2}{k_1^2+k_2^2}g_i^i(\vec{r},s^*) \end{bmatrix}\mathrm{d}\zeta'. \tag{6.100}$$

The transient voltage is obtained by applying the IFT.

### 6.3.3 Modified Transmission Line Method Approach

Within the framework of the MTL approach, one neglects the transverse propagation effects and the grounding system is simulated by means of a complex network [20]. The corresponding coupling equations for the scalar potential and the current in the TD for one-dimensional case are reported using the Agrawal model [20]:

$$\frac{\partial u^s(l,t)}{\partial l} + R\, i(l,t) + L\,\frac{\partial\, i(l,t)}{\partial t} = E_l^e(l,t), \tag{6.101}$$

$$\frac{\partial\, i(l,t)}{\partial l} + G\, u^s(l,t) + C\,\frac{\partial u^s(l,t)}{\partial t} = 0, \tag{6.102}$$

where $l = x$ or $y$.

$E_l^e(l,t)$ is the tangential component of the electric field excitation.

Combining two telegrapher's equations (6.101) and (6.102) the induced current or voltage can be eliminated and the second order partial differential equation for either voltage or current is obtained. If the propagation occurs in two directions, $x$ and $y$, the corresponding two-dimensional partial differential equation for transient voltage is given by

$$\frac{\partial^2 u^s}{\partial x^2} + \frac{\partial^2 u^s}{\partial y^2} - 2RGu^s - 2(RC+LG)\frac{\partial u^s}{\partial t} - 2LC\frac{\partial^2 u^s}{\partial t^2} = \frac{\partial E_x^e}{\partial x} + \frac{\partial E_y^e}{\partial y}, \tag{6.103}$$

where the electric field constitutes the excitation source, and $R$, $L$, $C$, and $G$ are per-unit-length parameters of the interconnected conductors.

For the grounding grid, the per-unit-length parameters are calculated taking into account the soil–air interface effects. There are various approaches for the assessment of these parameters, e.g. using the formulas suggested by Sunde [22] or by Liu [23].

### 6.3.4 Finite Difference Solution of the Potential Differential Equation for Transient Induced Voltage

Partial differential equation (6.103) is solved numerically using the finite difference technique. The spatial discretization of the second order differential

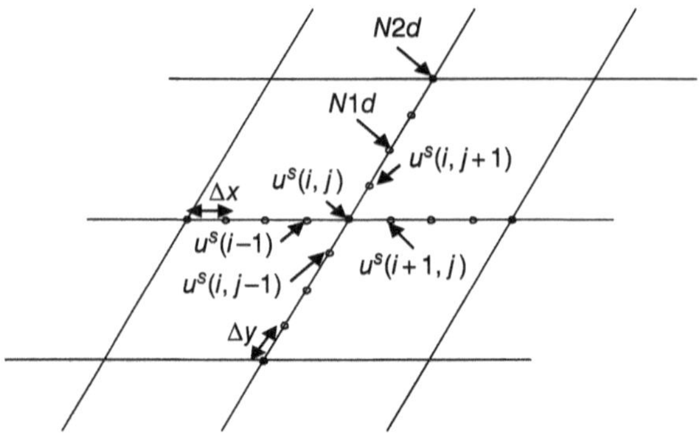


**Figure 6.51** Spatial discretization of the square grid.

operator at certain point $(i, j)$ using the finite difference approximation is shown in Figure 6.51.

The finite difference approximation of spatial and temporal derivatives at certain point $(i, j)$ can be written as

$$\frac{\partial^2 u^s}{\partial x^2} = \frac{1}{\Delta x^2}[((u^s)^n_{i+1,j} - 2(u^s)^n_{i,j} + (u^s)^n_{i-1,j})], \tag{6.104}$$

$$\frac{\partial^2 u^s}{\partial y^2} = \frac{1}{\Delta y^2}[(u^s)^n_{i,j+1} - 2(u^s)^n_{i,j} + (u^s)^n_{i,j-1}], \tag{6.105}$$

$$\frac{\partial u^s}{\partial t} = \frac{1}{\Delta t}[(u^s)^n_{i,j} - (u^s)^{n-1}_{i,j}], \tag{6.106}$$

$$\frac{\partial^2 u^s}{\partial t^2} = \frac{1}{\Delta t^2}[(u^s)^n_{i,j} - 2(u^s)^{n-1}_{i,j} + (u^s)^{n-2}_{i,j}], \tag{6.107}$$

$$\frac{\partial E^e_x}{\partial x} = \frac{1}{2\Delta x}[(E^e_x)^n_{i+1,j} - (E^e_x)^n_{i-1,j}], \tag{6.108}$$

$$\frac{\partial E^e_y}{\partial y} = \frac{1}{2\Delta y}[(E^e_y)^n_{i,j+1} - (E^e_y)^n_{i,j-1}]. \tag{6.109}$$

Substituting (6.104)–(6.109) into (6.103) yields the following relation:

$$\begin{aligned}
&\left[-\frac{2}{(\Delta x)^2} - \frac{2}{(\Delta y)^2} - 2RG - \frac{2(RC+LG)}{\Delta t} - \frac{2LC}{(\Delta t)^2}\right](u^s)^n_{i,j} \\
&+ \left[\frac{1}{(\Delta x)^2}\right](u^s)^n_{i+1,j} + \left[\frac{1}{(\Delta x)^2}\right](u^s)^n_{i-1,j} + \left[\frac{1}{(\Delta y)^2}\right](u^s)^n_{i,j+1} \\
&+ \left[\frac{1}{(\Delta y)^2}\right](u^s)^n_{i,j-1,k} = \left[-\frac{2(RC+LG)}{\Delta t} - \frac{4LC}{(\Delta t)^2}\right](u^s)^{n-1}_{i,j} \\
&+ \frac{2LC}{(\Delta t)^2}(u^s) + \frac{1}{2\Delta x}[(E^e_x)^n_{i+1,j} - (E^e_x)^n_{i-1,j}] + \frac{1}{2\Delta y}[(E^e_y)^n_{i,j+1} - (E^e_y)^n_{i,j-1}],
\end{aligned} \tag{6.110}$$

which can also be expressed in the matrix form:

$$\begin{bmatrix} A_{11} & \cdots & A_{1k} & \cdots & A_{1N} \\ \vdots & \ddots & \vdots & \ddots & \vdots \\ A_{1k} & \cdots & A_{kk} & \cdots & A_{kN} \\ \vdots & \ddots & \vdots & \ddots & \vdots \\ A_{N1} & \cdots & A_{Nk} & \cdots & A_{NN} \end{bmatrix} \begin{bmatrix} U_1 \\ \vdots \\ U_k \\ \vdots \\ U_N \end{bmatrix} = \begin{bmatrix} B_1 \\ \vdots \\ B_k \\ \vdots \\ B_N \end{bmatrix}, \tag{6.111}$$

where $[A]$ denotes the coefficients matrix, $[u^s]$ is the unknown voltage vector, $[B]$ is the entire right-hand side, and $N$ is the total number of nodes.

The diagonal elements of matrix $[A]$ are given by

$$A_{kk} = -\frac{2}{(\Delta x)^2} - \frac{2}{(\Delta y)^2} - 2RG - \frac{2(RC + LG)}{\Delta t} - \frac{2LC}{(\Delta t)^2}, \tag{6.112}$$

while the non-diagonal elements of matrix $[A]$ are as follows:

$$A_{kl} = \frac{1}{(\Delta x)^2} \text{ if 1 is the adjacent node } k \text{ in the } x \text{ direction,} \tag{6.113a}$$

$$A_{kl} = \frac{1}{(\Delta y)^2} \text{ if 1 is the adjacent node } k \text{ in the } y \text{ direction,} \tag{6.113b}$$

$$A_{kl} = 0 \text{ elsewhere.} \tag{6.113c}$$

The elements of vector $[B]$ are

$$B_k = \left[ -\frac{2(RC + LG)}{\Delta t} - \frac{4LC}{(\Delta t)^2} \right] (u^s)^{n-1}_{i,j} + \frac{2LC}{(\Delta t)^2}(u^s)$$
$$+ \frac{1}{2\Delta x}[(E^e_x)^n_{i+1,j} - (E^e_x)^n_{i-1,j}] + \frac{1}{2\Delta y}[(E^e_y)^n_{i,j+1} - (E^e_y)^n_{i,j-1}]. \tag{6.114}$$

The solution of the partial differential equation (6.103) requires knowledge of the conditions at the grid edges, as indicated in Figure 6.52.

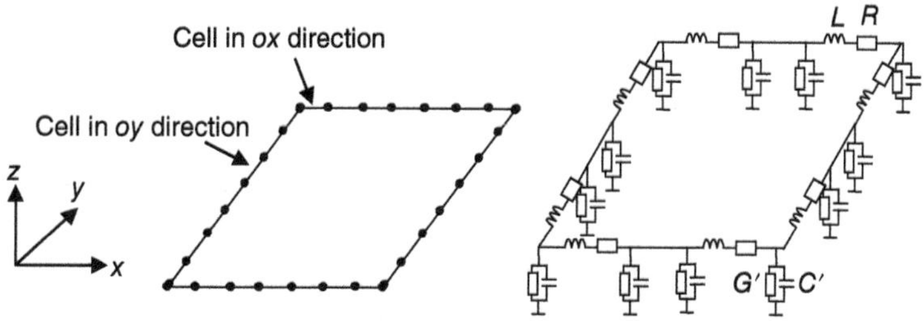


(a) border of the grid. (b) representation by $\pi$-Cell

**Figure 6.52** (a) Conditions at the grid edges; (b) equivalent electrical network of grounding grid.

In the case of an indirect lightning strike, at a certain point on the border of the grid in Figure 6.52 the following equation is to be used [20]:

$$u^s(l,t) = -(G')^{-1}I(l,t-\Delta t) + \int_0^h E_z^{es}(l,t)\mathrm{d}z, \tag{6.115}$$

where $G'$ is the equivalent conductance of corresponding nodes, while $I(l,t-\Delta t)$ is the transversal current known at instant $(t-\Delta t)$ and $E_z^{es}$ is the $z$ component of the electric field in soil. The presence of a two-media configuration is taken into account calculating the linear parameters of the electrical circuit for the case of the grounding electrode [22, 23]. Such a treatment of non-homogeneous media is identical to the case of a transmission line with ground return path.

At each calculation step the determination of induced voltages provides the evaluation of the currents induced along interconnected conductors of grounding grid by numerical integration of the telegrapher's equation (6.101). Note that in the case of the direct impact of the lightning strike the presented procedure is used by simply imposing $E^e = 0$. Therefore, in this case the voltage differential equation (6.103) in the FD becomes

$$\frac{\partial^2 u}{\partial x^2} + \frac{\partial^2 u}{\partial y^2} - 2[RG + j\omega(RC + LG) - LC\omega^2]u = 0, \tag{6.116}$$

where $u$ is potential in any point of the grid.

In this case, at the injection node the value of the current is known (lightning strike generator), enabling the assessment of the corresponding voltage.

The input impedance is defined as

$$Z_{\text{in}} = \frac{u(k,t)}{I(k,t)}, \tag{6.117}$$

where $k$ is the injection node index.

### 6.3.5 Computational Examples: Grounding Grids and Rings

In all computational examples the lightning current is expressed by the double exponential function:

$$i(t) = I_0(\mathrm{e}^{-at} - \mathrm{e}^{-bt}), \tag{6.118}$$

with parameters $I_0 = 1.1043$ kA, $a = 0.07924 \times 10^6\ \mathrm{s}^{-1}$, $b = 0.07924 \times 10^6\ \mathrm{s}^{-1}$.

Figure 6.53 shows the transient voltage at the feeding point calculated by the AT and TL approach, respectively, for all four grid configuration scenarios and soil conductivity $\sigma_1 = 1\ \mathrm{mS\ m^{-1}}$, while the related transient impedance of the grounding systems is shown in Figure 6.54.

The results obtained by different approaches for grid type 1 and 2 agree rather satisfactorily, and relatively good agreement is achieved for type 3, while major

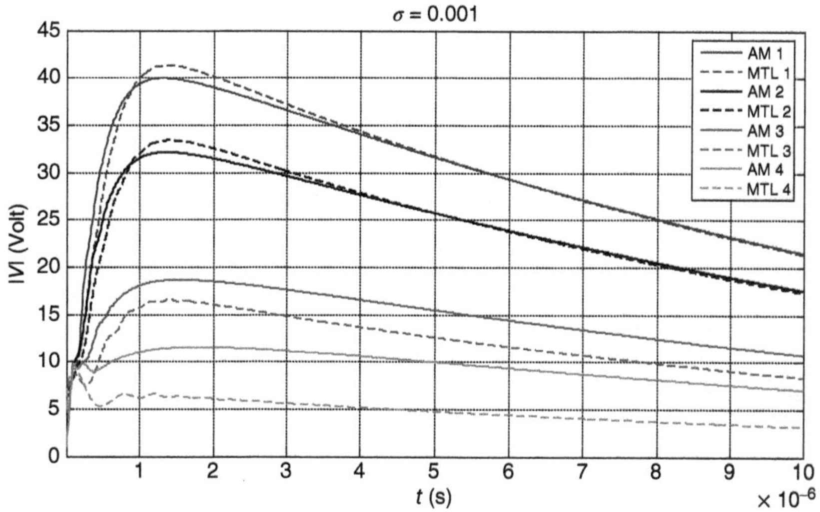


**Figure 6.53** Transient feeding point voltage for dry soil.

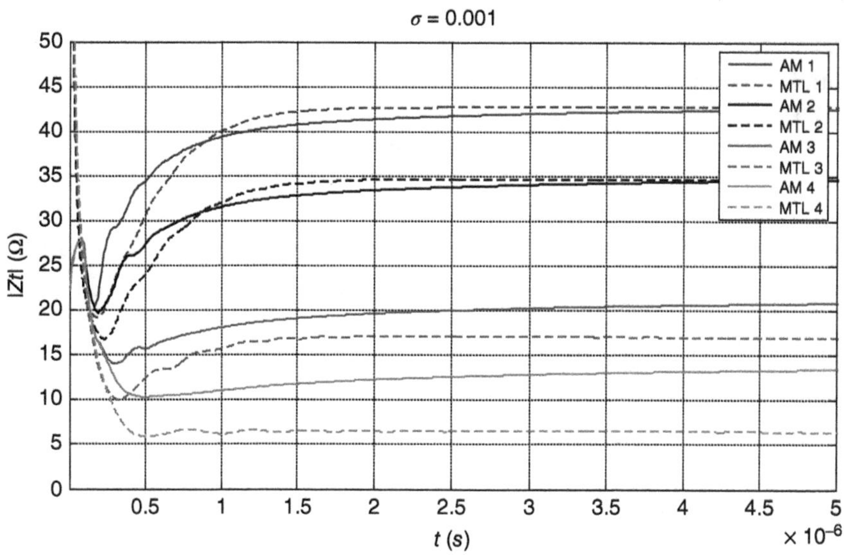


**Figure 6.54** Transient impedance for dry soil.

differences appear for type 4 grid. Some differences occur at very early time instants (from $10^{-8}$ to $10^{-7}$ s) corresponding to the high frequency content of the input signal spectrum, which cannot be accurately predicted by the MTL method. Consequently, there are some differences in rather early time instants, as is visible from Figure 6.53, i.e. MTL fails to accurately predict the early time behavior of grounding grids.

A similar conclusion can be drawn for the transient impedance, shown in Figure 6.53, as well. Greater the grid size, worse the agreement achieved between the results. MTL method would be expected to work better if the wires are longer, but it is not the case for this particular grid configuration.

Such a behavior can be explained by the fact that the parts of the grid behave as single antennas and there are many reflections from discontinuities that MTL fails to take into account. This effect is more evident for low conductive soils than for higher ones and what can be observed in Figures 6.55 and 6.56, which show the transient voltages and impedances at injection point calculated for ground conductivity $\sigma_2 = 0.01$ S m$^{-1}$.

The agreement between the results obtained via different approaches is found to be satisfactory, particularly for later time instants corresponding to the lower frequency part of the spectrum. For very early time instants the results are alike, although the transient voltage peaks, for all cases of grid configuration, are somewhat higher.

Comparing the results for both values of ground conductivities it can be noticed that the peak value of the voltage is advanced for the case of higher

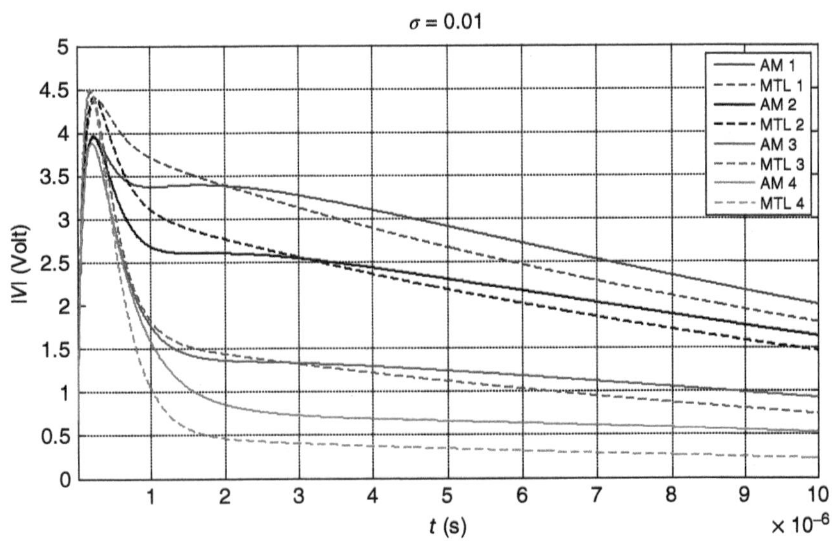


**Figure 6.55** Transient feeding point voltage for wet soil.

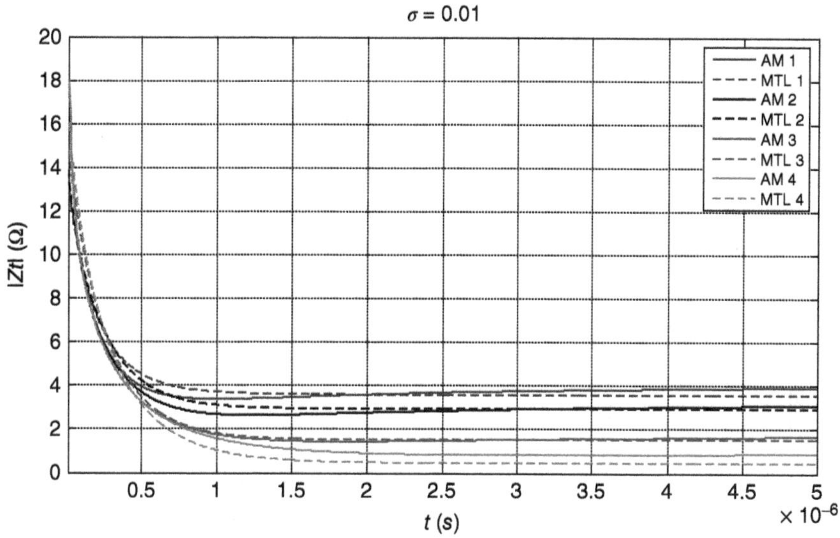


**Figure 6.56** Transient impedance for wet soil.

conductivity. Also, the values of the peak voltages are pretty much alike regardless of the grid size. It is well known that for very early time instants the higher frequency part of the impedance spectrum is important. As the grid density (mesh) remains the same for all grid configurations that portion of the frequency spectrum is unchanged regardless of the grid size, as shown in Figure 6.57. Depending on the conductivity, that part of spectrum starts at different frequency values. In the case of $\sigma_2 = 0.01\ \mathrm{S\ m^{-1}}$ the frequency is about 3 MHz (Figure 6.57), while for $\sigma_1 = 0.001\ \mathrm{S\ m{-}1}$ it is above 30 MHz (Figure 6.58). Thus, in high conductivity environment effective length of the grounding wires becomes very short at higher frequencies.

Therefore, very early time behavior is similar for all configurations as the portion of the very high frequency spectrum is pretty much the same. Also, it should be noted that the analysis of results presented in Figures 6.57 and 6.58, respectively is related to the results obtained by applying the antenna approach only, as the MTL results are obtained directly in time.

Some illustrative computational examples related to the assessment of transient voltage induced due to central and corner injection, respectively, of a grounding grid, as shown in Figure 6.59, are given as well.

The grounding system of interest is composed of grid 60m × 60m (6 by 6 $10\ \mathrm{m^2}$ meshes); wire radius 0.007 m; depth: 0.5 m; $\rho = 100\ \Omega\mathrm{m}$; $\varepsilon_r = 36$. The grounding grid is energized at certain points by the double exponential current source with $I_0 = 1.2\ \mathrm{kA}$, $a = 0.0142 \times 10^{-6}$, $b = 1.073 \times 10^{-6}$.

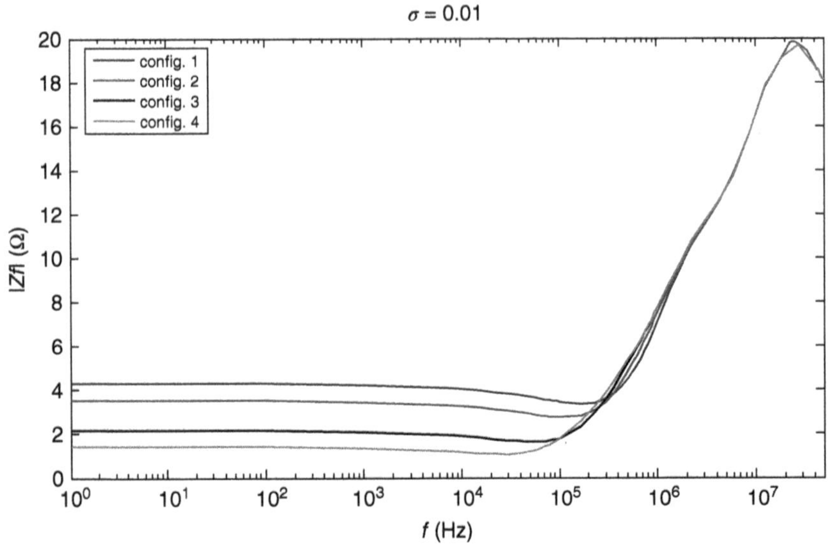


**Figure 6.57** The impedance spectrum for $\sigma_2 = 0.01$ S m$^{-1}$.

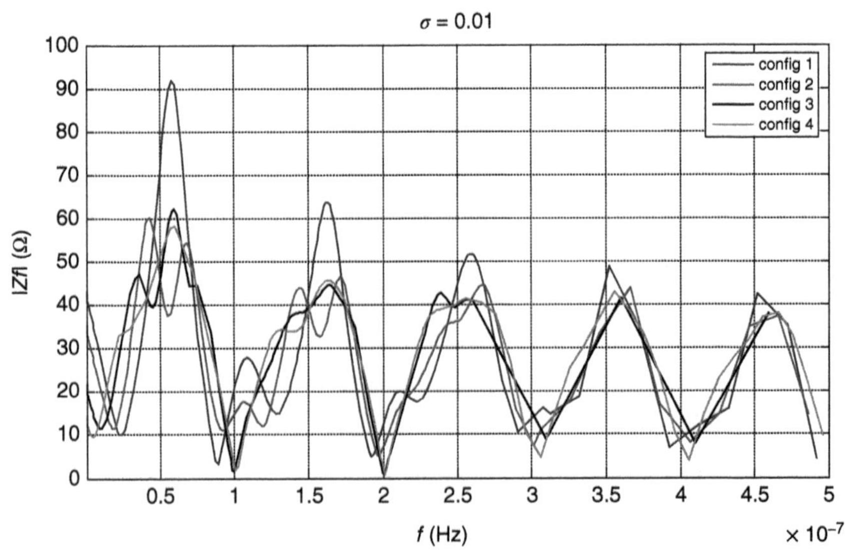


**Figure 6.58** The impedance spectrum for $\sigma_1 = 0.001$ S m$^{-1}$.

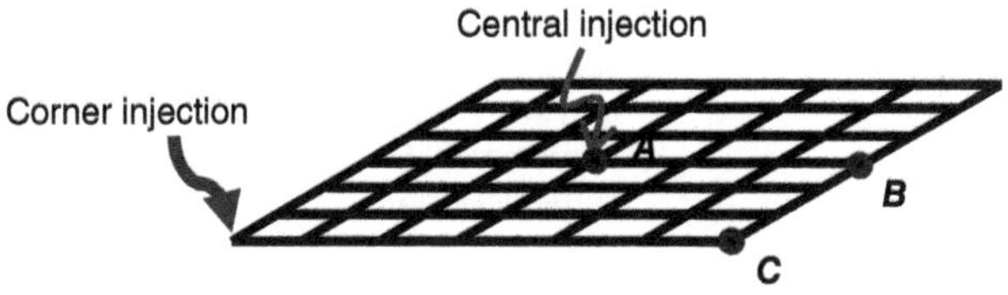


**Figure 6.59** Grounding grid under central injection of the current source (double exponential excitation).

**Figure 6.60** Spatial distribution of the voltage induced along the grounding grid (center injection) at $T = 0.5\,\mu s$ computed via (a) AT approach; (b) FDTD–TL approach.

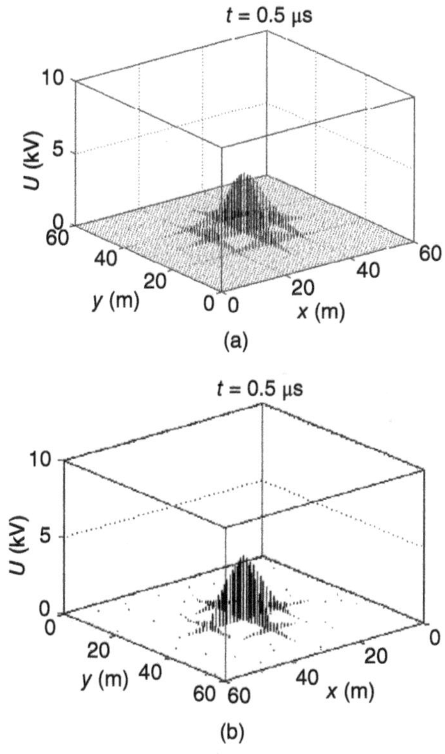


Figure 6.60 shows the spatial distribution of the voltage induced along the grounding grid at $T = 0.5\,\mu s$, due to a central current source excitation, computed via AT approach (Figure 6.60a) and TL approach, featuring the finite difference time domain (FDTD) approach, respectively (Figure 6.60b).

Furthermore, Figure 6.61 represents the spatial distribution of the voltage induced along the grid at $T = 0.5\,\mu s$, due to a corner current source excitation, computed via AT approach (Figure 6.61a) and TL approach, featuring the FDTD approach (Figure 6.61b), respectively.

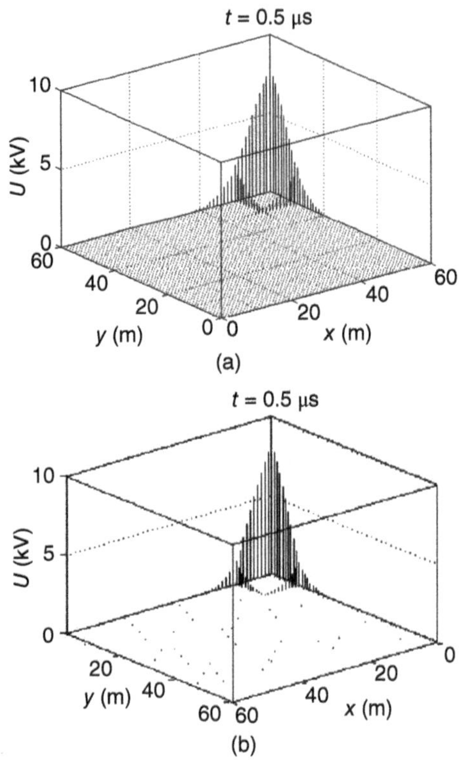


**Figure 6.61** Spatial distribution of the voltage induced along the grounding grid at $T = 0.5$ μs computed via (a) AT approach (b) FDTD–TL approach; corner injection.

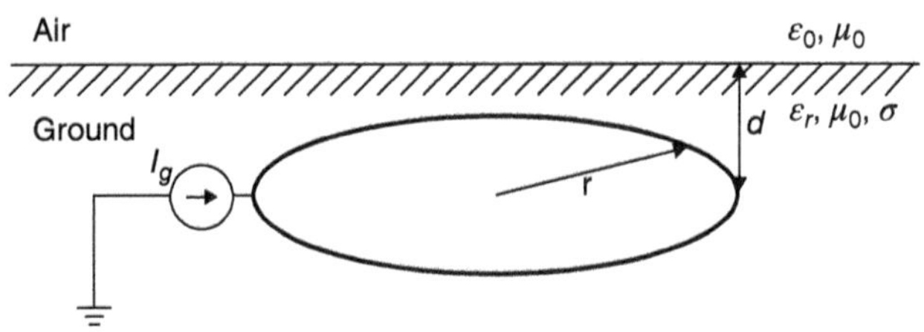


**Figure 6.62** Ring grounding system energized by a current source $I_g$.

The results obtained via different approaches agree relatively satisfactorily for both central and corner injection.

Next geometry of interest (Figure 6.62) is a ring-shaped grounding electrode of ring radius $r$ and wire radius $a$, buried in a lossy medium, characterized by permittivity $\varepsilon_r$, permeability $\mu_0$, and conductivity $\sigma$, at depth $d$, excited by a current source.

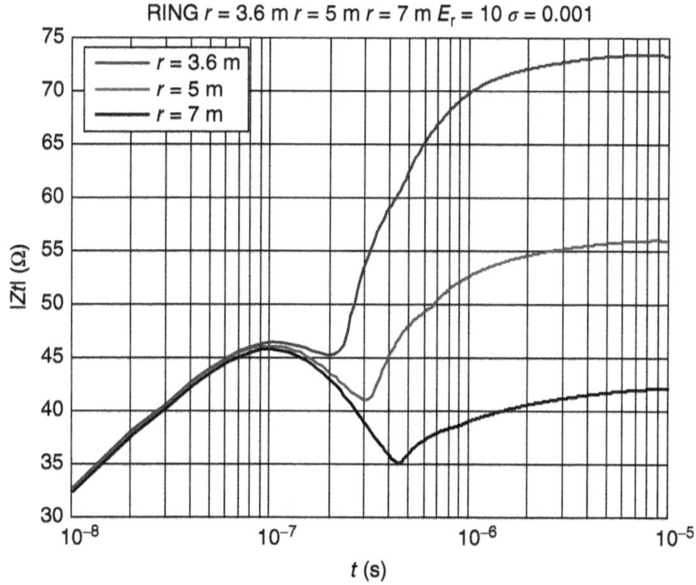


**Figure 6.63** Transient impedance of ring electrode calculated for various ring radiuses.

Figure 6.63 shows a transient impedance ring electrode buried in the ground at depth $d = 1$ m, with $\varepsilon_r = 10$, and $\sigma = 0.001$ S m$^{-1}$, for the case of different radius of the ring (3.6 m, 5 m, 7 m).

It can be observed that transient impedance varies from zero toward certain steady state value (corresponding to power frequency impedance), which decreases with the increase of the ring size. The early time behavior of the impedance remains almost the same.

### 6.3.6 Computational Examples: Grounding Systems for WTs

WTs are extremely vulnerable to lightning strikes due to their special shape and isolated locations mainly in high altitude areas. Lightning strikes may cause serious damages to WTs and available relevant statistical data indicate that between 4% and 8% of wind power systems in Europe suffer damages due to lightning strikes each year [24]. This situation is even worse in Southern Europe, due to an increased number of thunder storms and relatively low soil conductivities. Consequently, the development and installation of integral lightning protection system for WTs is of particular interest [24–31]. Although the methodology for WT lightning protection has been proposed in [24], a number of issues pertaining to transient behavior of grounding system, in the case of lightning strike, are still open research areas.

The grounding methodology reported in IEC 61400-24:2010 [24] is entirely subjected to the IEC 62305-3:2006 [32], which handles lightning protection for general structures including houses and buildings. The foundation and grounding system of a WT are generally much smaller compared to the grounding systems for buildings of the same height.

Therefore, there are two rather important EMC issues to be addressed:

- A study on WT's transient behavior due to a direct lightning strike, which involves the assessment of the transient current distribution along WT configuration
- Analysis and design of an efficient low impedance grounding system as a major prerequisite for an effective protection of WT from lightning strikes.

The first issue, an impact of lightning discharge to WT, particularly lightning strike to tall structures, has generally attracted the attention of many prominent researchers in the last few decades (e.g. [33]). Several models have been proposed for the assessment of the current distribution along the structure and lightning channel. Most of the representations are based on the extension of certain return stroke models initially developed for the case of return stroke initiated at the ground level. The presence of a tall structure has been included in two classes of return stroke models: engineering models and AT models [34]. The presence of a tall object within engineering models has been considered by representing the object as a uniform, lossless TL [35, 36]. The AT models [37–42] have been mostly used in the analysis of lightning strikes to CN tower in Toronto, e.g. [41], or similar towers, e.g. [42]. The formulation can be posed in either the time [37, 42] or the FD [38–41], respectively. Within the framework of the FD formulation a numerical electromagnetic code (NEC) is commonly used for current distribution calculation assuming the ground to be perfectly conducting (PEC) [37].

On the other hand, grounding systems, such as buried vertical or horizontal electrodes and large grounding grids, are important not only for safety of personnel but also for the protection of electrical equipment in industrial and power plants. The principal task of such grounding systems is to suppress the values of transient step and touch voltages, respectively, under the level that could cause adverse health effects. The secondary purpose of grounding systems is to provide a common reference voltage for all connected electrical and electronic systems.

Therefore, a low impedance, proper grounding for the protection of the WT should be designed to reach the grounding resistance of preferably less than 10 Ω (for an isolated WT), without taking into account the entire grounding system of a wind farm [24]. This requirement is rather difficult to meet in the case of the high specific resistance of the soil. It is worth noting that the standards for grounding systems [34] are based on the steady state or low frequency analysis. Consequently, almost all practical aspects of grounding systems design rely on the steady state analysis. However, such studies do not account for a transient behavior of a grounding system during lightning strikes.

On the other hand, transient analysis is of great importance as impulse currents increase the grounding system potential related to zero ground during the transient state, which could be dangerous to humans, installations, and equipment.

One of the most important parameters arising from the transient analysis of a grounding system is the transient impedance. Generally, grounding systems can be modeled using the simple circuit theory models [43, 44], the transmission line model (TLM) [1, 23, 45], or the antenna (full-wave) model (AM) [46–49]. While the circuit theory models are often considered to be oversimplified, the TL models are more accurate, still simple enough, and with relatively low computational cost. However, although valid for long horizontal conductors, a simplified TL approach is not convenient for vertical and interconnected conductors. Generally, the TL-based solutions are limited to a certain upper frequency, depending on the electrical properties of the ground and configuration of the particular grounding system [50]. On the other hand, the rigorous electromagnetic models based on the AT are regarded as the most accurate. The AT approach is based on the solution of Pocklington's integro- differential equation for the half-space problems [8, 20].

In the last decade, several researchers have raised important questions about WT grounding. Some papers on the subject [51–54] are based on the use of commercial software packages (EMPT and CEDGS) for WT grounding analysis. There are also papers reporting the use of MoM (method of moments) [55, 56], the FDTD [57], the FEM (Finite Element Method) [58], EMPT [59], and CEDGS [60, 61].

A review of not only a direct lightning strike to the WT but also transient analysis and design of realistic WT grounding systems is given in [48], where WT is represented by a simple PEC wire configuration, consisting of a tower and three blades, while the lightning channel is represented by a lossy vertical wire attached to WT. The lightning return stroke current is energized by an ideal voltage source at the tip of the WT blade. The current distribution along the WT and lightning channel is obtained by solving the set of Pocklington integro-differential equations in the FD by means of the GB-IBEM [8].

Furthermore, in [48] the transient impedance of a typical WT grounding system placed in a low conductivity soil is determined. Since the standard [32] provides very little information about installation and influence of additional electrodes attached to the original grounding system, special attention is focused on the influence of additional vertical and horizontal electrodes, respectively. The influence of a grounding wire placed in a cable trench on the transient behavior is also studied.

Contrary to the analyses reported in [51–61] the analysis presented in [48] deals with the antenna model presented in [20]. The model is based on the set of homogeneous integro-differential equations of Pocklington type, with ground–air interface effects being taken into account via the exact Sommerfeld integral formulation. Furthermore, instead of applying the usual approach featuring the Moment Method [48, 49] (also implemented in CEDGS), the current distribution along the grounding system is obtained by solving the set of Pocklington's integro-differential equations in the FD via the GB-IBEM [8] featuring the linear isoparametric elements. Finally, the corresponding transient response is obtained by means of the IFFT algorithm.

The problem of interest, i.e. WT connected to a corresponding grounding system, is shown in Figure 6.64. The WT grounding system is subjected to a transient current due to the direct lighting strike at a certain point. The influence of WT itself (tower, blades, etc.) is neglected.

The corresponding practical grounding system is shown in Figure 6.65.

The grounding system consists of a square of galvanized steel flanges (Fe/Zn 30 × 3.5 mm – gray line in Figure 6.65) at 2 m depth, two copper ring wires (Cu 70 mm$^2$ – black line in Figure 6.65) at different levels (smaller one of 3.25 m radius at 5 cm depth and the larger with 6.8 m radius buried at 55 cm depth),

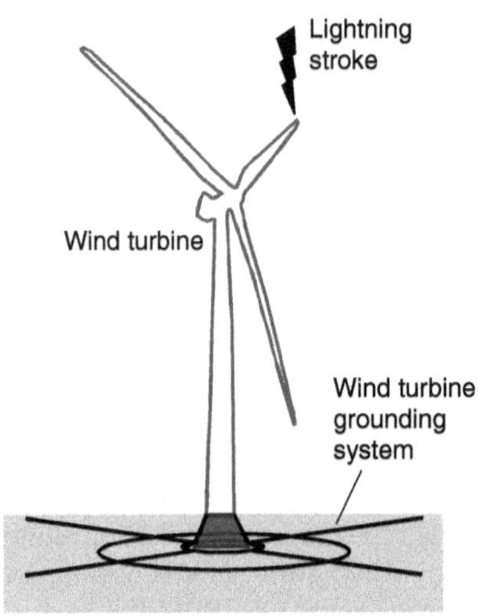


**Figure 6.64** WT subjected to a lightning strike.

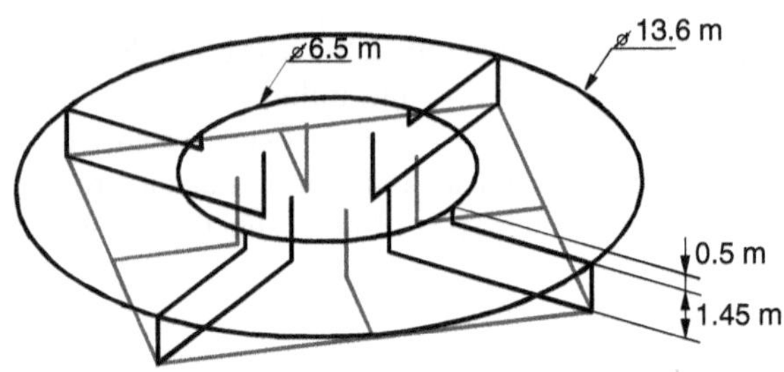


**Figure 6.65** Typical WT grounding system arrangement.

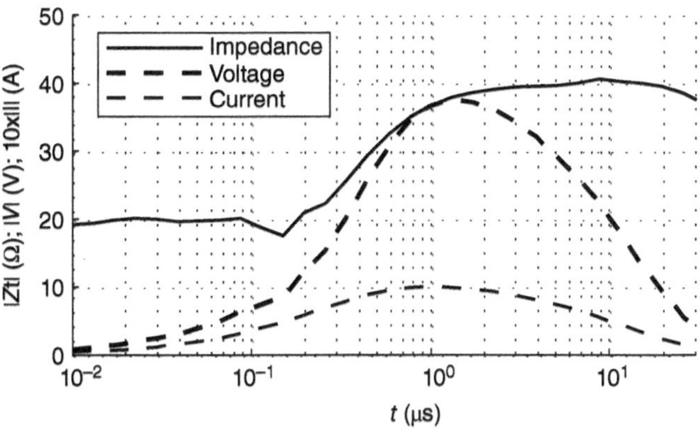


**Figure 6.66** Transient behavior of the basic grounding system.

and additional four copper wires connecting the rhombus with the tower. All these parts of the grounding system are connected by aluminothermy welding.

The grounding system is buried in a homogeneous soil of relatively high specific resistance of $\rho = 1200\ \Omega\cdot\mathrm{m}^{-1}$ while the relative dielectric constant is assumed to be $\varepsilon_r = 9$.

In all computational examples to follow the lightning current is expressed by the double exponential function (6.118) with $I_0 = 1.1043$ A, $a = 0.07924 \times 10^6\ \mathrm{s}^{-1}$, $b = 0.07924 \times 10^6\ \mathrm{s}^{-1}$ (the so-called 1/10 μs impulse).

Figure 6.66 shows the transient response of the basic grounding system. The dashed line represents a ten times higher input current waveform for comparison.

The maximal value of voltage is around 37 V and this value is reached slightly after current peak value. Transient impedance continuously increases from zero to the maximal value of 40 Ω being equivalent to the steady state condition.

As the transient impedance is relatively high, i.e. four times higher than proposed by the standards in the steady state, it follows that the actual grounding system does not meet the safety standards. Therefore, the original grounding system should be improved, which could be accomplished by adding horizontal and/or vertical electrodes.

To investigate the influence of additional horizontal electrodes, the WT grounding system has been upgraded with four 5 or 15 m long horizontal electrodes (Figure 6.67) placed at 2 m depth.

Figures 6.68 and 6.69 show the transient voltage induced at the injection point and the transient impedance, respectively, for different lengths of additional horizontal electrodes. A significant decrease in the maximal induced voltage and consequently the transient impedance can be observed with the increase of the horizontal electrode's length. Compared to the configuration without

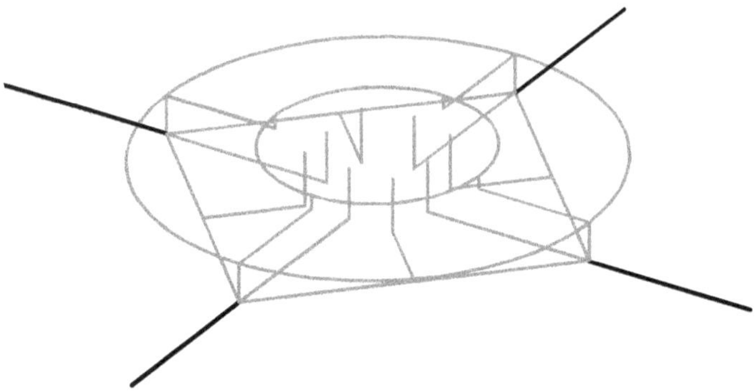

Figure 6.67 Additional horizontal electrodes on WT grounding system.

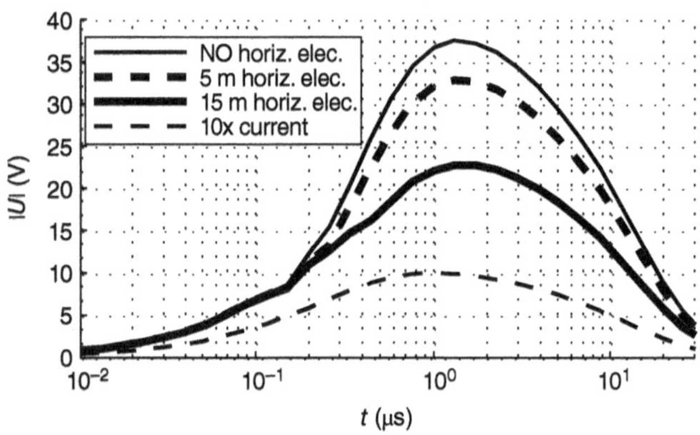


Figure 6.68 Induced feeding point transient voltage for different lengths of additional horizontal electrodes.

horizontal electrodes, additional four 5 m horizontal electrodes reduce maximal transient voltage by 13% while four 15 m electrodes reduce it by 40%. Similar behavior can be noticed for the transient impedance in Figure 6.69.

No difference in transient behavior occurs until 0.15 μs; therefore, the additional horizontal electrodes do not affect very early time behavior.

The influence of vertical electrodes on WT grounding system has been analyzed by investigating vertical electrodes of various length (3, 5.5, and 15 m).

Arrangement of vertical electrodes is shown on Figure 6.70.

The results given in Figures 6.71 and 6.72 show the influence of additional vertical electrodes of various lengths on the transient behavior of the WT grounding system. It is clear that additional, relatively short, vertical electrodes do not significantly decrease the value of the transient impedance with respect to the case without vertical electrodes (5% with four 3 m electrodes, and 12% with four 5.5 m electrodes).

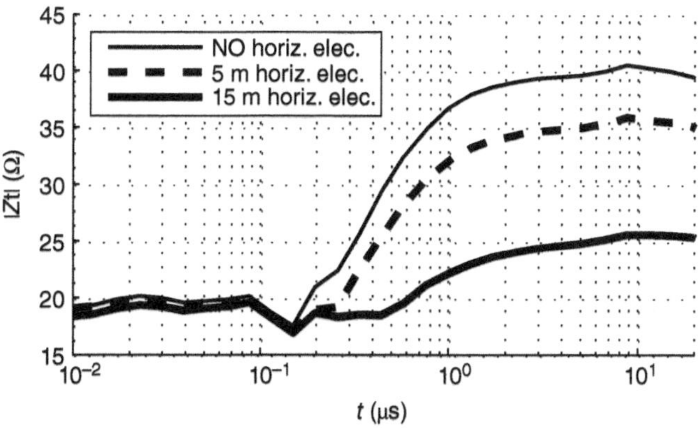


**Figure 6.69** Transient impedance for different lengths of additional horizontal electrodes.

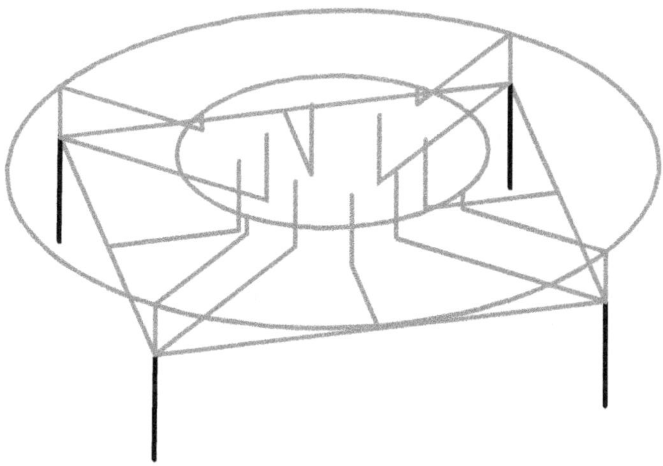

**Figure 6.70** Additional vertical electrodes on WT grounding system.

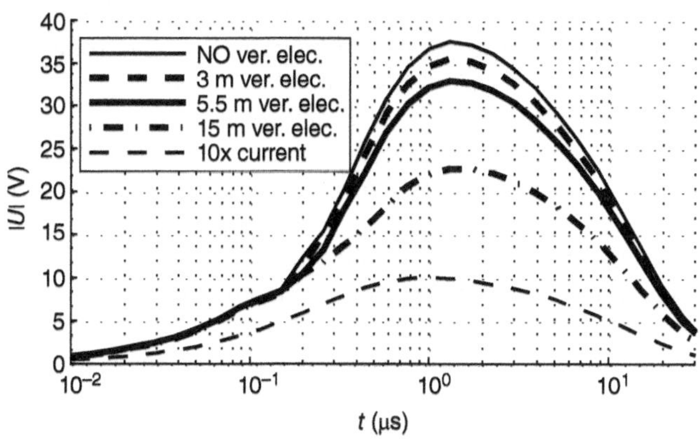


**Figure 6.71** Induced feeding point transient voltage for different lengths of additional vertical electrodes.

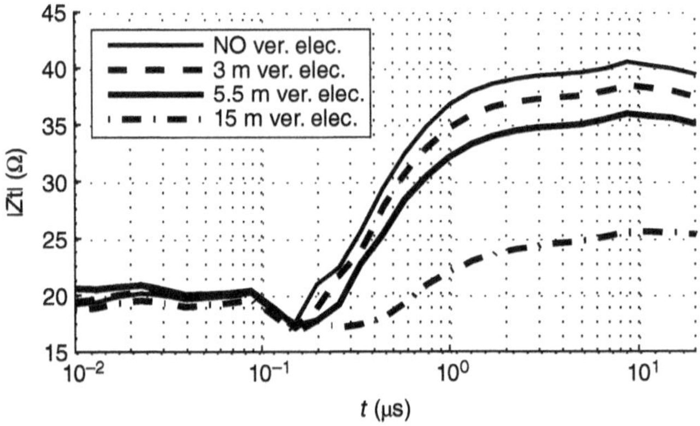


**Figure 6.72** Transient impedance for different lengths of additional vertical electrodes.

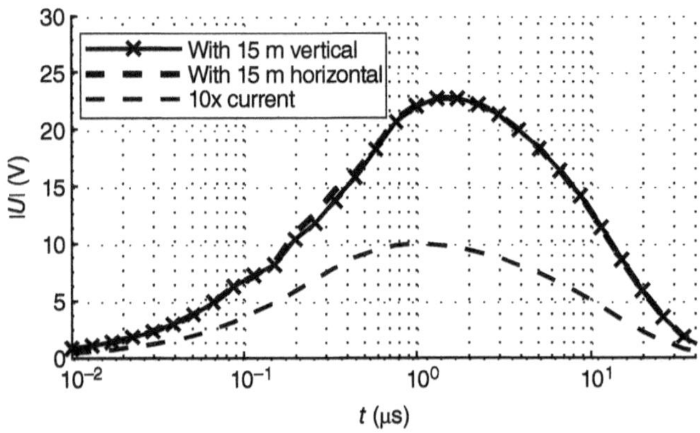


**Figure 6.73** Comparison of the induced feeding point transient voltage in the case of added 15 m horizontal or vertical electrodes, respectively.

On the other hand, in the case of long vertical electrodes such influence is significant (around 40% with respect to the case without vertical electrodes).

Figures 6.73 and 6.74 show the transient behavior of the grounding system with added four 15 m horizontal or vertical electrodes, respectively.

Comparing those two cases no significant difference can be found. This conclusion is important as installation of horizontal electrodes is appreciably cheaper than placing vertical ones, especially in a rocky terrain.

Therefore, the principal parameter that determines the overall grounding system performance is the electrode length. On the other hand, the installation of the vertical electrodes is plausible in the case of highly conductive earth layer under the main grounding system, which is rarely the case in the rocky terrain.

It is also important to analyze the impact of the grounding wire in cable trench on the behavior of grounding system in the impulse mode. The

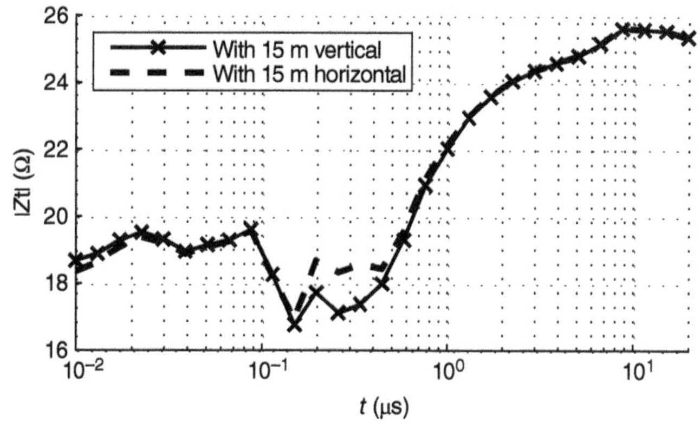


**Figure 6.74** Comparison of transient in the case of added 15 m horizontal or vertical electrodes, respectively.

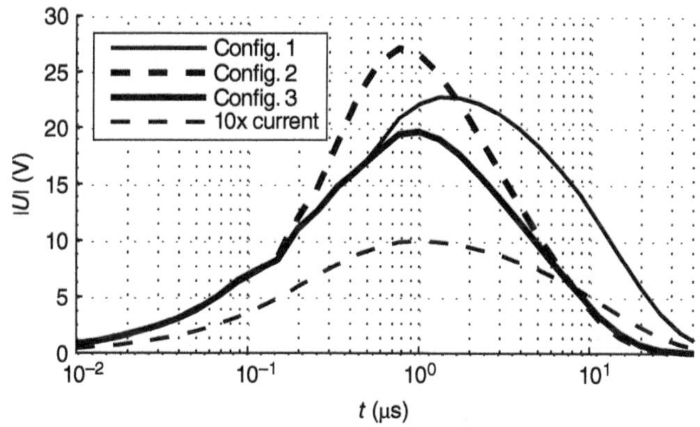


**Figure 6.75** Influence of grounding wire in a cable trench on the induced feeding point transient voltage.

numerical results are compared (Figures 6.75 and 6.76) for three different configurations:

- Grounding system with additional four 15 m horizontal electrodes without grounding wire in cable trench (Configuration No. 1).
- Grounding system without additional horizontal electrodes with 200 m grounding wire in cable trench (Configuration No. 2).
- Grounding system with additional three 15 m horizontal electrodes with 200 m grounding wire in cable trench (Configuration No. 3).

All configurations do not contain vertical electrodes. The grounding wire is modeled as a 200 m horizontal electrode.

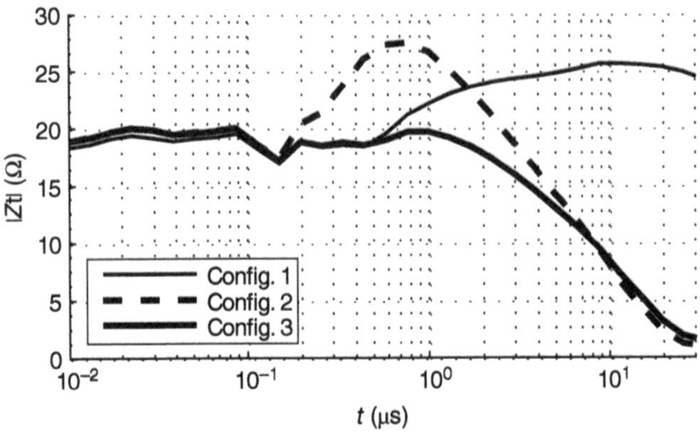


**Figure 6.76** Influence of the grounding wire in a cable trench on the transient impedance.

It is clear from the obtained numerical results that the addition of grounding wire in a cable trench reduces the grounding impedance in steady state. A significant increase in maximal induced voltage and related transient impedance is also evident in the case of configuration without additional horizontal electrodes.

Consequently, added long grounding wire in a cable trench significantly influences only the steady state behavior. Therefore, the configuration with both horizontal electrodes and grounding wire in cable trench seems to be optimal.

## 6.4 Time Domain Analysis of Horizontal Grounding Electrodes

A direct TD analysis of a horizontal grounding electrode is of continuous interest in studies of LPS. The proposed geometry is of interest as a simple grounding system itself in some engineering applications and it could be also useful as a benchmark for testing other solution techniques.

This section deals with a transient analysis of a horizontal grounding wire immersed in a lossy half-space by using a rigorous AT and an approximate TL approach, respectively. The AT formulation is based on the homogeneous space–time Pocklington's integro-differential equation. The presence of the earth–air interface is, as previously shown, taken into account via the simplified RC arising from the MIT. The TL approach is based on the corresponding set of space–time telegrapher's equations.

What follows is the numerical and analytical solution of Pocklington's equation while the telegrapher's equations have been numerically solved via the FDTD method.

The direct TD electromagnetic modeling of horizontal grounding electrode by means of the AT approach is presented in [14], [8]. The space–time Pocklington's integro-differential equation, arising from the AT approach, is numerically solved by using the GB-IBEM. The analytical solution has been reviewed in [14]. The obtained numerical results for the current distribution and scattered voltage induced along the horizontal grounding electrode agree satisfactorily with the results calculated via other solution methods.

### 6.4.1 Homogeneous Integral Equation Formulation in the Time Domain

The geometry of interest, shown in Figure 6.77, is related to the horizontal grounding electrode of length $L$ and radius $a$, buried in a lossy medium of permittivity $\varepsilon$ and conductivity $\sigma$ at depth $d$. As in the FD analysis, the electrode is energized at one end with an equivalent current source.

Direct TD analysis of the horizontal grounding electrode is based on the following homogeneous space–time Pocklington's integro-differential equation [14]:

$$\left(\frac{\partial^2}{\partial x^2} - \mu\sigma\frac{\partial}{\partial t} - \mu\varepsilon\frac{\partial^2}{\partial t^2}\right) \cdot \begin{bmatrix} \frac{\mu}{4\pi}\int_0^L I\left(x', t - \frac{R}{v}\right)\frac{\mathrm{e}^{-\frac{1}{\tau_g}\frac{R}{v}}}{R}\mathrm{d}x' \\ -\frac{\mu}{4\pi}\int_0^t \int_0^L \Gamma_{\mathrm{ref}}^{\mathrm{MIT}}(\tau) \\ I\left(x', t - \frac{R^*}{v} - \tau\right)\frac{\mathrm{e}^{-\frac{1}{\tau_g}\frac{R^*}{v}}}{R^*}\mathrm{d}x'\,\mathrm{d}\tau \end{bmatrix} = 0, \tag{6.119}$$

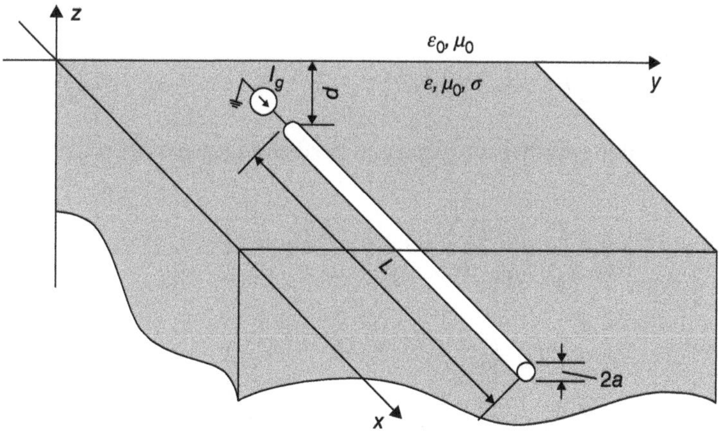


**Figure 6.77** Horizontal grounding electrode buried in a lossy medium.

where $I\left(x', t - \frac{R}{v}\right)$ is the space–time current along the electrode, $v$ is the propagation velocity in a lossy medium, and $\Gamma_{\text{ref}}^{\text{MIT}}$ is the RC given by [14]

$$\Gamma_{\text{ref}}(t) = -\left[\frac{\tau_1}{\tau_2}\delta(t) + \frac{1}{\tau_2}\left(1 - \frac{\tau_1}{\tau_2}\right)\mathrm{e}^{-t/\tau_2}\right], \tag{6.120}$$

while $R$ and $R^*$ are the distances from the source and image wire (in the wire axis), respectively to the observation point at the wire surface.

Time constants $\tau_g$, $\tau_1$ and $\tau_2$ are given by [14]

$$\tau_g = \frac{2\varepsilon}{\sigma}, \quad \tau_1 = \frac{\varepsilon_r - 1}{\sigma}\varepsilon_0, \quad \tau_2 = \frac{\varepsilon_r + 1}{\sigma}\varepsilon_0. \tag{6.121}$$

The ICS is included into the integral equation formulation in terms of the boundary condition

$$I(0, t) = I_g, \tag{6.122}$$

which is inserted subsequently in the global matrix system [8].

The initial conditions are related to the fact that the electrode is not excited before a certain instant, $t = 0$ [8].

### 6.4.2 Numerical Solution Procedure for Pocklington's Equation

The application of GB-IBEM to the solution of the Pocklington's equation suffers from numerical instabilities. The origin of these instabilities is the existence of the space–time differential operator [8].

For the sake of simplicity, this section analyzes only the case of an infinite lossy medium.

The space–time dependent current along the electrode can be expressed as follows:

$$I(x', t - R/v) = \sum_{i=1}^{N} I(t - R/v) f_i(x'). \tag{6.123}$$

Utilizing the weighted residual approach and performing space discretization yields

$$\sum_{i=1}^{N} I_i(t - \tau_{ij}) \left[\frac{\mu}{4\pi}\int_{\Delta l_j}\int_{\Delta l_i} \frac{\partial f_j(x)}{\partial x}\frac{\partial f_i(x')}{\partial x'}\frac{\mathrm{e}^{-\frac{\sigma}{2\varepsilon v}R}}{R}\mathrm{d}x'\,\mathrm{d}x + \right.$$
$$\frac{1}{v^2}\frac{\partial^2}{\partial t^2}\int_{\Delta l_j}\int_{\Delta l_i} f_j(x) f_i(x')\frac{\mathrm{e}^{-\frac{\sigma}{2\varepsilon v}R}}{R}\mathrm{d}x'\,\mathrm{d}x +$$
$$\left. + \frac{\sigma}{\varepsilon}\frac{\partial}{\partial t}\int_{\Delta l_j}\int_{\Delta l_i} f_j(x) f_i(x')\frac{\mathrm{e}^{-\frac{\sigma}{2\varepsilon v}R}}{R}\mathrm{d}x'\,\mathrm{d}x\right] = 0,$$
$$j = 1, 2, \ldots, N. \tag{6.124}$$

Performing the discretization in the TD the following set of TD differential equations is obtained:

$$[M]\frac{\partial^2}{\partial t^2}\{I(t')\} + [C]\frac{\partial}{\partial t}\{I(t')\} + [K]\{I(t')\} = 0. \tag{6.125}$$

The corresponding space dependent matrices are given by

$$M_{ji} = \frac{1}{v^2}\int_{\Delta l_j}\int_{\Delta l_i}\{f\}_j\{f\}_i^T\frac{e^{-\frac{T}{\tau}}}{R}\mathrm{d}x'\,\mathrm{d}x, \tag{6.126}$$

$$C_{ji} = \frac{\sigma}{\varepsilon}\int_{\Delta l_j}\int_{\Delta l_i}\{f\}_j\{f\}_i^T\frac{e^{-\frac{T}{\tau}}}{R}\mathrm{d}x'\,\mathrm{d}x, \tag{6.127}$$

$$K_{ji} = \frac{\mu}{4\pi}\int_{\Delta l_j}\int_{\Delta l_i}\{D\}_j\{D\}_i^T\frac{e^{-\frac{T}{\tau}}}{R}\mathrm{d}x'\,\mathrm{d}x, \tag{6.128}$$

where $\{D\}$ contains the shape functions derivatives and the corresponding time constants are given by $\tau = \frac{2\varepsilon}{\sigma}$ and $T = \frac{R}{v}$.

The set of differential equations (6.125) is solved by using the marching-on-in-time procedure [47]:

$$\sum_{i=1}^{n}[M_{ji} + \beta\Delta t^2 K_{ji}]I_i^k = -\sum_{i=1}^{n}\left[-2M_{ji} + \left(\frac{1}{2} - 2\beta + \gamma\right)\Delta t^2 K_{ji}\right]I_i^{k-1}$$
$$-\sum_{n=1}^{n}\left[M_{ji} + \left(\frac{1}{2} + \beta - \gamma\right)\Delta t^2 K_{ji}\right]I_i^{k-2}, \tag{6.129}$$

where $\Delta t$ is the time increment, and the stability of the procedure is ensured by choosing $\gamma = 1/2$ and $\beta = 1/4$ [8].

### 6.4.3 Numerical Results for Grounding Electrode

Computational example deals with the transient response of the electrode with length $L = 10$ m, radius $a = 5$ mm, immersed in the lossy ground with $\varepsilon_r = 10$, and $\sigma = 0.001\mathrm{S\,m^{-1}}$. The electrode is energized with the double exponential current pulse (6.118) with $I_0 = 1.1043$ A, $a = 0.07924 \times 10^7\ \mathrm{s^{-1}}$, $b = 4.0011 \times 10^7\ \mathrm{s^{-1}}$.

The transient current at the center of the electrode obtained via the direct TD approach and the indirect FD approach GB-IBEM with IFFT is shown in Figure 6.78.

Satisfactory agreement between the results obtained via different approaches can be observed.

### 6.4.4 Analytical Solution of Pocklington's Equation

Assuming the excitation function in terms of the equivalent current generator representing the lightning strike current (6.118) the space–time current flowing

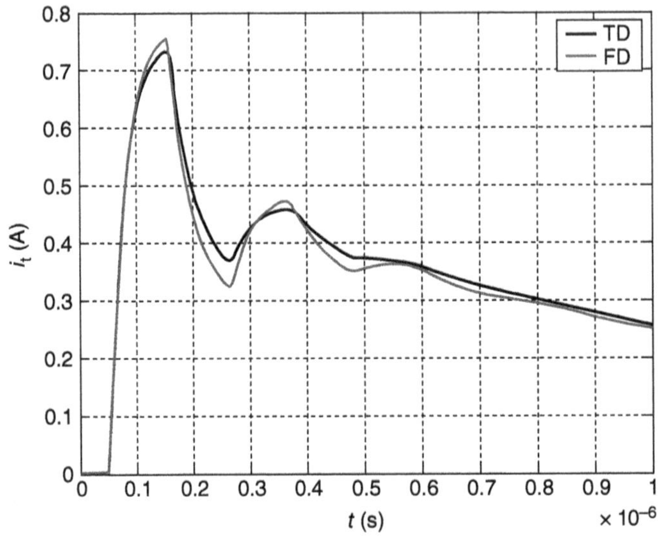


**Figure 6.78** Transient current induced at the center of the grounding electrode.

along the electrode, obtained by analytically solving (6.119), is given by [14]

$$I(x,t) = \frac{2\pi I_0}{\mu\varepsilon L^2}\sum_{n=1}^{\infty}\frac{(-1)^{n-1}n}{\pm\sqrt{b^2-4c_n}}$$
$$\sin\frac{n\pi(L-x)}{L}\cdot\left(\frac{e^{s_{1,2n}t}-e^{-\alpha t}}{s_{1,2n}+\alpha}-\frac{e^{s_{1,2n}t}-e^{-\beta t}}{s_{1,2n}+\beta}\right), \tag{6.130}$$

where

$$s_{1,2n} = \frac{1}{2}(-b\pm\sqrt{b^2-4c_n}),$$
$$b = \frac{\sigma}{\varepsilon},$$
$$c_n = \frac{n^2\pi^2}{\mu\varepsilon L^2},\quad n = 1,2,3,\ldots \tag{6.131}$$

More mathematical details regarding the analytical solution of (6.119) are available in [14].

### 6.4.5 Transmission Line Model

Using the framework of the TL theory, the current and voltage induced along the electrode can be obtained by solving the set of telegrapher's equations [14]:

$$\frac{\partial v(x,t)}{\partial x} + Ri(x,t) + L\frac{\partial i(x,t)}{\partial t} = 0, \tag{6.132}$$

$$\frac{\partial i(x,t)}{\partial x} + Gv(x,t) + C\frac{\partial v(x,t)}{\partial t} = 0, \tag{6.133}$$

where $v(x,t)$, $i(x,t)$ are induced voltage and current along the conductor, respectively, and $R$ is the per-unit-length series resistance. $L$, $G$, and $C$ are the effective per-unit-length inductance, conductance, and capacitance of the conductor, respectively.

The per-unit lines parameters of buried horizontal wires can be calculated using the following expressions [14]:

$$L = \frac{\mu_0}{2\pi}\ln\left(\frac{2l}{\sqrt{2ad}} - 1\right), \tag{6.134}$$

$$G = \frac{2\pi\sigma}{\ln\left(\frac{2l}{\sqrt{2ad}} - 1\right)}, \tag{6.135}$$

$$C = \frac{2\pi\varepsilon_0\varepsilon_r}{\ln\left(\frac{2l}{\sqrt{2ad}} - 1\right)}. \tag{6.136}$$

As in AT, the electrode is assumed to be PEC, i.e. $R = 0$. Note that $l$, $a$, and $d$ are the length, radius, and depth, respectively, of the buried horizontal electrode, while $\mu_0$, $\varepsilon_r$ and $\sigma$ are the permeability of air, the relative permittivity, and the conductivity of the ground, respectively.

The solution of telegrapher's equations (6.132) and (6.133) is carried out using the FDTD method and is outlined in [14].

### 6.4.6 FDTD Solution of Telegrapher's Equations

Discretizing the electrode into $N_s$ segments, with segment length $\Delta x$ and the time interval of interest into $nt$ sections, with increment $\Delta t$, the set of telegrapher's equations (6.136) and (6.137) become

$$v_k^{n+1} = D\left[E \cdot v_k^n - \left(i_k^{n+1/2} - i_{k-1}^{n+1/2}\right)\right], \quad \text{for } k = 2, 3, \dots, N, \tag{6.137}$$

$$i_k^{n+3/2} = A\left[B \cdot i_k^{n+1/2} - (v_{k+1}^{n+1} - v_k^{n+1})\right], \quad \text{for } k = 1, 2, \dots, N, \tag{6.138}$$

where

$$v_k^n = v[(k-1)\Delta x, n\Delta t], \qquad \text{for } n = 0, 1, \dots, nt, \tag{6.139}$$

$$i_k^n = i\left[(k - 1/2)\,\Delta x, n\Delta t\right], \quad \text{for } n = 0, 1, \dots, n, \tag{6.140}$$

and

$$A = \left(\frac{\Delta x}{\Delta t} l + \Delta x \frac{r}{2}\right)^{-1}, \tag{6.141}$$

$$B = \left(\frac{\Delta x}{\Delta t} l - \Delta x \frac{r}{2}\right)^{-1}, \tag{6.142}$$

$$D = \left(\frac{\Delta x}{\Delta t} c + \Delta x \frac{g}{2}\right)^{-1}, \tag{6.143}$$

$$E = \left(\frac{\Delta x}{\Delta t} c - \Delta x \frac{g}{2}\right)^{-1}. \tag{6.144}$$

Incorporating the appropriate boundary conditions for voltages at the wire ends, $v_1^n, v_{nx+1}^n$ are evaluated from the following expressions:

$$v_1^{n+1} = D\left[E \cdot v_1^n - 2i_1^{n+1/2} + (i_g^{n+1} - i_g^n)\right], \tag{6.145}$$

$$v_{nx+1}^{n+1} = D\left(E \cdot v_{nx+}^n + 2i_{nx}^{n+1/2}\right). \tag{6.146}$$

The current excitation $I_g^n$ at the time instant $t = n\Delta t$ is given by (6.118).

### 6.4.7 The Leakage Current

One of the parameters that can be useful in the analysis of the grounding electrode properties is the leakage current density [14].

The concept of leakage current density flowing radially from the electrode is depicted in Figure 6.79.

This current density can be evaluated as the product of the radial field component $E_\rho$

$$E_\rho = -\frac{1}{j\omega\varepsilon_{\text{eff}} \cdot 2\pi\rho} \frac{\partial I(z)}{\partial z}, \tag{6.147}$$

and the soil conductivity $\sigma$.

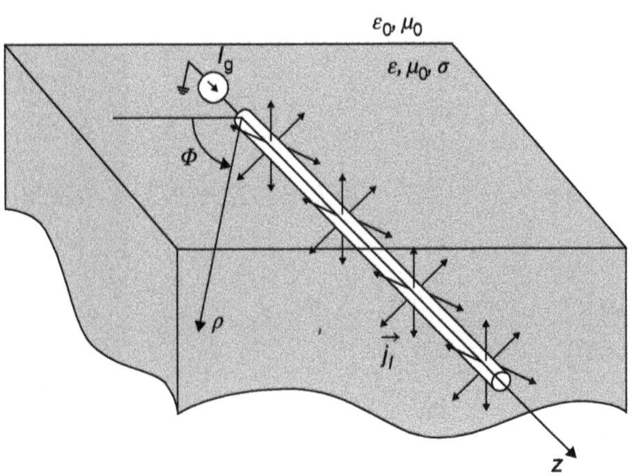


**Figure 6.79** The leakage current density.

Thus, the leakage current density is given by [14]

$$J_l = \sigma \cdot E_\rho = -\frac{\sigma}{j\omega\varepsilon_{\text{eff}} \cdot 2\pi\rho}\frac{\partial I(z)}{\partial z}. \tag{6.148}$$

The total current flowing out of the electrode can be obtained by integrating the current density along the cylinder:

$$I_{l,\text{tot}} = \int_S \vec{J}_l \cdot \text{d}\vec{S} = -\frac{\sigma}{j\omega\varepsilon_{\text{eff}} \cdot 2\pi}\int_0^{2\pi}\int_0^L \frac{1}{\rho}\frac{\partial I(z)}{\partial z}\rho \text{d}\phi \,\text{d}z, \tag{6.149}$$

and is given by

$$I_{l,\text{tot}} = \frac{\sigma}{j\omega\varepsilon_{\text{eff}}} I(0). \tag{6.150}$$

In the Laplace domain (6.150) becomes

$$I_{l,\text{tot}}(s) = \frac{1}{1+sT} I_{\text{g}}(s), \tag{6.151}$$

where $T = \varepsilon/\sigma$.

It is convenient to rewrite (6.151), as follows:

$$I_{l,\text{tot}}(s) = H(s) \cdot I_{\text{g}}(s), \tag{6.152}$$

where $H(s)$ can be written in the form

$$H(s) = \frac{\frac{1}{T}}{s + \frac{1}{T}}. \tag{6.153}$$

The product from (6.151) in the Laplace domain corresponds to the convolution in the TD:

$$i_{l,\text{tot}}(t) = \int_0^t h(t-\tau) i_{\text{g}}(\tau) \text{d}\tau, \tag{6.154}$$

where $h(t)$ is the TD counterpart of (6.153) given by

$$h(t) = \frac{1}{T}\text{e}^{-\frac{t}{T}}. \tag{6.155}$$

Inserting (6.118) and (6.155) into (6.154) yields

$$i_{l,\text{tot}}(t) = \int_0^t \frac{1}{T}\text{e}^{-\frac{t-\tau}{T}} I_0(\text{e}^{-\alpha\tau} - \text{e}^{-\beta\tau}) \text{d}\tau. \tag{6.156}$$

The solution of (6.156) is straightforward and can be written in the form

$$i_{l,\text{tot}}(t) = \frac{I_0}{T}\text{e}^{-\frac{t}{T}} \left\{ \frac{1}{\alpha - \frac{1}{T}}\left[1 - \text{e}^{-t\left(\alpha - \frac{1}{T}\right)}\right] - \frac{1}{\beta - \frac{1}{T}}\left[1 - \text{e}^{-t\left(\beta - \frac{1}{T}\right)}\right] \right\}. \tag{6.157}$$

### 6.4.8 Computational Examples for the Horizontal Grounding Electrode

All examples are related to the horizontal electrode of radius $a = 5$ mm, buried in a lossy half space with relative permittivity $\varepsilon_r = 10$ at depth $d = 0.5$ m. The grounding electrode is energized with the double exponential current source with $I_0 = 100$ kA. The transient current is determined by (6.130). The obtained analytical results are compared to the numerical results obtained by solving the telegrapher's equations (6.21) and (6.22) via FDTD method.

Figure 6.80 shows the transient response at the center of the 1 m long electrode buried in a lossy medium ($\sigma = 10\,\text{mS m}^{-1}$) and excited by the

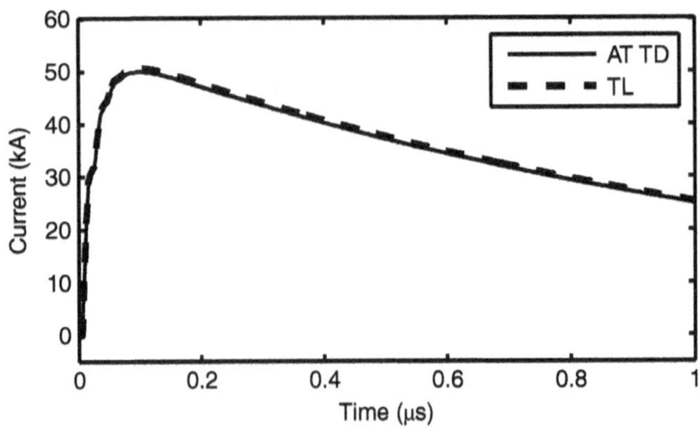


**Figure 6.80** Transient current at the center of the electrode, $L = 1$ m, $\sigma = 10$ mS m$^{-1}$. 0.1/1 μs pulse.

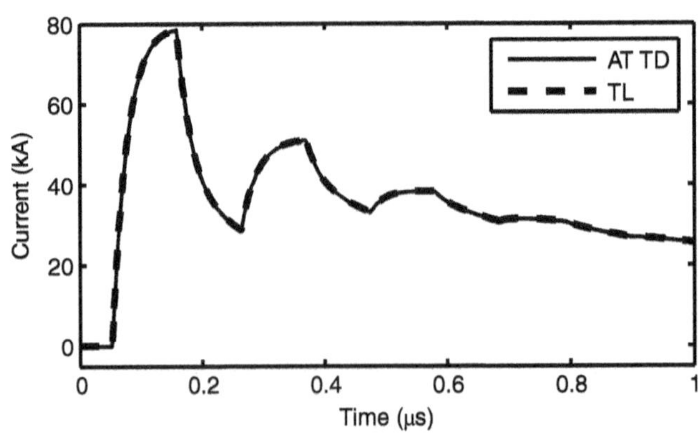


**Figure 6.81** Transient current at the center of the electrode, $L = 10$ m, $\sigma = 1$ mS m$^{-1}$. 0.1/1 μs pulse.

0.1/1 μs pulse defined by the following set of parameters: $\alpha = 0.07924 \times 10^7\ \mathrm{s}^{-1}$, $\beta = 4.0011 \times 10^7\ \mathrm{s}^{-1}$.

Figures 6.81 and 6.82 show the transient response at the center of the 10 m long electrode for the soil conductivity $\sigma = 1\ \mathrm{mS\,m^{-1}}$ and $\sigma = 10\ \mathrm{mS\,m^{-1}}$, respectively.

The transient response at the center of the 20 m long electrode buried in the lossy medium with the conductivity $\sigma = 1\ \mathrm{mS\,m^{-1}}$ is shown in Figure 6.83.

It can be observed that the results obtained by means of the AT and TL approach, respectively, agree rather satisfactorily.

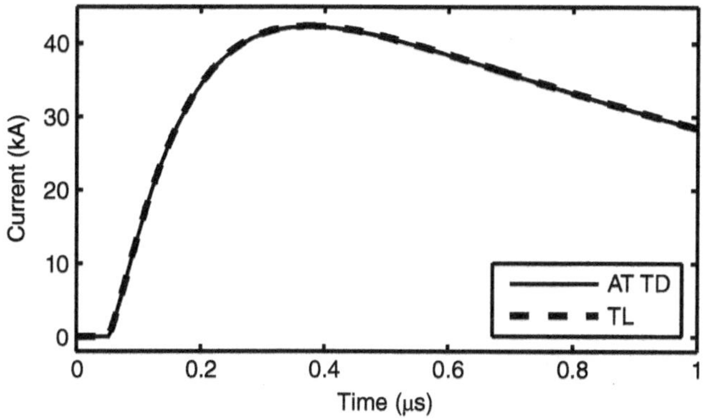


**Figure 6.82** Transient current at the center of the electrode, $L = 10$ m, $\sigma = 10\ \mathrm{mS\,m^{-1}}$. 0.1/1 μs pulse.

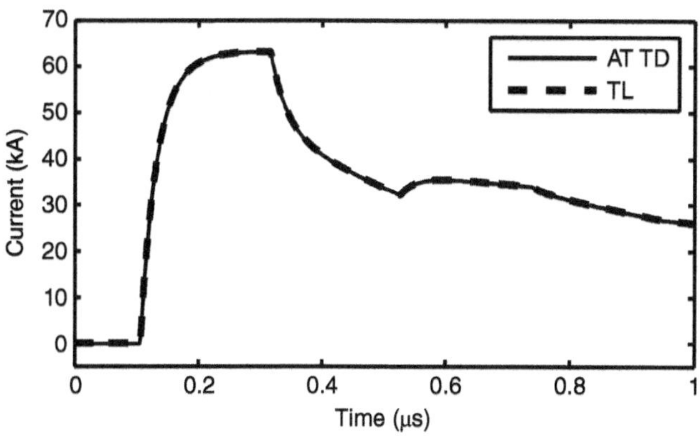


**Figure 6.83** Transient current at the center of the electrode, $L = 20$ m, $\sigma = 1\ \mathrm{mS\,m^{-1}}$. 0.1/1 μs pulse.

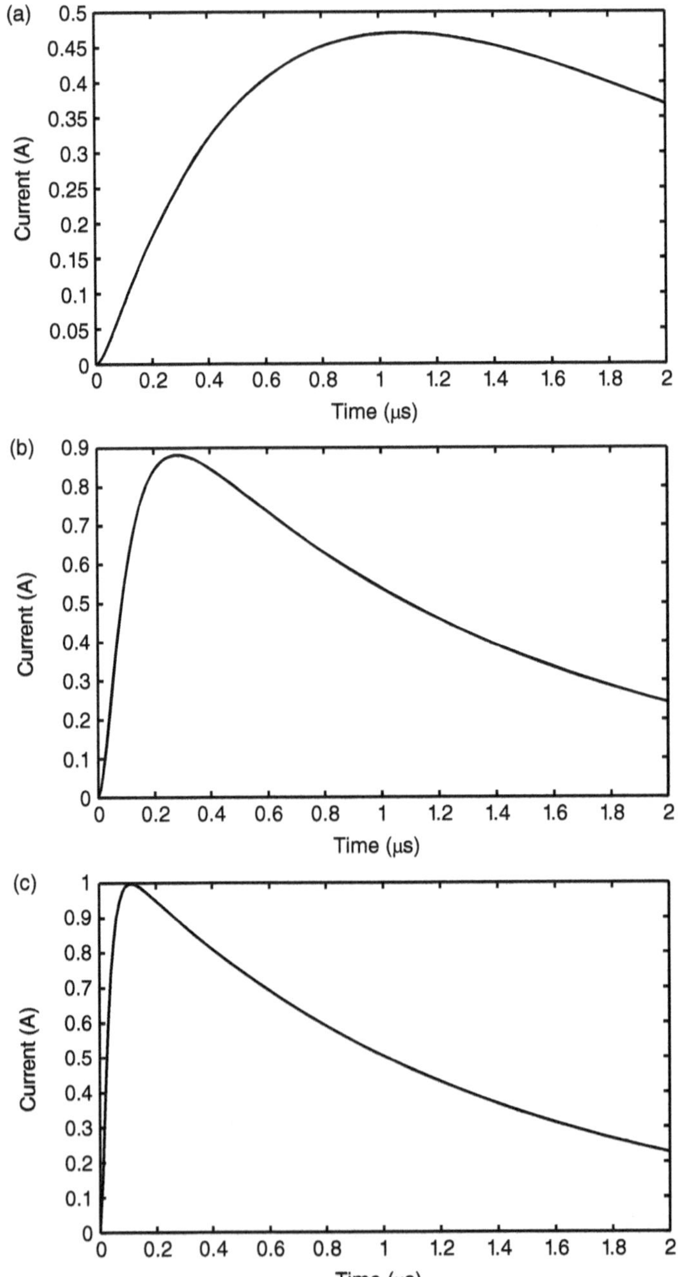


**Figure 6.84** Total leakage current versus time. (a) $\sigma = 0.1$ mS m$^{-1}$. (b) $\sigma = 1$ mS m$^{-1}$. (c) $\sigma = 10$ mS m$^{-1}$.

Therefore, it can be concluded that the approximations adopted within the analytical solution procedure of the integro-differential equation (6.119) to a certain extent correspond to the approximations adopted in the TL model itself.

Figure 6.84 shows a time-dependent total leakage current flowing out of the electrode in the normal direction, for the ground permittivity $\varepsilon_r = 10$ and different values of ground conductivity. The double exponential excitation (6.118) is related to the 0.1/1/ μs, 1 A pulse.

It is obvious that higher the conductivity the faster is the die-off of the total leakage current curve.

## References


1 Liu, Y., Zitnik, M., and Thottappillil, R. (2001). An improved transmission line model of grounding system. *IEEE Trans. EMC* **43** (3): 348–355.

2 Ala, G. and Di Silvestre, M.L. (2003). A simulation model for electromagnetic transients in lightning protection systems. *IEEE Trans. EMC* **44** (4): 539–534.

3 Lorentzou, M.I., Hatziargyriou, N.D., and Papadias, C. (2003). Time domain analysis of grounding electrodes impulse response. *IEEE Trans. Power Delivery* **2**: 517–524.

4 Grcev, L. and Dawalibi, F. (1990). An electromagnetic model for transients in grounding systems. *IEEE Trans. Power Delivery* **4**: 1773–1781.

5 Grcev, L.D. and Menter, F.E. (1996). Transient electro-magnetic fields near large earthing systems. *IEEE Trans. Magnetics* **32**: 1525–1528.

6 Poljak, D. and Roje, V. (1997). The Integral equation method for ground wire impedance. In: *Integral Methods in Science and Engineering*, vol. I (ed. C. Constanda, J. Saranen and S. Seikkala), 139–143. UK: Longman.

7 Tesche, F., Ianoz, M., and Carlsson, F. (1997). *EMC Analysis Methods and Computational Models.* New York: Wiley.

8 Poljak, D. (2007). *Advanced Modelling in Computational Electromagnetic Compatibility.* New York: Wiley.

9 Poljak, D. (2009). Generalized form of telegrapher's equations for the electromagnetic field coupling to buried wires of finite length. *IEEE Trans. EMC.*

10 Poljak, D., Rachidi, F., and Tkachenko, S. (2007). Generalized form of telegrapher's equations for the electromagnetic field coupling to finite length lines above a lossy ground. *IEEE Trans. EMC* **49** (3): 689–697.

11 Olsen, R.G. and Willis, M.C. (1996). A comparison of exact and Quasi-static methods for evaluating grounding systems at high frequencies. *IEEE Trans. Power Delivery* **11** (2): 1071–1081.

12 Poljak, D., Drissi, K., Kerroum, K., and Sesnić, S. (2011). Comparison of analytical and boundary element modeling of electromagnetic field coupling to overhead and buried wires. *Engineering Analysis with Boundary Elements* **35** (3): 555–563.

13 Poljak, D., Sesnic, S., and Goic, R. (2010). Analytical versus boundary element modelling of horizontal ground electrode. *Engineering Analysis with Boundary Elements* **34**: 307–314.

14 Poljak, D., Sesnić, S., El-Khamlichi Drissi, K. et al. (2016). Transient electromagnetic field coupling to buried thin wire configurations: antenna model versus transmission line approach in the time domain. *International Journal of Antennas and Propagation* 3943754-1–3943754-11.

15 Poljak, D. and Doric, V. (2006). Wire antenna model for transient analysis of simple grounding systems. Part II: the horizontal grounding electrode. *Progress in Electromagnetics Research* **64**: 167–189.

16 Poljak, D., Cavka, D., and Rachidi, F. (2017). On the use of magnetic current loop source model in lightning electromagnetics. URSI GASS, Montreal (August 2017).

17 Poljak, D., Sesnic, S., El-Khamlichi Drissi, K., and Kerroum, K. (2014). Transient response of grounding electrode using the wire antenna theory approach. In: *Conference Series: Materials Science and Engineering* (ed. V. Monebhurrun), 1–4. IOP Publishing 2011.

18 Poljak, D., Sesnić, S., Cavka, D., and El-Khamlichi Drissi, K. (2015). On the use of the vertical straight wire model in electromagnetics and related boundary element solution. *Engineering Analysis with Boundary Elements* **50**: 19–28.

19 Poljak, D. and Doric, V. (2006). Wire antenna model for transient analysis of simple grounding systems. Part I: the vertical grounding electrode. *Progress in Electromagnetics Research* **64**: 149–166.

20 Poljak, D., El-Khamlichi Drissi, K., and Nekhoul, B. (2013). Electromagnetic field coupling to arbitrary wire configurations buried in a lossy ground: a review of antenna model and transmission line approach. *International Journal of Computational Methods and Experimental Measurements* **1** (2): 142–163.

21 Poljak, D., Cavka, D., Nekhoul, B., et al. (2013). Transient voltage induced along the grounding system using the antenna theory approach. ICECOM 2013, Conference Proceedings.

22 Sunde, E.D. (1968). *Earth Conducting Effects in Transmission Systems*. New York, NY, Dover Publications, Inc.

23 Liu, Y., Theethayi, N., and Thottappillil, R. (2005). An engineering model for transient analysis of grounding system under lightning strikes: nonuniform transmission-line approach. *IEEE Trans. On Power Delivery* **20** (2): 722–730.

24 IEC International STANDARD IEC 61400-24 (2010). *Wind turbine generation system – 24: lightning protection*. Geneva: International Electro-Technical Commission.

25 IEA (1997). Recommended practices for wind turbine testing and evaluation, 9. Lightning Protection for Wind Turbine Installations. Ed. 1997.

26 IEE Professional Group S1 (1997). (New concepts in the generation, distribution and use of electrical energy): half-day colloquium on "Lightning protection of wind turbines", (1997, 11).

27 Sorensen, T., Sorensen, J.T., and Nielsen, H. (1998). Lightning damages to power generating wind turbines. Proceedings of 24th International Conference on Lightning Protection (ICLP98), 176–179.

28 McNiff, B. (2002). Wind Turbine Lightning Protection Project 1999–2001. NREL Subcontractor report. *SR-500-31115*.

29 Rachidi, F., Rubinstein, M., Montanya, J. et al. (2008). A review of current issues in lightning protection of new-generation wind turbine blades. *IEEE Transactions on Industrial Electronics* **55** (6): 2489–2496.

30 Yoh, Y., Toshiaki, F., and Toshiaki, U. (2007). How does ring earth electrode effect to wind turbine? 42nd International Universities Power Engineering Conference, UPEC 2007, 796–799.

31 Glushakow, B. (2007). Effective lightning protection for wind turbine generators. *IEEE Transactions on Energy Conversion* **22** (1): 214–222.

32 IEC International Standard IEC 62305-3 (2006). Protection against lightning – Part 3: Physical damage to structures and life hazard. Geneva: International Electro-Technical Commission.

33 Rakov, V.A. (2001). Transient response of a tall object to lightning. *IEEE Transactions on Electromagnetic Compatibility* **43**: 654–661.

34 Rakov, V.A. and Uman, M.A. (1998). Review and evaluation of lightning return stroke models including some aspects of their application. *IEEE Transactions on Electromagnetic Compatibility* **40**: 403–426.

35 Rachidi, F., Rakov, V.A., Nucci, C.A., and Bermudez, J.L. (2002). The effect of vertically-extended strike object on the distribution of current along the lightning channel. *Journal of Geophysical Research* **107** (D23): 4699.

36 Pavanello, D., Rachidi, F., Rakov, V.A. et al. (2007). Return stroke current profiles and electromagnetic fields associated with lightning strikes to tall towers: comparison of engineering models. *Journal of Electrostatics* **65**: 316–321.

37 Podgorski, S. and Landt, J.A. (1987). Three dimensional time domain modeling of lightning. *IEEE Transactions on Power Delivery* **2**: 931–938.

38 Petrache, E., Rachidi, F., Pavanello, D., et al. (2005). Lightning strikes to elevated structures: influence of grounding conditions on currents and electromagnetic fields. Presented at IEEE International Symposium on Electromagnetic Compatibility, Chicago.

**39** Petrache, E., Rachidi, F., Pavanello, D. et al. (2005). Influence of the finite ground conductivity on the transient response to lightning of a tower and its grounding. Presented at 28th General Assembly of International Union of Radio Science (URSI), New Delhi, India.

**40** Podgorski, S. and Landt, J.A. (1985). Numerical analysis of the lightning-CN tower interaction. Presented at 6th Symposium and Technical Exhibition on Electromagnetic Compatibility, Zurich, Switzerland.

**41** Baba, Y. and Ishii, M. (2001). Numerical electromagnetic field analysis of lightning current in tall structures. *IEEE Transactions on Power Delivery* **16**: 324–328.

**42** Kordi, B., Moini, R., Janischewskyj, W. et al. (2003). Application of the antenna theory model to a tall tower struck by lightning. *Journal of Geophysical Research* **108**.

**43** Meliopoulos, A.P. and Moharam, M.G. (1983). Transient analysis of grounding systems. *IEEE Transactions on Power Apparatus and Systems* **102** (2): 389–399.

**44** Ramamoorty, M., Narayanan, M.M.B., Parameswaran, S., and Mukhedkar, D. (1989). Transient performance of grounding grids. *IEEE Transactions on Power Delivery* **4** (4): 2053–2059.

**45** Lorentzou, M.I., Hatziargyriou, N.D., and Papadias, B.C. (2003). Time domain analysis of grounding electrodes impulse response. *IEEE Trans Power Delivery* (2): 517–524.

**46** Poljak, D., Doric, V., El Khamlichi Drissi, K. et al. (2008). Comparison of wire antenna and modified transmission line approach to the assessment of frequency response of horizontal grounding electrodes. *Engineering Analysis with Boundary Elements* **32** (8): 676–681.

**47** Poljak, D. (2016). Frequency domain and time domain response of the horizontal grounding electrode using the antenna theory approach. In: *Engineering Mathematics I Electromagnetics, Fluid Mechanics, Material Physics and Financial Engineering*, 1–12. Cham, Switzerland: Springer.

**48** Poljak, D. and Cavka, D. (2015). Electromagnetic compatibility issues of wind turbine analysis and design. *International Journal of Computational Methods and Experimental Measurements* **3** (3): 250–268.

**49** Grcev, L. and Heimbach, M. (1997). Frequency dependent and transient characteristics of substation grounding systems. *IEEE Trans. on Power Delivery* **12** (1): 172–178.

**50** Cavka, D., Harrat, B., Poljak, D. et al. (2011). Wire antenna versus modified transmission line approach to the transient analysis of grounding grid. *Engineering Analysis with Boundary Elements* **3**: 1101–1108.

**51** Hatziargvriou, N., Lorentzou, M., Cotton, I., and Jenkins, N. (1997). Wind farm earthing. Proceedings of IEE Half-Day Colloquium on Lightning Protection of Wind Turbines, No. 6.

52 Cotton, I. and Jenkins, N. (1997). The effects of lightning on structures and establishing the level of risk. Proceedings of IEE Half-Day Colloquium on Lightning Protection of Wind Turbines, No. 3.

53 Cotton, I. and Jenkins, N. (1999). Wind farm earthing. Proceedings of European Wind Energy Conference (EWEC1999), 725–728.

54 Lorentzou, M., Hatziargyriou, N., and Papadias, B.C. (2000). Analysis of wind turbine grounding systems. Proceeding of 10th Mediterranean Electrotechnical Conference (MELECON2000), Cyprus, 936-939.

55 Lewke, B., Krug, F., and Kindersberger, J. (2006). Risk of lightning strike to wind turbines for maintenance personnel inside the hub. Proceedings of 28th International Conference on Lightning Protection (ICLP2006), No.XI-9, Kanazawa.

56 Ukar, O. and Zamora, I. (2011). Wind farm grounding system design for transient currents. *Renewable Energy* **36**: 2004–2010. doi: 10.1016/j.renene.2010.12.026.

57 Yasuda, Y., Fuji, T., and Ueda, T. (2007). Transient analysis of ring earth electrode for wind turbine. Proceedings of European Wind Energy Conference (EWEC2007), No. BL3.212, Milan.

58 Muto, A., Suzuki, J., and Ueda, T. (2010). Performance comparison of wind turbine blade receptor for lightning protection. Proceedings of 30th International Conference on Lightning Protection (ICLP2010), No.9A-1263, Cagliari.

59 Yasuda, Y., Uno, N., Kobayashi, H., and Funabashi, T. (2008). Surge analysis on wind farm when winter lightning strikes. *IEEE Transactions on Energy Conversion* **23** (1): 257–262.

60 Kontargyri, V.T., Gonos, I.F., and Stathopulos, I.A. (2005). Frequency response of grounding systems for wind turbine generators. Proceedings of the 14th International Symposium on High-Voltage Engineering (ISH 2005) No.B-13, Beijing.

61 Elmghairbi, A., Haddad, A., and Griffiths, H. (2009). Potential rise and safety voltages of wind turbine earthing systems under transient conditions. Proceedings of 20th International Conference on Electricity Distribution (CIRED2009), 8–11.

# Part II

# Advanced Models in Bioelectromagnetics

# 7

# Human Exposure to Electromagnetic Fields – General Aspects

In the last half a century the presence of artificially created electromagnetic fields in the environment due to the tremendous growth of power grids, radio and television stations, radars, base stations, cellular phones, numerous domestic appliances, and appliances at workplaces has dramatically increased. There is also continuing public concern associated with the possible adverse health effects due to human exposure to these fields, particularly exposure to high voltage power lines and radiation from cellular base stations and mobile phones [1–3].

The electromagnetic spectrum extends from extremely low frequencies (ELF) and very low frequencies (VLF) to radio frequencies (RF), infrared radiation, visible light, ultraviolet (UV), X-rays, and $\gamma$-ray frequencies exceeding $10^{24}$ Hz. All types of electromagnetic radiation have the same physical properties such as interference, coherence, or polarization, but differ in terms of energy. Depending on the frequency, electromagnetic radiation is classified as either nonionizing or ionizing. The separation between ionizing and nonionizing radiation is at wavelengths around 10 nm in the far-UV region.

Nonionizing radiation is a general term for the part of electromagnetic spectrum with weak photon energy insufficient for breaking atomic bonds in the irradiated material. Natural sources of nonionizing radiation are the sun, distant radio stars, other cosmic sources, and terrestrial sources such as lightning. These sources are extremely weak, and with the tremendous growth of electricity applications, the density of artificial electromagnetic energy in the environment is much higher than natural levels.

Above $10^{17}$ Hz, ionizing radiation contains energy to physically change the molecules or atoms it strikes, changing them into charged particles, i.e. ions, being chemically more active than their electrically neutral forms. Chemical changes occurring in biological systems may be cumulative and detrimental or even fatal.

Nonionizing electromagnetic fields are split into two main categories: low frequencies (LF, up to ~30 kHz) and high frequencies (HF, from 30 kHz to 300 GHz). Above this frequency lie the infrared, visible light, ultraviolet, X-ray,

*Computational Methods in Electromagnetic Compatibility: Antenna Theory Approach versus Transmission line Models*, First Edition. Dragan Poljak and Khalil El Khamlichi Drissi.

and $\gamma$-ray spectra, following an ascending order of frequencies, which all belong to the ionizing radiation spectrum.

At ELF (up to 3 kHz), the wavelengths are very long (6000 km at 50 Hz and 5000 km at 60 Hz) so the fields are not radiating in nature. In other words, electric and magnetic fields could be analyzed separately. Since the wavelength of 50/60 Hz fields is appreciably higher than the relevant distances from the field source, the near-field terms, nonradiative in nature, are considerably larger than the radiative terms. While ELF fields are generally used for power utilities (transmission, distribution, and applications) and for strategic global communications with submarines submerged in conducting seawater, RF electromagnetic fields lie in the frequency range from 3 kHz to 300 GHz and are used for radio and television, radar, and other RF/microwave applications.

The LF fields may cause excitation of sensory, nerve, and muscle cells. Humans are particularly sensitive to HF electromagnetic fields as the body absorbs the radiated energy, and the related heating effects become dominant. The human body absorbs a great deal of energy at certain frequencies, since the body acts as an equivalent antenna if the dimensions of the body parts are comparable to the field wavelength. When the body size is half the wavelength, the resonant frequency is reached and a large amount of energy is absorbed from the field at frequencies between 30 and 300 MHz. Children have a higher resonant frequency than adults. HF fields generate the internal temperature rise and, it is either not noticed or noticed too late because the heat is basically sensed by the skin. Eventually, the temperature control mechanisms may be affected.

Some research has been undertaken with respect to other effects, such as irritation of the nervous system and long-term consequences. Nevertheless, the importance of these in connection with health and safety is not yet well established.

Protective measures include engineering and administrative controls, personal protection programs, and medical surveillance. As a fundamental step, engineering controls should be undertaken wherever possible to reduce device emissions of fields to acceptable levels.

## 7.1 Dosimetry

The fundamentals of interaction of electromagnetic fields with materials were established in the nineteenth century in the form of Maxwell's equations. However, the application of these basic laws of electromagnetics and Maxwell's equations to biological systems is an extremely difficult task due to their high complexity and multiple organization levels.

Since human experimentation in the high dose range or for long-term exposures is not possible, irradiation experiments can be performed only on phantoms, tissue probes, and animals. Theoretical models are required to interpret and confirm the experiment, develop an extrapolation process, and thereby establish safety guidelines and exposure limits for humans. The related mathematical complexity has led researchers to investigate simple models such as plane slabs, cylinders, homogeneous and layered spheres, and prolate spheroids. Spherical models are still being used to study the power deposition characteristics of the heads of humans and animals. Sophisticated numerical modeling is required to successfully predict distribution of internal fields [4–6]. Today, realistic computational models comprising of cubical cells mostly apply finite difference time domain (FDTD) methods, e.g. [7]. In some recent studies, the finite element method (FEM), e.g. [8], is considered to be a more accurate method than the FDTD, and a more sophisticated and versatile tool as well, particularly if the treatment of irregular or curved shape domains is of interest. Recent research has also demonstrated that the boundary element method (BEM), fast multipole techniques, and wavelet techniques can be used to reduce the computational task, e.g. [9].

### 7.1.1 Low Frequency Exposures

The induced currents and fields in human organs may give rise to thermal and nonthermal effects. When humans are exposed to LF radiation, the thermal effects seem to be negligible, and possible nonthermal effects may occur at the cellular level.

Knowledge of the current density induced inside the human body is the key to understanding the interaction of the human being with LF fields. The current density inside the human body can be induced due to an external electric or magnetic field. Internal current densities induced by external electric fields have axial character, whereas induced currents due to external magnetic fields form loops.

The internal current density $J$ due to an external electric field is defined by the constitutive equation

$$J = \sigma E, \tag{7.1}$$

where $\sigma$ is tissue conductivity and $E$ the corresponding internal electric field.

Since 2010, ICNIRP Guidelines [10] propose a basic restriction in terms of internal electric field instead of current density proposed by the 1998 ICNIRP Guidelines [11].

The current density induced in the human body due to an external ELF magnetic field forms circular loops, and this is given by Hand [2]:

$$J = \sigma \pi r f B, \tag{7.2}$$

where $B$ is the corresponding magnetic induction normal to the human body, $f$ the operating frequency of the external magnetic field, and $r$ the radius of the circular loop.

### 7.1.2 High Frequency Exposures

The key point in HF bioelectromagnetics is how much of electromagnetic energy is absorbed by a biological body where it is deposited, and which biochemical reactions are disturbed and electron transport flow interfered. The basic dosimetric quantity for HF fields is the specific absorption rate (SAR), which denotes the rate of energy $W$ absorbed by or dissipated in a unit mass of the body [10, 11]:

$$\mathrm{SAR} = \frac{\mathrm{d}P}{\mathrm{d}m} = \frac{\mathrm{d}}{\mathrm{d}m}\frac{\mathrm{d}W}{\mathrm{d}t} = C\frac{\mathrm{d}T}{\mathrm{d}t}, \tag{7.3}$$

expressed in watts per kilogram of tissue (W kg$^{-1}$), where $C$ is the specific heat capacity of tissue, $T$ the temperature, and $t$ the time.

In tissue, SAR is proportional to the square of the internal electric field:

$$\mathrm{SAR} = \frac{\mathrm{d}P}{\mathrm{d}m} = \frac{\mathrm{d}P}{\rho\,\mathrm{d}V} = \frac{\sigma}{\rho}|E|^2, \tag{7.4}$$

where $E$ is the root mean square value of the electric field, $\rho$ the tissue density, and $\sigma$ the tissue conductivity.

Thus, the localized SAR is directly related to the internal field, and the main task of dosimetry involves the assessment of the electric field distribution inside the biological body.

The distribution of SAR can be calculated using computational models or estimated from laboratory measurements. Generally, SAR values depend on the incident field parameters, the characteristics of the exposed body, ground effects, and reflector effects. Namely, the effects of reflection, transmission, and attenuation of electromagnetic waves in the presence of finitely conducting objects have to be taken into account.

When the electric field is oriented parallel to the long axis of the human body, the whole-body SAR reaches maximal values.

On the basis of the evaluated electromagnetic energy absorbed by the human body due to the HF radiation exposure, a corresponding thermal response of the body can be determined.

## 7.2 Coupling Mechanisms

The interaction of electromagnetic fields with biological materials is studied through microscopic or macroscopic models. Since the analysis of interaction on a macroscopic level with charges in the material is very difficult, at LF this process is usually described macroscopically through the polarization of bound charges, the orientation of permanent electric dipoles, and the drift of conduction charges.

In the RF range, biological tissues behave like solutions of electrolytes containing polar molecules, that is, RF fields interact with living organisms via ionic conduction and rotation of polar molecules of water and protein relaxation. Absorbed RF energy is then transformed into kinetic energy of molecules, which is directly associated with a temperature rise of the irradiated tissue.

There are three established basic coupling mechanisms through which time-varying electromagnetic fields interact with the biological body:

- Coupling to LF electric fields
- Coupling to LF magnetic fields
- Absorption of energy from electromagnetic radiation

### 7.2.1 Coupling to LF Electric Fields

The interaction of LF electric fields with humans results in electric current, formation of electrical dipoles, and the reorientation of the already presented electric dipoles in tissue. The intensity of these effects depends on the electrical properties of the body that vary with the type of tissue and also on the frequency of the applied field. External electric fields induce a shift of surface charges on the body resulting in induced currents in the body, the distribution of which varies with the size and shape of the body.

### 7.2.2 Coupling to LF Magnetic Fields

The interaction of an LF magnetic field with the human body results in induced electric fields and currents flowing in circular loops inside the body. The magnitudes of the induced field and the current density are proportional to the loop radius, the tissue conductivity, and the rate of change and magnitude of the magnetic flux density. For a specified magnitude and frequency of magnetic field, the strongest electric fields are induced where the loop dimensions are greatest. The path and the magnitude of the current induced in any part of the body depend on the tissue conductivity.

### 7.2.3 Absorption of Energy from Electromagnetic Radiation

Although exposures to LF electric and magnetic fields result in negligible energy absorption and no measurable temperature rise in the human body, exposure to electromagnetic radiation at frequencies above around 100 kHz can lead to significant absorption of energy and consequently temperature increases. Generally, exposure to a plane wave electromagnetic field can result in a highly nonuniform deposition and distribution of the energy within the body, which has to be determined by computational dosimetry and measurement procedures. Regarding energy absorption by the human body, electromagnetic fields are divided into four different ranges:

- Frequencies from approximately 100 kHz to less than approximately 20 MHz, at which absorption in the trunk decreases rapidly with decreasing frequency, and significant absorption may occur in the neck and legs.
- Frequencies in the range from approximately 20 to 300 MHz, at which relatively high absorption can occur in the whole body, and to even higher values if partial body (e.g., head) resonances are considered.
- Frequencies in the range from approximately 300 MHz to several GHz, at which significant local, nonuniform absorption occurs.
- Frequencies above approximately 10 GHz, at which energy absorption occurs primarily at the body surface.

The amount of energy absorbed depends on a number of factors, including the size of the exposed body. The resonant absorption frequency of the ungrounded human body is around 70 MHz. For taller persons, the resonant frequency is somewhat lower. However, for shorter adults, children, babies, and seated persons, it could be around 100 MHz. For grounded humans, resonant frequencies are lower by a factor of about 2.

The near-field exposures can lead to a high local SAR in the head, wrists, and ankles. The local SAR and whole-body SAR strongly depend on the separation distance between the radiation source and the body.

At frequencies above approximately 10 GHz, the depth of the field penetration into tissues is small, and SAR is not a convenient measure for determining the energy absorption in the body; the incident power density of the field is a more appropriate parameter.

### 7.2.4 Indirect Coupling Mechanisms

There are two established coupling mechanisms:

- Contact currents that appear when the human body comes into contact with an object at a different electric potential.
- Coupling of electromagnetic field to medical devices worn by, or implanted in, a person.

## 7.3 Biological Effects

Biological effects occur when exposures to electromagnetic fields cause any noticeable or detectable physiological change in a biological body [10, 11]. Sometimes, such effects may result in physiological change exceeding a normal range for a brief period of time, that is, in adverse health effects. Such health conditions occur when the biological effect lies outside the normal range for the human body to compensate.

These adverse health effects are often the result of accumulated biological effects over time and depend on exposure dose.

### 7.3.1 Effects of ELF Fields

Exposure limits of ELF fields are mostly expressed in terms of magnetic fields, as electric fields are greatly diminished by many orders of magnitude inside biological tissues from their values in air external to the tissues. This is due to the boundary conditions arising from the electromagnetic theory that require internal current density to be approximately equal to the displacement current density outside the body. However, the magnetic field inside the body is the same as the field outside the body since biological tissues are nonmagnetic materials. The electric fields induced inside the human bodies are generally less than approximately $10^{-7}$ times field outside the body and rarely exceed approximately $10^{-4}$ times the external field.

Within the ELF range, biological materials can be regarded as conducting. A possible effect of electromagnetic fields on living systems has been theorized to involve the ability, through magnetic induction, to stimulate eddy currents at cell membranes and in tissue fluids, which circulate in a closed loop that lies in a plane normal to the direction of the magnetic field. Accurate calculation of this induced current and related field inside the body is only possible using numerical simulation.

Provided the properties of the biological system are constant, the induced current is directly proportional to the frequency of the applied field. Although at very high frequencies (above 100 kHz), heat generated due to induced currents can cause thermal damage to the exposed biological tissue, at ELF fields, heating of tissue is insignificant.

However, if the level of the induced current is too high, there is a risk of stimulating electrically excitable cells such as neurons or muscle cells. At frequencies below approximately 100 kHz, currents necessary to heat biological systems are greater than currents necessary to stimulate neurons and other electrically excitable cells.

Some of the experimental findings moderately support a causal relationship between ELF environmental levels and changes in biological function. Furthermore, there is evidence for a relationship between ELF fields and cancer, in particular childhood leukemia. The International Agency for Research in Cancer (IARC) evaluation of ELF fields, published in June 2002, classified power frequency fields as possibly carcinogenic.

### 7.3.2 Effects of HF Radiation

The interaction of HF fields with living systems and their related bioeffects can be considered at various levels, including the molecular, subcellular, organ, or system level, or the entire body. SAR, a measure of radiation power converted to heat, is used as the basis for setting limiting values. The parts of the human body are split into three sections regarding absorption: low, medium, and high absorption regions [1]. The degree to which parts of the body are vulnerable to heat from HF field depends on the circulation and on the ability of heat conduction. The lens of the eye and the kneecap are the most susceptible parts. On the contrary, the heart, lungs, and skin are relatively insensitive, due to the high blood circulation.

For the case of modulated field, such as GSM, cell irritation may occur in addition to the thermal effects. There is increasing evidence that pulsed radiation (such as radar or digital mobile radio telephones) has a greater effect on biosystems than nonpulsed signals. Bioeffects due to the RF radiation can be classified as

- high level (thermal) effects;
- intermediate level (athermal) effects;
- low level (nonthermal) effects.

An obvious consequence of HF energy absorbed by the body is heating where the core temperature of the body rises a few degrees despite the process of thermoregulation of the body. Therefore, potential harmful thermal effects can be defined as energy deposition higher than the thermoregulatory capacity of the human body. In other words, thermally harmful effects can occur if the total power absorbed by the body is large enough to cause the protective mechanisms for heat control to break down, resulting in uncontrolled rise in the body temperature (hyperthermia).

However, there is a controversy associated with biological effects of intermediate and low level HF radiation: first, whether RF radiation at such levels can cause harmful effects in the absence of demonstrable thermal effects and, second, whether effects can occur from HF radiation when thermoregulation maintains the temperature of the body at the normal level despite the HF energy deposition, or when thermoregulation is not challenged and there is no significant temperature change.

Biological systems alter their functions as a result of temperature change. The most adverse health effects due to HF exposure between 1 MHz and 10 GHz are associated with responses to induced heating, which results in a temperature rise in the tissue higher than 1 °C.

The human body generates heat from metabolism. The basal metabolic rate (BMR) is defined as the heat production of a human in a thermoneutral environment (33 °C) at mental and physical rest more than 12 h after the last meal. The standard basal metabolic rate for a 70 kg man is approximately 1.2 W kg$^{-1}$, but it can be altered by changes in active body mass, diets, and endocrine levels.

When humans are exposed to heating from an external thermal source at a much greater rate, thermal damage can occur. However, exposure levels comparable to the BMR might produce thermal effects due to the induction of thermoregulation.

Thermal effects imposed on the body by a given SAR are significantly affected by ambient temperature, relative humidity, and airflow. The human body attempts to regulate temperature increase due to the thermal effect through perspiration and heat exchange via blood circulation.

Certain areas with limited blood circulatory ability, such as the lens of the eye and the testes, run a particularly high risk of being damaged.

Other thermal effects may show up around electrically conducting objects, either implanted (nails, screws, artificial hip joints, etc.) or external (watches, bows of spectacles, etc.). For adverse health effects, such as eye cataracts and skin burns, to occur from exposure to RF fields at HF, power densities above 1000 W $m^{-2}$ are needed. Such densities exist in close proximity to powerful transmitters such as radars.

Furthermore, during some exposures (e.g., when cell phones are used), a nonuniform distribution of absorbed RF power is possible, that is, only certain parts of the body absorb RF power, which results in nonuniform heating. These points in the body are usually referred to as hot spots. Localized temperatures above 41.6 °C may cause protein denaturation, coagulation of protein, increased permeability of cell membranes, or the liberation of toxins in the immediate vicinity where the hot spots exists.

Moreover, the severity of the physiologic effects produced by a localized temperature increase can be expected to be appreciably enhanced in critical organs such as the brain or heart.

With the recent widespread use of the mobile phones, there is growing public concern regarding the possible health hazards due to the temperature rise in the head caused by the HF near-field exposure.

It is generally accepted that, in common use, approximately half of the cell-phone-radiated electromagnetic energy is absorbed by tissues on the head side closest to the handset.

Some studies indicate that the steady state has been achieved after approximately 30 min of exposure and that a call duration of 3 min causes a temperature rise over 60% of the steady state value. A call duration of 6–7 min yields a temperature rise of approximately 90% of the steady state value. It is known that the thermal time constant for humans is above 6 min; that is, for all of the tissues, the thermal time constants were found to be between 6 and 8 min.

If the field intensity increases, the human body can compensate these effects by its defense mechanisms if the time interval is short enough. For example, if the considered biological effect is tissue heating, then the defense response of the body is a thermoregulatory reaction (sweating and increase of blood flow).

Calculations revealed a deeper penetration into the brain and a 50% increase of the 1 g averaged spatial SAR in a model of the head of a 5-year-old child as compared to adults, whereas other studies reported no significant differences in the absorption between adults and children. At present, dosimetry for children is not satisfactorily clarified, due to the lack of a database of the electric and thermal properties. However, with the rapidly increasing use of cellular phones by children, a more detailed dosimetry for children is considered to be an urgent issue.

The hypothalamus, located below the thalamus as part of the diencephalon, is considered to be the central part of the body temperature regulation system. A temperature rise exceeding a threshold of 0.3 °C for the hypothalamus would result in certain temperature regulation actions such as extensions of blood vessels in the skin, a blood flow increase, and evaporation by sweat.

An important biological effect of cell phone radiation is related to the permeability of the blood–brain barrier (BBB). The BBB maintains the physiochemical microenvironment of cells and neurons and of the brain within strict narrow limits, which are essential for life. When high SAR values are induced due to HF radiation, some studies have confirmed that BBB permeability increases for substances normally excluded from the brain parenchyma.

Some researchers have reported that exposure to cell phone radiation can affect mental processes such as attention, short-term memory, information manipulation, or response-reaction times. The propagation velocity of nerve impulses is known to increase or decrease due to small temperature increases: 0.3–0.6 °C. Therefore, the increase in responsiveness or decrease in choice reaction time of humans is found to be consistent with the effects of mild localized heating of the underlying nervous tissue.

Some results implied that pulse-modulated radiation from cell phones may reduce sleep onset latency and modify the sleep EEG.

So far, nobody knows for sure what the long-term effect of RF radiation is and if it is cumulative in nature. However, if cumulative effects exist, these would be of paramount importance for health effects. The long-term cumulative exposure is the product of duration and mean exposure.

It is worth noting that to reliably estimate cancer risks, human epidemiology studies require an observation period of at least 15–20 years.

To sum up, HF biological effects are proved to be hazardous only if the radiation intensity is rather high. In the case of most environmental RF exposures, particularly radio base station antennas and cellular phones, the intensity is usually not very high; that is, it does not exceed the adopted exposure limits.

## 7.4 Safety Guidelines and Exposure Limits

Safety guidelines for exposure to electromagnetic fields rely on well-established effects based on experimental data from biological systems, on epidemiological and human studies, as well as on understanding of the various mechanisms of interaction.

A safety limit is considered as a threshold below which exposure is safe according to the available scientific knowledge. Nevertheless, the safety limit is not an exact demarcation between safety and hazard, and the possible risk to human health increases with higher exposure levels.

For efficient protection against the risk of harmful effects of exposure, regulatory agencies, in addition to setting safety limits, need to incorporate a safety margin to allow for uncertainty.

Another critical issue refers to an international effort to secure various standard-setting bodies, health agencies, governments, and international organizations to cooperate on safety standard development. It does not necessarily mean that the world should have only one accepted standard, but it does mean that the reasons for the existing differences should be made available.

The United States, Canada, several countries of the EU, Russia, Australia, and many Asian countries as well as international organizations have issued radiation protection guidelines. The reason for the large number of guidelines is the manner in which they are defined, for example, by frequency, exposure duration, and periodicity of exposure and other factors.

The International Commission on Non-Ionizing Radiation Protection (ICNIRP), an independent scientific commission, approved radiation protection guidelines for limiting exposure to electric and magnetic fields, current density, SAR, and power density up to 300 GHz.

Exposure limits are based on the established short-term adverse effects: stimulation of electrically excitable cells in muscle and nervous tissue and heating of the tissue.

Limiting values of the induced current density in the body $J$ (A m$^{-2}$), power density $S$ (W m$^{-3}$), and SAR (W kg$^{-1}$) are the basic restrictions established for interaction of the human body with electromagnetic radiation. This means that adverse effects are directly related to these quantities, and all other quantities (reference levels) are to be derived from basic restrictions. The reference levels are provided for practical exposure assessment purposes to estimate if the basic restrictions are not exceeded. The derived quantities are *electric field strength* $E$ (V m$^{-1}$), *magnetic field strength* $H$ (A m$^{-1}$) or flux density $B$ (T), and *power density* $S$ (W m$^{-2}$).

The fundamental quest in protection of humans exposed to electromagnetic radiation is compliance with the given basic restrictions. If it is not feasible to calculate or measure quantities associated with the basic restrictions, then comparison with reference levels can be used for the compliance and safety tests.

If the reference levels are exceeded, it does not necessarily mean that the basic restrictions are exceeded as well. However, when the reference levels are

**Table 7.1** Basic restrictions for SAR according to the ICNIRP guidelines.

| Frequency range: 10 MHz–10 GHz | Whole body SAR (W kg$^{-1}$) | Localized SAR (head and trunk) (W kg$^{-1}$) | Localized SAR (limbs) (W kg$^{-1}$) |
|---|---|---|---|
| Workers | 0.4 | 10 | 20 |
| Public | 0.08 | 2 | 4 |

exceeded, it is necessary to test compliance with the relevant basic restrictions and to figure out if additional protective measures are necessary.

Exposure limits are defined for two categories: occupationally exposed population and general public. The basic restrictions for whole-body average SAR and localized SAR for the frequency range between 10 MHz and 10 GHz are presented in Table 7.1 [11].

The ICNIRP guidelines for the general public and the occupationally exposed population differ by a factor of 5 for SAR levels. The reason for this approach was the possibility that some members of the general public might be exceptionally sensitive to RF radiation.

After establishing the limits for induced current density and SAR, the values of corresponding incident fields, reference levels, which are necessary to reach such basic restrictions, are determined by mathematical modeling and by extrapolation from the results of laboratory investigations at specific frequencies.

The human body is most sensitive in the range between 10 and 400 MHz for whole-body exposure, which implies the most rigorous exposure limits within this range of frequencies. In addition to the knowledge of the field strength at a specific frequency, important parameters for the estimation of potential adverse effects are exposure period, radiation type (continuous or modulated), and simultaneous exposure to multiple frequency fields.

For pulse-modulated sources, beside time averaging, it is necessary to check the peak value for short pulse durations. For example, according to the guidelines for frequencies exceeding 10 MHz, it is suggested to limit peaks to 1000 times the corresponding limit value of the power density at the specific frequency.

Regarding thermal effects, application of the ANSI/IEEE safety guidelines [12], restricting the 1 g averaged spatial peak SAR to 1.6 W kg$^{-1}$, results in a maximum temperature rise of 0.06 °C in the brain, and application of the ICNIRP/Japan Safety guidelines, restricting the 10 g averaged spatial peak SAR to 2 W kg$^{-1}$, results in a maximum temperature rise in the brain of 0.11 °C. These values have a margin of 17 for the ANSI/IEEE safety guidelines and 9 for ICNIRP/Japan safety guidelines, in comparison with the temperature rise of 1 °C, at which thermal effects will probably occur.

IEEE standards contain some of the characteristics of the current ICNIRP guidelines, but also differ in certain aspects.

## 7.5 Some Remarks

The presence of electromagnetic fields in the environment and their potential hazard to humans represent a controversial scientific, technical, and often public issue.

The starting point in the analysis of possible health risk is the incident field dosimetry (including the evaluation of incident fields from various electromagnetic sources in the near- and far-field regions) and internal field dosimetry (including the various techniques for the determination of internal electromagnetic field).

The investigation of biological effects integrates several aspects of electromagnetic fields such as the biological, medical, biochemical, epidemiological, and environmental aspects, risk assessment, and health policy. Research groups that can fully grasp all of these subjects are still rare.

Most of the recent research results have suggested that the determined field levels in the environment due to ELF sources (power lines, substations) and HF sources (base stations) are in principle lower than internationally established limits of exposure.

## References

1 Habash, R.W.Y. (2002). *Electromagnetic Fields and Radiation*. New York: Marcel Dekker.

2 Hand, J.W. (2008). Modeling the interaction of electromagnetic fields (10 MHz–10 GHz) with the human body: methods and applications. *Physics in Medicine and Biology* **53** (16): 243–286.

3 Poljak, D. (2011). Electromagnetic fields: environmental exposure. In: *Encyclopedia of Environmental Health*, vol. **2** (ed. J.O. Nriagu), 259–268. Burlington: Elsevier.

4 Nielsen, N.F., Michelsen, J., Michalsen, J.A., and Schneider, T. (1996). Numerical calculation of electrostatic field surrounding a human head in visual display environments. *Journal of Electrostatics* **36**: 209–223.

5 Gonzalez, C., Peratta, A., and Poljak, D. (2007). Boundary element modelling of the realistic human body exposed to extremely-low-frequency (ELF) electric fields: computational and geometrical aspects. *IEEE Transactions on Electromagnetic Compatibility* **49**: 153–162.

6 Singh, K.D., Longan, N.S., and Gilmartin, B. (2006). Three dimensional modeling of the human eye based on magnetic resonance imaging. *Investigative Opthamology and Visual Science* **47**: 2272–2279.

7 Fujimoto, M., Hirata, A., Wang, J. et al. (2006). FDTD-derived correlation of maximum temperature increase and peak SAR in child and adult head models due to dipole antenna. *IEEE Transactions on Electromagnetic Compatibility* **48** (1): 240–247.
8 Cavka, D., Poljak, D., and Peratta, A. (2011). Comparison between finite and boundary element methods for analysis of electrostatic field around human head generated by video display units. *Journal of Communications Software and Systems* 7: 22–28.
9 Poljak, D., Čavka, D., Dodig, H. et al. (2014). On the se of boundary element analysis in bioelectromagnetics. *Engineering Analysis with Boundary Elements*, (Special issue on Bioelectromagnetics) **49**: 2–14.
10 International Commission on Non-Ionizing Radiation Protection (2010). Guidelines for limiting exposure to time-varying electric and magnetic fields (1 HZ – 100 kHZ). *Health Physics* **99** (6): 818–836.
11 International Commission on Non-Ionizing Radiation Protection (1998). Guidelines for limiting exposure to time-varying electric, magnetic, and electromagnetic fields (up to 300 GHz). *Health Physics* **74** (4): 494–522.
12 American National standards Institute (ANSI). 1992 *Safety levels with respect to human exposure to radio frequency electromagnetic fields, 3kHz to 300GHz*, ANSI/IEEE C.95.1-1992.

# 8

# Modeling of Human Exposure to Static and Low Frequency Fields

Assessment of human exposure to static and low frequency (LF) fields involves the calculation of internal current density and internal electric field [1–3]. Measurement of internal fields is generally not possible and human exposure assessment is undertaken by means of sophisticated, anatomically based computational models [4–7] mostly related to the use of the finite difference methods (FDM). The finite element method (FEM), boundary element method (BEM), method of moments (MoM), and some other methods are used to a somewhat lesser extent.

Integral equation approaches, using the Green integral representation, are based on the fundamental solution of the leading operator for the governing equation, thus being competitive with other well-established methods, such as FDM, FDTD (suffering from the so-called staircasing approximation), or FEM in terms of accuracy and efficiency [5]. Effects of numerical artifacts in LF dosimetry are studied by working group 2 (WG 2) of IEEE/ICES TC95 SC6 EMF Dosimetry Modeling [8].

Within the framework of integral approaches, such as BEM, it is possible to avoid the staircasing error (by use of isoparametric elements), implementation of absorbing boundary conditions, and volume meshes for large-scale problems.

Some disadvantages of integral methods, such as BEM are more complex formulation (particularly for nonhomogeneous domains) and the related numerical implementation (dense matrices, Green function singularities/quasi-singularities).

The use of some integral methods in static and LF electromagnetic dosimetry has been discussed elsewhere, e.g. in [9–11], and outlined in this chapter. Thus, Laplace equation formulation/BEM scheme for the exposure of human head to electrostatic fields and whole body to LF electric fields is presented.

To sum up, several classes of numerical methods are available for solving various problems in static and LF bioelectromagnetics such as FDM, FEM, and BEM. While the main advantage of FDM is simplicity and robustness of the algorithm, it evidently suffers from staircasing approximation error [8]. On the

*Computational Methods in Electromagnetic Compatibility: Antenna Theory Approach versus Transmission line Models,* First Edition. Dragan Poljak and Khalil El Khamlichi Drissi.

other hand, FEM and BEM sufficiently treat domains with complex geometries. Furthermore, the application of FEM leads to sparse and usually symmetric matrix systems, which can be stored with relatively low memory requirements.

The principle advantage of BEM is the reduction of the problem dimensionality [5, 12].

An important issue arising from the use of FEM or BEM is a trade-off regarding the accuracy and efficiency. Generally BEM has a better accuracy, as it is based on the fundamental solution of the leading partial differential equation (Green's functions), and the use of approximations is restricted to the boundary. Furthermore, in BEM one has to deal with two types of variables, such as potentials and fluxes, while in FEM only the potential is introduced as a degree of freedom. The subsequent need to differentiate the potential in the case of FEM in order to compute fluxes considerably reduces the accuracy of the solution.

Since the approximations in BEM are restricted to the boundary, boundary element meshes should not be compared to the finite element meshes with the internal nodes removed. To achieve comparable accuracy finite element meshes would need more boundary divisions than the equivalent boundary element mesh.

The exposure of high intensity electrostatic fields, detected from several types of VDUs, could be possibly related with skin rashes [13, 14], but the main concern is related to particle transport and deposition [15].

On the other hand, extremely low frequency (ELF) (~Hz) electromagnetic and electrostatic fields might be associated with certain skin diseases, suppression of melatonin, or induction of phosphenes in the eyes, despite the fact that there is no strong evidence of adverse health effects from domestic levels of ELF electromagnetic fields [12, 13, 16].

The realistic, three-dimensional, anatomically based model of the human head exposed to electrostatic field from VDU featuring the use of FEM and BEM has been reported in [10]. Contrary to the more common approach featuring the use of the FDM [13], the electrostatic field around the human head is assessed using both FEM and BEM. The mathematical formulation is based on the Laplace equation for electrostatic potential.

Exposure of pregnant woman/fetus to ELF electric field has been documented elsewhere, e.g. in [3] and [11].

## 8.1 Exposure to Static Fields

Assuming the charge density to be negligible in the space between the human head and the display, the mathematical description of the 3D electrostatic field between a VDU and the head is given by the Laplace equation for scalar electric potential $\varphi$ [10]:

$$\nabla^2 \varphi = 0, \tag{8.1}$$

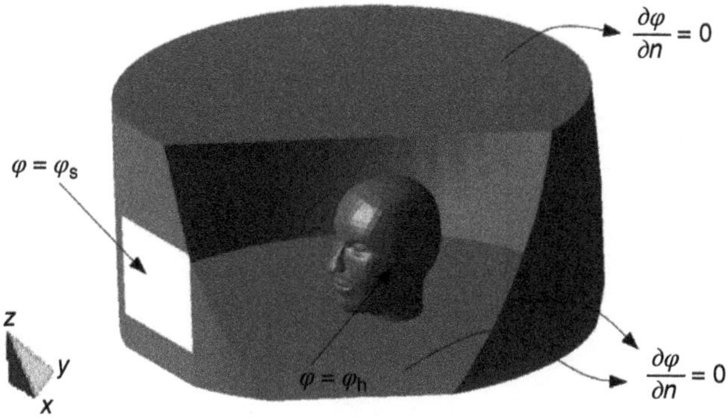


**Figure 8.1** Geometry and boundary conditions for numerical 3D model of a human seated in front of a VDU.

with the associated boundary conditions:

$$\varphi = \varphi_s \text{ on the display,} \tag{8.2}$$

$$\varphi = \varphi_h \text{ on the head,} \tag{8.3}$$

$$\nabla \varphi \cdot \vec{n} = 0 \text{ on the far field boundaries.} \tag{8.4}$$

As shown in Figure 8.1, the Dirichlet boundary conditions (8.2) and (8.3) are specified on the face and the display, respectively, while Neumann conditions (8.4) are imposed on the rest of the outer boundary. The head is considered to be a perfect conductor, thus being itself an equipotential surface with potential $\varphi_h$.

The standard conditions for the person in front of the display are defined as follows: $l_s = 40$ cm, $d_s = 42.5$ cm (17 in), $\varphi_s = 15$ kV and $\varphi_h = 0$ kV. It is presumed that the monitor is of the 4 : 3 format, meaning that width of the screen is 34.3 cm (13.6 in) and height is 25.7 cm (10.2 in). Also, the diameter of the computational domain is around 1 m, with a height of around 0.6 m. The size of the head follows the standards proposed by NASA [17], i.e. the head length is about 21 cm and the head breadth about 16.5 cm.

Furthermore, the eyebrows are assumed to have the same potential as the face. Other parameters such as temperature, humidity, and conductivity of the screen glass surface are not taken into account in order to simplify the model.

Two different face geometries are shown in Figure 8.2: person 1 (Figure 8.2a) and person 2 (Figure 8.2b).

The preprocessing and the geometry implementation represent a major problem that has been handled by a customizable geometry modeler, a preprocessor, and mesh generator (GID) [10].

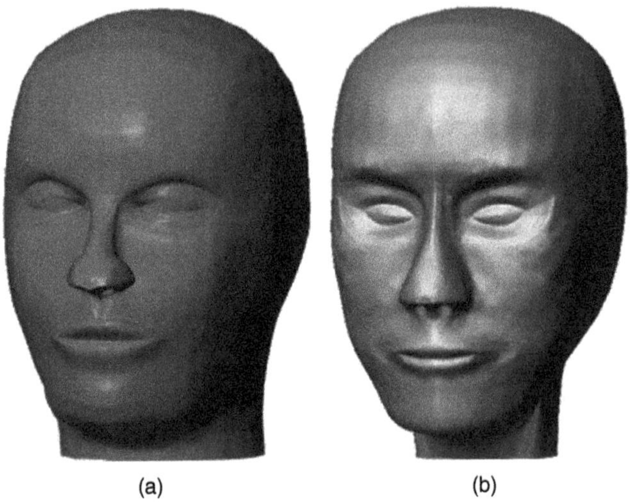

Figure 8.2 Head models: (a) person 1; (b) person 2.

### 8.1.1 Finite Element Solution

Applying the weighted residual approach to (8.1) yields [10]

$$\int_{\Omega} \nabla^2 \varphi W_j \, d\Omega = 0, \tag{8.5}$$

where $W_j$ denotes the set of test functions.

The weak formulation of the problem yields

$$\int_{\Omega} \nabla \varphi \cdot \nabla f_j \, d\Omega = \int_{\Gamma} \frac{\partial \varphi}{\partial n} f_j \, d\Gamma. \tag{8.6}$$

The unknown potential over the finite element is expressed in terms of the linear combination of four shape functions [10]:

$$\varphi^e = \sum_{i=1}^{4} \alpha_i f_i, \tag{8.7}$$

where $\alpha_i$ represent unknown coefficients of the solution, while $f_i$ are shape functions, which in matrix notation becomes

$$\varphi^e = \{f\}^T \{\alpha\}. \tag{8.8}$$

For three-dimensional problems the shape functions $f_i$ are given by

$$f_i(x, y, z) = \frac{1}{D}(V_i + a_i x + b_i y + c_i z), \quad i = 1, 2, 3, 4. \tag{8.9}$$

Expressions for $a_i$, $b_i$, $c_i$ could be found elsewhere, i.e. in [18].

Then, the potential gradient defined as

$$\nabla\varphi = \frac{\partial\varphi}{\partial x}\vec{e}_x + \frac{\partial\varphi}{\partial y}\vec{e}_y + \frac{\partial\varphi}{\partial z}\vec{e}_z \tag{8.10}$$

can be written in terms of shape functions as follows:

$$\nabla\varphi = \begin{bmatrix} \frac{\partial\varphi}{\partial x} \\ \frac{\partial\varphi}{\partial y} \\ \frac{\partial\varphi}{\partial z} \end{bmatrix} = \begin{bmatrix} \frac{\partial f_1}{\partial x} & \frac{\partial f_2}{\partial x} & \frac{\partial f_3}{\partial x} & \frac{\partial f_4}{\partial x} \\ \frac{\partial f_1}{\partial y} & \frac{\partial f_2}{\partial y} & \frac{\partial f_3}{\partial y} & \frac{\partial f_4}{\partial y} \\ \frac{\partial f_1}{\partial z} & \frac{\partial f_2}{\partial z} & \frac{\partial f_3}{\partial z} & \frac{\partial f_4}{\partial z} \end{bmatrix} \begin{bmatrix} \alpha_1 \\ \alpha_2 \\ \alpha_3 \\ \alpha_4 \end{bmatrix}. \tag{8.11}$$

Having performed FEM discretization one obtains the matrix equation

$$[a]\{\alpha\} = \{Q\}, \tag{8.12}$$

where $\{Q\}$ represents the flux vector, and the global matrix $[a]$ is composed of local finite element matrices $[a]_{ji}^{e}$.

Hence, inserting (8.11) into (8.6) the finite element matrix becomes

$$[a]^e = \int_{\Omega_e} \begin{bmatrix} \frac{\partial f_1}{\partial x} & \frac{\partial f_1}{\partial y} & \frac{\partial f_1}{\partial z} \\ \frac{\partial f_2}{\partial x} & \frac{\partial f_2}{\partial y} & \frac{\partial f_2}{\partial z} \\ \frac{\partial f_3}{\partial x} & \frac{\partial f_3}{\partial y} & \frac{\partial f_3}{\partial z} \\ \frac{\partial f_4}{\partial x} & \frac{\partial f_4}{\partial y} & \frac{\partial f_4}{\partial z} \end{bmatrix} \begin{bmatrix} \frac{\partial f_1}{\partial x} & \frac{\partial f_2}{\partial x} & \frac{\partial f_3}{\partial x} & \frac{\partial f_4}{\partial x} \\ \frac{\partial f_1}{\partial y} & \frac{\partial f_2}{\partial y} & \frac{\partial f_3}{\partial y} & \frac{\partial f_4}{\partial y} \\ \frac{\partial f_1}{\partial z} & \frac{\partial f_2}{\partial z} & \frac{\partial f_3}{\partial z} & \frac{\partial f_4}{\partial z} \end{bmatrix} d\Omega. \tag{8.13}$$

Once the scalar potential $\varphi$ is obtained, the electric field $E$ can be calculated from the potential gradient

$$\vec{E} = -\nabla\varphi \tag{8.14}$$

as indicated in Ref. [10].

### 8.1.2 Boundary Element Solution

Applying the weighted residual approach, (8.1) is integrated over the calculation domain $\Omega$ [10, 11, 13]:

$$\int_{\Omega} \nabla^2\phi \cdot \psi \, d\Omega = 0, \tag{8.15}$$

where $\psi$ is the weighting function.

Using Green identities, and applying the generalized Gauss theorem, it follows that

$$\int_{\Omega} \psi \nabla^2 \varphi \, d\Omega = \int_{\Gamma} \psi \frac{\partial \varphi}{\partial n} d\Gamma - \int_{\Gamma} \varphi \frac{\partial \psi}{\partial n} d\Gamma + \int_{\Omega} \varphi \nabla^2 \psi \, d\Omega. \tag{8.16}$$

Then, $\psi$ is chosen to be the fundamental solution of the differential equation

$$\nabla^2 \psi - \delta(\vec{r} - \vec{r}') = 0, \tag{8.17}$$

which, for 3D problems becomes

$$\psi = \frac{1}{4\pi R}, \tag{8.18}$$

where $\delta$ is the Dirac delta function, $\vec{r}$ denotes the observation points, $\vec{r}'$ denotes the source points, and $R = |\vec{r} - \vec{r}'|$ is the distance between them.

Now, the domain integral from the left-hand side of (8.16) becomes

$$\int_{\Omega} \phi \nabla^2 \psi \, d\Omega = - \int_{\Omega} \phi \delta(\vec{r} - \vec{r}') d\Omega = -\phi_i \tag{8.19}$$

for any point inside the domain.

Now, combining (8.15)–(8.19) yields

$$\phi_i = \int_{\Gamma} \psi \frac{\partial \phi}{\partial n} d\Gamma - \int_{\Gamma} \phi \frac{\partial \psi}{\partial n} d\Gamma, \tag{8.20}$$

which can be regarded as the Green representation of $\varphi$.

When the observation point $i$ is located on the boundary $\Gamma$, the boundary integral becomes singular as $R$ approaches zero. In order to deal with this singularity a small sphere for 3D problems is considered. By sorting out the surface integrals over the small sphere centered in the singular point, the following integral formulation is obtained:

$$c_i \phi_i = \int_{\Gamma} \Psi \frac{\partial \phi}{\partial n} d\Gamma - \int_{\Gamma} \phi \frac{\partial \Psi}{\partial n} d\Gamma, \tag{8.21}$$

where

$$c_i = \begin{cases} 1, & i \in \Omega, \\ 1/2, & i \in \Gamma \text{ (smooth boundary)}. \end{cases} \tag{8.22}$$

Discretizing the boundary $\Gamma$ into $N_e$ boundary elements (8.21) becomes

$$c_i u_i = \sum_{j=1}^{N_e} \int_{\Gamma_j} \psi \frac{\partial \varphi}{\partial n} d\Gamma - \sum_{j=1}^{N_e} \int_{\Gamma_j} \varphi \frac{\partial \psi}{\partial n} d\Gamma. \tag{8.23}$$

Each boundary element contains a number ($N_{fn}$) of subjacent collocation nodes, in which the potential or fluxes are evaluated. Thus, the values of the potential or its normal derivative at any point defined by the local coordinates

$\boldsymbol{\xi} = (\xi_1, \xi_2)$ on a given boundary element can be defined in terms of their values at the collocation nodes, and the $N_{\mathrm{fn}}$ interpolation functions $\Phi_k$ with $k = 1, N_{\mathrm{fn}}$ as follows:

$$\varphi(\boldsymbol{\xi}) = \sum_{k=1}^{N_{\mathrm{fn}}} \Phi_k(\boldsymbol{\xi})\,\varphi_k; \qquad \frac{\partial \varphi(\boldsymbol{\xi})}{\partial n} = \sum_{k=1}^{N_{\mathrm{fn}}} \Phi_k(\boldsymbol{\xi}) \left.\frac{\partial \varphi}{\partial n}\right|_k. \tag{8.24}$$

Thus, (8.23) can now be rewritten in the following form:

$$c_i u_i = \sum_{j=1}^{N_e} \sum_{k=1}^{N_{\mathrm{fn}}} \left( \int_{\Gamma_j} \psi\, \Phi_k \, \mathrm{d}\Gamma_j \right) \frac{\partial \varphi_{jk}}{\partial n} - \sum_{j=1}^{N_e} \sum_{k=1}^{N_{\mathrm{fn}}} \left( \int_{\Gamma_j} \frac{\partial \psi}{\partial n} \Phi_k \, \mathrm{d}\Gamma_j \right) \varphi_{jk} \tag{8.25}$$

while the corresponding matrix notation is

$$[H]\{\varphi\} = [G][\partial_n \varphi], \tag{8.26}$$

where the corresponding matrix coefficients are

$$H_{il} = \delta_{il} c_i + \int_{\Gamma_j} \left.\frac{\partial \psi_i}{\partial n_j}\right|_{\xi} \Phi_k(\boldsymbol{\xi}) \mathrm{d}\Gamma_j, \tag{8.27}$$

$$G_{il} = \int_{\Gamma_j} \psi_i(\xi)\, \Phi_k(\boldsymbol{\xi}) \mathrm{d}\Gamma_j. \tag{8.28}$$

Index $l$, identifying a collocation node within the domain, can be calculated in terms of indices $j$ and $k$, by means of the nodal connectivity of the mesh, i.e. index $l = 1, \ldots, M$. and $M$, is the total number of collocation nodes in the domain. Basically, the index $l$ is used to identify one of the adjacent freedom (collocation) nodes from a global point of view, and is given as a function of the indicator of element ($j$), and the local collocation node of that element ($k$). In the case of discontinuous collocation nodes it follows that

$$M = \sum_{j=1}^{N_e} f_j. \tag{8.29}$$

The boundary differential $\mathrm{d}\Gamma_j$ can be expressed in terms of the domain local coordinates ($\boldsymbol{\xi}$) via the Jacobian of the transformation $|J|$:

$$\mathrm{d}\Gamma_j = |J| \mathrm{d}\xi_1 \, \mathrm{d}\xi_2. \tag{8.30}$$

Finally, the application of the prescribed boundary conditions through boundary discretization and further collocation into $N_{\mathrm{fs}}$ degrees of freedom yields an algebraic linear system of equations of the form [10]:

$$[A]\{x\} = \{b\}, \tag{8.31}$$

where $[A]$ is an $N_{\mathrm{fe}} \times N_{\mathrm{fe}}$ matrix that involves the coefficients of the single and double layer potential operators, i.e. the coefficients of $[H]$ and $[G]$ matrices given by (8.27) and (8.28); the 1-column array of unknowns $x$ contains the potentials and normal fluxes that were not prescribed as boundary conditions, and the right-hand side term involves boundary conditions.

### 8.1.3 Numerical Results

Figure 8.3 shows the electrostatic field induced at the surface of the female face calculated via FEM and BEM, respectively. The details of the head model can be found elsewhere, e.g. in [5] and [10].

Comparison of the FEM and BEM results for specific points on the head are shown in Figure 8.4.

The details of the FEM solution can be found elsewhere, e.g. in [10]. Differences on certain points, (nose tip in particular) between the FEM and BEM solution arise due to difficulties in accomplishing sufficiently refined mesh around complex geometries, such as the human face. Generally, BEM is more accurate and versatile than the FEM, allowing for a better representation of the shape of the human face. Nevertheless, the numerical differences between BEM and FEM (15%) are generally expected to be smaller than the fluctuations due to slight changes of geometry from face to face.

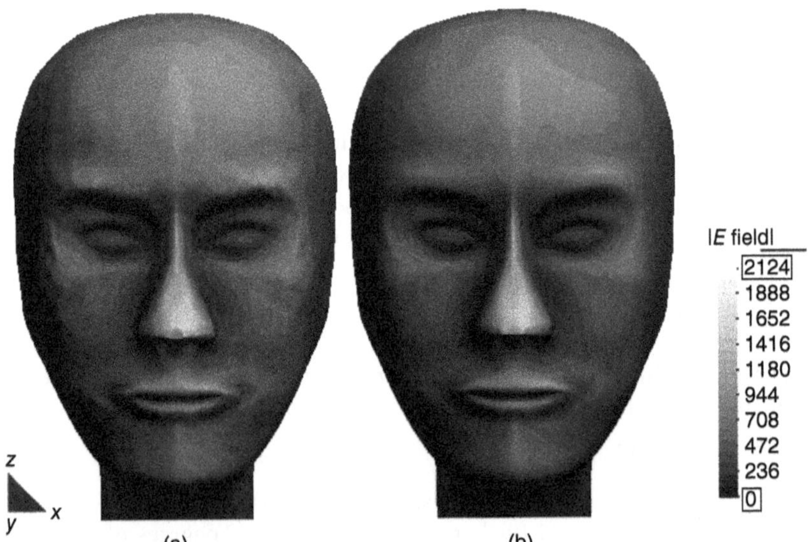


**Figure 8.3** Electrostatic field on the female face: (a) FEM solution; (b) BEM solution.

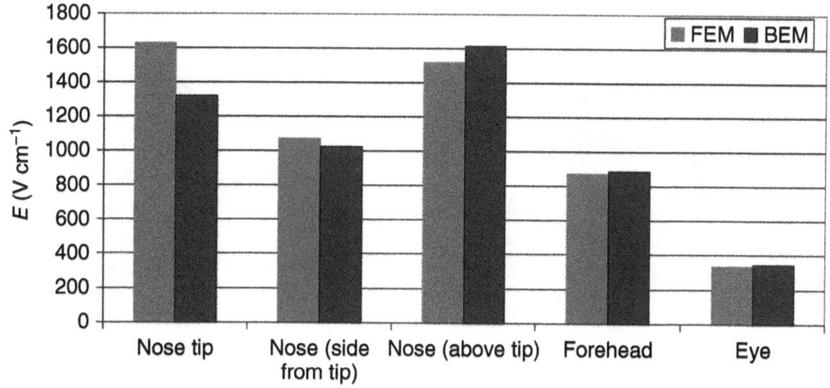


**Figure 8.4** Electrostatic field at the specific locations of the female face.

## 8.2 Exposure to Low Frequency (LF) Fields

The quasi-static formulation for LF exposures used in [5, 10] is based on the Laplace (quasi-static) version of the continuity equation for scalar potential $\varphi$ given by

$$\nabla \cdot [(\sigma + j\omega\varepsilon)\nabla\varphi] = 0, \tag{8.32}$$

where $\omega$ is the frequency of the incident field, and $\sigma$ and $\varepsilon$ are the conductivity and permittivity of the material.

As the body is dominantly conducting at LF exposures, expression (8.32) for the body is given by

$$\nabla \cdot (\sigma\nabla\varphi) = 0. \tag{8.33}$$

For the ambient air, (8.31) simplifies into

$$\nabla^2\varphi = 0. \tag{8.34}$$

Knowing the scalar potential along the body, the induced current density is computed from the differential form of Ohm's Law:

$$\vec{J} = (\sigma + j\omega\varepsilon)\nabla\phi. \tag{8.35}$$

The electric field $E$ can be readily obtained from (8.14). The boundary integral representation of (8.32) is given by (8.21) [5]. Note that the BEM procedure is the same as presented in Section 8.1.2. More mathematical details can be found elsewhere, e.g. in [19].

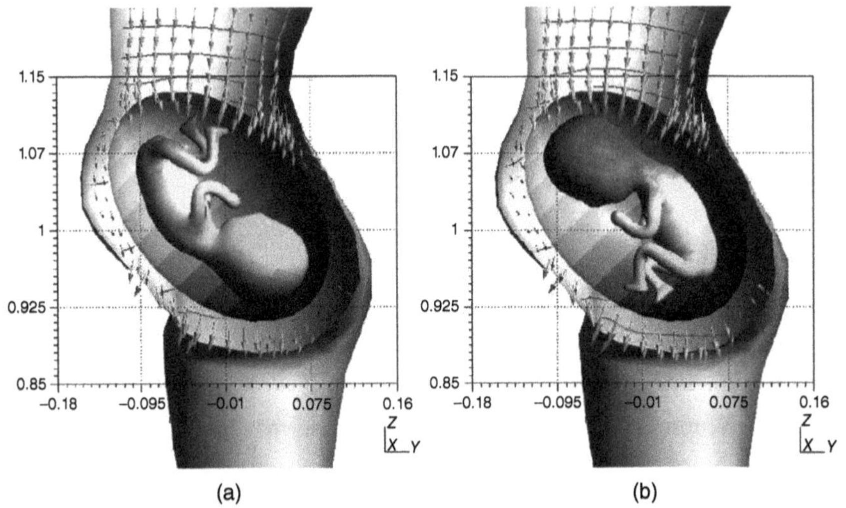


**Figure 8.5** Lateral view of the pregnant woman at 26th gestational week: (a) fetus in the cephalic presentation; (b) fetus in the breach presentation.

### 8.2.1 Numerical Results

An illustrative computational example is the pregnant woman/fetus exposure to power line electric field $E = 10\,\text{kV m}^{-1}$. The amniotic fluid (AF) has the highest conductivity, which varies depending on the period of gestation. Kidney, muscle, cortical bone, bladder, spleen, and skin have conductivity very close to $0.1\,\text{S m}^{-1}$, while the ovary and cartilage conductivity is around $0.2\,\text{S m}^{-1}$. More details on the electrical properties of the pregnant woman tissue can be found elsewhere, e.g. in [11].

Figure 8.5 shows the sliced model of the pregnant woman at 26th gestational week with the electric field/scalar potential lines for the fetus in cephalic and breach presentation, respectively.

It is visible that the uterus, due to the higher conductivity of the AF compared to the maternal tissue, tends to concentrate the field lines.

The maximal value of current density in the fetus occurs during the 8th week, while the maximum current density induced in the fetus is $7.4\,\text{mA m}^{-2}$ for an external field exposure of $10\,\text{kV m}^{-1}$. Since the restriction recommended for public exposure by ICNRP [2] is $2\,\text{mA m}^{-2}$, through modeling it is possible to set restrictions on external field exposure, which translates to a maximum external field of $2.7\,\text{kV m}^{-1}$ in order to limit the current density exposure in the earlier months of development [5].

## References


1 Poljak, D. (2011). Electromagnetic fields: environmental exposure. In: *Encyclopedia of Environmental Health*, vol. **2** (ed. J.O. Nriagu), 259–268. Burlington: Elsevier.

2 International Commission on Non-Ionizing Radiation Protection (1998). Guidelines for limiting exposure to time-varying electric, magnetic and electromagnetic fields (up to 300 GHZ). *Health Physics* **74** (4): 494–522.

3 International Commission on Non-Ionizing Radiation Protection (2010). Guidelines for limiting exposure to time-varying electric and magnetic fields (1 HZ – 100 kHZ). *Health Physics* **99** (6): 818–836.

4 Hand, J.W. (2008). Modeling the interaction of electromagnetic fields (10 MHz–10 GHz) with the human body: methods and applications. *Physics in Medicine and Biology* **53** (16): 243–286.

5 Poljak, D., Cavka, D., Dodig, H. et al. (2014). On the use of boundary element analysis in bioelectromagnetics. *Engineering Analysis with Boundary Elements*(Special issue on Bioelectromagnetics) **49**: 2–14.

6 Singh, K.D., Longan, N.S., and Gilmartin, B. (2006). Three dimensional modeling of the human eye based on magnetic resonance imaging. *Investigative Opthamology and Visual Science* **47**: 2272–2279.

7 Hirata, A. (2005). Temperature increase in human eyes due to near-field and far-field exposures at 900 MHz, 1.5 GHz, and 1.9 GHz. *IEEE Transactions on Electromagnetic Compatibility* **47** (1): 68–76.

8 Reilly, J. and Hirata, A. (2016). Low-frequency electrical dosimetry: research agenda of the IEEE international committee on electromagnetic safety. *Physics in Medicine and Biology* **61** (12): R138.

9 Poljak, D., Cvetkovic, M., Peratta, A., et al. (2016). On some integral approaches in electromagnetic dosimetry. The Joint Annual Meeting of The Bioelectromagnetics Society and the European BioElectromagnetics Association – BioEM 2016, Ghent, Belgium, 289–295.

10 Čavka, D., Poljak, D., and Peratta, A. (2011). Comparison between finite and boundary element methods for analysis of electrostatic field around human head generated by video display units. *Journal of Communications Software and Systems* 7: 22–28.

11 Poljak, D. (2013). On the assessment of pregnant woman/foetus exposure to electromagnetic fields by using numerical methods. *Paediatria Croatica* **1** (Suppl. 57): 1–9.

12 Institute of Electrical and Electronics Engineers (1997). Biological and health effects of electric and magnetic fields from video display terminals. *IEEE Engineering in Medicine and Biology Magazine* **16** (3): 87–92.

13 Nielsen, N.F., Michelsen, J., Michalsen, J.A., and Schneider, T. (1996). Numerical calculation of electrostatic field surrounding a human head in visual display environments. *Journal of Electrostatics* **36**: 209–223.
14 Australian Radiation Protection and Nuclear Safety Agency (2003). *Radiation Emission from Video Display Terminals.* Australian Radiation Protection and Nuclear Safety Agency.
15 Nielsen, N.F. and Schneider, T. (1998). Particle deposition onto a human head: influence of electrostatic and wind fields. *Bioelectromagnetics* **19**: 246–258.
16 International Commission on Non-Ionizing Radiation Protection (ICNIRP) and International Labour Organization (1994). *Visual Display Units: Radiation Protection Guidelines.* Geneva: International Labour Office.
17 NASA (1995). *Man-systems integration standards: anthropometry and biomechanics.* Washington: NASA.
18 Čavka, D., Poljak, D., and Peratta, A. (2006). Finite element model of the human head exposed to electrostatic field generated by Video Display Units. SoftCOM 2006, SYM1/I-6122-2909, Split, Croatia.
19 Gonzalez, M.G., Peratta, A., and Poljak, D. (2007). Boundary element modeling of the realistic human body exposed to extremely-low-frequency (ELF) electric fields: computational and geometrical aspects. *IEEE Transactions on Electromagnetic Compatibility* **49** (1): 153–162.

# 9

# Modeling of Human Exposure to High Frequency (HF) Electromagnetic Fields

Tremendous growth of wireless and mobile communication systems has increased public concern regarding possible adverse health effects. As the dominant and well-established biological effect of high frequency (HF) radiation is tissue heating, HF exposure assessment is related to the determination of SAR distribution and related temperature increase in the tissue. Of particular interest is the exposure of the head, i.e. the eye and brain, respectively [1–5]. As measurement of internal fields and related temperature rise is not possible theoretical models for exposure assessment are necessary to simulate various exposure scenarios, and thereby establish safety guidelines and exposure limits for humans [6, 7]. Computational models for HF exposures can be classified as either realistic models of the body (or particular organs) mostly based on magnetic resonance imaging (MRI), e.g. [4], or simplified models, computationally much less demanding but failing to provide accurate results in most of the exposure scenarios, e.g. [8].

Modern realistic, anatomically based computational models are related to the use of the finite difference time domain (FDTD) method or the finite element method (FEM), applied to structured meshes comprising of cubical cells or voxels. The conformal FEM, boundary element method (BEM), and method of moments (MoM) are, on the other hand, being used somewhat less frequently [8, 9].

FDTD is rather robust method and has the definitive advantage of algorithm simplicity and efficiency in analyzing heterogeneous domains, but it suffers from the staircasing errors related to curved geometries and does not provide an elegant way to efficiently truncate the computational domain to the region of interest.

FEM, on the other hand, is well suited for modeling complex objects, providing accurate geometrical representation of structures with irregular boundaries, while also allowing elementwise variation in dielectric properties. However, the discretization of an entire domain and the use of absorbing boundary conditions are needed for the use of FEM, thus requiring rather large computational resources when finer discretization is used.

*Computational Methods in Electromagnetic Compatibility: Antenna Theory Approach versus Transmission line Models*, First Edition. Dragan Poljak and Khalil El Khamlichi Drissi.

The advantage of the conformal methods based on corresponding integral formulation of the problem, such as BEM and MoM, is the accurate representation of the domain boundary, avoiding the staircasing errors due to the use of voxel elements. Furthermore, the nature of Green integral representation enables one to carry out the exact treatment of exterior field problem such as the human head exposed to electromagnetic radiation.

However, MoM and BEM are not well suited for modeling nonhomogeneous domains, e.g. complex and arbitrary shaped dielectric objects. Also, their numerical implementation results in dense matrices being computationally far more expensive than is the case when FDTD and FEM approaches are used.

The advantages of the domain/boundary techniques can be combined, if one makes an additional effort in coupling them into hybrid methods, e.g. exploiting the ability of FEM to deal with arbitrary material properties, and to accurately model curved geometries, and the suitability of BEM for the open boundary problems. Therefore, the hybrid approach is considered to be convenient for the accurate characterization of the bioelectromagnetic phenomena compared to the already existing and well-established approaches [5, 9–13].

The hybrid FEM/BEM approach [9] for the human head exposure to HF electromagnetic radiation is reported in [11], thus extending the approach presented in [9] pertaining to the realistic three-dimensional model of the human eye.

This chapter presents the use of integral, differential, and hybrid approaches in bioelectromagnetics and thermal dosimetry mostly reported in [9–12], featuring the use of the hybrid BEM/FEM, the MoM, and FEM. The obtained maximal values of SAR are compared to the exposure limits proposed by ICNIRP [6]. Computational examples presented in this chapter are related to the head, eye, and brain exposure to HF electromagnetic fields and to the calculation of the related temperature rise.

## 9.1 Internal Electromagnetic Field Dosimetry Methods

### 9.1.1 Solution by the Hybrid Finite Element/Boundary Element Approach

Electromagnetic wave incident on the human eye or head represents an unbounded scattering problem. Using the Stratton–Chu integral expression, the time harmonic electric field in the domain exterior to the head is expressed in terms of the following boundary integral equation [9–13]:

$$\alpha \vec{E}'_{\text{ext}} = \vec{E}'_{\text{inc}} + \oint_{\partial V} \vec{n} \times -(\nabla \times \vec{E}'_{\text{ext}}) G(\vec{r}, \vec{r}') \mathrm{d}S + \oint_{\partial V} [(\vec{n} \times \vec{E}'_{\text{ext}}) \times \nabla G(\vec{r}, \vec{r}') + (\vec{n} \cdot \vec{E}'_{\text{ext}}) \nabla G(\vec{r}, \vec{r}')] \mathrm{d}S, \tag{9.1}$$

where $\vec{n}$ is an outer normal to surface $\partial V$ bounding the volume $V$ and $\alpha$ is the solid angle subtended at the observation point, $\vec{E}'_{ext}$ and $\vec{E}'_{inc}$ are the total and the incident electric field, respectively, while $G(\vec{r},\vec{r}')$ denotes the Green's function for the free space given by

$$G(\vec{r},\vec{r}') = \frac{e^{-jk|\vec{r}-\vec{r}'|}}{4\pi|\vec{r}-\vec{r}'|}, \tag{9.2}$$

where $|\vec{r}-\vec{r}'|$ is the distance from the observation point to the source point, and $k$ denotes the wave number.

Performing some further mathematical manipulations, Equation (9.1) can be written in terms of tangential components of the electric field $E$ and magnetic field $H$ on the boundary surface $\partial V$:

$$\begin{aligned}\alpha\vec{E}'_{ext} = \vec{E}'_{inc} - j\omega\mu \oint_{\partial V} (\vec{n}\times\vec{H}_{ext})G(\vec{r},\vec{r}')dS \\ + \oint_{\partial V} \Bigg[(\vec{n}\times\vec{E}'_{ext})\times\nabla G(\vec{r},\vec{r}') \\ - \frac{1}{\sigma + j\omega\mu}\nabla_s\cdot(\vec{n}\times\vec{H}_{ext})\nabla G(\vec{r},\vec{r}')\Bigg] dS.\end{aligned} \tag{9.3}$$

The interior region is given by following differential equation:

$$\nabla\times\left(\frac{j}{\omega\mu}\nabla\times\vec{E}_{int}\right) - (\sigma + j\omega\varepsilon)\vec{E}_{int} = 0, \tag{9.4}$$

which needs to be coupled with (9.3).

The fields $E$ and $H$ are approximated using the edge elements [11] that preserve the tangential continuity of the fields on the boundary:

$$\vec{E} = \sum_{i=1}^{n} \delta_i \vec{w}_i e_i, \tag{9.5}$$

$$\vec{H} = \sum_{i=1}^{n} \delta_i \vec{w}_i h_i. \tag{9.6}$$

The unknown coefficients $e_i$ and $h_i$, respectively, associated with each edge of the model, are determined from the global system of equations. The coefficient $\delta_i = 1$ if the local edge direction coincides with the chosen global edge direction; otherwise, it is $-1$.

Applying the weighted residual approach to (9.4), by taking the dot product of (9.4) with test function $w_i$, through the Galerkin–Bubnov procedure, it follows that

$$\int_V \left[\nabla\times\left(\frac{j}{\omega\mu}\nabla\times\vec{E}_{int}\right) - (\sigma + j\omega\varepsilon)\vec{E}_{int}\right]\delta_i\vec{w}_i\, dV = 0. \tag{9.7}$$

Having applied some standard vector identities, followed by the divergence theorem, the weak formulation is obtained:

$$\int_V [\nabla \times \delta_i \vec{w}_i \cdot \vec{E}_{\text{int}} - (\sigma + j\omega\varepsilon)\delta_i \vec{w}_i \cdot \vec{E}_{\text{int}}]_i \mathrm{d}V = \oint_{\partial V} \delta_i \vec{w}_i \cdot \mathrm{d}\vec{S} \times \vec{H}_{\text{int}}. \tag{9.8}$$

Now FEM/BEM coupling can be undertaken by ensuring the tangential components of the electric and magnetic fields to be continuous across the surface $\partial V$. Thus, it can be written that

$$\oint_{\partial V} \delta_i \vec{w}_i \cdot \mathrm{d}\vec{S} \times \vec{E}'_{\text{int}} = \oint_{\partial V} \delta_i \vec{w}_i \cdot \mathrm{d}\vec{S} \times \vec{E}'_{\text{ext}}. \tag{9.9}$$

Now, substituting $E_{\text{ext}}$ and $H_{\text{ext}}$ in (9.3) with $E_{\text{int}}$ and $H_{\text{int}}$, respectively, and after inserting (9.3) into (9.10), the following double surface integral is obtained:

$$\begin{aligned}\oint_{\partial V} \delta_i \vec{w}_i \cdot \mathrm{d}\vec{S} \times \alpha_i \vec{E}'_{\text{ext}} = &\oint_{\partial V} \delta_i \vec{w}_i \cdot \mathrm{d}\vec{S} \times \vec{E}'_{\text{inc}} \\ &- j\omega\mu \oint_{\partial V} \delta_i \vec{w}_i \cdot \mathrm{d}\vec{S} \times \oint_{\partial V} \vec{n} \\ &\times \vec{H}_{\text{int}} G(\vec{r}, \vec{r}')\mathrm{d}\vec{S} + \oint_{\partial V} \delta_i \vec{w}_i \cdot \mathrm{d}\vec{S} \\ &\times \oint_{\partial V} (\vec{n} \times \vec{E}_{\text{int}}) \times \nabla G(\vec{r}, \vec{r}')\mathrm{d}\vec{S} \\ &- \frac{1}{\sigma + j\omega\mu} \oint_{\partial V} \delta_i \vec{w}_i \cdot \mathrm{d}\vec{S} \\ &\times \oint_{\partial V} \nabla_s \cdot (\vec{n} \times \vec{H}_{\text{ext}}) \times \nabla G(\vec{r}, \vec{r}')\mathrm{d}\vec{S}.\end{aligned} \tag{9.10}$$

Now, inserting (9.5) into (9.9) results in the following system of equations for the edges of the boundary surface:

$$[E_{\text{BEM}}]\{e_{\text{BEM}}\} = \{e_{\text{inc}}\} + [H_{\text{BEM}}]\{h_{\text{BEM}}\}, \tag{9.11}$$

$$[E_{\text{FEM}}]\{e_{\text{FEM}}\} = [H_{\text{FEM}}]\{h_{\text{FEM}}\}, \tag{9.12}$$

where $\{e_{\text{BEM}}\}$ and $\{h_{\text{BEM}}\}$ are unknown coefficients associated with boundary surface of the scattering problem, $\{e_{\text{inc}}\}$ are known coefficients calculated from the incident field, matrices $[E_{\text{BEM}}]$ and $\{h_{\text{BEM}}\}$ arise from boundary integral equation (9.9), while matrices $[E_{\text{FEM}}]$ and $[H_{\text{FEM}}]$ stem from (9.7).

### 9.1.2 Numerical Results for the Human Eye Exposure

As the human eye is a complex organ consisting of many fine parts, it is important to use a detailed model of the eye. The cross section of the human eye presented in [12] is shown in Figure 9.1.

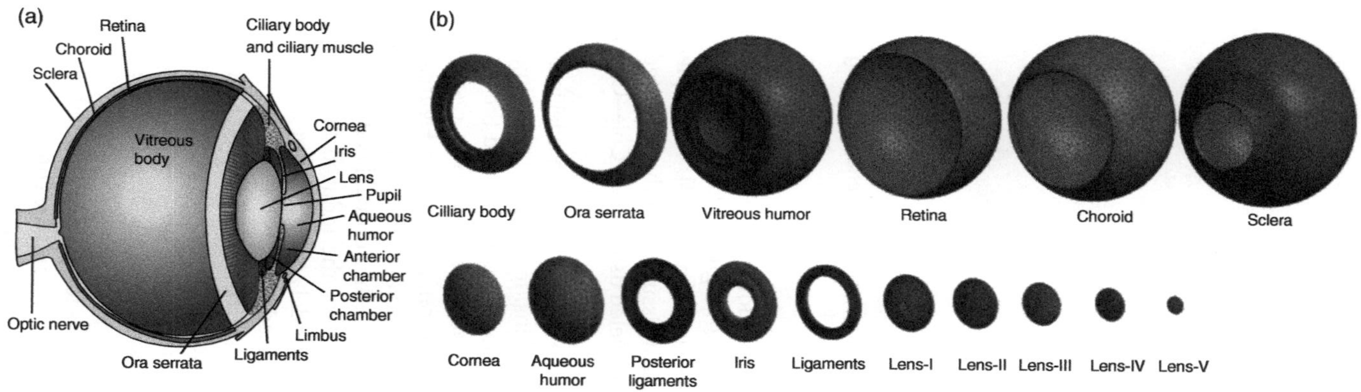


**Figure 9.1** Cross section of the human eye (a) and various eye tissues (b) from the compound and the extracted eye model.

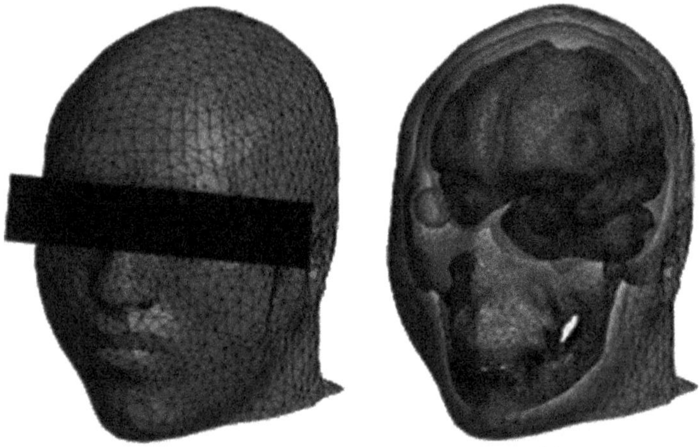


**Figure 9.2** Model of the human head (a), and the overlay depicting various head tissues; (b), surrounding the compound eye model.

The extracted compound eye model is a part of the head model composed of various head tissues as shown in Figure 9.2. The model of the human head is built from the MRI of a 24-year-old male [14]. The implemented model uses 7 tissues in addition to 16 ocular tissues from the extracted eye model. The head tissue parameters are also modeled using the 4-Cole–Cole method [15] and are given in Table 9.1.

The boundary surface of the complete head model is discretized using triangular elements, while the head inside is discretized using tetrahedral elements.

The numerical results for the electric field and the related SAR, induced in the extracted and the compound model of the eye, respectively, obtained using the hybrid FEM/BEM formulation, are given in Figures 9.3–9.5. The incident plane wave of 1 GHz frequency is vertically polarized, and directed toward the corneal surface, perpendicular to the coronal head/eye cross section.

Figure 9.3 shows the results for the induced field on the transverse cross sections of the extracted eye model and the compound eye model, respectively, due to 1 GHz vertically polarized plane wave.

It is seen that the higher values of the induced electric field in the compound eye model, being incorporated in the whole head model, are obtained in the superficial region of the eye, i.e. in the corneal and scleral regions. More details on the electric field distribution on the surface of both eye models can be found in Figure 9.4.

The obtained numerical results for the SAR computed on the transverse cross sections of the extracted eye model, and the compound eye model, respectively are presented in Figure 9.5. Again, higher values for the SAR are obtained in the compound eye model in the right part of the corneal and scleral regions.

**Table 9.1** Tissue parameters [12].

| Tissue | $\sigma$ (S m$^{-1}$) | $\varepsilon$ (−) | $\rho$ (g m$^{-3}$) |
|---|---|---|---|
| Brainstem | 0.622 | 38.577 | 1043 |
| Cerebellum | 1.308 | 48.858 | 1039 |
| Head skin | 0.899 | 40.936 | 1050 |
| Liquor | 1.667 | 68.875 | 1035 |
| Skull | 0.364 | 20.584 | 1900 |
| Mandible | 0.364 | 20.584 | 1900 |
| Gray matter | 0.985 | 52.282 | 1039 |
| Anterior chamber | 1.667 | 68.875 | 1003 |
| Choroid | 0.729 | 44.561 | 1000 |
| Ciliary body | 0.978 | 54.811 | 1040 |
| Cornea | 1.438 | 54.835 | 1076 |
| Iris | 0.978 | 54.811 | 1040 |
| Ligaments | 0.760 | 45.634 | 1000 |
| Ora serrata | 0.882 | 45.711 | 1000 |
| Posterior chamber | 1.667 | 68.875 | 1000 |
| Retina | 1.206 | 55.017 | 1039 |
| Sclera | 1.206 | 55.017 | 1076 |
| Vitreous body | 1.667 | 68.875 | 1009 |
| Lens-I | 0.824 | 46.399 | 1100 |
| Lens-II | 0.824 | 47.011 | 1100 |
| Lens-III | 0.824 | 47.694 | 1100 |
| Lens-IV | 0.824 | 48.383 | 1100 |
| Lens-V | 0.824 | 49.076 | 1100 |

*Source:* From [12]. Reproduced with permission of SpliTech.

This nonsymmetric distribution of SAR in the compound eye model, contrary to symmetrical distribution obtained in the extracted eye model, will result in a somewhat different distribution.

The next example deals with the exposure of the entire head. The model of the human head including the various head tissues is shown in Figure 9.6. The model is built with the aid of the MRI of a 24-year-old male [14]. The current implementation of the model consists of eight tissues, whose frequency dependent dielectric parameters, modeled using the 4-Cole–Cole method [15], are presented in Table 9.2.

The surface of the head is discretized using 6.838 triangular elements, while the interior domain of the head is discretized using 1.034.641 tetrahedral elements.

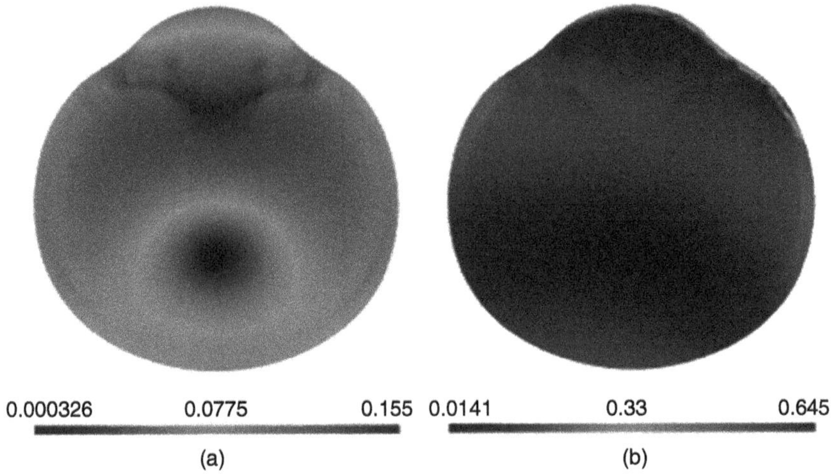


**Figure 9.3** Induced electric field due to 1 GHz vertically polarized plane wave in the transverse cross section of (a) the extracted eye model and (b) the compound eye model.

The computational examples are given for the head exposed to incident plane wave of 900 and 1800 MHz. Incident field $E\text{inc} = 1\,\text{V m}^{-1}$ is horizontally and vertically polarized, respectively, and directed toward the nose, perpendicular to the coronal head cross section. Numerical results for the horizontal polarization are denoted by HP, while vertical polarization is denoted by VP.

The electric field induced on the surface of the head model is shown in Figure 9.7 for 900 MHz horizontally polarized EM wave. The highest value of the field is obtained at the tip of the nose.

Figures 9.8–9.10 show the results for the induced field on the sagittal, coronal, and transverse cross sections of the head, respectively, due to 900 and 1800 MHz, vertically, i.e. horizontally polarized plane wave.

As is evident from Figures 9.8–9.10, the highest values of the induced field are formed in regions around the nose and the eyes, suggesting possible formation of hot spots.

Finally, considering the influence of the polarization of the incident electromagnetic wave on the induced field, the results show somewhat higher values for the vertical polarization. The study in [16] attributed this fact to the component of the surface area perpendicular to the incident electric field. The results obtained in [11] suggest similar behavior.

### 9.1.3 Solution by the Method of Moments

The lossy dielectric model of the human brain exposed to HF radiation is based on the surface integral equation (SIE) approach [9]. The formulation itself is usually derived from the equivalence theorem and by using the appropriate

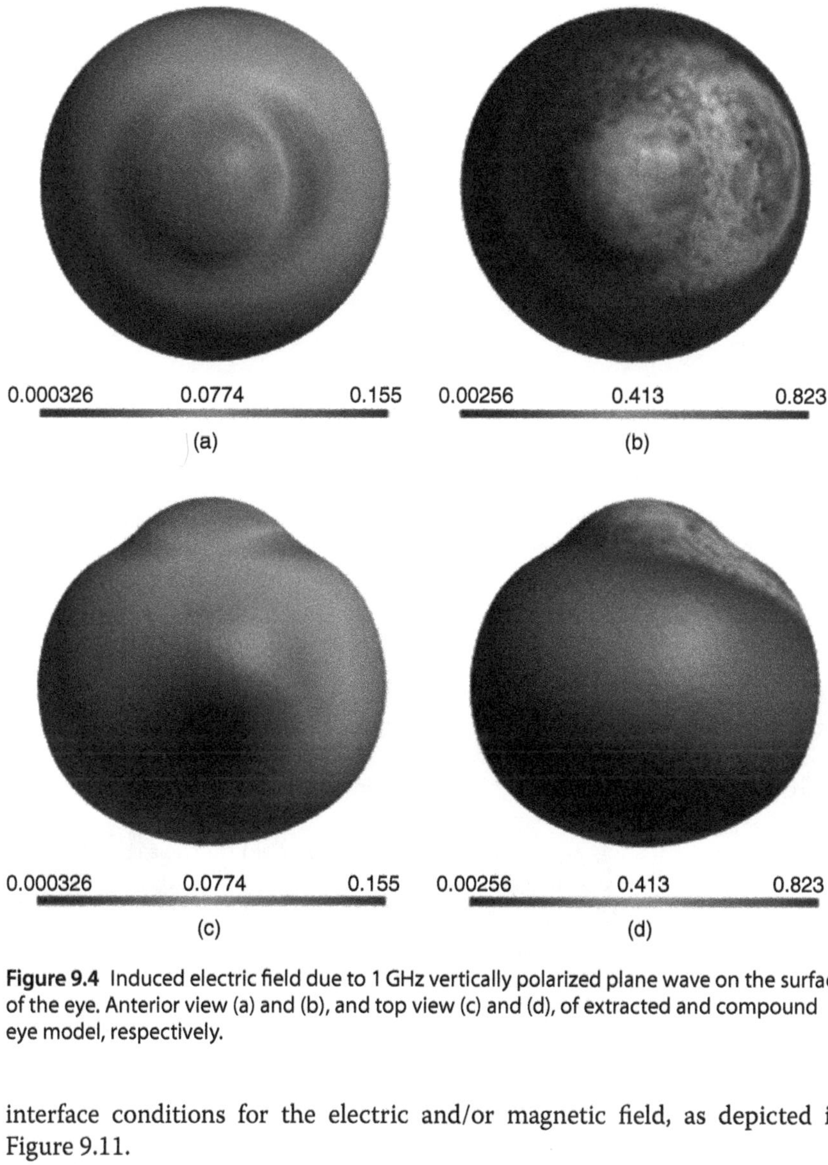


**Figure 9.4** Induced electric field due to 1 GHz vertically polarized plane wave on the surface of the eye. Anterior view (a) and (b), and top view (c) and (d), of extracted and compound eye model, respectively.

interface conditions for the electric and/or magnetic field, as depicted in Figure 9.11.

The lossy homogeneous object representing the brain is illuminated by the incident electromagnetic field ($\vec{E}^{inc}$; $\vec{H}^{inc}$).

Using the equivalence theorem, two problems are formulated, in terms of the equivalent electric and magnetic current densities $\vec{J}$ and $\vec{M}$ existing at the surface $S$, one for the region 1 (exterior to dielectric) and other for the region 2 (inside the dielectric) [9].

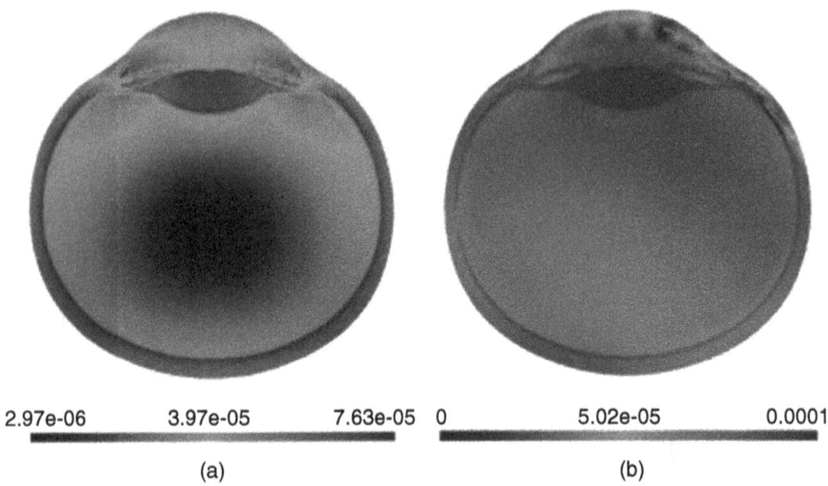


**Figure 9.5** Induced SAR due to 1 GHz vertically polarized plane wave in the transverse cross section of the (a) extracted eye model and (b) compound eye model.

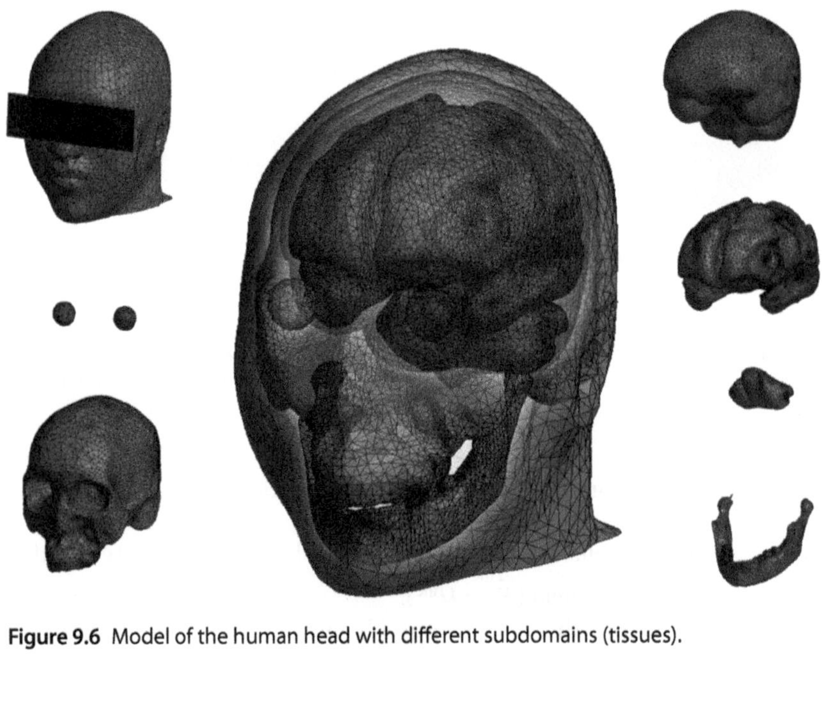

**Figure 9.6** Model of the human head with different subdomains (tissues).

**Table 9.2** Tissue dielectric parameters according to the 4-Cole–Cole model described in [15].

| | **900 MHz** | | **1800 MHz** | |
|---|---|---|---|---|
| **Tissue** | $\sigma$ **(S m$^{-1}$)** | $\varepsilon$ **(–)** | $\sigma$ **(S m$^{-1}$)** | $\varepsilon$ **(–)** |
| Brainstem | 0.591 | 38.886 | 0.915 | 37.011 |
| Cerebellum | 1.263 | 49.444 | 1.709 | 46.114 |
| Eye (vitreous) | 1.636 | 68.902 | 2.032 | 68.573 |
| Head skin | 0.867 | 41.405 | 1.185 | 38.872 |
| Skull and mandible | 0.339 | 20.788 | 0.588 | 19.343 |
| Gray matter | 0.942 | 52.725 | 1.391 | 50.079 |
| Muscle tissue | 0.943 | 55.032 | 1.341 | 53.549 |

*Source:* From [15]. Reproduced with permission of King's College London (United Kingdom) Department of Physics.

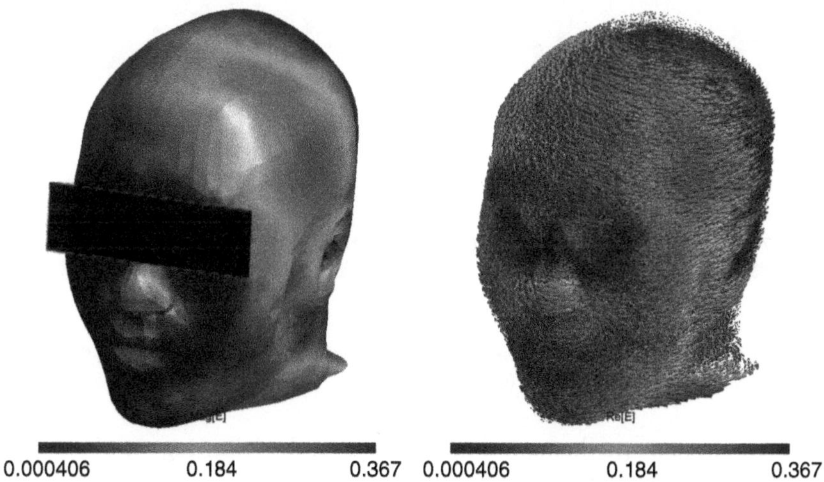


(a) (b)

**Figure 9.7** Electric field induced on the surface of the human head model due to 900 MHz horizontally polarized plane wave: (a) field magnitude and (b) field direction.

The boundary conditions at the surface $S$ are satisfied by introducing equivalent surface currents $\vec{J}_2 = -\vec{J}_1$ and $\vec{M}_2 = -\vec{M}_1$. Using the same procedure for the interior equivalent problem yields another homogeneous domain introducing the equivalent surface currents $\vec{J}_1$ and $\vec{M}_1$.

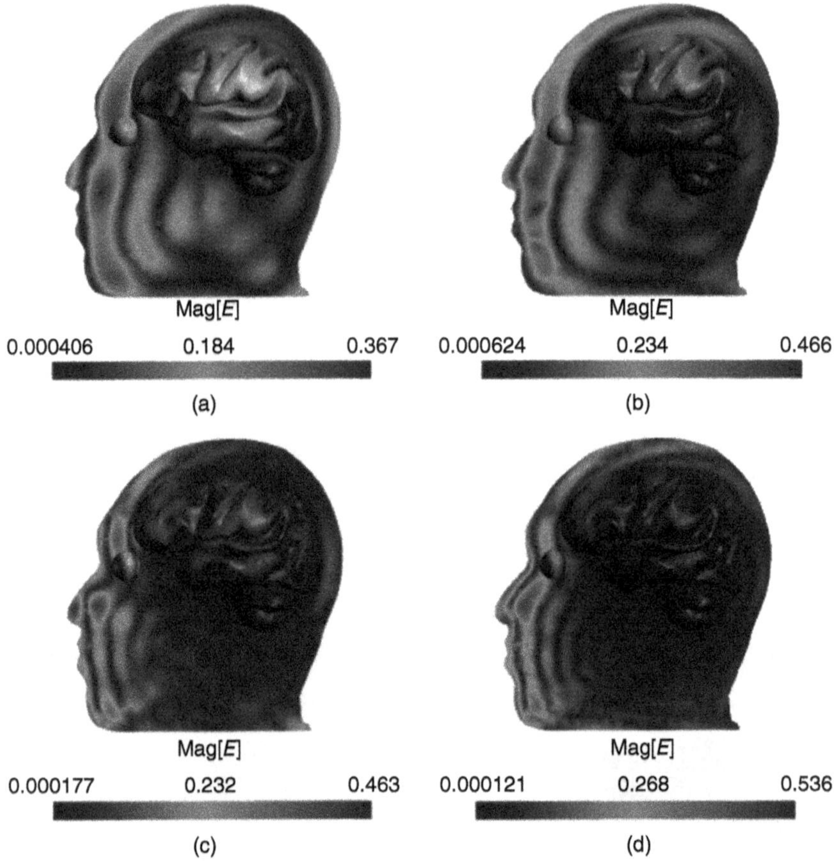


**Figure 9.8** Induced electric field in the sagittal cross section of the head due to (a) 900 MHz HP, (b) 900 MHz VP, (c) 1800 MHz HP, (d) 1800 MHz VP.

Performing some mathematical manipulations, the following set of integral equations is obtained [9]:

$$j\omega\mu_n \iint_S \vec{J}(\vec{r}')G_n(\vec{r},\vec{r}')\,\mathrm{d}S' - \frac{j}{\omega\varepsilon_n}\iint_S \nabla_S' \cdot \vec{J}(\vec{r}')\nabla G_n(\vec{r},\vec{r}')\,\mathrm{d}S' + \iint_S \vec{M}(\vec{r}') \times \nabla' G_n(\vec{r},\vec{r}')\,\mathrm{d}S' = \begin{cases} \vec{E}^{\mathrm{inc}}, & n = 1 \\ 0, & n = 2 \end{cases}, \tag{9.13}$$

where the Green's function for the homogeneous medium is given by

$$G_n(\vec{r},\vec{r}') = \frac{\mathrm{e}^{-jk_nR}}{4\pi R}; \quad R = |\vec{r} - \vec{r}'| \tag{9.14}$$

and $R$ is the distance from the source to observation point, respectively while $kn$ is the wave number of a medium $n$, $(n = 1; 2)$.

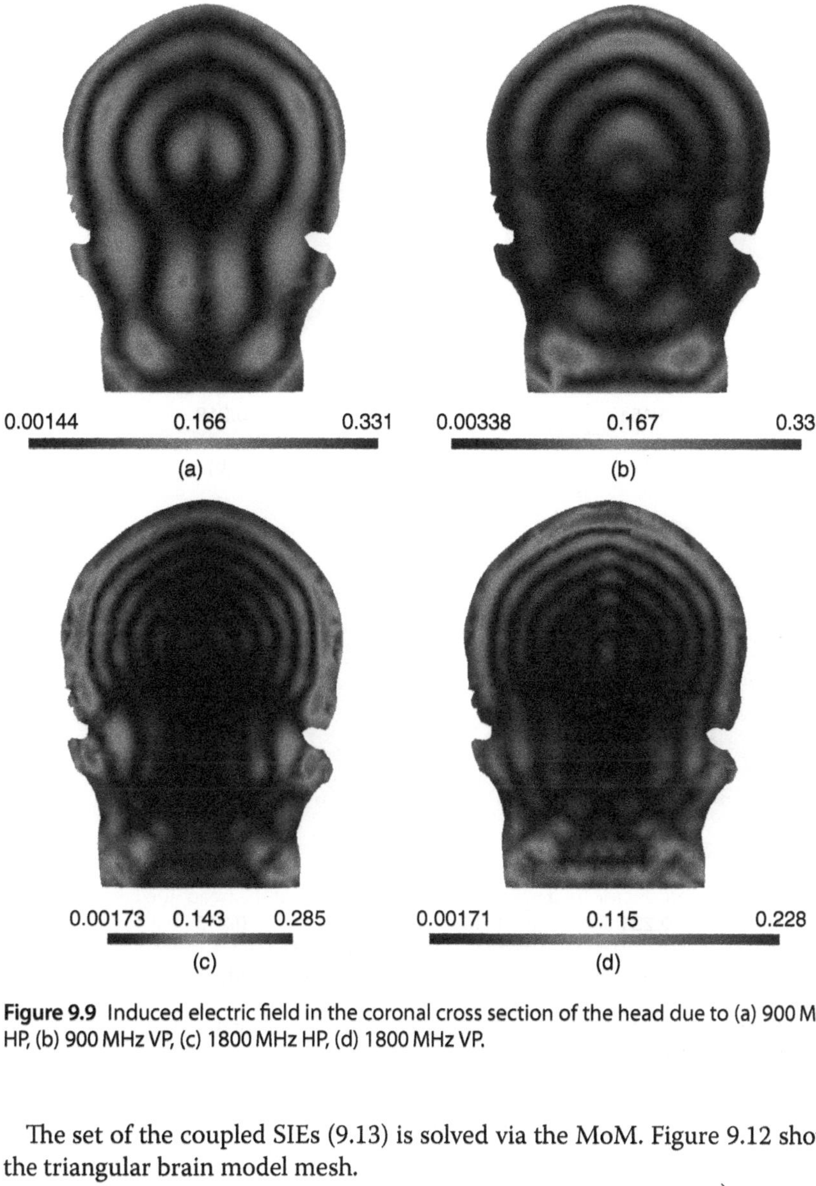


**Figure 9.9** Induced electric field in the coronal cross section of the head due to (a) 900 MHz HP, (b) 900 MHz VP, (c) 1800 MHz HP, (d) 1800 MHz VP.

The set of the coupled SIEs (9.13) is solved via the MoM. Figure 9.12 shows the triangular brain model mesh.

As a first step, the equivalent electric and magnetic currents $\vec{J}$ and $\vec{M}$ in (9.13) are expressed in terms of a linear combination of basis functions $\vec{f}_n$ and $\vec{g}_n$, respectively.

$$\vec{J}(\vec{r}) = \sum_{n=1}^{N} J_n \vec{f}_n(\vec{r}), \tag{9.15}$$

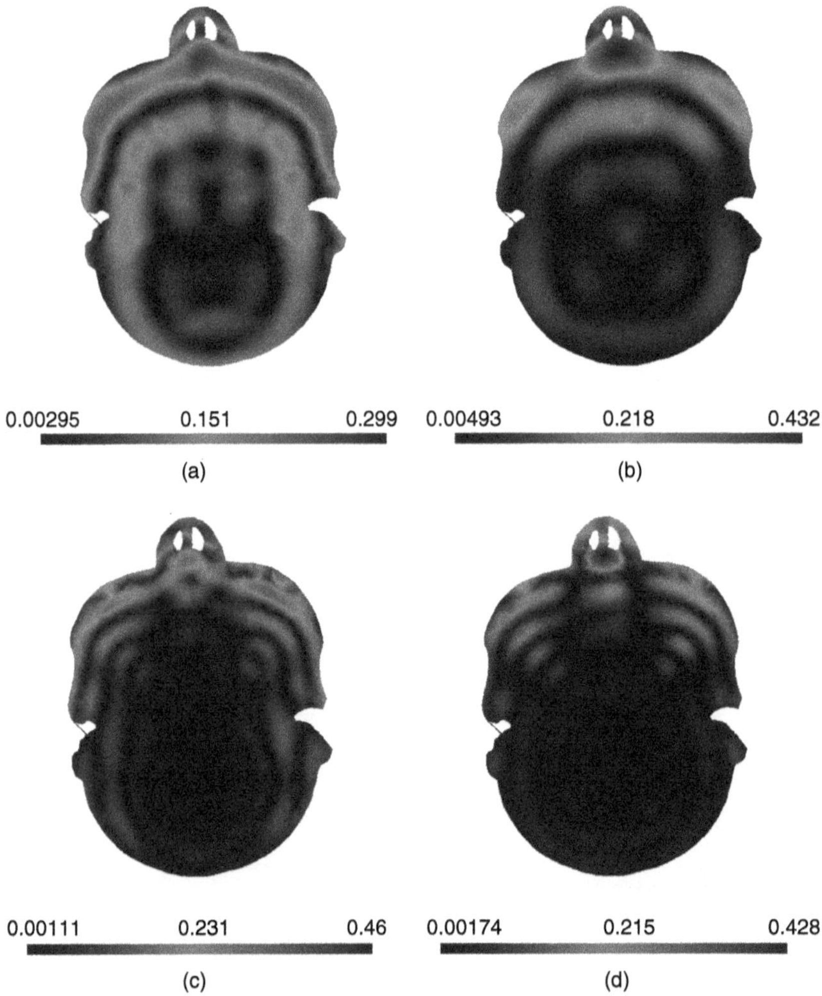


**Figure 9.10** Induced electric field in the transverse cross section of the head due to (a) 900 MHz HP, (b) 900 MHz VP, (c) 1800 MHz HP, (d) 1800 MHz VP.

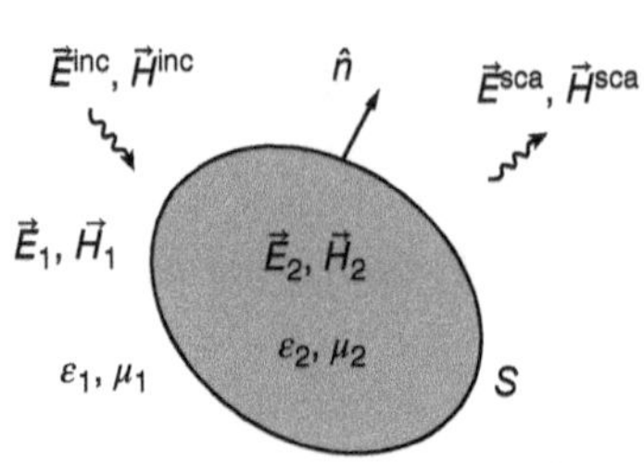


**Figure 9.11** The brain represented by a lossy homogeneous dielectric.

**Figure 9.12** Triangular mesh of the homogeneous brain model.

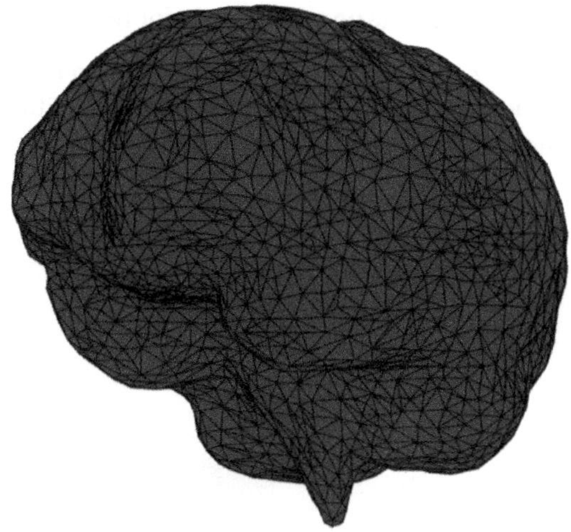

$$\vec{M}(\vec{r}) = \sum_{n=1}^{N} M_n \vec{g}_n(\vec{r}), \tag{9.16}$$

where $J_n$ and $M_n$ are unknown coefficients, while $N$ is the total number of triangular elements.

Applying the weighted residual approach, i.e. multiplying (9.13) by the set of a test functions $\vec{f}_m$ and integrating over the surface $S$, provided some mathematical manipulations are carried out, yields

$$\begin{aligned}
& j\omega\mu_i \sum_{n=1}^{N} J_n \iint_S \vec{f}_m(\vec{r}) \cdot \iint_{S'} \vec{f}_n(\vec{r}') G_i \,\mathrm{d}S' \,\mathrm{d}S \\
& + \frac{j}{\omega\varepsilon_i} \sum_{n=1}^{N} J_n \iint_S \nabla_S \cdot \vec{f}_m(\vec{r}) \iint_{S'} \nabla'_S \cdot \vec{f}_n(\vec{r}') G_i \,\mathrm{d}S' \,\mathrm{d}S + \\
& \pm \sum_{n=1}^{N} M_n \iint_S \vec{f}_m(\vec{r}) \cdot [\hat{n} \times \vec{g}_n(\vec{r}')] \,\mathrm{d}S \\
& + \sum_{n=1}^{N} M_n \iint_S \vec{f}_m(\vec{r}) \cdot \iint_{S'} \vec{g}_n(\vec{r}') \\
& \times \nabla' G_i \,\mathrm{d}S' \,\mathrm{d}S = \begin{cases} \iint_S \vec{f}_m(\vec{r}) \cdot \vec{E}^{\mathrm{inc}} \,\mathrm{d}S, & i = 1, \\ 0, & i = 2, \end{cases}
\end{aligned} \tag{9.17}$$

where subscript $i$ denotes the index of the medium. The details of the numerical solution procedure could be found elsewhere, e.g. in [9] or [17].

### 9.1.4 Computational Example for the Brain Exposure

Figure 9.13 shows the *SAR* distribution in the brain at $f = 900$ MHz due to the vertically polarized incident plane wave with power density $P = 5\,\text{mW cm}^{-2}$. The brain electrical parameters are $\varepsilon_r = 46$, $\sigma = 0.8\,\text{S m}^{-1}$ [9].

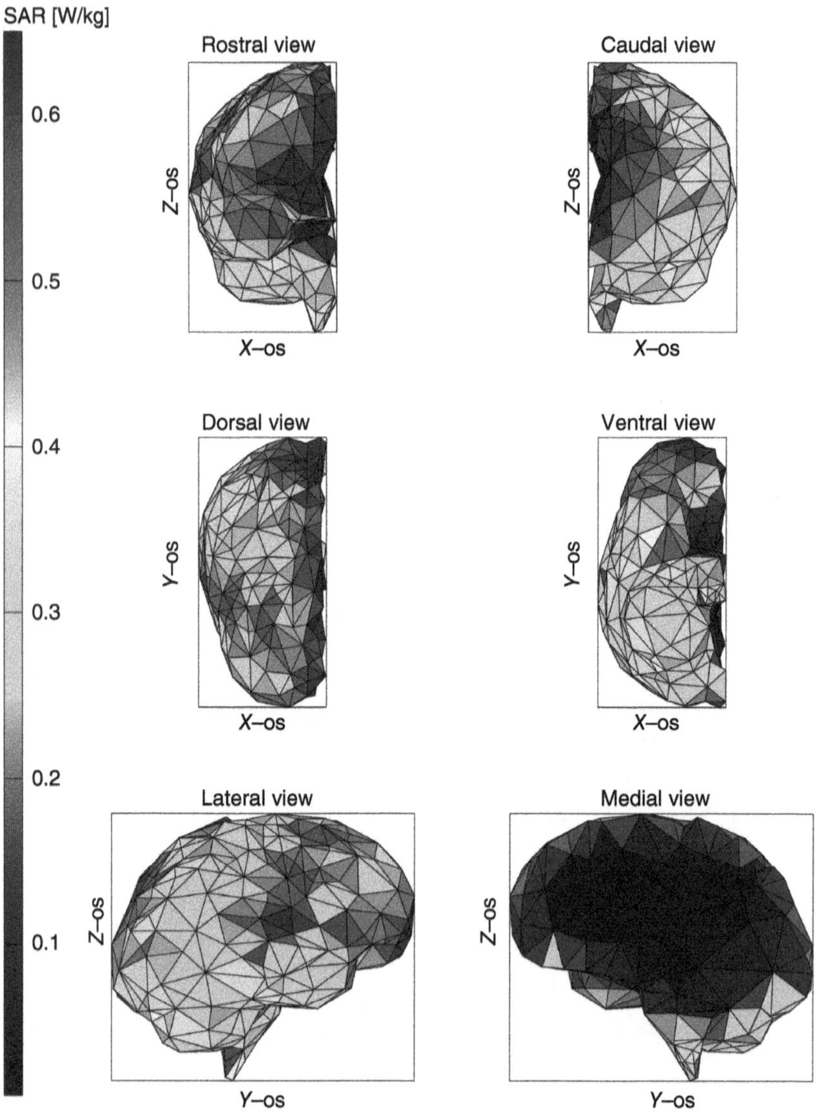


**Figure 9.13** SAR distribution in the brain at $f = 900$ MHz.

The obtained numerical results for peak and average SAR value are $SAR_{max} = 0.866$ W kg$^{-1}$ and $SAR_{avg} = 0.158$ W kg$^{-1}$, respectively.

## 9.2 Thermal Dosimetry Procedures

Provided the distribution of SAR inside tissues of interest is determined, the related temperature increase could be calculated.

### 9.2.1 Finite Element Solution of Bio-Heat Transfer Equation

The temperature increase in a tissue exposed to external HF fields is obtained by solving the Pennes bio-heat transfer equation [9]

$$\nabla \cdot (k\nabla T) + \rho_b c_b w(T_a - T) + Q_m + \rho \cdot \mathrm{SAR} = 0 \tag{9.18}$$

via FEM.

The integral formulation of (9.18) convenient for FEM solution is given by [9]

$$\int_{V'} [k\nabla f_j \cdot \nabla T + \rho_b c_b w \cdot T \cdot f_j] \mathrm{d}V' = \int_{V'} (\rho_b c_b w \cdot T_a + Q_m + \rho \cdot \mathrm{SAR}) \cdot f_j \mathrm{d}V' + \oint_{\partial V'} k f_j \nabla T \cdot \mathrm{d}\vec{S}'. \tag{9.19}$$

The appropriate boundary condition at the interface between skin and air is

$$q = H(T_s - T_a), \tag{9.20}$$

where $q$ denotes the heat flux density,

$$q = -\lambda \frac{\partial T}{\partial n}, \tag{9.21}$$

while $H$, $T_s$, and $T_a$ denote, respectively, the convection coefficient, the temperature of the skin, and the temperature of the air.

The standard finite element discretization of Helmholtz equation yields the following matrix equation:

$$[K]\{T\} = \{M\} + \{P\}, \tag{9.22}$$

where $[K]$ is the finite element matrix of the form

$$K_{ji} = \int_{\Omega_e} \nabla f_j (\lambda \nabla f_i) \mathrm{d}\Omega_e + \int_{\Omega_e} W_b C_{pb} f_j f_i \, \mathrm{d}\Omega_e, \tag{9.23}$$

while $\{M\}$ denotes the flux vector

$$M_j = \int_{\Gamma_e} \lambda \frac{\partial T}{\partial n} f_j \mathrm{d}\Omega_e \tag{9.24}$$

and $\{P\}$ stands for the source vector

$$P_j = \int_{\Omega_e} (W_b C_{pb} T_a + Q_m + \rho \cdot \text{SAR}) \cdot f_j d\Omega_e. \tag{9.25}$$

More details can be found elsewhere, e.g. in [9, 13, 18].

### 9.2.2 Numerical Results

A plane wave incident on the corneal part of the eye (Figure 9.14) can be treated as an unbounded scattering problem, using the hybrid method presented in Section 9.1.1. The eye is discretized to 36 027 tetrahedral elements.

Figure 9.15 shows numerical results for the SAR in the eye exposed to plane wave with power density of 10 W m$^{-2}$ at different frequencies. The key physical properties of the eye model are available in [9].

The maximal values of SAR at $f = 1$ GHz is 0.34 W kg$^{-1}$ and at $f = 2$ GHz is 0.62 W kg$^{-1}$.

The results show that as frequency increases SAR distribution becomes more localized, promoting the formation of "hot spots." Nevertheless, the obtained values of whole eye averaged SAR values are below the ICNIRP exposure limits for localized SAR in the head for the general public population (2 W kg$^{-1}$) [6]. It is worth noting that the SAR results presented in this chapter are in good agreement with the results reported in [5].

Figure 9.16 shows the related temperature increase in the eye due to the induced SAR presented in Figure 9.15.

The obtained maximal temperature increase is less than 0.1 °C while the hot spot region is within the vitreous body as the absorbed energy is focused in the vitreous region and there is also a lack of the blood perfusion within the vitreous body.

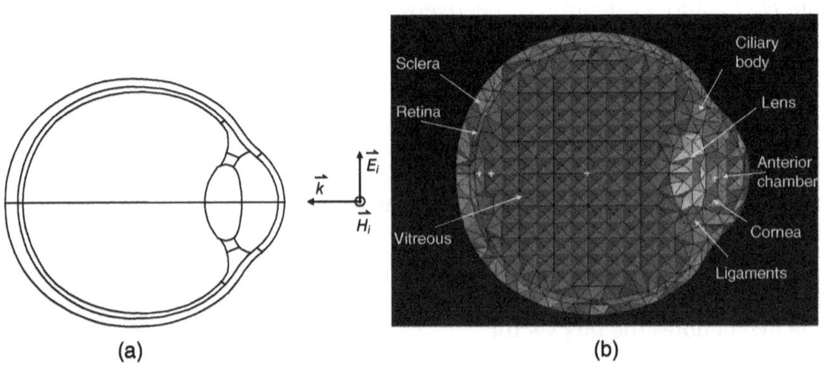


**Figure 9.14** (a) Model of the eye exposed to plane wave. (b) Meshing detail.

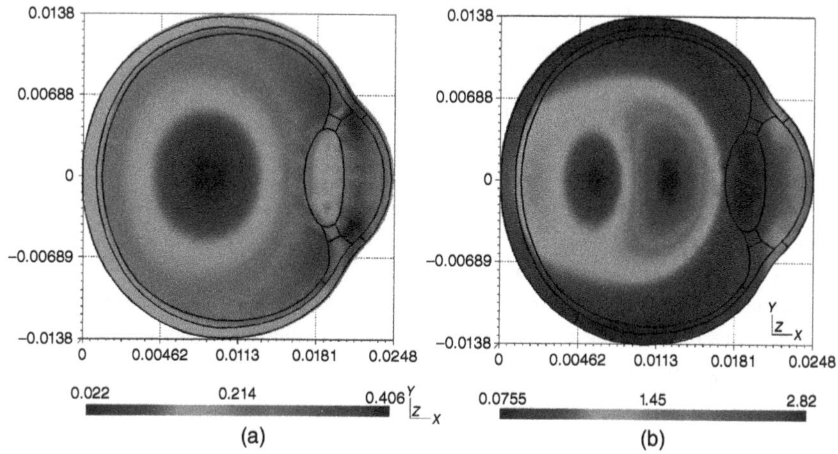


**Figure 9.15** SAR in the eye exposed to plane wave of power density 10 W m$^{-2}$ at frequency (a) 1 GHz, (b) 2 GHz.

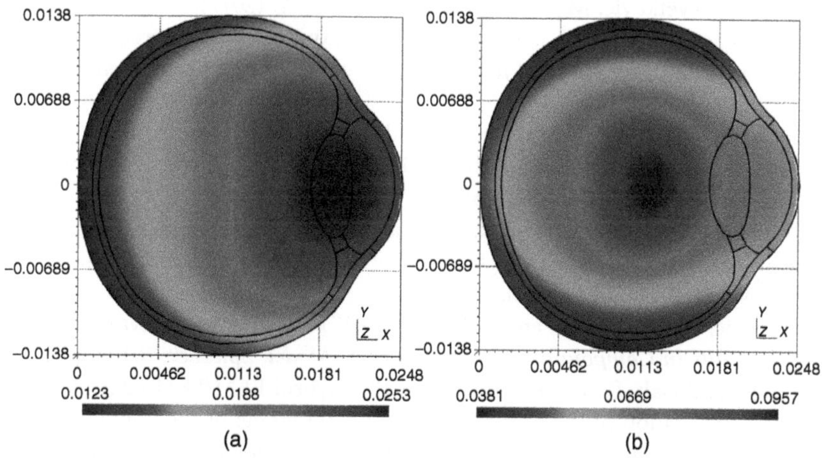


**Figure 9.16** Temperature increase in the eye exposed to plane wave of power density $P = 10$ W m$^{-2}$ at frequency (a) 1 GHz, (b) 2 GHz.

# References

1 Hand, J.W. (2008). Modeling the interaction of electromagnetic fields (10 MHz–10 GHz) with the human body: methods and applications. *Physics in Medicine and Biology* **53** (16): 243–286.

2 Poljak, D. (2011). Electromagnetic fields: environmental exposure. In: *Encyclopedia of Environmental Health*, vol. **2** (ed. J.O. Nriagu), 259–268. Burlington: Elsevier.

3 Poljak, D., Cavka, D., Dodig, H. et al. (2014). On the use of boundary element analysis in bioelectromagnetics. *Engineering Analysis with Boundary Elements (Special issue on Bioelectromagnetics)* **49**: 2–14.
4 Singh, K.D., Longan, N.S., and Gilmartin, B. (2006). Three dimensional modeling of the human eye based on magnetic resonance imaging. *Investigative Opthamology and Visual Science* **47**: 2272–2279.
5 Hirata, A. (2005). Temperature increase in human eyes due to near-field and far-field exposures at 900 MHz, 1.5 GHz, and 1.9 GHz. *IEEE Transactions on Electromagnetic Compatibility* **47** (1): 68–76.
6 International Commission on Non-Ionizing Radiation Protection (1998). Guidelines for limiting exposure to time-varying electric, magnetic and electromagnetic fields (up to 300 GHZ). *Health Physics* **74** (4): 494–522.
7 International Commission on Non-Ionizing Radiation Protection (2010). Guidelines for limiting exposure to time-varying electric and magnetic fields (1 HZ – 100 kHZ). *Health Physics* **99** (6): 818–836.
8 Poljak, D. (2003). *Human Exposure to Electromagnetic Fields.* Southampton-Boston: WIT Press.
9 Poljak, D., Cvetković, M., Dodig, H., and Peratta, A. (2017). Electromagnetic-thermal analysis for human exposure to high frequency (HF) radiation. *International Journal of Design and Nature and Ecodynamics* **12** (1): 55–67.
10 Poljak, D., Cvetkovic, M., Cavka, D., et al. (2017). Boundary integral methods in bioelectromagnetics and biomedical applications of electromagnetic fields. BEM/MRM 40, Southampton.
11 Dodig, H., Cvetkovic, M., Poljak, D., et al. (2017). Hybrid FEM/BEM for human head exposed to high frequency electromagnetic radiation. BEM/MRM 40, Southampton.
12 Cvetkovic, M., Dodig, H., and Poljak, D. (2017). *Comparison of Numerical SAR Results in Compound and Extracted Eye Model.* SpliTech 2017, Split.
13 Dodig, H., Poljak, D. and Peratta, A. (2012). Hybrid BEM/FEM edge element computation of the thermal rise in the 3D model of the human eye induced by high frequency EM waves. 2012 International Conference on Software, Telecommunications and Computer Networks, Split.
14 Laakso, I., Tanaka, S., Koyama, S. et al. (2015). Intersubject variability in electric fields of motor cortical tDCS. *Brain Stimulation* **8** (5): 906–913.
15 Gabriel, C. and Gabriel, S. (1996). Compilation of the Dielectric Properties of Body Tissues at RF and Microwave Frequencies. Technical report, King's College London, *AL/OE-TR-1996-0037*.
16 Hirata, A., Ito, N., and Fujiwara, O. (2009). Influence of electromagnetic polarization on the whole-body averaged sar in children for plane-wave exposures. *Physics in Medicine and Biology* **54** (4): N59.

17 Cvetković, M. and Poljak, D. (2014). An efficient integral equation based dosimetry model of the human brain. Proceedings of the 2014 International Symposium on Electromagnetic Compatibility (EMC Europe 2014), Gothenburg, Sweden (1–4 September 2014), 375–380.
18 Poljak, D., Dodig, H., Cavka, D., and Peratta, A. (2012). Some numerical methods of thermal dosimetry for applications in bioelectromagnetics. Proceedings of the Heat Transfer 2012, Split, Croatia (September 2012), 271–280.

# 10

# Biomedical Applications of Electromagnetic Fields

While human exposure to electromagnetic fields generated by different sources has initiated many questions regarding potential adverse effects, particularly for the brain and eye on exposure to high frequency (HF) radiation, some biomedical applications of electromagnetic fields are also of great interest. Thus, electromagnetic fields are applied in medical diagnostics and for therapy purposes featuring the use of techniques such as transcranial magnetic stimulation (TMS) [1–3], percutaneous electrical nerve stimulation (PENS), transcutaneous electrical nerve stimulation (TENS) [4–6], or intraoperational methods such as transcranial electrical stimulation (TES) and direct cortical stimulation (DCS). TMS is a noninvasive and painless technique for excitation or inhibition of brain regions, and in the last few decades became an important tool in preoperative neurosurgical diagnosis/evaluation of patients [7–9]. TMS is also used for therapeutic purposes (e.g. depression treatment), and is the subject of interest in neurophysiologic research. Various efficiency aspects of TMS stimulation have been stressed out primarily due to differences in relevant stimulation parameters such as pulse waveform, frequency, and intensity of treatment. The choice of optimal stimulation intensity is still investigated in many TMS studies.

The coil orientation and positioning appreciably influence misalignment from the targeted brain region, thus reducing TMS efficiency, although by using navigated TMS, this problem could be somewhat alleviated. In addition to the stimulation parameters, being adjustable to the requirements of the TMS operator, to a certain extent the difference in individual brain morphology due to age, gender, or health status and the biological tissue parameters appreciably influence the distribution of the induced fields in the brain. Most of the parameters are obtained under different measurements on *ex vivo* animal and human tissues, and usually exhibit large variations from their average values. The level of uncertainty in the values of brain conductivity and permittivity is even more pronounced at low frequencies.

Modeling and computer simulation of TMS phenomena could be helpful in determining the exact location of stimulation, in the interpretation of

*Computational Methods in Electromagnetic Compatibility: Antenna Theory Approach versus Transmission line Models*, First Edition. Dragan Poljak and Khalil El Khamlichi Drissi.

experimental results as well as in designing some more efficient stimulation setups. Realistic TMS models can also provide a more reliable prediction of the distribution of internal fields and currents by taking into account the variability of the various input parameters.

Furthermore, techniques such as PENS, electro-acupuncture, or TENS are widely used in the treatment of neurological disorders. Basically, there are two types of electric potential occurring in the stimulated nerve cells: the electrotonic potential and the action potential. The electrotonic potential existing due to local changes in ion conductivity decays along the fiber, and the passive membrane then shows linear nature, which satisfies Ohm's law. The action potential is initiated when the threshold potential is achieved. Studies on electrical excitation of nerves, among other aspects involve nerve excitation using stimulating electrodes, nerve conduction velocity tests, or noninvasive stimulation of nerves. The exceptional complexity of the nervous system, particularly with nerve cells as its basis, and widespread neurological disorders and a need for better insight into the complex functioning of the nervous system motivate researchers to develop efficient and accurate nerve fiber models. Thus, such computational models of nerve fibers could provide a study of the nerve fiber response to different stimulus waveforms, often used within various electrotherapy techniques, particularly electro-acupuncture and PENS. Such a model can be a beneficial and versatile tool for interpreting experimental results and analyzing variables for which there could be difficulties in the laboratory implementation.

This chapter deals with the integral equation approach to study TMS induced fields and nerve fiber excitation used in therapeutic procedures such as electro-acupuncture and/or PENS reported in [10] by providing more mathematical details and presenting some additional computational examples. The work reported in [11], together with [10, 12], has been carried out through tasks and activities within the framework of IEEE/ICES TC95 SC6 EMF Dosimetry Modeling and COST Action BM1309.

## 10.1 Modeling of Induced Fields due to Transcranial Magnetic Stimulation (TMS) Treatment

The lossy dielectric model of the brain exposed to TMS electromagnetic field is based on the set of coupled surface integral equations (SIEs) that could be derived from the equivalence theorem and by forcing the corresponding interface conditions for the electric and/or magnetic field [1], as indicated in Figure 10.1.

The lossy homogeneous object representing the brain is illuminated by the electromagnetic wave characterized by incident electric field $\vec{E}^{inc}$ and incident magnetic field $\vec{H}^{inc}$.

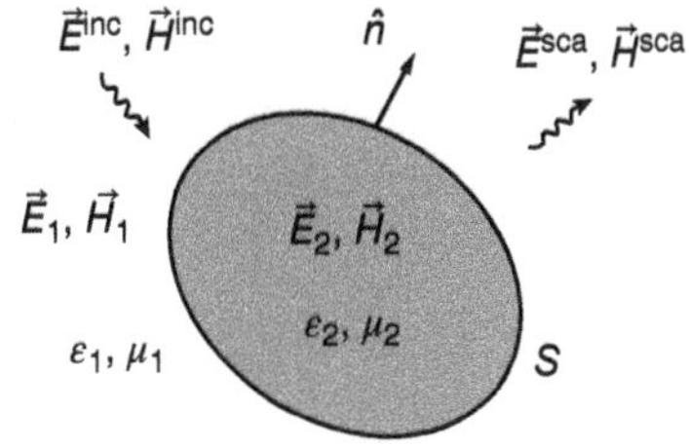


**Figure 10.1** The lossy homogeneous dielectric brain model.

Performing some mathematical manipulations, the set of coupled SIEs is obtained [1]:

$$j\omega\mu_n \iint_S \vec{J}(\vec{r}')G_n(\vec{r},\vec{r}')\ \mathrm{d}S' - \frac{j}{\omega\varepsilon_n}\iint_S \nabla'_S\cdot\vec{J}(\vec{r}')\nabla G_n(\vec{r},\vec{r}')\ \mathrm{d}S' + \iint_S \vec{M}(\vec{r}')\times\nabla' G_n(\vec{r},\vec{r}')\ \mathrm{d}S' = \begin{cases}\vec{E}^{inc}, & n=1,\\ 0, & n=2,\end{cases} \tag{10.1}$$

where $\vec{J}$ and $\vec{M}$ represent equivalent electric and magnetic current density, respectively, and $G_n$ is the interior/exterior Green function given by [1]

$$\left| G_n(\vec{r}\vec{r}') = \frac{e^{-jk_nR}}{4\pi R};\quad R = |\vec{r}-\vec{r}'| \right| \tag{10.2}$$

and $R$ is the distance from the source to observation point, respectively, while $k_n$ denotes the wave number of a considered medium $n$.

The brain model is generated from a freely available Google Sketchup rendering of the human brain, as shown in Figure 10.2.

The set of integral equations (10.1) is solved by means of an efficient method of moments (MoM) scheme reported elsewhere, e.g. in [1]. For the sake of completeness, the MoM procedure is outlined in this section as well.

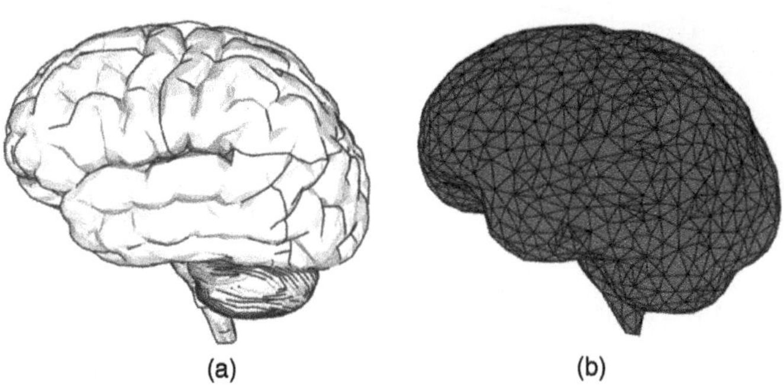

**Figure 10.2** The human brain model for SIE formulation. (a) Detailed 3-D model from Google Sketchup. (b) Final model discretized using the triangular elements.

As a first step, the equivalent electric and magnetic currents $\vec{J}$ and $\vec{M}$ in (10.1) are expressed by means of a linear combination of basis functions $\vec{f}_n$ and $\vec{g}_n$, respectively:

$$\vec{J}(\vec{r}) = \sum_{n=1}^{N} J_n \vec{f}_n(\vec{r}), \tag{10.3}$$

$$\vec{M}(\vec{r}) = \sum_{n=1}^{N} M_n \vec{g}_n(\vec{r}), \tag{10.4}$$

where $J_n$ and $M_n$ are unknown coefficients, while $N$ is the total number of triangular elements.

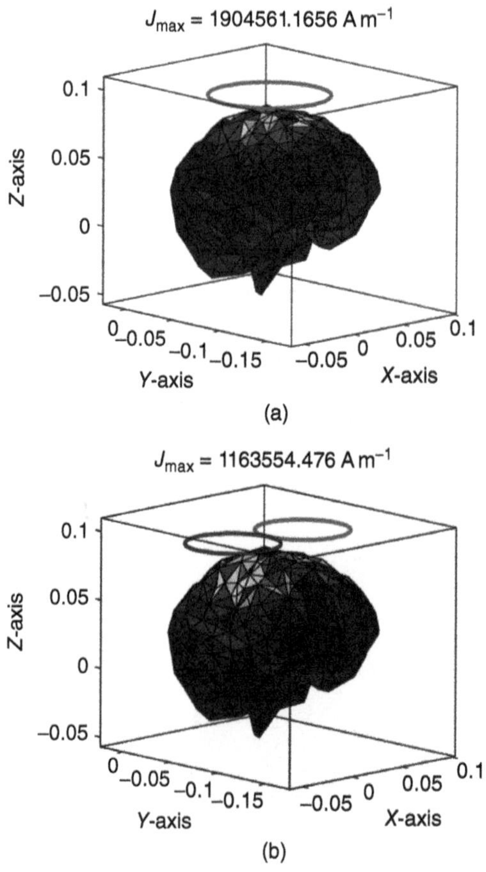


**Figure 10.3** Equivalent electric current density for (a) circular coil, (b) figure-8 coil.

Applying the weighted residual approach, i.e. multiplying (10.1) by the set of a test functions $\vec{f}_m$ and integrating over the surface $S$, after some mathematical manipulations, it follows that

$$
\begin{aligned}
&j\omega\mu_i \sum_{n=1}^{N} J_n \iint_S \vec{f}_m(\vec{r}) \cdot \iint_{S'} \vec{f}_n(\vec{r}')G_i \; \mathrm{d}S'\,\mathrm{d}S \\
&+\frac{j}{\omega\varepsilon_i} \sum_{n=1}^{N} J_n \iint_S \nabla_S \cdot \vec{f}_m(\vec{r}) \iint_{S'} \nabla'_S \cdot \vec{f}_n(\vec{r}')G_i \; \mathrm{d}S'\,\mathrm{d}S \\
&\pm \sum_{n=1}^{N} M_n \iint_S \vec{f}_m(\vec{r}) \cdot [\hat{n} \times \vec{g}_n(\vec{r}')] \; \mathrm{d}S \\
&+ \sum_{n=1}^{N} M_n \iint_S \vec{f}_m(\vec{r}) \cdot \iint_{S'} \vec{g}_n(\vec{r}') \\
&\times \nabla' G_i \; \mathrm{d}S'\,\mathrm{d}S = \begin{cases} \iint_S \vec{f}_m(\vec{r}) \cdot \vec{E}^{\mathrm{inc}} \; \mathrm{d}S, & i = 1, \\ 0, & i = 2, \end{cases}
\end{aligned} \tag{10.5}
$$

where subscript $i$ denotes the index of the medium. The details of the numerical solution procedure could be found elsewhere, e.g. in [12].

### 10.1.1 Numerical Results

Figures 10.3 and 10.4 show numerical results for the equivalent electric and magnetic current densities for the case of (a) circular coil and (b) figure-8 coil. The actual coil insulation and casing are not taken into account. Each coil is discretized into 80 linear segments. The coils are driven by a sinusoidal current of amplitude $I = 2843$ A, $f = 2.44$ kHz. The radius of the circular coil and figure-8 coil is 4.5 cm and 3.5 cm, respectively, while the number of windings for the circular coil is 14 and for figure-8 is 15 turns.

Figures 10.5 and 10.6 show the numerical results for the sagittal and transversal cross sections of the internal and external field, respectively, due to the circular coil and figure-8 coil. In both cases, the coils are positioned tangentially to the brain surface, 1 cm over the primary motor cortex area, i.e. the distance between the coil geometric center and brain surface is 1 cm.

The results shown in Figures 10.5 and 10.6 confirm that the maximum electric field of figure-8 coil is induced directly under the coil geometric center, while for the case of circular coil maximum field is induced under the coil windings. In addition to the induced electric field value, it is important to consider the orientation of the field vector. The results show the obvious importance of the correct TMS coil placement with respect to the brain surface.

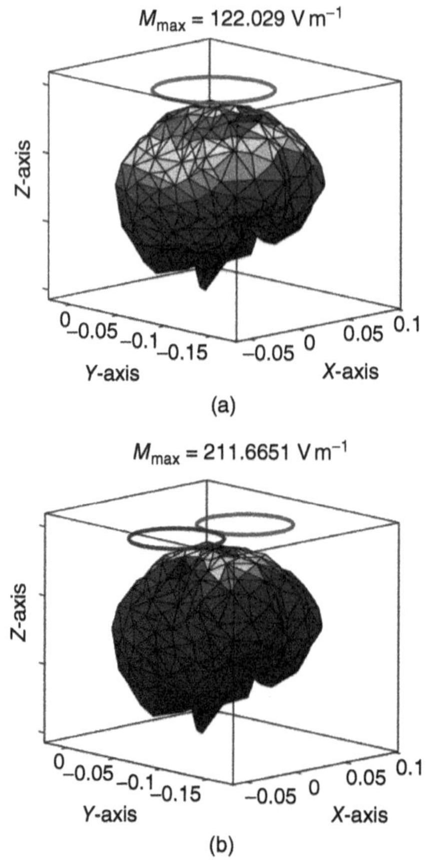


**Figure 10.4** Equivalent magnetic current density for (a) circular coil, (b) figure-8 coil.

## 10.2 Modeling of Nerve Fiber Excitation

Figure 10.7 shows the myelinated nerve fiber with an arbitrary number of Ranvier's nodes represented by a straight thin wire antenna [6, 13].

Note that $L$ denotes the fiber length, $a$ is the inner axon radius and $b$ is the radius including the myelin sheath. Electrode nerve fiber stimulation is taken into account by means of the equivalent current source $I_g$ located at the fiber beginning. The current generator $I_g$ represents the nerve fiber stimulation used in electro-acupuncture or PENS, which both make use of thin needles injected through the skin. The fiber, oriented along the $x$ axis, is assumed to be immersed in an unbounded homogeneous lossy medium.

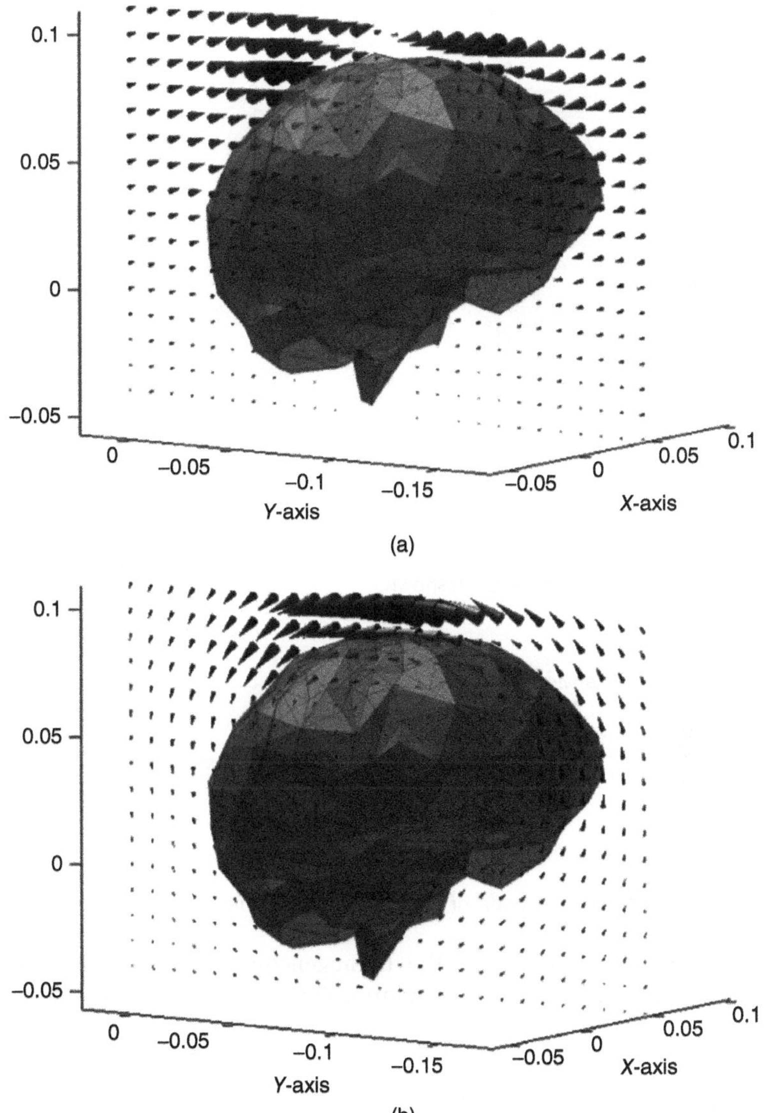


**Figure 10.5** Sagittal cross section of the electric field induced in the brain model by (a) circular coil, (b) figure-8 coil. Coils not shown.

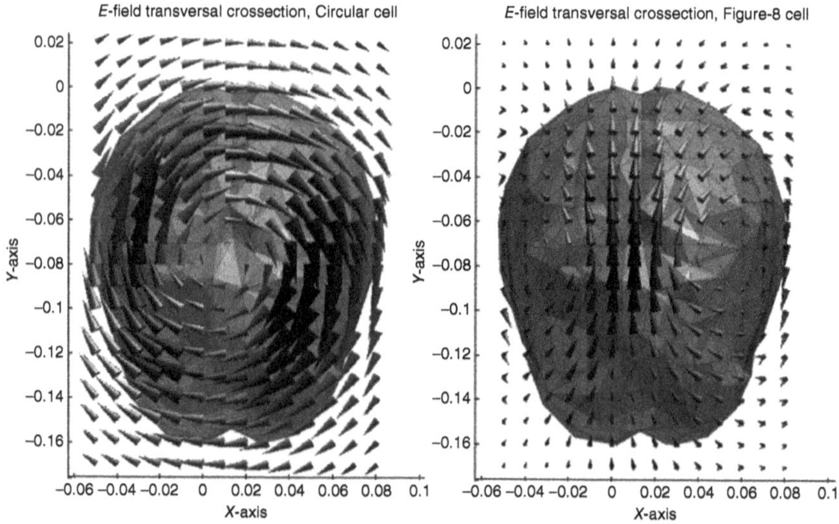


**Figure 10.6** Transversal cross section of the electric field induced in the brain model by (a) circular coil, (b) figure-8 coil. Coils not shown.

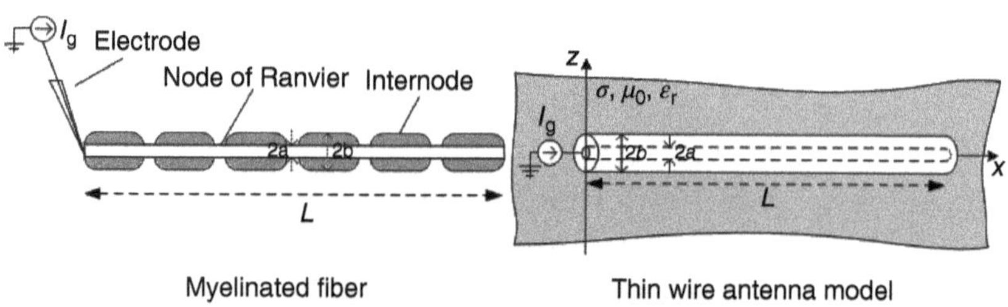


**Figure 10.7** Thin wire antenna model of the myelinated nerve fiber.

The formulation is based on the homogeneous Pocklington's integro-differential equation for the unknown intracellular current $I_a$ along the nerve fiber [6, 13]:

$$-\frac{1}{j4\pi\omega\varepsilon_{\text{eff}}}\int_0^L\left(\frac{\partial^2}{\partial x^2}-\gamma^2\right)g(x,x')I_a(x')\mathrm{d}x' = 0, \tag{10.6}$$

while $g(x, x')$ is the lossy medium Green function defined as

$$g(x,x') = \frac{\mathrm{e}^{-\gamma R}}{R}, \tag{10.7}$$

where $R$ denotes the distance from the source to the observation point, respectively, and $\gamma$ is the complex propagation constant given by

$$\gamma = \sqrt{j\omega\mu\sigma - \omega^2\mu\varepsilon_0\varepsilon_r} \tag{10.8}$$

and $\varepsilon_{\text{eff}}$ is the complex permittivity of a lossy medium given by

$$\varepsilon_{\text{eff}} = \varepsilon_0\varepsilon_{\text{r}} - j\frac{\sigma}{\omega} \tag{10.9}$$

with $\sigma$, $\varepsilon_{\text{r}}$, and $\omega$ being the conductivity and relative permittivity of the medium, respectively, and $\omega$ is the angular frequency.

The current generator $I_{\text{g}}$ is taken into account via the conditions at the nerve fiber ends:

$$I_a(0) = I_{\text{g}}, \quad I_a(L) = 0. \tag{10.10}$$

The properties of the lossy medium, nerve fiber membrane, and myelin sheath are taken into account via the conductivity and relative permittivity, which are obtained from the cable equation and the transmission line equation, respectively, as described in [6, 13].

The Helmholtz equation for the corresponding transmembrane voltage is given by [6, 13]

$$\frac{\partial^2 V}{\partial x^2} - \gamma_m^2 V = 0, \tag{10.11}$$

where

$$\gamma_m = \sqrt{\frac{1 + j\omega\tau}{\lambda^2}} \tag{10.12}$$

and $V$ is a transmembrane voltage, while $\lambda$ is the length constant defined by [13]

$$\lambda = \sqrt{\frac{r_{\text{m}}a}{2\rho_{\text{a}}}} \tag{10.13}$$

and $\tau$ is the time constant given by [13]

$$\tau = r_{\text{m}}c_{\text{m}}. \tag{10.14}$$

$\rho_{\text{a}}$ is the resistivity of axoplasm, $r_{\text{m}}$ the myelin layer resistance for unit area, and $c_{\text{m}}$ is the capacitance of the Ranvier's node membrane per unit area.

The corresponding relations for the conductivity and relative permittivity are obtained via the procedure reported in [6]. Homogeneous Pocklington's integro-differential equation (10.6) is solved via the Galerkin–Bubnov indirect boundary element method (GB-IBEM). The details could be found elsewhere, e.g. in [13]. For the sake of completeness, the procedure is outlined in this section as well.

Thus, using the boundary element formalism, Equation (10.6) is transformed into the set of linear equations, which can be written using the following matrix notation [13, 14]:

$$\sum_{i=1}^{n} [Z]_{ji}^{e}\{\alpha\}_{i}^{e} = 0, \quad j = 1, 2, \ldots, n, \tag{10.15}$$

where $n$ is the number of wire segments, $[Z]_{ji}^{e}$ is the mutual impedance matrix representing the interaction of the observation segment $j$ with the source segment $i$, and $\{\alpha\}_{i}^{e}$ is the solution vector, expressed in global nodes. The mutual impedance matrix is

$$[Z]_{ji}^{e} = -\frac{1}{j4\pi\omega\varepsilon_{\text{eff}}}$$

$$\left[\int_{\Delta l_j}\int_{\Delta l_i}\{D\}_j\{D'\}_i^T g(x,x') - \gamma^2 \int_{\Delta l_j}\int_{\Delta l_i}\{f\}_j\{f'\}_i^T g(x,x')\mathrm{d}x\,\mathrm{d}x'\right], \quad (10.16)$$

where matrices $\{f\}$ and $\{f'\}$ contain linear shape functions and matrices $\{D\}$ and $\{D'\}$ contain their derivatives. $\Delta l_j$ and $\Delta l_i$ are the observation and source segment lengths, respectively. The nerve fiber stimulation, in the form of an equivalent current generator, is incorporated into the matrix system via boundary conditions (11).

### 10.2.1 Passive Nerve Fiber

When modeling the passive nerve fiber, the ionic current in each Ranvier's node is to be taken into account. Thus, the additional current sources in the nonactivated Ranvier's nodes are impressed. Figure 10.8 shows the passive nerve fiber model having three Ranvier's nodes and four internodes.

Analyzing the CRRSS (Chiu, Ritchie, Rogert, Stagg, and Sweeney) model [5, 15, 16], it follows that the ionic current in each nonactivated Ranvier's node is approximately equal to $-0.6\ I_a$, where $I_a$ is the intracellular current flowing into the observed Ranvier's node. Intracellular current $I_{a2}$, which leaves the first Ranvier's node is equal to $0.4\ I_{a1}$. By analogy, the intracellular current $I_{a3}$ flowing out of the second Ranvier's node is equal to $0.4\ I_{a2}$ and the intracellular current $I_{a4}$ that leaves the third Ranvier's node is $0.4\ I_{a3}$.

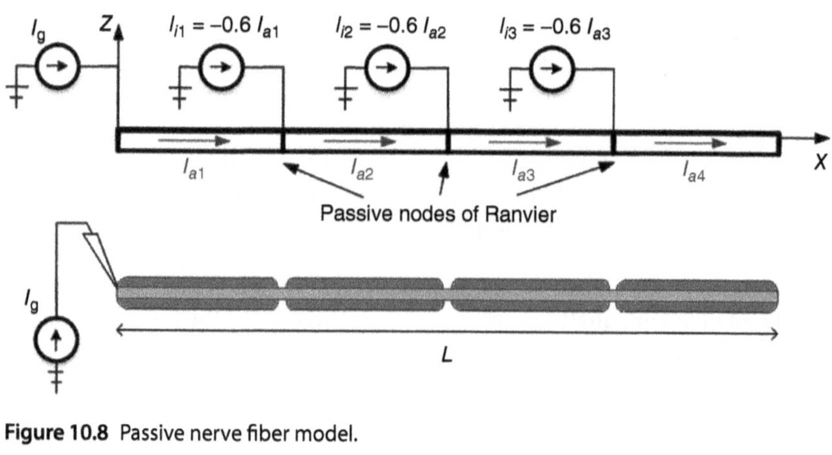


Figure 10.8 Passive nerve fiber model.

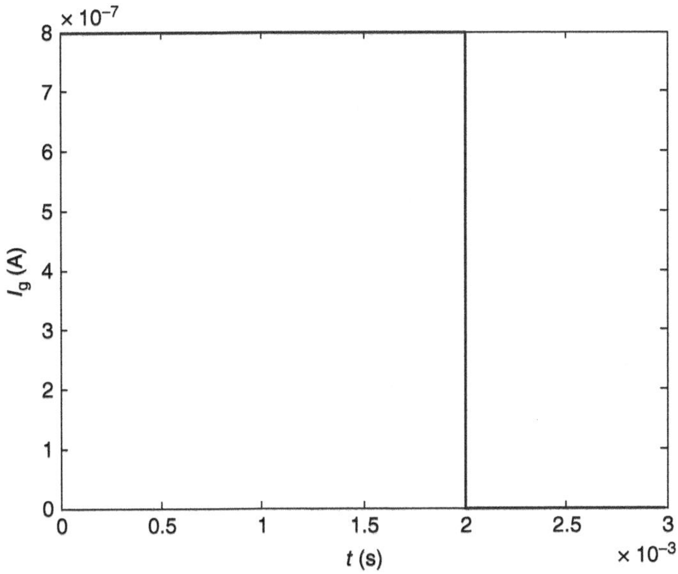


**Figure 10.9** Rectangular subthreshold current pulse.

### 10.2.2 Numerical Results for Passive Nerve Fiber

The intracellular current is calculated for the nerve fiber of 2 cm, with 9 Ranvier's nodes and 10 internodes. The fiber is stimulated at its beginning, using a subthreshold rectangular current pulse, plotted in Figure 10.9.

Ranvier's nodes are located at $x = 2, 4, 6, 8, 10, 12, 14, 16, 18$ mm along the fiber, respectively. The intracellular current distribution along the passive nerve fiber is plotted in Figure 10.10. The result is compared to the CRRSS model [5, 15, 16], and calculated for the time instant $t = 1$ ms.

Figures 10.11 and 10.12 show the subthreshold intracellular current versus time in the Ranvier's nodes 2, and 6, respectively.

The results for the intracellular current obtained by the antenna model seem to be in rather satisfactory agreement with the results obtained using the CRRSS model. Note that the intracellular current waveform in Ranvier's nodes follows the excitation pulse form, as due to the subthreshold current stimulation, the voltage gated channels are not activated.

### 10.2.3 Active Nerve Fiber

Model of an active nerve fiber is presented on a myelinated nerve fiber with three active Ranvier's nodes and four internodes, as shown in Figure 10.13.

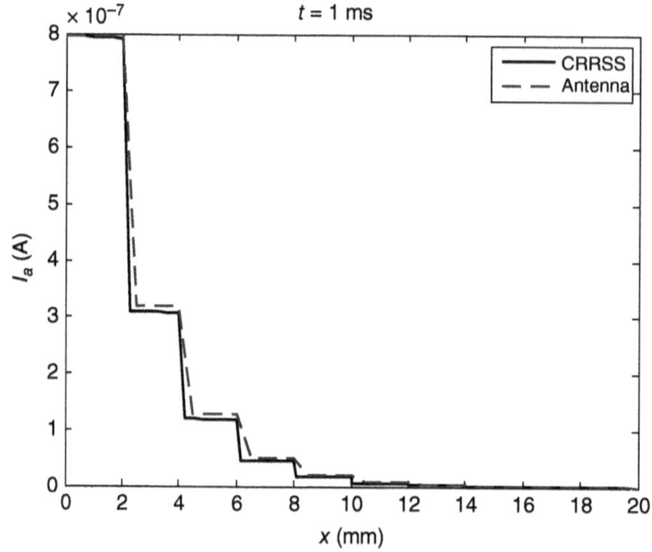


**Figure 10.10** Intracellular current along the passive nerve fiber, $t = 1$ ms, $L = 2$ cm.

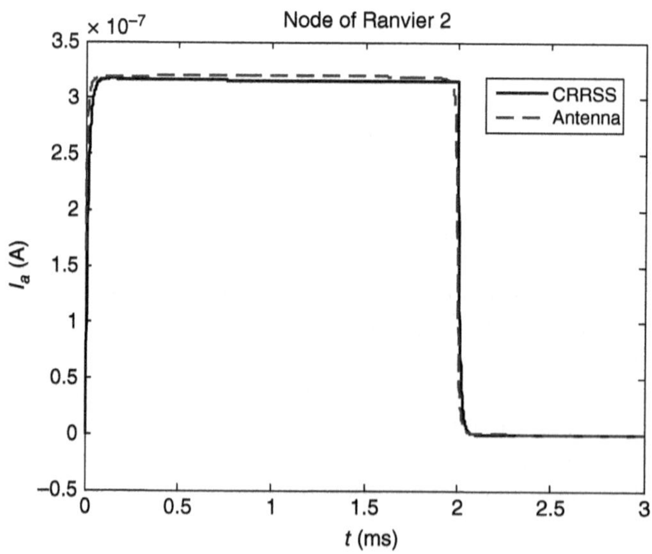


**Figure 10.11** Intracellular current in the passive Ranvier's node 2, $L = 2$ cm.

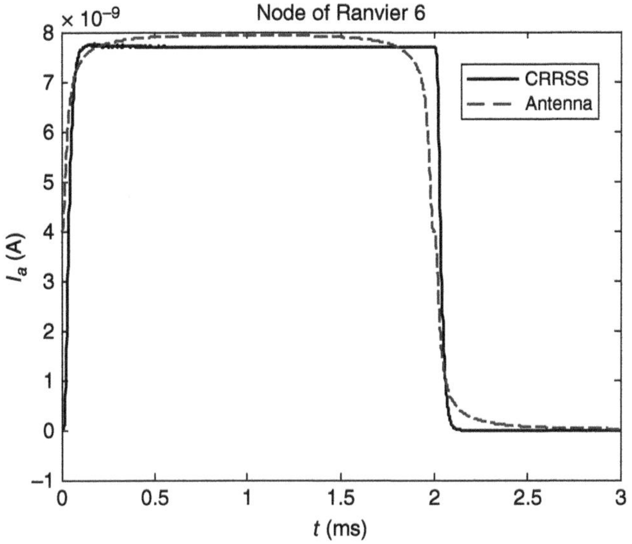


**Figure 10.12** Intracellular current in the passive Ranvier's node 6, $L = 2$ cm

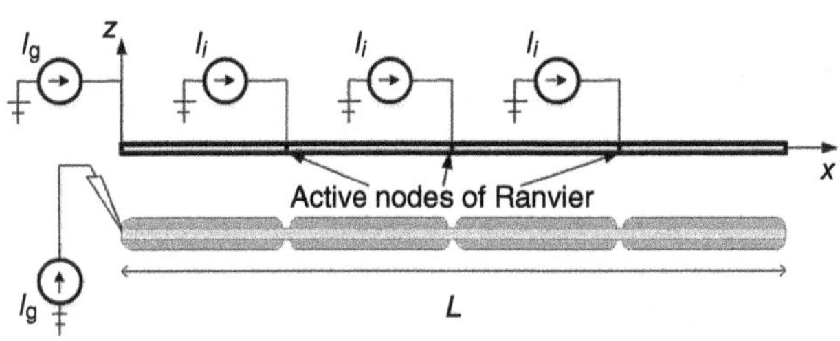


**Figure 10.13** Active nerve fiber model.

Each node is represented by a wire junction of two thin wires representing an active Ranvier's node. Thus, three additional current sources at the active Ranvier nodes represent an ionic current $I_i$ of the activated node. The ionic current is determined by analyzing the *CRRSS* model and the analytical expression for the ionic current is obtained by curve fitting procedure [10]:

$$I_i(t) = Au(t)[e^{-Bt} - e^{-Dt}] - Eu(t - t_1)e^{-G(t-t_2)^2}, \tag{10.17}$$

where $t_1 = 0.031$ ms, $t_2 = 0.252$ ms, $A = 0.02075$ mA ms$^{-1}$, $B = 199.9$ (ms)$^{-1}$, $D = 46.91$ (ms)$^{-1}$, $E = 0.000664$ mA ms$^{-1}$, $G = 60$ (ms)$^{-2}$. The ionic current is depicted in Figure 10.14.

A thin wire junction model of the active node can be represented by two separate thin wires, as shown in Figure 10.15.

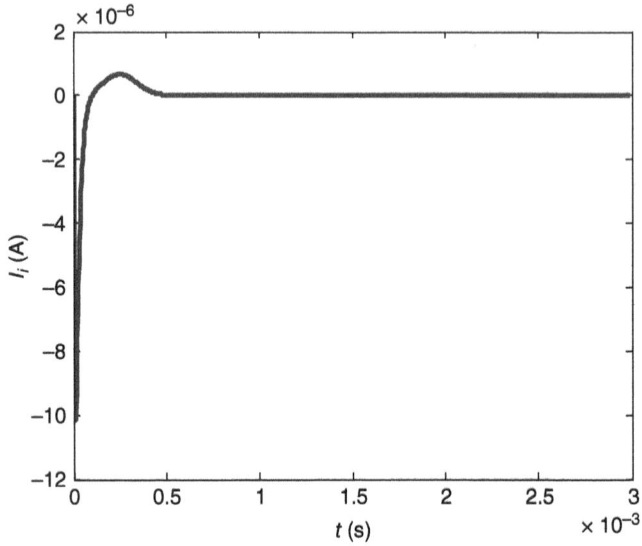


**Figure 10.14** Ionic current of activated node of Ranvier.

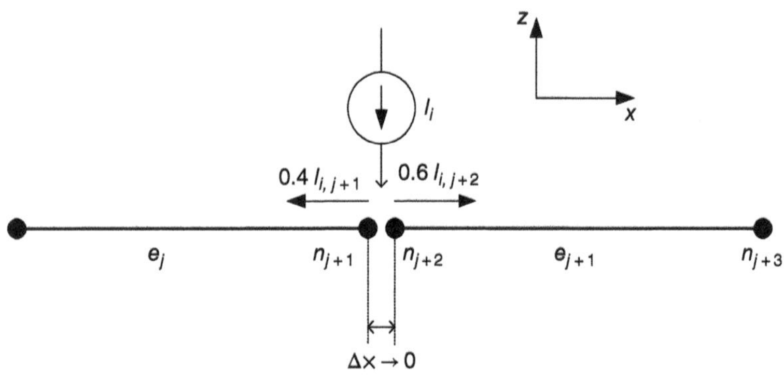


**Figure 10.15** Two wire junction representation of the active node.

Kirchhoff's law has to be satisfied at the junction [10]:

$$I_i = 0.4I_{i,j+1} + 0.6I_{i,j+2} \tag{10.18}$$

where $I_{i,j+1}$ represents the ionic current value flowing out of the junction in one direction and $I_{i,j+2}$ is the current flowing out of the junction in the opposite direction.

First, the nerve fiber is stimulated by a current generator $I_g$ at a fiber beginning, as in the case of the passive nerve fiber. Furthermore, if the stimulating current exceeds the threshold in the corresponding Ranvier's node, the second current source, representing the ionic current, activates. The intracellular current for the activated fiber is a summation of the node activation current and the ionic current of the activated node. The resulting current, flowing out of the activated node, then activates the next node. The same procedure is repeated for each node.

The threshold for nerve fiber activation depends on the strength and duration of the stimulus [17–19].

### 10.2.4 Numerical Results for Active Nerve Fiber

The intracellular current for the active nerve fiber is calculated with 9 Ranvier's nodes and 10 internodes. The rectangular current pulse used to depolarize the nerve fiber membrane is plotted in Figure 10.16.

The current pulse shown in Figure 10.16 stimulates the nerve fiber at the fiber beginning. Once a threshold current of 1.5 μA reaches the first Ranvier's node, the other current source activates. Summation of the intracellular current activating the node and the ionic current of the second current source yields the resulting intracellular current in the first activated Ranvier's node. The resulting current then activates the second Ranvier's node. The related intracellular current plot is shown in Figure 10.17.

The results for the intracellular current obtained via the antenna and CRRSS model show a rather satisfactory agreement to a certain extent. Some minor

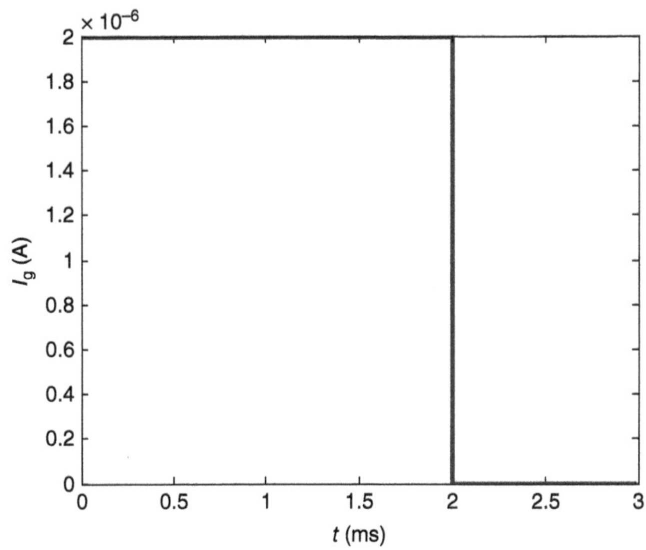


**Figure 10.16** Rectangular superthreshold current pulse.

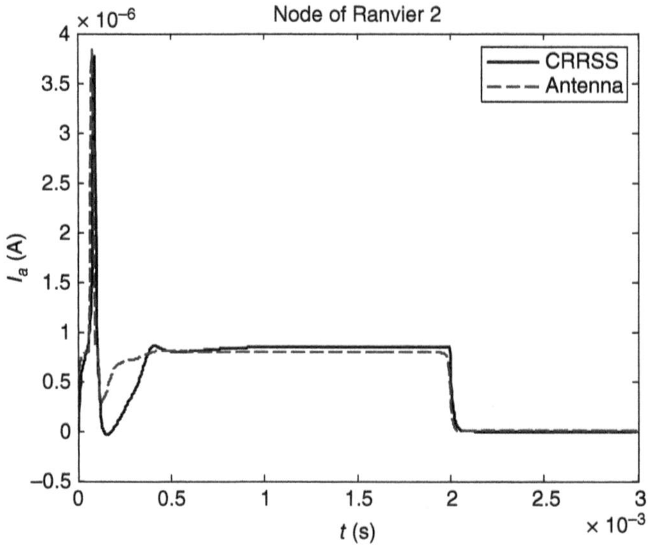


**Figure 10.17** Intracellular current in the active Ranvier's node 2, $L = 2$ cm.

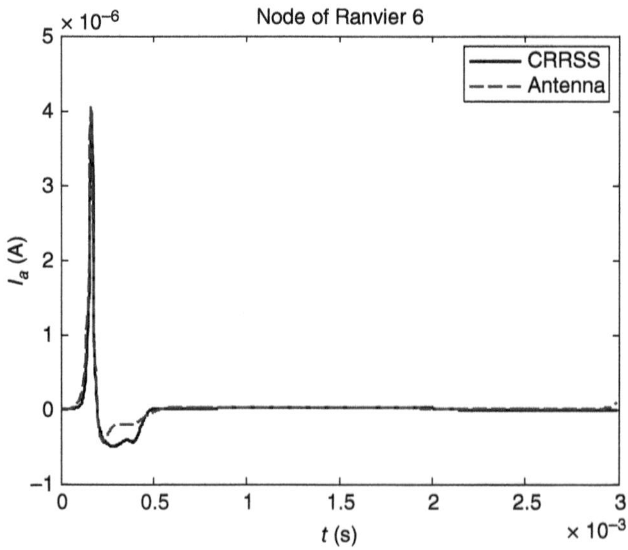


**Figure 10.18** Intracellular current in the active Ranvier's node 6, $L = 2$ cm.

discrepancies appear due to the ionic current approximation. The intracellular current part pertaining to the rectangular current pulse is in very good agreement.

Finally, Figure 10.18 shows the intracellular current for the fourth Ranvier's node.

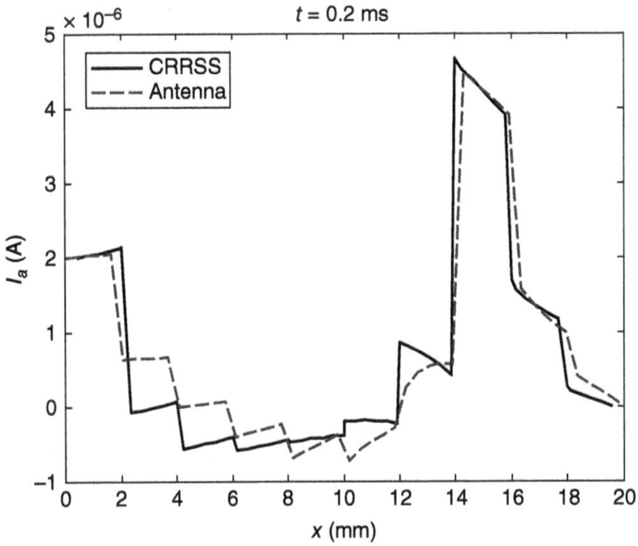


**Figure 10.19** Intracellular current along the active nerve fiber, $t = 0.2$ ms, $L = 2$ cm.

Obviously, the rectangular current pulse decays along the fiber, while the intracellular current part keeps almost the same amplitude at the activated node. The signal of the activated nerve propagates along the fiber almost without attenuation.

Finally, Figure 10.19 shows the intracellular current distribution along the nerve fiber, plotted for the time instant of 0.2 ms obtained by the antenna and CRRSS model, respectively.

There is satisfactory agreement between the results, with some visible discrepancies in amplitude of the intracellular current traveling along the fiber, mainly due to the instability of numerical procedure in the antenna model.

## References

1 Cvetković, M., Poljak, D., and Haueisen, J. (2015). Analysis of transcranial magnetic stimulation based on the surface integral equation formulation. *IEEE Transactions on Biomedical Engineering* **62** (6): 1535–1545.

2 Garvey, M.A. and Mall, V. (2008). Transcranial magnetic stimulation in children. *Clinical Neurophysiology* **119** (5): 973–984.

3 Rajapakse, T. and Kirton, A. (2013). Non-invasive brain stimulation in children: applications and future directions. *Translational Neuroscience* **4**: 1–29.

4 Frijns, J.H. and ten Kate, J.H. (1994). A model of myelinated nerve fibres for electrical prosthesis design. *Medical & Biological Engineering & Computing* **32**: 391–398.

5 Rattay, F. (1999). The basic mechanism for the electrical stimulation of the nervous system. *Neuroscience* **89** (2): 335–346.
6 Zulim, I., Dorić, V., Poljak, D., and El Khamlichi Drissi, K. (2015). Antenna model for passive myelinated nerve fiber. SoftCOM 2015. 23rd International Conference on Software, Telecommunications and Computer Networks.
7 Picht, T., Krieg, S.M., Sollmann, N. et al. (2013). A comparison of language mapping by preoperative navigated transcranial magnetic stimulation and direct cortical stimulation during awake surgery. *Neurosurgery* **72** (5): 808–819.
8 Deletis, V., Rogić, M., Fernández-Conejero, I. et al. (2014). Neurophysiologic markers in laryngeal muscles indicate functional anatomy of laryngeal primary motor cortex and premotor cortex in the caudal opercular part of inferior frontal gyrus. *Clinical Neurophysiology* **125** (9): 1912–1922.
9 Rogić, M., Deletis, V., and Fernández-Conejero, I. (2014). Inducing transient language disruptions by mapping of Broca's area with modified patterned repetitive transcranial magnetic stimulation protocol. *Journal of Neurosurgery* **120** (5): 1033–1041.
10 Poljak, D., Cvetković, M., Dorić, V., et al. (2016). Integral equation models in some biomedical applications of electromagnetic fields: transcranial magnetic stimulation (TMS), nerve fiber stimulation. Computer and Energy Science (SpliTech), International Multidisciplinary Conference, Split, Croatia, 1–6.
11 Poljak, D., Cvetković, M., Dorić, V. et al. (2018). Integral equation formulations and related numerical solution methods in some biomedical applications of electromagnetic fields: transcranial magnetic stimulation (TMS), nerve fiber stimulation, which was submitted for inclusion in the title. *International Journal of E-Health and Medical Communications (IJEHMC)* **9** (1).
12 Poljak, D., Cvetković, M., Peratta, A., et al. (2016). On some integral equation approaches in electromagnetic dosimetry. BioEM 2016, Ghent, Belgium.
13 Zulim, I., Dorić, V., Poljak, D., and El Khamlichi Drissi, K. (2016). *Antenna Model for Passive Myelinated Nerve Fiber*. SoftCOM.
14 Poljak, D. (2007). *Advanced Modeling in Computational Electromagnetic Compatibility*. Hoboken, NJ: Wiley-Interscience.
15 Chiu, S.Y., Ritchie, J.M., Rogart, R.B., and Stagg, D. (1979). A quantitative description of membrane currents in rabbit myelinated nerve. *The Journal of Physiology* **292**: 149–166.
16 Sweeney, J.D., Mortimer, J.T., and Durand, D. (1987). Modeling of mammalian myelinated nerve for functional neuromuscular electrostimulation. Proceedings of the 9th Annual Conference of the IEEE EMBS, 1577–1578.
17 Geddes, L.A. and Bourland, J.D. (1985). The strength-duration curve. *IEEE Transactions on Biomedical Engineering* **BME-32** (6): 458–459.

**18** Boinagrov, D., Loudin, J., and Palanker, D. (2010). Strength–duration relationship for extracellular neural stimulation: numerical and analytical models. *Journal of Neurophysiology* **104**: 2236–2248.

**19** Ashley, Z., Sutherland, H., Lanmuller, H. et al. (2005). Determination of the chronaxie and rheobase of denervated limb muscles in conscious rabbits. *Artificial Organs* **29** (3): 212–215.

# Index



*Computational Methods in Electromagnetic Compatibility: Antenna Theory Approach versus Transmission line Models*, First Edition. Dragan Poljak and Khalil El Khamlichi Drissi.